Theory and Applications of OFDM and CDMA

Theory and Applications of OFDM and CDMA

Wideband Wireless Communications

Henrik Schulze
and
Christian Lüders

Both of
Fachhochschule Südwestfalen
Meschede, Germany

John Wiley & Sons, Ltd

Telephone (+44) 1243 779777

Email (for orders and customer service enquiries): cs-books@wiley.co.uk
Visit our Home Page on www.wiley.com

Reprinted with corrections November 2006.

Other Wiley Editorial Offices

John Wiley & Sons Inc., 111 River Street, Hoboken, NJ 07030, USA

Jossey-Bass, 989 Market Street, San Francisco, CA 94103-1741, USA

Wiley-VCH Verlag GmbH, Boschstr. 12, D-69469 Weinheim, Germany

John Wiley & Sons Australia Ltd, 42 McDougall Street, Milton, Queensland 4064, Australia

John Wiley & Sons (Asia) Pte Ltd, 2 Clementi Loop #02-01, Jin Xing Distripark, Singapore 129809

John Wiley & Sons Canada Ltd, 22 Worcester Road, Etobicoke, Ontario, Canada M9W 1L1

Wiley also publishes its books in a variety of electronic formats. Some content that appears in print may not be available in electronic books.

British Library Cataloguing in Publication Data

A catalogue record for this book is available from the British Library

ISBN-13 978-0-470-85069-5 (HB)
ISBN-10 0-470-85069-8 (HB)

Typeset in 10/12pt Times by Laserwords Private Limited, Chennai, India.

Contents

Preface

Wireless communication has become increasingly important not only for professional applications but also for many fields in our daily routine and in consumer electronics. In 1990, a mobile telephone was still quite expensive, whereas today most teenagers have one, and they use it not only for calls but also for data transmission. More and more computers use wireless local area networks (WLANs), and audio and television broadcasting has become digital.

Many of the above-mentioned communication systems make use of one of two sophisticated techniques that are known as *orthogonal frequency division multiplexing* (OFDM) and *code division multiple access* (CDMA).

The first, OFDM, is a digital multicarrier transmission technique that distributes the digitally encoded symbols over several *subcarrier* frequencies in order to reduce the symbol clock rate to achieve robustness against long echoes in a multipath radio channel. Even though the spectra of the individual subcarriers overlap, the information can be completely recovered without any interference from other subcarriers. This may be surprising, but from a mathematical point of view, this is a consequence of the *orthogonality* of the base functions of the Fourier series.

The second, CDMA, is a multiple access scheme where several users share the same physical medium, that is, the same frequency band at the same time. In an ideal case, the signals of the individual users are *orthogonal* and the information can be recovered without interference from other users. Even though this is only approximately the case, the concept of *orthogonality* is quite important to understand why CDMA works. It is due to the fact that pseudorandom sequences are approximately orthogonal to each other or, in other words, they show good correlation properties. CDMA is based on *spread spectrum*, that is, the spectral band is *spread* by multiplying the signal with such a pseudorandom sequence. One advantage of the enhancement of the bandwidth is that the receiver can take benefit from the multipath properties of the mobile radio channel.

OFDM transmission is used in several digital audio and video broadcasting systems. The pioneer was the European DAB (Digital Audio Broadcasting) system. At the time when the project started in 1987, hardly any communication engineers had heard about OFDM. One author (Henrik Schulze) remembers well that many practical engineers were very suspicious of these rather abstract and theoretical underlying ideas of OFDM. However, only a few years later, the DAB system became the leading example for the development of the digital terrestrial video broadcasting system, DVB-T. Here, in contrast to DAB, coherent higher-level modulation schemes together with a sophisticated and powerful channel estimation technique are utilized in a multipath-fading channel. High-speed WLAN systems like IEEE 802.11a and IEEE 802.11g use OFDM together with very similar channel coding and modulation. The European standard HIPERLAN/2 (High Performance Local Area Network, Type 2) has the same OFDM parameters as these IEEE systems and differs only in a few options concerning channel coding and modulation. Recently, a broadcasting system called DRM (Digital Radio Mondiale) has been developed to replace the antiquated analog AM radio transmission in the frequency bands below 30 MHz. DRM uses OFDM together with a sophisticated multilevel coding technique.

The idea of spread spectrum systems goes back to military applications, which arose during World War II, and were the main field for spread spectrum techniques in the following decades. Within these applications, the main benefits of spreading are to hide a signal, to protect it against eavesdropping and to achieve a high robustness against intended interference, that is, to be able to separate the useful signal from the strong interfering one. Furthermore, correlating to a spreading sequence may be used within radar systems to obtain reliable and precise values of propagation delay for deriving the position of an object.

A system where different (nearly orthogonal) spreading sequences are used to separate the signals transmitted from different sources is the Global Positioning System (GPS) developed in about 1970. Hence, GPS is the first important system where code division multiple access (CDMA) is applied. Within the last 10 years, CDMA has emerged as the most important multiple access technique for mobile communications. The first concept for a CDMA mobile communication system was developed by Qualcomm Incorporated in approx 1988. This system proposal was subsequently refined and released as the so-called IS-95 standard in North America. In the meantime, the system has been rebranded as cdmaOne, and there are more than 100 millions of cdmaOne subscribers in more than 40 countries. Furthermore, cdmaOne has been the starting point for cdma2000, a third-generation mobile communication system offering data rates of up to some Mbit/s. Another very important third-generation system using CDMA is the Universal Mobile Telecommunications System (UMTS); UMTS is based on system proposals developed within a number of European research projects. Hence, CDMA is the dominating multiple access technique for third generation mobile communication systems.

This book has both theoretical and practical aspects. It is intended to provide the reader with a deeper understanding of the concepts of OFDM and CDMA. Thus, the theoretical basics are analyzed and presented in some detail. Both of the concepts are widely applied in practice. Therefore, a considerable part of the book is devoted to system design and implementation aspects and to the presentation of existing communication systems.

The book is organized as follows. In Chapter 1, we give a brief overview of the basic principles of digital communications and introduce our notation. We represent signals as vectors, which often leads to a straightforward geometrical visualization of many seemingly abstract mathematical facts. The concept of orthogonality between signal vectors is a key to the understanding of OFDM and CDMA, and the Euclidean distance between signal vectors is an important concept to analyze the performance of a digital transmission system. Wireless communication systems often have to cope with severe multipath fading in a mobile radio channel. Chapter 2 treats these aspects. First, the physical situation of multipath propagation is analyzed and statistical models of the mobile radio channel are presented. Then, the problems of digital transmission over these channels are discussed and the basic principles of Chapter 1 are extended for those channels. Digital wireless communication over fading channels is hardly possible without using some kind of error protection or channel coding. Chapter 3 gives a brief overview of the most important channel coding techniques that are used in the above-mentioned communication systems. Convolutional codes are typically used in these systems, and many of the systems have very closely related (or even identical) channel coding options. Thus, the major part of Chapter 3 is dedicated to convolutional codes as they are applied in these systems. A short presentation of Reed–Solomon Codes is also included because they are used as outer codes in the DVB-T system, together with inner convolutional codes. Chapter 4 is devoted to OFDM. First, the underlying ideas and the basic principles are explained by using the basic principles presented in Chapter 1. Then implementation aspects are discussed as well as channel estimation and synchronization aspects that are relevant for the above-mentioned systems. All these systems are designed for mobile radio channels and use channel coding. Therefore, we give a comprehensive discussion of system design aspects and how to fit all these things together in an optimal way for a given channel. Last but not least, the transmission schemes for DAB, DVB-T and WLAN systems are presented and discussed. Chapter 5 is devoted to CDMA, focusing on its main application area – mobile communications. This application area requires not only sophisticated digital transmission techniques and receiver structures but also some additional methods as, for example, a soft handover, a fast and exact power control mechanism as well as some special planning techniques to achieve an acceptable radio network performance. Therefore, the first section of Chapter 5 discusses these methods and some general principles of CDMA and mobile radio networks. CDMA receivers may be simple or quite sophisticated, thereby making use of knowledge about other users. These theoretically involved topics are treated in the following three subsections. As examples of CDMA applications we discuss the most important systems already mentioned, namely, GPS, cdmaOne (IS-95), cdma2000 and UMTS with its two transmission modes called Wideband CDMA and Time Division CDMA. Furthermore, Wireless LAN systems conforming to the standard IEEE 802.11 are also included in this section as some transmission modes of these systems are based on spreading.

This book is supported by a companion website on which lecturers and instructors can find electronic versions of the figures contained within the book, a solutions manual to the problems at the end of each chapter and also chapter summaries. Please go to ftp://ftp.wiley.co.uk/pub/books/schulze

1

Basics of Digital Communications

1.1 Orthogonal Signals and Vectors

The concept of *orthogonal signals* is essential for the understanding of OFDM (orthogonal frequency division multiplexing) and CDMA (code division multiple access) systems. In the normal sense, it may look like a miracle that one can separately demodulate overlapping carriers (for OFDM) or detect a signal among other signals that share the same frequency band (for CDMA). The concept of orthogonality unveils this miracle. To understand these concepts, it is very helpful to interpret signals as vectors. Like vectors, signals can be added, multiplied by a scalar, and they can be expanded into a base. In fact, signals fit into the mathematical structure of a vector space. This concept may look a little bit abstract. However, vectors can be visualized by geometrical objects, and many conclusions can be drawn by simple geometrical arguments without lengthy formal derivations. So it is worthwhile to become familiar with this point of view.

1.1.1 The Fourier base signals

To visualize signals as vectors, we start with the familiar example of a Fourier series. For reasons that will become obvious later, we do not deal with a periodic signal, but cut off outside the time interval of one period of length T. This means that we consider a well-behaved (e.g. integrable) real signal $x(t)$ inside the time interval $0 \leq t \leq T$ and set $x(t) = 0$ outside. Inside the interval, the signal can be written as a Fourier series

$$x(t) = \frac{a_0}{2} + \sum_{k=1}^{\infty} a_k \cos\left(2\pi \frac{k}{T} t\right) - \sum_{k=1}^{\infty} b_k \sin\left(2\pi \frac{k}{T} t\right). \tag{1.1}$$

The Fourier coefficients a_k and b_k are given by

$$a_k = \frac{2}{T} \int_0^T \cos\left(2\pi \frac{k}{T} t\right) x(t)\, \mathrm{d}t \tag{1.2}$$

Theory and Applications of OFDM and CDMA Henrik Schulze and Christian Lüders

and

$$b_k = -\frac{2}{T} \int_0^T \sin\left(2\pi \frac{k}{T} t\right) x(t)\, \mathrm{d}t. \tag{1.3}$$

These coefficients are the amplitudes of the cosine and (negative) sine waves at the respective frequencies $f_k = k/T$. The cosine and (negative) sine waves are interpreted as a *base* for the (well-behaved) signals inside the time interval of length T. Every such signal can be expanded into that base according to Equation (1.1) inside that interval. The underlying mathematical structure of the Fourier series is similar to the expansion of an N-dimensional vector $\mathbf{x} \in \mathcal{R}^N$ into a base $\{\mathbf{v}_i\}_{i=1}^N$ according to

$$\mathbf{x} = \sum_{i=1}^{N} \alpha_i \mathbf{v}_i. \tag{1.4}$$

The base $\{\mathbf{v}_i\}_{i=1}^N$ is called *orthonormal* if two different vectors are orthogonal (perpendicular) to each other and if they are normalized to length one, that is,

$$\mathbf{v}_i \cdot \mathbf{v}_k = \delta_{ik},$$

where δ_{ik} is the Kronecker Delta ($\delta_{ik} = 1$ for $i = k$ and $\delta_{ik} = 0$ otherwise) and the dot denotes the usual scalar product

$$\mathbf{x} \cdot \mathbf{y} = \sum_{i=1}^{N} x_i y_i = \mathbf{x}^T \mathbf{y}$$

for real N-dimensional vectors. In that case, the coefficients α_i are given by

$$\alpha_i = \mathbf{v}_i \cdot \mathbf{x}.$$

For an orthonormal base, the coefficients α_i can thus be interpreted as the projections of the vector $\mathbf{x}$ onto the base vectors, as depicted in Figure 1.1 for $N = 2$. Thus, α_i can be interpreted as the *amplitude* of $\mathbf{x}$ in the direction of $\mathbf{v}_i$.

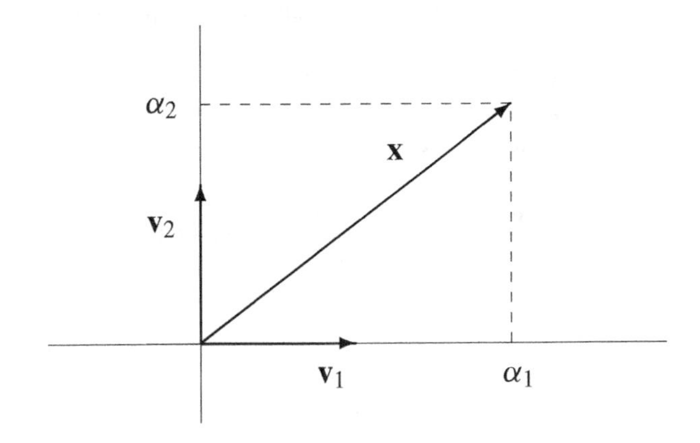

Figure 1.1 A signal vector in two dimensions.

The Fourier expansion (1.1) is of the same type as the expansion (1.4), except that the sum is infinite. For a better comparison, we may write

$$x(t) = \sum_{i=0}^{\infty} \alpha_i v_i(t)$$

with the normalized base signal vectors $v_i(t)$ defined by

$$v_0(t) = \sqrt{\frac{1}{T}}\, \Pi\left(\frac{t}{T} - \frac{1}{2}\right)$$

and

$$v_{2k}(t) = \sqrt{\frac{2}{T}} \cos\left(2\pi \frac{k}{T} t\right) \Pi\left(\frac{t}{T} - \frac{1}{2}\right)$$

for even $i > 0$ and

$$v_{2k+1}(t) = -\sqrt{\frac{2}{T}} \sin\left(2\pi \frac{k}{T} t\right) \Pi\left(\frac{t}{T} - \frac{1}{2}\right)$$

for odd i with coefficients given by

$$\alpha_i = \int_{-\infty}^{\infty} v_i(t) x(t)\, \mathrm{d}t,$$

that is,

$$\alpha_{2k} = \sqrt{T/2} a_k$$

and

$$\alpha_{2k+1} = \sqrt{T/2} b_k.$$

Here we have introduced the notation $\Pi(x)$ for the rectangular function, which takes the value one between $x = -1/2$ and $x = 1/2$ and zero outside. Thus, $\Pi(x - 1/2)$ is the rectangle between $x = 0$ and $x = 1$. The base of signals $v_i(t)$ fulfills the orthonormality condition

$$\int_{-\infty}^{\infty} v_i(t) v_k(t)\, \mathrm{d}t = \delta_{ik}. \tag{1.5}$$

We will see in the following text that this just means that the Fourier base forms a set of orthogonal signals. With this interpretation, Equation (1.5) says that the base signals for different frequencies are orthogonal and, for the same frequency $f_k = k/T$, the sine and cosine waves are orthogonal.

We note that the orthonormality condition and the formula for α_i are very similar to the case of finite-dimensional vectors. One just has to replace sums by integrals. A similar geometrical interpretation is also possible; one has to regard signals as vectors, that is, identify $v_i(t)$ with $\mathbf{v}_i$ and $x(t)$ with $\mathbf{x}$. The interpretation of α_i as a projection on $\mathbf{v}_i$ is obvious. For only two dimensions, we have $x(t) = \alpha_1 v_1(t) + \alpha_2 v_2(t)$, and the signals can be adequately described by Figure 1.1. In this special case, where $v_1(t)$ is a cosine signal and $v_2(t)$ is a (negative) sine signal, the figure depicts nothing else but the familiar phasor diagram. However, this is just a special case of a very general concept that applies to many other scenarios in communications.

Before we further discuss this concept for signals by introducing a scalar product for signals, we continue with the complex Fourier transform. This is because complex signals are a common tool in communications engineering.

Consider a well-behaved complex signal $s(t)$ inside the time interval $[0, T]$ that vanishes outside that interval. The complex Fourier series for that signal can be written as

$$s(t) = \sum_{k=-\infty}^{\infty} \alpha_k v_k(t) \tag{1.6}$$

with the (normalized) complex Fourier base functions

$$v_k(t) = \sqrt{\frac{1}{T}} \exp\left(j2\pi \frac{k}{T} t\right) \Pi\left(\frac{t}{T} - \frac{1}{2}\right). \tag{1.7}$$

The base functions are orthonormal in the sense that

$$\int_{-\infty}^{\infty} v_i^*(t) v_k(t)\, \mathrm{d}t = \delta_{ik}. \tag{1.8}$$

holds. The Fourier coefficient α_k will be obtained from the signal by the Fourier analyzer. This is a detection device that performs the operation

$$\alpha_k = \int_{-\infty}^{\infty} v_i^*(t) s(t)\, \mathrm{d}t. \tag{1.9}$$

This coefficient is the complex amplitude (i.e. amplitude and phase) of the wave at frequency f_k. It can be interpreted as the component of the signal vector $s(t)$ in the direction of the base signal vector $v_k(t)$, that is, we interpret frequency components as vector components or vector *coordinates*.

Example 1 (OFDM Transmission) *Given a finite set of complex numbers s_k that carry digitally encoded information to be transmitted, we may use the complex Fourier series for this purpose and transmit the signal*

$$s(t) = \sum_{k=0}^{K} s_k v_k(t). \tag{1.10}$$

The transmitter performs a Fourier synthesis. *In an ideal transmission channel with perfect synchronization and no disturbances, the transmit symbols s_k can be completely recovered at the receiver by the Fourier analyzer that plays the role of the detector. One may send K new complex symbols during every time slot of length T by performing the Fourier synthesis for that time slot. At the receiver, the Fourier analysis is done for every time slot. This method is called* orthogonal frequency division multiplexing (OFDM). *This name is due to the fact that the transmit signals form an orthogonal base belonging to different frequencies f_k. We will see in the following text that other – even more familiar – transmission setups use orthogonal bases.*

1.1.2 The signal space

A few mathematical concepts are needed to extend the concept of orthogonal signals to other applications and to represent the underlying structure more clearly. We consider (real or complex) signals of finite energy, that is, signals $s(t)$ with the property

$$\int_{-\infty}^{\infty} |s(t)|^2 \, \mathrm{d}t < \infty. \tag{1.11}$$

The assumption that our signals should have finite energy is physically reasonable and leads to desired mathematical properties. We note that this set of signals has the property of a vector space, because finite-energy signals can be added or multiplied by a scalar, resulting in a finite-energy signal. For this vector space, a scalar product is given by the following:

Definition 1.1.1 (Scalar product of signals) *In the vector space of signals with finite energy, the scalar product of two signals $s(t)$ and $r(t)$ is defined as*

$$\langle s, r \rangle = \int_{-\infty}^{\infty} s^*(t) r(t) \, \mathrm{d}t. \tag{1.12}$$

Two signals are called orthogonal if their scalar product equals zero. The Euclidean norm of the signal is defined by $\|s\| = \sqrt{\langle s, s \rangle}$, and $\|s\|^2 = \langle s, s \rangle$ is the signal energy. $\|s - r\|^2$ is called the squared Euclidean distance *between $s(t)$ and $r(t)$.*

We add the following remarks:

- This scalar product has a structure similar to the scalar product of vectors $\mathbf{s} = (s_1, \ldots, s_K)^T$ and $\mathbf{r} = (r_1, \ldots, r_K)^T$ in a K-dimensional complex vector space given by

$$\mathbf{s}^\dagger \mathbf{r} = \sum_{k=1}^{K} s_k^* r_k,$$

  where the dagger (†) means conjugate transpose. Comparing this expression with the definition of the scalar product for continuous signals, we see that the sum has to be replaced by the integral.

- It is a common use of notation in communications engineering to write a function with an argument for the function, that is, to write $s(t)$ for a signal (which is a function of the time) instead of s, which would be the mathematically correct notation. In most cases, we will use the engineer's notation, but we write, for example, $\langle s, r \rangle$ and not $\langle s(t), r(t) \rangle$, because this quantity does not depend on t. However, sometimes we write s instead of $s(t)$ when it makes the notation simpler.

- In mathematics, the vector space of square integrable functions (i.e. finite-energy signals) with the scalar product as defined above is called the *Hilbert space* $L^2(\mathcal{R})$. It is interesting to note that the Hilbert space of finite-energy signals is the same as the Hilbert space of wave functions in quantum mechanics. For the reader who is interested in details, we refer to standard text books in mathematical physics (see e.g. (Reed and Simon 1980)).

Without proof, we refer to some mathematical facts about that space of signals with finite energy (see e.g. (Reed and Simon 1980)).

- Each signal $s(t)$ of finite energy can be expanded into an orthonormal base, that is, it can be written as

$$s(t) = \sum_{k=1}^{\infty} \alpha_k v_k(t) \tag{1.13}$$

with properly chosen orthonormal base signals $v_k(t)$. The coefficients can be obtained from the signal as

$$\alpha_k = \langle v_k, s \rangle \,. \tag{1.14}$$

The coefficient α_k can be interpreted as the component of the signal vector s in the direction of the base vector v_k.

- For any two finite energy signals $s(t)$ and $r(t)$, the Schwarz inequality

$$|\langle s, r \rangle| \leq \|s\| \; \|r\|$$

holds. Equality holds if and only if $s(t)$ is proportional to $r(t)$.

- The Fourier transform is well defined for finite-energy signals. Now, let $s(t)$ and $r(t)$ be two signals of finite energy, and $S(f)$ and $R(f)$ denote their Fourier transforms. Then,

$$\langle s, r \rangle = \langle S, R \rangle$$

holds. This fact is called *Plancherel theorem* or *Rayleigh theorem* in the mathematical literature (Bracewell 2000). The above equality is often called *Parseval's equation*. As an important special case, we note that the signal energy can be expressed either in the time or in the frequency domain as

$$E = \int_{-\infty}^{\infty} |s(t)|^2 \, \mathrm{d}t = \int_{-\infty}^{\infty} |S(f)|^2 \, \mathrm{d}f.$$

Thus, $|S(f)|^2 \, \mathrm{d}f$ is the energy in an infinitesimal frequency interval of width $\mathrm{d}f$, and $|S(f)|^2$ can be interpreted as the spectral density of the signal energy.

In communications, we often deal with subspaces of the vector space of finite-energy signals. The signals of finite duration form such a subspace. An appropriate base of that subspace is the Fourier base. The Fourier series is then just a special case of Equation (1.13) and the Fourier coefficients are given by Equation (1.14). Another subspace is the space of strictly band-limited signals of finite energy. From the sampling theorem we know that each such signal $s(t)$ that is concentrated inside the frequency interval between $-B/2$ and $B/2$ can be written as a series

$$s(t) = \sum_{k=-\infty}^{\infty} s(k/B) \operatorname{sinc}(Bt - k) \tag{1.15}$$

with $\operatorname{sinc}(x) = \sin(\pi x)/(\pi x)$.

We define a base as follows:

Definition 1.1.2 (Normalized sinc base) *The orthonormal sinc base for the bandwidth $B/2$ is given by the signals*

$$\psi_k(t) = \sqrt{B}\,\mathrm{sinc}\,(Bt - k)\,. \tag{1.16}$$

We note that $\sqrt{B}\,\psi_0(t)$ is the impulse response of an ideal low-pass filter of bandwidth $B/2$, so that the sinc base consists of delayed and normalized versions of that impulse response. From the sampling theorem, we conclude that these $\psi_k(t)$ are a base of the subspace of strictly band-limited functions. By looking at them in the frequency domain, we easily see that they are orthonormal. From standard Fourier transform relations, we see that the Fourier transform $\Psi_k(f)$ of $\psi_k(t)$ is given by

$$\Psi_k(f) = \frac{1}{\sqrt{B}}\Pi\,(f/B)\,\mathrm{e}^{-j2\pi kf/B}.$$

Thus, $\Psi_k(f)$ is just a Fourier base function for signals concentrated inside the frequency interval between $-B/2$ and $B/2$. This base is known to be orthogonal. Thus, we rewrite the statement of the sampling theorem as the expansion

$$s(t) = \sum_{k=-\infty}^{\infty} s_k \psi_k(t)$$

into the orthonormal base $\psi_k(t)$. From the sampling theorem, we know that

$$s_k = \frac{1}{\sqrt{B}} s(k/B)$$

holds. The Fourier transform of this signal is given by

$$S(f) = \sum_{k=-\infty}^{\infty} s_k \Psi_k(f).$$

This is just a Fourier series for the spectral function that is concentrated inside the frequency interval between $-B/2$ and $B/2$. The coefficients are given by

$$s_k = \langle \psi_k, s \rangle = \langle \Psi_k, S \rangle\,.$$

This relates the coefficients of a Fourier expansion of a signal $S(f)$ in the frequency domain to the samples of the corresponding signal $s(t)$ in the time domain. As we have seen from this discussion, the Fourier base and the sinc base are related to each other by interchanging the role of time and frequency.

1.1.3 Transmitters and detectors

Any linear digital transmission setup can be characterized as follows: As in the OFDM Example 1, for each transmission system we have to deal with a synthesis of a signal (at the transmitter site) and the analysis of a signal (at the receiver site). Given a finite set $\{s_k\}_{k=1}^{K}$ of coefficients that carry the information to be transmitted, we choose a base $g_k(t)$ to transmit the information by the signal

$$s(t) = \sum_{k=1}^{K} s_k g_k(t). \tag{1.17}$$

Definition 1.1.3 (Transmit base, pulses and symbols) *In the above sum, each signal $g_k(t)$ is called a transmit pulse, the set of signals $\{g_k(t)\}_{k=1}^K$ is called the transmit base, and s_k is called a transmit symbol. The vector $\mathbf{s} = (s_1, \ldots, s_K)^T$ is called the transmit symbol vector.*

Note that in the terminology of vectors, the transmit symbols s_k are the *coordinates* of the signal vector $s(t)$ corresponding to the base $g_k(t)$.

If the transmit base is orthonormal, then, for an ideal transmission channel, the information symbols s_k can be recovered completely as the projections onto the base. These scalar products

$$s_k = \langle g_k, s \rangle = \int_{-\infty}^{\infty} g_k^*(t) s(t)\, \mathrm{d}t \tag{1.18}$$

can also be written as

$$s_k = \left[\int_{-\infty}^{\infty} g_k^*(\tau - t) s(\tau)\, \mathrm{d}\tau \right]_{t=0} = \left[g_k^*(-t) * s(t) \right]_{t=0}. \tag{1.19}$$

This means that the *detection* of the information s_k transmitted by $g_k(t)$ is the output of the filter with impulse response $g_k^*(-t)$ sampled at $t = 0$. This filter is usually called *matched filter*, because it is matched to the transmit pulse $g(t)$.

Definition 1.1.4 (Detector and matched filter) *Given a transmit pulse $g(t)$, the corresponding matched filter is the filter with the impulse response $g^*(-t)$. The detector $\mathcal{D}_g$ for $g(t)$ is defined by the matched filter output sampled at $t = 0$, that is, by the detector output $\mathcal{D}_g[r]$ given by*

$$\mathcal{D}_g[r] = \int_{-\infty}^{\infty} g^*(t) r(t)\, \mathrm{d}t \tag{1.20}$$

for any receive signal $r(t)$. If two transmit pulses $g_1(t)$ and $g_2(t)$ are orthogonal, then the corresponding detectors are called orthogonal.

Thus, a detector extracts a (real or complex) number. This number carries the information. The detector may be visualized as depicted in Figure 1.2.

The matched filter has an interesting property: if a transmit pulse is corrupted by white (not necessarily Gaussian) noise, then the matched filter is the one that maximizes the signal-to-noise ratio (SNR) for $t = 0$ at the receiver end (see Problem 6).

We add the following remarks:

- For a finite-energy receive signal $r(t)$, $\mathcal{D}_g[r] = \langle g, r \rangle$ holds. However, we usually have additive noise components in the signal at the receiver, which are typically not of finite energy, so that the scalar product is not defined.

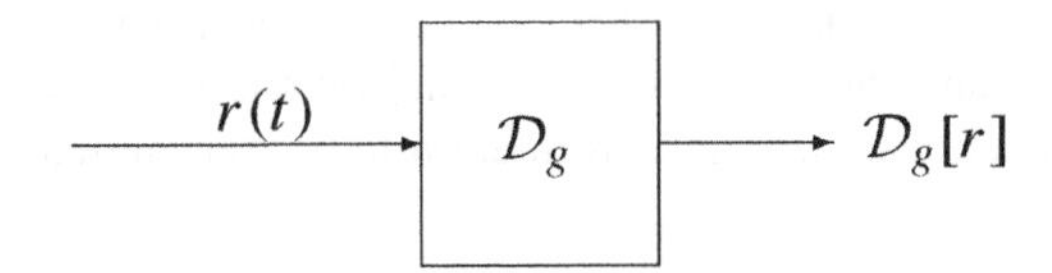

Figure 1.2 Detector.

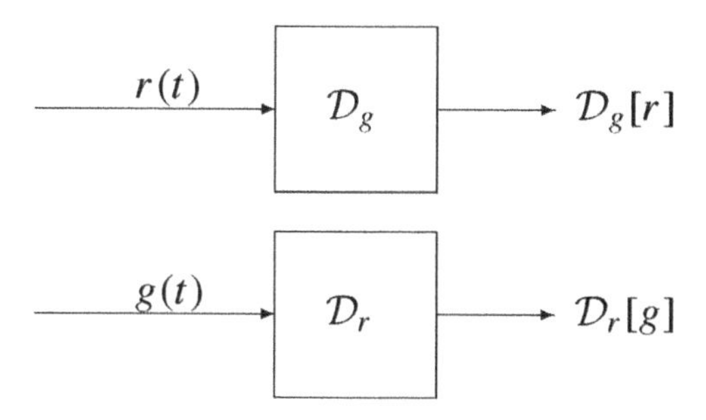

Figure 1.3 Input and detector pulses.

- The detector $\mathcal{D}_g[r]$ compares two signals: in principle, it answers the question 'How much of $r(t)$ looks like $g(t)$?'. The role of receive signal and the pulse can be interchanged according to

$$\mathcal{D}_g[r] = \mathcal{D}_r^*[g],$$

  as Figure 1.3 depicts.

- The Fourier analyzer given by Equation (1.9) is a set of orthogonal detectors. The sinc base is another set of orthogonal detectors.

- For an orthonormal base, the energy of the transmit signal equals the energy of the vector of transmit symbols, that is,

$$E = \int_{-\infty}^{\infty} |s(t)|^2 \, dt = \sum_{k=1}^{K} |s_k|^2,$$

  or more compactly written as $E = \|s\|^2 = \|\mathbf{s}\|^2$. For the proof, we refer to Problem 1.

The following example shows that the familiar Nyquist pulse shaping can be understood as an orthogonality condition.

Example 2 (The Nyquist Base) *Consider a transmission pulse $g(t)$ with the property*

$$\int_{-\infty}^{\infty} g^*(t)g(t - kT_S) \, dt = \delta[k] \tag{1.21}$$

for a certain time interval T_S, that is, the pulse $g(t)$ and its versions time-shifted by kT_S build an orthonormal base. This property also means that the pulse shape $h(t) = g^*(-t) * g(t)$ is a so-called Nyquist pulse that satisfies the first Nyquist criterion $h(kT_S) = \delta[k]$ and allows a transmission that is free of intersymbol interference (ISI). For the Fourier transforms $G(f)$ and $H(f)$ of $g(t)$ and $h(t)$, the relation $H(f) = |G(f)|^2$ holds. Therefore, $g(t)$ is often called a sqrt-Nyquist pulse. If we define $g_k(t) = g(t - kT_S)$ as the pulse transmitted at time kT_S, the condition (1.21) is equivalent to

$$\langle g_i, g_k \rangle = \delta_{ik}. \tag{1.22}$$

We then call the base formed by the signals $g_k(t)$ a (sqrt-) Nyquist base. Equation (1.22) means that if the pulse is transmitted at time kT_S, the detector for the pulse transmitted at time iT_S has output zero unless $i = k$. Thus, the pulses do not interfere with each other.

We note that in practice the pulse $g(t)$ is the impulse response of the transmit filter, that is, the pulse-shaping filter that will be excited by the transmit symbols s_k. The corresponding matched filter $g^*(-t)$ is the receive filter. Its output will be sampled once in every symbol clock T_S to recover the symbols s_k. The impulse response of the whole transmission chain $h(t) = g^*(-t) * g(t)$ must be a Nyquist pulse to ensure symbol recovery without intersymbol interference.

The Nyquist criterion

$$h(kT_S) = \delta[k]$$

in the time domain can be equivalently written in the frequency domain as

$$\sum_{k=-\infty}^{\infty} H\left(f - \frac{k}{T_S}\right) = T_S,$$

where $H(f)$ is the Fourier transform of $h(t)$. Familiar Nyquist pulses like *raised cosine* (RC) pulses are usually derived from this criterion in the frequency domain (see e.g. (Proakis 2001)). In the following text, we shall show a simple method to construct Nyquist pulses in the time domain.

Obviously, every pulse of the shape

$$h(t) = u(t) \cdot \operatorname{sinc}(t/T_S)$$

satisfies the Nyquist criterion in the time domain. $u(t)$ should be a function that improves the decay of the pulse. In the frequency domain, this is equivalent to

$$H(f) = T_S U(f) * \Pi(fT_S),$$

where $U(f)$ is the Fourier transform of $u(t)$. The convolution with $U(f)$ smoothens the sharp flank of the rectangle. Writing out the convolution integral leads to

$$H(f) = T_S \left(V\left(f + \frac{1}{2T_S}\right) - V\left(f - \frac{1}{2T_S}\right)\right)$$

with

$$V(f) = \int_{-\infty}^{f} U(x)\,\mathrm{d}x.$$

$V(f)$ is a function that describes the flank of the filter given by $H(f)$.

One possible choice for $U(f)$ is

$$U(f) = \frac{\pi}{2\alpha} T_S \cos\left(\frac{\pi}{\alpha} fT_S\right) \Pi\left(\frac{1}{\alpha} fT_S\right), \quad 0 \le \alpha \le 1$$

with a constant α between that we call the *rolloff factor*. The corresponding time domain function obviously given by

$$u(t) = \frac{\pi}{4}\left(\operatorname{sinc}\left(\alpha\frac{t}{T_S} + \frac{1}{2}\right) + \operatorname{sinc}\left(\alpha\frac{t}{T_S} - \frac{1}{2}\right)\right),$$

which equals the expression

$$u(t) = \frac{\cos(\pi \alpha t/T_S)}{1 - (2\alpha t/T_S)^2}.$$

The filter flank is obtained by integration as

$$V(f) = \begin{cases} 0 & : \quad fT_S \leq -\alpha/2 \\ \frac{1}{2}\left(1 + \sin\left(\frac{\pi}{\alpha} fT_S\right)\right) & : \quad -\alpha/2 \leq fT_S \leq \alpha/2 \\ 1 & : \quad fT_S \geq \alpha/2. \end{cases} \tag{1.23}$$

The shape of the filter is then given by the familiar expression

$$H(f) = \begin{cases} T_S & : \quad 2|f|T_S \leq 1 - \alpha \\ \frac{T_S}{2}\left(1 - \sin\left(\frac{\pi}{\alpha}\left(|f|T_S - \frac{1}{2}\right)\right)\right) & : \quad 1 - \alpha \leq 2|f|T_S \leq 1 + \alpha \\ 0 & : \quad 2|f|T_S \geq 1 + \alpha. \end{cases} \tag{1.24}$$

The corresponding pulse is

$$h(t) = \operatorname{sinc}(t/T_S) \cdot \frac{\cos(\pi \alpha t/T_S)}{1 - (2\alpha t/T_S)^2}. \tag{1.25}$$

Other Nyquist pulses than these so-called *raised cosine* (RC) pulses are possible. A Gaussian

$$u(t) = \exp\left(-(\beta t/T_S)^2\right)$$

guarantees an exponential decay in both the time and the frequency domain. β is a shaping parameter similar to the rolloff factor α. The Fourier transform $U(f)$ of $u(t)$ is given by

$$U(f) = \sqrt{\pi T_S^2/\beta^2} \exp\left(-(\pi T_S f/\beta)^2\right),$$

and the filter flank is

$$V(f) = \frac{1}{2}\left(1 + \operatorname{erf}\left(\frac{\pi}{\beta} T_S f\right)\right).$$

The filter curve is then given by

$$H(f) = \frac{1}{2} T_S \left(\operatorname{erf}\left(\frac{\pi}{\beta}\left(T_S f + \frac{1}{2}\right)\right) - \operatorname{erf}\left(\frac{\pi}{\beta}\left(T_S f - \frac{1}{2}\right)\right)\right). \tag{1.26}$$

The RC pulse and the Gaussian Nyquist pulse

$$h(t) = \operatorname{sinc}(t/T_S) \cdot \exp\left(-(\beta t/T_S)^2\right) \tag{1.27}$$

have a very similar shape if we relate α and β in such a way that their flanks $V(f)$ have the same first derivative at $f = 0$. This is the case for

$$\beta = \frac{2\alpha}{\pi}.$$

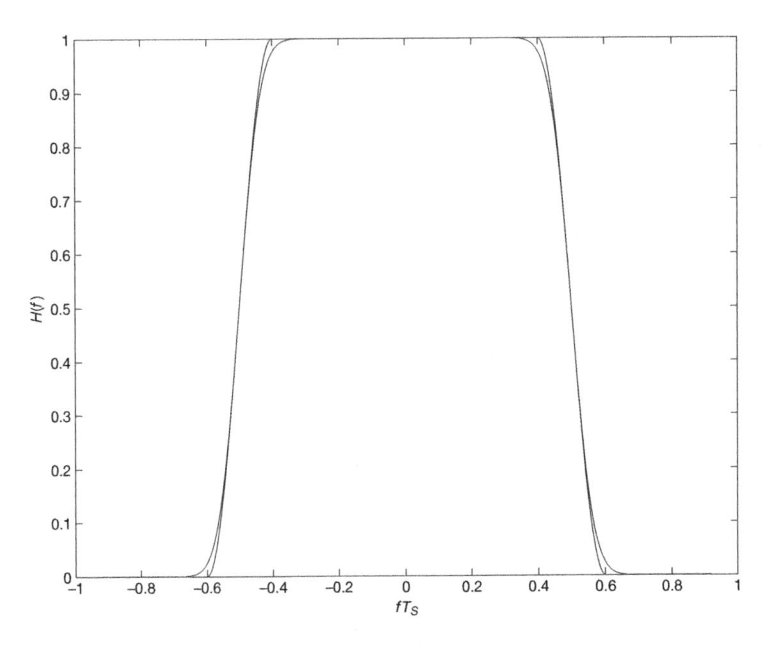

Figure 1.4 RC and Gaussian Nyquist filter shape for $\alpha = 0.2$.

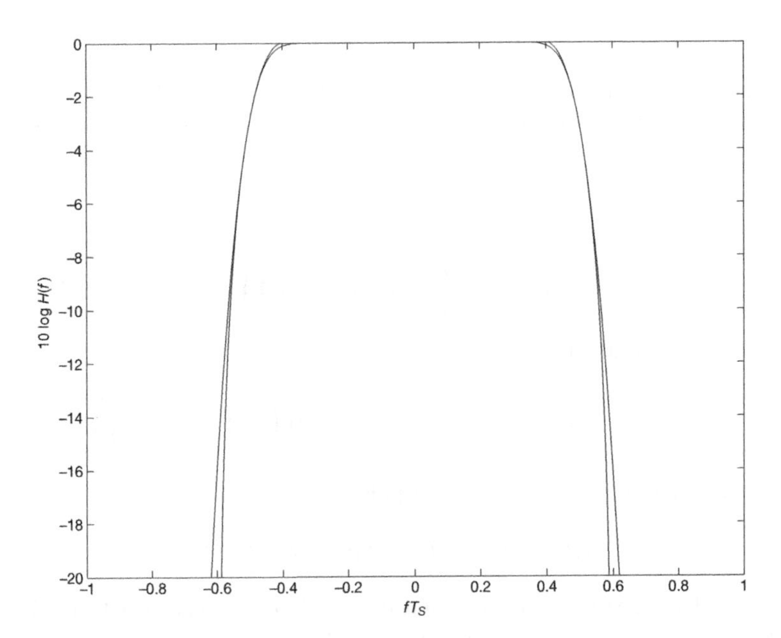

Figure 1.5 RC and Gaussian Nyquist filter shape for $\alpha = 0.2$ (decibel scale).

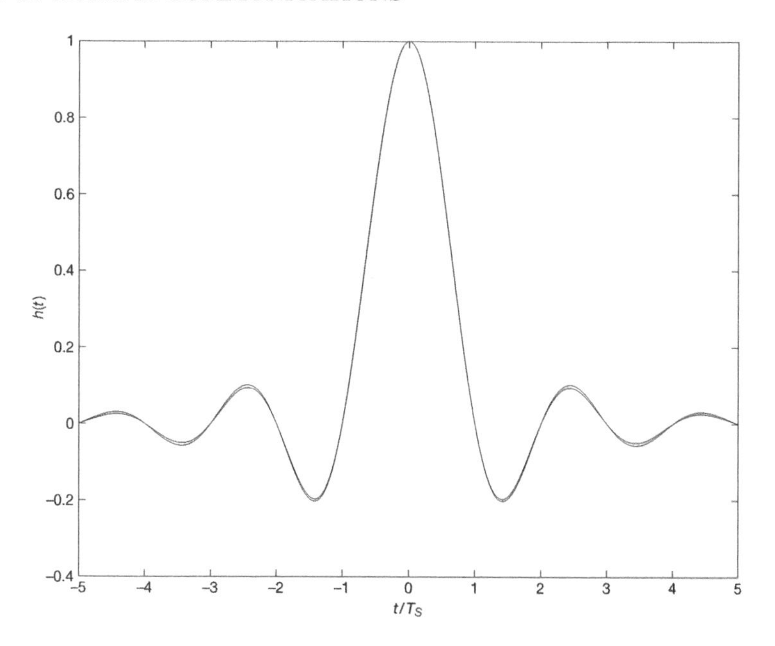

Figure 1.6 RC and Gaussian Nyquist pulse shape for $\alpha = 0.2$.

For this case and $\alpha = 0.2$, both filter shapes are depicted in Figure 1.4, and with a logarithmic scale, in Figure 1.5. The Gaussian shaping is slightly broader at the outer flanks, but the difference is very small even on the logarithmic scale. Figure 1.6 shows the corresponding pulses. The Gaussian pulse has slightly smaller amplitudes, but the difference is very small. Of course, both curves differ significantly in their asymptotic behavior (which can be seen on a logarithmic scale), but this is in a region where pulses are practically zero.

1.1.4 Walsh functions and orthonormal transmit bases

In this subsection, we introduce the Walsh functions that play an important role in CDMA signaling. We regard them as an example to discuss an orthonormal base that can be interpreted as a base of signals or as a base in an Euclidean space.

The $M \times M$ Walsh–Hadamard (WH) matrices $\mathbf{H}_M$, where M is a power of two, are defined by $\mathbf{H}_1 = 1$ and the recursive relation

$$\mathbf{H}_M = \begin{bmatrix} \mathbf{H}_{M/2} & \mathbf{H}_{M/2} \\ \mathbf{H}_{M/2} & -\mathbf{H}_{M/2} \end{bmatrix}.$$

For example, the matrix $\mathbf{H}_4$ is given by

$$\mathbf{H}_4 = \begin{bmatrix} 1 & 1 & 1 & 1 \\ 1 & -1 & 1 & -1 \\ 1 & 1 & -1 & -1 \\ 1 & -1 & -1 & 1 \end{bmatrix}.$$

The column vectors of the WH matrix are pairwise orthogonal but not normalized to one. We may divide them by their length $\sqrt{M}$ to obtain the normalized WH vectors $\mathbf{g}_k$ related to $\mathbf{H}_M$ by

$$\sqrt{M}\,\mathbf{H}_M = \mathbf{G} = \left[\mathbf{g}_1, \ldots, \mathbf{g}_M\right]. \tag{1.28}$$

The column vectors are orthonormal, that is,

$$\mathbf{g}_i \cdot \mathbf{g}_k = \delta_{ik}.$$

For a given value of M, the Walsh functions $g_k(t)$, $k = 1, \ldots, M$ are functions defined on a time interval $t \in [0, T_S]$ that are piecewise constant on time sub-intervals (called *chips*) of duration $T_c = T_S/M$. The sign of the function on the ith time sub-interval ($i = 1, \ldots, M$) is given by the ith component h_{ik} of the kth column vector in the WH matrix $\mathbf{H}_M$. The absolute value is normalized to $1/\sqrt{T_S}$ to obtain an orthonormal signal base, that is,

$$\langle g_i, g_k \rangle = \delta_{ik}.$$

Outside the interval $t \in [0, T_S]$, the Walsh functions vanish. For $M = 8$, the Walsh functions are depicted in Figure 1.7.

We can write the normalized Walsh functions as

$$g_k(t) = \sum_{i=1}^{M} g_{ik} c_i(t), \tag{1.29}$$

where $g_{ik} = h_{ik}/\sqrt{M}$, and the chip pulse $c_i(t)$ is defined by $c_i(t) = 1/\sqrt{T_c}$ in the ith chip interval and zero outside. Obviously

$$\langle c_i, c_k \rangle = \delta_{ik}$$

holds, that is, the chip pulses are orthonormal. Given the transmit base $\{g_k(t)\}_{k=1}^{M}$, we may transmit information carried by the symbols $\{s_k\}_{k=1}^{M}$ by using the signal

$$s(t) = \sum_{k=1}^{M} s_k g_k(t). \tag{1.30}$$

Instead of $\{g_k(t)\}_{k=1}^{M}$, we can use the equivalent discrete base $\{\mathbf{g}_k\}_{k=1}^{M}$ and transmit the vector

$$\mathbf{s} = \sum_{k=1}^{M} s_k \mathbf{g}_k. \tag{1.31}$$

This representation allows the geometrical interpretation of the transmit signal as a vector $\mathbf{s}$ in an M-dimensional Euclidean space. The transmit symbols s_k are the components of the vector $\mathbf{s}$ in the orthonormal Walsh–Hadamard base $\{\mathbf{g}_k\}_{k=1}^{M}$, which is a rotated version of the canonical Euclidean base $\mathbf{e}_1 = (1, 0, 0, \ldots, 0)^T$, $\mathbf{e}_2 = (0, 1, 0, \ldots, 0)^T, \ldots, \mathbf{e}_M = (0, \ldots, 0, 1)^T$ with the rotation given by the matrix $\mathbf{G}$. Equation (1.29) is the coordinate transform corresponding to the base transform from the base of chip pulses to the base of normalized Walsh functions. Thus, the chip pulse base can be identified with the canonical Euclidean base. For an ideal channel, the components s_k of the vector $\mathbf{s}$ can be completely

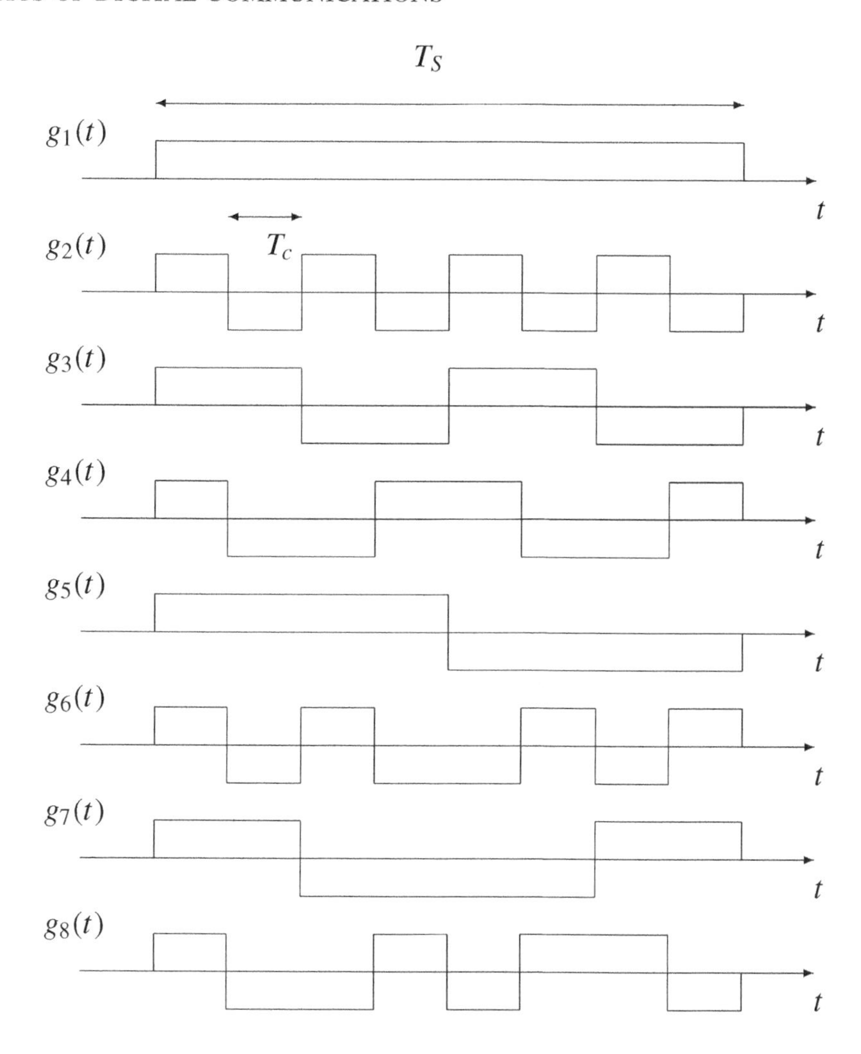

Figure 1.7 Walsh functions for $M = 8$.

recovered by projecting the receive vector on the base vectors $\mathbf{g_k}$. For $M = 2$, the situation is depicted in Figure 1.8.

The orthogonality of the base can be interpreted as the existence of M independent channels without cross talk as illustrated in Figure 1.9 for $M = 2$. The orthogonal detectors $\mathcal{D}_{g_i}$ correspond to the projections on the base vectors. If the symbols $\{s_k\}_{k=1}^{M}$ are part of the same data stream, M symbols are transmitted in parallel via M channels during the time interval T_S. Orthogonality means absence of ISI between these channels. Another possibility is to use the M channels to transmit M independent data streams, each of them with the rate of one symbol transmitted during the time interval T_S . In that case, the orthogonality is used for *multiplexing* or *multiple access*, as it is the situation in CDMA (code division multiple access). Each Walsh function corresponds to another code that may be allocated to

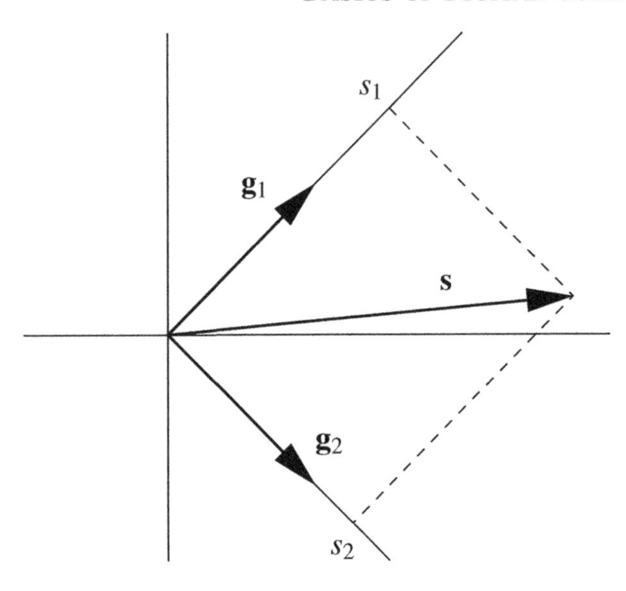

Figure 1.8 The orthonormal Walsh–Hadamard base.

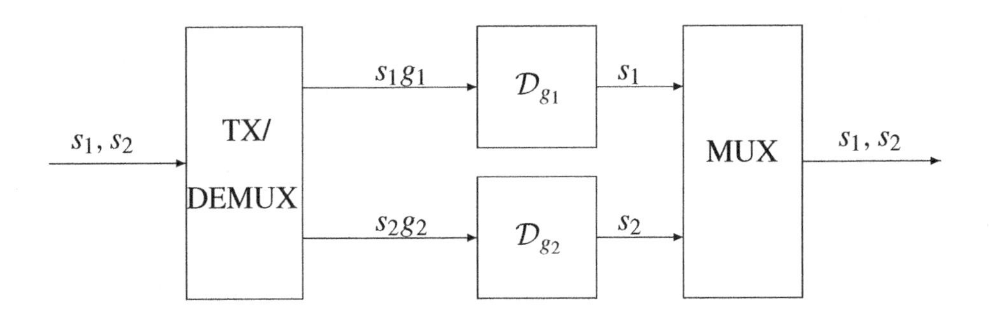

Figure 1.9 Orthogonal channels.

a certain user in a mobile radio scenario. The downlink of the Qualcomm CDMA system IS-95 (now called cdmaOne, see Subsection 5.5.6) is an example for such a setup (see the discussion in Chapter 5).

Walsh functions may also be used for orthogonal signaling. In that case the term *Walsh modulation* is often used. This transmission scheme is very robust against noise, but it leads to a significant spreading of the bandwidth. For CDMA systems, both properties together are desirable. This modulation scheme is used in the uplink of the Qualcomm system IS-95. In a setup with orthogonal Walsh modulation, only one base pulse $g_k(t)$ will be transmitted during the time period T_S. One can transmit $\log_2(M)$ bits during T_S by selecting one of the M base pulses. The transmit signal during this time period is simply given by

$$s(t) = \sqrt{E_S}\, g_k(t)$$

or

$$\mathbf{s} = \sqrt{E_S}\,\mathbf{g}_k,$$

where $\sqrt{E_S}$ is the energy of the signal in the time period of length T_S.

We note that all the properties discussed in this subsection hold not only for the piecewise constant chip pulse but also for any set of chip pulses $c_i(t)$ that satisfies the orthonormality condition $\langle c_i, c_k \rangle = \delta_{ik}$, and the base signal given by Equation (1.30) holds. One may, for example, choose any Nyquist base for the pulses $c_i(t)$ with smoother pulse shape than the rectangular pulses in order to get better spectral properties. It is also possible to use the Fourier base pulse for $c_i(t)$ in Equation (1.30), resulting in multicarrier (MC-) CDMA. Then, a chip may be interpreted as a frequency pulse rather than a time pulse.

1.1.5 Nonorthogonal bases

Orthonormal transmit bases have desirable properties for simple detection. However, sometimes it is necessary to deal with nonorthogonal bases. This may be the case when a transmission channel corrupts the orthogonality. For example, the channel may introduce ISI, so the ISI-free Nyquist base used for transmission will be convolved by the channel impulse response resulting in nonorthogonal base vectors. In such a case, the channel must be regarded as a part of the transmit setup.

Now, let the pulses $b_k(t)$ be such a base of nonorthogonal, but linearly independent, transmit pulses and let x_k, $k = 1, \ldots, K$ be the finite set of transmit symbols. The transmitted signal is then given by

$$s(t) = \sum_{k=1}^{K} x_k b_k(t). \tag{1.32}$$

There exists an orthonormal base $\{\psi_k\}_{k=1}^{K}$ for the finite-dimensional vector space spanned by the transmit pulses $b_k(t)$, which can be obtained using the Gram–Schmidt algorithm. The two bases are related by the base transform

$$b_k(t) = \sum_{i=1}^{K} b_{ik} \psi_i(t) \tag{1.33}$$

with $b_{ik} = \langle \psi_i, b_k \rangle$. We take the scalar product of Equation (1.32) with the vector $\psi_i(t)$ and obtain

$$s_i = \sum_{k=1}^{K} x_k b_{ik},$$

where we have defined $s_i = \langle \psi_i, s \rangle$. This is a coordinate transform of the coordinates x_k of the signal corresponding to the nonorthogonal base $b_k(t)$ to the coordinates s_k corresponding to the orthonormal base $\psi_k(t)$ by writing

$$s(t) = \sum_{k=1}^{K} s_k \psi_k(t).$$

Defining $\mathbf{x} = (x_1, \ldots, x_k)^T$ and $\mathbf{s} = (s_1, \ldots, s_k)^T$, the coordinate transform can be written in the more compact matrix notation as

$$\mathbf{s} = \mathbf{B}\mathbf{x}$$

with the matrix $\mathbf{B}$ of entries b_{ik}. Because this coordinate transform is invertible, the transmit symbols x_k can be recovered from the s_k by $\mathbf{x} = \mathbf{B}^{-1}\mathbf{s}$. Thus, in a channel without noise, the receiver may use the detector outputs corresponding to the orthonormal base and transform them by using the inverse matrix. However, in a noisy channel, this inversion is not optimal for the detection, because it causes noise enhancement.

For a receive signal $r(t)$, the detector outputs in the nonorthogonal base

$$y_k = \mathcal{D}_{b_k}[r] = \int_{-\infty}^{\infty} b_k^*(t) r(t)\, \mathrm{d}t$$

are related to the detector outputs

$$r_k = \mathcal{D}_{\psi_k}[r] = \int_{-\infty}^{\infty} \psi_k^*(t) r(t)\, \mathrm{d}t$$

in the orthogonal base by

$$y_k = \sum_{i=1}^{K} b_{ik}^* r_i$$

or, in matrix notation, by

$$\mathbf{y} = \mathbf{B}^{\dagger}\mathbf{r}.$$

Because this transform is invertible, the detector outputs r_k can be recovered from the y_k by $\mathbf{r} = \left(\mathbf{B}^{\dagger}\right)^{-1}\mathbf{y}$.

1.2 Baseband and Passband Transmission

It is convenient to describe a digitally modulated passband signal by a so-called *complex equivalent low-pass* or *complex baseband* signal. Let $\tilde{s}(t)$ be a strictly band-limited[1] passband signal of bandwidth B centered around a carrier frequency f_0. We have chosen the sign $\tilde{}$ for the signal $\tilde{s}(t)$ to indicate a wave for the RF signal. It is possible to completely describe $\tilde{s}(t)$ by its equivalent low-pass signal $s(t)$ of bandwidth $B/2$. Both signals are related by

$$\tilde{s}(t) = \sqrt{2}\,\Re\{s(t)\mathrm{e}^{j2\pi f_0 t}\}. \tag{1.34}$$

This one-to-one correspondence between both signals $\tilde{s}(t)$ and $s(t)$ is easy to visualize if we look at them in the frequency domain. Writing $\tilde{S}(f)$ and $S(f)$ for their respective Fourier transforms, Equation (1.34) is equivalent to

$$\tilde{S}(f) = \frac{1}{\sqrt{2}}\left(S\left(f - f_0\right) + S^*\left(-f - f_0\right)\right).$$

This is because, in Equation (1.34), multiplication with the exponential corresponds to a frequency shift by f_0, and taking the real part corresponds (up to a factor of 2) to adding

[1] We note that strictly band-limited signals do not exist in reality. This is due to the fact that a strictly band-limited signal cannot be time limited, which should be the case for signals in reality. However, it is mathematically convenient to make this assumption, always keeping in mind that this is only an approximation and the accuracy of this model has to be discussed for any practical case.

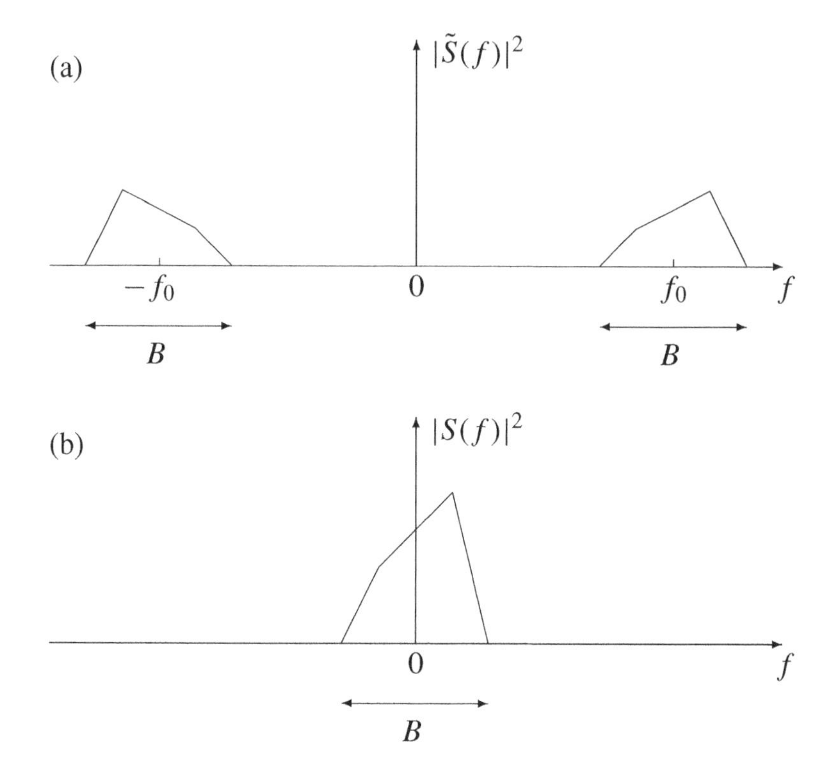

Figure 1.10 Equivalence of (a) passband and (b) complex baseband representation of the same signal.

the negative spectral component (see Problem 2). Figure 1.10 shows the energy spectral densities $|\tilde{S}(f)|^2$ and $|S(f)|^2$ for both signals. We have chosen a normalization such that the total energy of both signals is the same, that is,

$$\int_{-\infty}^{\infty} |s(t)|^2 \, dt = \int_{-\infty}^{\infty} |\tilde{s}(t)|^2 \, dt$$

and

$$\int_{-\infty}^{\infty} |S(f)|^2 \, df = \int_{-\infty}^{\infty} |\tilde{S}(f)|^2 \, df.$$

It is obvious from the figure that both signals $\tilde{S}(f)$ and $S(f)$ are only located at different frequencies, but they carry the same information. The signal $S(f)$ can be obtained from $\tilde{S}(f)$ by a frequency shift by $-f_0$ followed by an ideal low-pass filter of bandwidth $B/2$, and a scaling by the factor $\sqrt{2}$. Denoting the low-pass filter impulse response by $\Phi(t)$, this operation can be written in the time domain as

$$s(t) = \Phi(t) * \left[\sqrt{2} \exp\left(-j2\pi f_0 t\right) \tilde{s}(t)\right]. \tag{1.35}$$

We note that the upconversion from $s(t)$ to $\tilde{s}(t)$ as described by Equation (1.34) doubles the bandwidth of the two components (real and imaginary part) of the baseband signal, resulting in one signal of bandwidth B instead of two (real) signals of bandwidth $B/2$. We write $s(t) = x(t) + jy(t)$ and call the real part, $x(t)$, the I- *(inphase)* and the imaginary part $y(t)$ the Q- *(quadrature)* component. Both components together are called *quadrature components*. Equation (1.34) can then be written as

$$\tilde{s}(t) = \sqrt{2}\,x(t)\cos(2\pi f_0 t) - \sqrt{2}\,y(t)\sin(2\pi f_0 t), \tag{1.36}$$

which is the superposition of the cosine-modulated I-component and the (negative) sine-modulated Q-component. It is an important fact that the passband of width B is shared by two separable channels: one is the *cosine* and the other is the *sine* channel. We shall see that the I-component modulated by the cosine is orthogonal to the Q-component modulated by the sine. Thus, they behave like two different channels (without cross talk) that can be individually used for transmission. We will further see that both the I-modulation and the Q-modulation leave scalar products invariant. To make the following treatment simpler, we first introduce a formal shorthand notation for the quadrature modulator and demodulator.

1.2.1 Quadrature modulator

First we define the quadrature modulator as given by Equation (1.36), but separately for both components.

Definition 1.2.1 (I- and Q-modulator) *Let $x(t)$ and $y(t)$ be some arbitrary real signals. Let $\Phi(t)$ denote the impulse response of an ideal low-pass filter of bandwidth $B/2$. Let $f_0 > B/2$. We then define the modulated signals $\tilde{x}(t)$ and $\tilde{y}(t)$ as*

$$\tilde{x}(t) = \sqrt{2}\cos(2\pi f_0 t)\,[\Phi(t) * x(t)] \tag{1.37}$$

and

$$\tilde{y}(t) = -\sqrt{2}\sin(2\pi f_0 t)\,\left[\Phi(t) * y(t)\right]. \tag{1.38}$$

We write $\tilde{x}(t) = I\{x(t)\}$ and $\tilde{y}(t) = Q\{y(t)\}$ or, shorthand, $\tilde{x} = Ix$ and $\tilde{y} = Qy$ and call the time-variant systems given by I and Q the I-modulator *and the* Q-modulator, *respectively. The time-variant system that converts the pair of signals $x(t)$ and $y(t)$ to the passband signal $\tilde{s} = Ix + Qy$ is called* quadrature modulator.

In a practical setup, the signals $x(t)$ and $y(t)$ are already low-pass signals, and thus the convolution with $\Phi(t)$ at the input is obsolete. For mathematical convenience, we prefer to define this time-variant system for arbitrary (not only low-pass) signals as inputs.

The following theorem states the orthogonality of the outputs of the I- and Q-modulator and that both modulators leave scalar products invariant for band-limited signals.

Theorem 1.2.2 *Let $x(t)$ and $y(t)$ be arbitrary real signals of finite energy and let $\tilde{x}(t) = I\{x(t)\}$ and $\tilde{y}(t) = Q\{y(t)\}$. Then,*

$$\langle \tilde{x}, \tilde{y} \rangle = 0. \tag{1.39}$$

Furthermore, let $\tilde{u}(t) = I\{u(t)\}$ and $\tilde{v}(t) = Q\{v(t)\}$ be two other signals of finite energy. If x, y, u, v are real low-pass signals strictly band-limited to $B/2$, then

$$\langle \tilde{u}, \tilde{x} \rangle = \langle u, x \rangle\,, \ \langle \tilde{v}, \tilde{y} \rangle = \langle v, y \rangle \tag{1.40}$$

holds. As a special case, we observe that both I- and Q-modulator leave the signal energy unchanged, that is, $\|x\|^2 = \|\tilde{x}\|^2$ and $\|y\|^2 = \|\tilde{y}\|^2$.

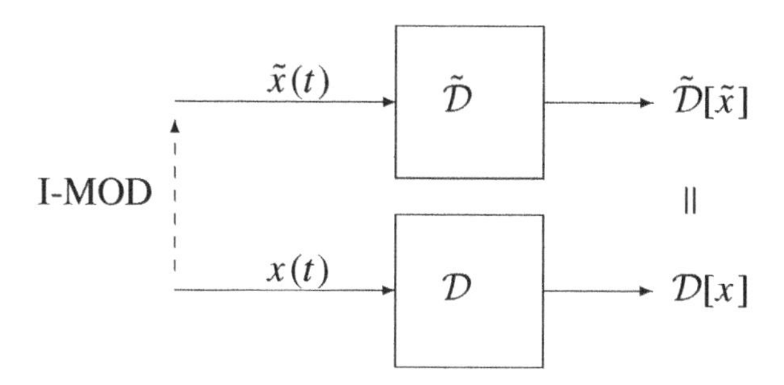

Figure 1.11 Equivalence of baseband and passband detection.

The proof of the theorem is left to Problem 3. Equation (1.39) states that the sine and the cosine channels are orthogonal. In particular, it means that a detector for a Q-modulated sine wave detects zero if only an I-modulated cosine wave has been transmitted and vice versa. For strictly band-limited baseband signals, Equation (1.40) states the following equivalence between baseband and passband transmission: we consider the detector for the baseband pulse $u(t)$ and denote it shorthand by $\mathcal{D} = \mathcal{D}_u$. For the passband detector corresponding to the I-modulated pulse $\tilde{u}(t) = I\{u(t)\}$, we write shorthand $\tilde{\mathcal{D}} = \mathcal{D}_{\tilde{u}}$. Then, as depicted in Figure 1.11, the baseband detector output for the baseband signal $x(t)$ is the same as the passband detector output for the I-modulated signal $\tilde{x}(t) = I\{x(t)\}$:

$$\tilde{\mathcal{D}}[\tilde{x}] = \mathcal{D}[x].$$

The same holds for the Q-modulation.

Now let $s(t) = x(t) + jy(t)$ and $z(t) = u(t) + jv(t)$ be strictly band-limited complex low-pass signals. The corresponding passband signals can be written as $\tilde{s} = Ix + Qy$ and $\tilde{z} = Iu + Qv$. Their scalar product is

$$\langle \tilde{z}, \tilde{s} \rangle = \langle Iu, Ix \rangle + \langle Qv, Qy \rangle + \langle Iu, Qy \rangle + \langle Qv, Ix \rangle .$$

As a consequence of the above theorem, the last two terms vanish and the first two can be converted, resulting in

$$\langle \tilde{z}, \tilde{s} \rangle = \langle u, x \rangle + \langle v, y \rangle .$$

We compare this expression with the scalar product of the two complex baseband signals

$$\langle z, s \rangle = \langle u, x \rangle + \langle v, y \rangle + j(\langle u, y \rangle - \langle v, x \rangle)$$

and find the relation

$$\langle \tilde{z}, \tilde{s} \rangle = \Re\{\langle z, s \rangle\}. \tag{1.41}$$

We note that there is a similar relation between the scalar products of vectors in the complex N-dimensional vector space $\mathcal{C}^N$ and in the real $2N$-dimensional space $\mathcal{R}^{2N}$. Let $\mathbf{s} = \mathbf{x} + j\mathbf{y}$ and $\mathbf{z} = \mathbf{u} + j\mathbf{v}$ be vectors in $\mathcal{C}^N$ and define the real vectors

$$\tilde{\mathbf{s}} = \begin{bmatrix} \mathbf{x} \\ \mathbf{y} \end{bmatrix}, \; \tilde{\mathbf{z}} = \begin{bmatrix} \mathbf{u} \\ \mathbf{v} \end{bmatrix}.$$

Then,

$$\tilde{\mathbf{z}} \cdot \tilde{\mathbf{s}} = \mathbf{u} \cdot \mathbf{x} + \mathbf{v} \cdot \mathbf{y}$$

and

$$\mathbf{z}^\dagger \mathbf{s} = \mathbf{u} \cdot \mathbf{x} + \mathbf{v} \cdot \mathbf{y} + j(\mathbf{u} \cdot \mathbf{y} - \mathbf{v} \cdot \mathbf{x})$$

and thus,

$$\tilde{\mathbf{z}} \cdot \tilde{\mathbf{s}} = \Re\{\mathbf{z}^\dagger \mathbf{s}\} \tag{1.42}$$

hold.

1.2.2 Quadrature demodulator

Consider again the detector $\mathcal{D} = \mathcal{D}_u$ and the detector $\tilde{\mathcal{D}} = \mathcal{D}_{\tilde{u}}$ for an I-modulated pulse $\tilde{u}(t) = I\{u(t)\}$. The detector output to an input signal $\tilde{s}(t)$ is given by $\tilde{\mathcal{D}}[\tilde{s}]$. We may ask for a time-variant system I^D called I-demodulator that maps $\tilde{s}(t)$ to a low-pass signal $x(t) = I^D\{\tilde{s}(t)\}$ which is defined by the property

$$\mathcal{D}[x] = \tilde{\mathcal{D}}[\tilde{s}],$$

(see Figure 1.12). Using simple integral manipulations, one can derive the explicit form of the I- (and similarly for the Q-) demodulator from this condition (see Problem 4). It turns out that I^D and Q^D are just given by the real and imaginary part of the Equation (1.35). We summarize the result in the following definition and theorem.

Definition 1.2.3 (I- and Q-demodulator) *Let $\tilde{s}(t)$ be a real signal. Let $\Phi(t)$ denote the impulse response of an ideal low-pass filter of bandwidth $B/2$. Let $f_0 > B/2$. We define the demodulated signals $x(t)$ and $y(t)$ as*

$$x(t) = \Phi(t) * \left[\sqrt{2}\cos(2\pi f_0 t)\, \tilde{s}(t)\right] \tag{1.43}$$

and

$$y(t) = -\Phi(t) * \left[\sqrt{2}\sin(2\pi f_0 t)\, \tilde{s}(t)\right] \tag{1.44}$$

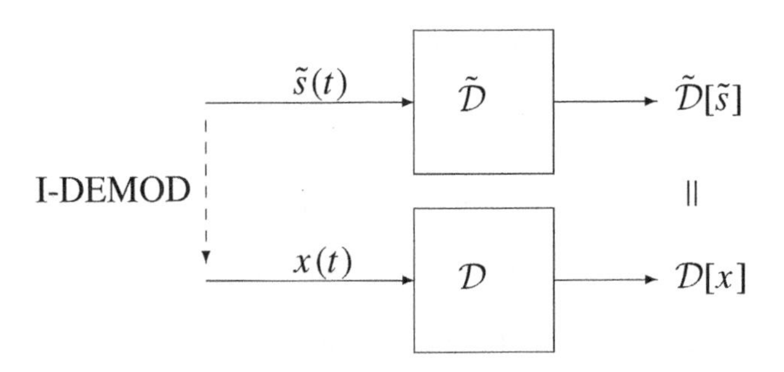

Figure 1.12 Characterization of the I-demodulator.

and write $x(t) = I^D\{\tilde{s}(t)\}$ and $y(t) = Q^D\{\tilde{s}(t)\}$ or, using shorthand, $x = I^D\tilde{s}$ and $y = Q^D\tilde{s}$. We call the time-variant systems given by I^D and Q^D the I-demodulator and the Q-demodulator, respectively. The time-variant system given by $I^D + jQ^D$ that converts $\tilde{s}$ to the complex signal

$$s(t) = \Phi(t) * \left[\sqrt{2}\exp(-j2\pi f_0 t)\,\tilde{s}(t)\right] \tag{1.45}$$

is called quadrature demodulator.

Theorem 1.2.4 *For real signals of finite energy, the I- and Q-demodulator have the following properties:*

$$\langle Iu, \tilde{s}\rangle = \left\langle u, I^D\tilde{s}\right\rangle, \quad \langle Qv, \tilde{s}\rangle = \left\langle v, Q^D\tilde{s}\right\rangle, \tag{1.46}$$

$$Q^D I x = 0, \quad I^D Q y = 0.$$

Conversely, Equation (1.46) uniquely determines the I- and Q-demodulation given by the above definition. Furthermore, let $x(t)$ and $y(t)$ be real signals of finite energy that are strictly band-limited to $B/2$. Then,

$$I^D I x = x, \quad Q^D Q y = y$$

holds. Thus, for band-limited signals, the I- (Q-)demodulator inverts the I- (Q-)modulator.

We may also write Equation (1.46) in the detector notation as

$$\mathcal{D}_{\tilde{u}}[\tilde{s}] = D_u[I^D\tilde{s}], \quad \mathcal{D}_{\tilde{v}}[\tilde{s}] = D_v[Q^D\tilde{s}] \tag{1.47}$$

with $\tilde{u}(t) = I\{u(t)\}$ and $\tilde{v}(t) = Q\{v(t)\}$. Without going into mathematical details, we note that if the detection pulses $u(t)$ and $v(t)$ are sufficiently well behaved, these equations – written as integrals – still make sense if the input signal is no longer of finite energy. This is the case, for example, for a sine wave, for a Dirac impulse, or for white noise, which is the topic of the next section.

1.3 The AWGN Channel

In reality, transmission is always corrupted by noise. The usual mathematical model is the AWGN (Additive White Gaussian Noise) channel. It is a very good model for the physical reality as long as the thermal noise at the receiver is the only source of disturbance. Nevertheless, because of its simplicity, it is often used to model man-made noise or multiuser interference. The AWGN channel model can be characterized as follows:

- The noise $w(t)$ is an *additive* random disturbance of the useful signal $s(t)$, that is, the receive signal is given by

$$r(t) = s(t) + w(t).$$

- The noise is *white*, that is, it has a constant power spectral density (psd). The one-sided psd is usually denoted by N_0, so $N_0/2$ is the two-sided psd, and BN_0 is the noise inside the (noise) bandwidth B, see part (a) of Figure 1.13. For thermal resistor noise, $N_0 = kT_0$, where k is the Boltzmann constant and T_0 is the absolute temperature. The

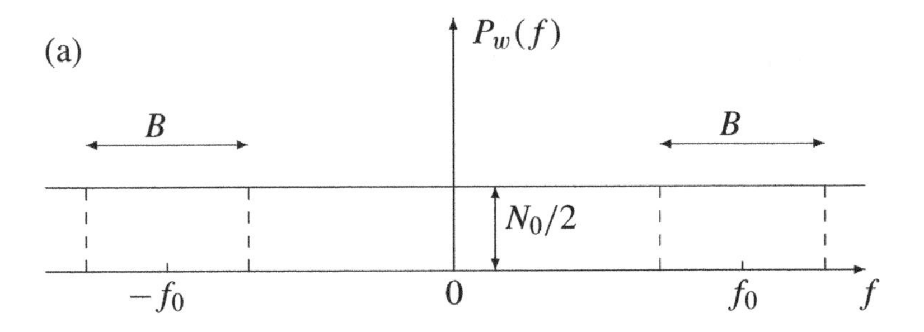

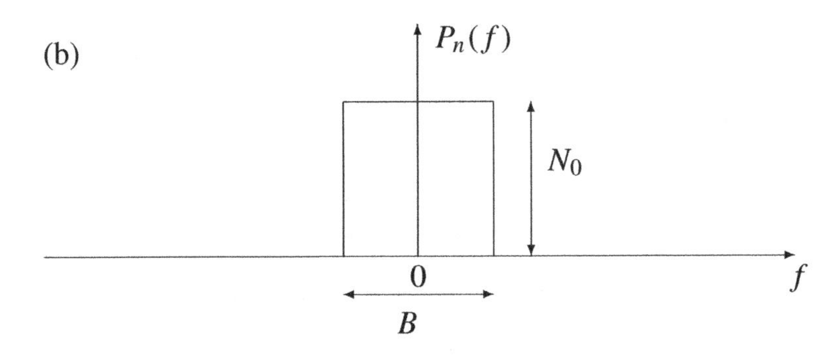

Figure 1.13 Equivalence of (a) wideband and (b) complex baseband AWGN for band-limited detection.

unit of N_0 is [W/Hz], which is the same as the unit [J] for the energy. Usually, N_0 is written as dBm/Hz. For $T_0 = 290$ K, $N_0 \approx -174$ dBm/Hz. However, this is only the ideal physical limit for an ideal receiver. In practice, some decibels according to the so-called *noise figure* have to be added. Typically, N_0 will be a value slightly above -170 dBm/Hz.

- The noise is a stationary and zero mean *Gaussian* random process. This means that the output of every (linear) noise measurement is a zero mean Gaussian random variable that does not depend on the time instant when the measurement is done.

One must keep in mind that the AWGN model is a mathematical fiction, because it implies that the total power (i.e. the psd integrated over all frequencies) is infinite. Thus, a time sample of the white noise has infinite average power, which is certainly not a physically reasonable property. It is known from statistical physics that the thermal noise density decreases exponentially at (very!) high frequencies. But to understand the physical situation in communications engineering it is better to keep in mind that every receiver limits the bandwidth as well as every physical noise measurement. So it makes sense to think of the noise process to be white, but it cannot be sampled directly without an input device. Each input device filters the noise and leads to a finite power.

1.3.1 Mathematical wideband AWGN

The mathematical AWGN random process $w(t)$ can be characterized as a zero mean Gaussian process with the autocorrelation function

$$\mathrm{E}\{w(t_1)w(t_2)\} = \frac{N_0}{2}\delta(t_1 - t_2). \tag{1.48}$$

We see that for $t_1 = t_2$, this expression is not well defined, because $\delta(t)$ is not well defined for $t = 0$. As for the δ-impulse, we must understand white noise as a *generalized function* that cannot be sampled directly, but it can be measured by a proper set of linear detectors. These linear detectors are the sampled outputs of linear filters. Thus, formally we can write the output of the detector for the (real) signal $\phi(t)$ of such a white noise measurement as $\mathcal{D}_\phi[w] = [\phi(-t) * w(t)]_{t=0}$ or

$$\mathcal{D}_\phi[w] = \int_{-\infty}^{\infty} \phi(t)w(t)\,\mathrm{d}t, \tag{1.49}$$

that is, $\phi(-t)$ is the impulse response of the measuring device. In the mathematical literature, $\phi(t)$ is called a *test function* (Reed and Simon 1980). Note that the integral in Equation (1.49) formally looks like the scalar product $\langle\phi, w\rangle$. Keeping in mind that the scalar product is well defined only for finite-energy signals, we have avoided such a notation. We can now characterize the white noise by the statistical properties of the outputs of linear detectors.

Definition 1.3.1 (White Gaussian Noise) *White Gaussian noise $w(t)$ is a random signal, characterized by the following properties: the output of a (finite-energy) linear detector $\mathcal{D}_\phi[w]$ is a Gaussian random variable with zero mean. Any two detector outputs $\mathcal{D}_{\phi_1}[w]$ and $\mathcal{D}_{\phi_2}[w]$ are jointly Gaussian with cross-correlation*

$$\mathrm{E}\left\{\mathcal{D}_{\phi_1}[w]\mathcal{D}_{\phi_2}[w]\right\} = \frac{N_0}{2}\langle\phi_1, \phi_2\rangle. \tag{1.50}$$

Since Gaussian random variables are completely characterized by their second-order properties (Papoulis 1991), all the statistical properties of $w(t)$ are fixed by this definition. Using the integral representation (1.49), it is easy to show that the characterization by detector outputs (1.50) is equivalent to (1.48) (see Problem 5).

Note that the AWGN outputs of orthogonal detectors are uncorrelated and thus, as Gaussian random variables, even statistically independent. In many transmission setups as discussed in the above examples, the transmission base and therefore the corresponding detectors are orthogonal. Thus, for an orthonormal transmission base, the detector outputs are both ISI-free and independent. Orthogonality thus means complete separability, which will lead us to the concept of *sufficient statistics* (see Subsection 1.4.1).

1.3.2 Complex baseband AWGN

Up to now, we have only considered wideband white Gaussian noise. The question now is what happens at the receiver when this additive disturbance of the useful signal is converted (together with the signal) to the complex baseband.

Consider a baseband detector for the real low-pass signal $\phi(t)$ and the detector that corresponds to the I-modulated version for the signal, $\tilde{\phi} = I\phi$. We know from Equation (1.47) that the output of the detector for $\tilde{\phi}$ for any signal r is the same as the output of ϕ for the I-demodulated version $I^D r$ of that signal. This same property also holds for white noise, that is, $\mathcal{D}_{\tilde{\phi}}[w] = \mathcal{D}_{\phi}[I^D w]$. Similarly, for the Q-demodulated noise, we have the property $\mathcal{D}_{\tilde{\psi}}[w] = \mathcal{D}_{\psi}[Q^D w]$ with $\tilde{\psi} = Q\psi$. Using this fact, together with the definition of AWGN and Theorem 1.2.2, we get the following Proposition.

Proposition 1.3.2 *Let $w(t)$ be additive white Gaussian noise and its I- and Q-demodulated versions be denoted by $I^D w$ and $Q^D w$. Let ϕ_1 and ϕ_2 be two strictly band-limited low-pass detectors. Then, the following properties hold:*

$$\mathrm{E}\left\{\mathcal{D}_{\phi_1}[I^D w]\mathcal{D}_{\phi_2}[I^D w]\right\} = \frac{N_0}{2}\langle\phi_1, \phi_2\rangle \tag{1.51}$$

$$\mathrm{E}\left\{\mathcal{D}_{\phi_1}[Q^D w]\mathcal{D}_{\phi_2}[Q^D w]\right\} = \frac{N_0}{2}\langle\phi_1, \phi_2\rangle \tag{1.52}$$

$$\mathrm{E}\left\{\mathcal{D}_{\phi_1}[I^D w]\mathcal{D}_{\phi_2}[Q^D w]\right\} = 0. \tag{1.53}$$

The last equation means that the I- and the Q-demodulator produce statistically independent outputs[2]. Since both demodulators include a low-pass filter, both $I^D w$ and $Q^D w$ are well-behaved random processes having finite average power $BN_0/2$. The samples (with sampling frequency B) of the low-pass white noise are given by the outputs of the detector corresponding to $\Phi(t)$, the impulse response of the ideal low-pass filter of bandwidth $B/2$. Thus, the low-pass white noise can be characterized by its detector outputs and by its samples as well.

We now define the complex baseband noise process $n(t)$ as the IQ-demodulated white noise

$$n = (I^D + jQ^D)w. \tag{1.54}$$

From the above proposition, we conclude

$$\mathrm{E}\{\mathcal{D}_{\phi_1}[n]\,\mathcal{D}^*_{\phi_2}[n]\} = N_0\langle\phi_1, \phi_2\rangle \tag{1.55}$$

and

$$\mathrm{E}\{\mathcal{D}_{\phi_1}[n]\,\mathcal{D}_{\phi_2}[n]\} = 0. \tag{1.56}$$

Complex random variables are characterized by these two types of covariances. Here the second one, the so-called *pseudocovariance*, has vanished. Gaussian processes with this property are called *proper* Gaussian. Nonproper Gaussian random variables have undesired properties for describing communication systems (Neeser and Massey 1993). The autocorrelation properties of $n(t)$ can simply be obtained by setting $\phi_1(t) = \Phi_{t_1}(t) = \Phi(t - t_1)$ and $\phi_2(t) = \Phi_{t_2}(t) = \Phi(t - t_2)$, where $\Phi(t)$ is the impulse response of the ideal low-pass filter of bandwidth $B/2$. Using $\langle\Phi_{t_1}, \Phi_{t_2}\rangle = \Phi(t_1 - t_2)$ we easily derive the following properties

$$\mathrm{E}\{n(t_1)\,n^*(t_2)\} = N_0\Phi(t_1 - t_2) \tag{1.57}$$

[2]Equivalently, we may say that the I- and the Q-component of the noise are statistically independent.

and

$$\mathrm{E}\{n(t_1)\, n(t_2)\} = 0. \tag{1.58}$$

$n(t)$ has similar properties as $w(t)$. It is white, but restricted to a bandwidth B, and the constant psd is N_0 instead of $N_0/2$. This can be understood because $n(t)$ is complex and the total power is the sum of the powers of the real and the imaginary part. Passband and complex baseband white noise psd is depicted in Figure 1.13.

Instead of dealing with complex white Gaussian noise with psd N_0 band-limited to $B/2$, many authors regard it as convenient to perform the limit $B \to \infty$, so that Equation (1.57) turns into

$$\mathrm{E}\{n(t_1)\, n^*(t_2)\} = N_0 \delta(t_1 - t_2),$$

that is, $n(t)$ becomes complex infinite-bandwidth white noise with *one-* sided psd N_0 (see e.g. (Kammeyer 2004; Proakis 2001)). This is reasonable if we think of downconverting with a low-pass filter of bandwidth much larger than the signal bandwidth. We wish to point out that the limit $B \to \infty$, that is, the wideband complex white noise does not reflect the physical reality but is only a mathematically convenient model. The equivalence between passband and baseband is only true for band-limited signals, especially $B/2 < f_0$ must hold.

Proposition 1.3.3 (Baseband stochastic processes) *Consider a (real-valued) stochastic process $\tilde{z}(t)$ that influences the useful signal in the air. We want to characterize the IQ-demodulator output*

$$z(t) = \Phi(t) * \left[\sqrt{2}\, \exp\left(-j2\pi f_0 t\right) \tilde{z}(t)\right]$$

by its second-order properties, that is, an autocorrelation function. Here $\Phi(t) = B\, \mathrm{sinc}\,(Bt)$ is the impulse response of the ideal low-pass filter of bandwidth $B/2$. We may think of white noise $w(t)$ as such a process, but also of an RF carrier that is broadened by the Doppler effect in a mobile radio environment (see Chapter 2). The following treatment is very general. We only assume that the random RF signal $\tilde{z}(t)$ has zero mean and it is wide-sense stationary (WSS), which means that the autocorrelation function

$$\tilde{\mathcal{R}}(\tau) = \mathrm{E}\left\{\tilde{z}(t+\tau)\tilde{z}(t)\right\}$$

of the process does not depend on t. We want to show that

$$\mathrm{E}\left\{z(t+\tau)z^*(t)\right\} = \Phi(\tau) * \left[2\, \exp\left(-j2\pi f_0 \tau\right)\, \tilde{\mathcal{R}}(\tau)\right] \tag{1.59}$$

and

$$\mathrm{E}\left\{z(t+\tau)z(t)\right\} = 0. \tag{1.60}$$

Obviously, for the special case of AWGN, this property is just given by the two Equations (1.57, 1.58) above.

To prove Equation (1.59), we apply the convolution in the definition of $z(t)$

$$\begin{aligned}
&\mathrm{E}\left\{z(t+\tau)z^*(t)\right\} \\
&\quad = 2\,\mathrm{E}\left\{\int_{-\infty}^{\infty} \mathrm{d}t_1\,\Phi(t+\tau-t_1)\mathrm{e}^{-j2\pi f_0 t_1}\tilde{z}(t_1)\int_{-\infty}^{\infty} \mathrm{d}t_2\,\Phi(t-t_2)\mathrm{e}^{j2\pi f_0 t_2}\tilde{z}(t_2)\right\} \\
&\quad = 2\,\mathrm{E}\left\{\int_{-\infty}^{\infty} \mathrm{d}t_1 \int_{-\infty}^{\infty} \mathrm{d}t_2\,\Phi(t+\tau-t_1)\Phi(t-t_2)\mathrm{e}^{-j2\pi f_0(t_1-t_2)}\tilde{\mathcal{R}}(t_1-t_2)\right\} \\
&\quad = 2\,\mathrm{E}\left\{\int_{-\infty}^{\infty} \mathrm{d}t_1 \int_{-\infty}^{\infty} \mathrm{d}t_2 \int_{-\infty}^{\infty} \mathrm{d}f\;\Phi(t+\tau-t_1)\Phi(t-t_2)\mathrm{e}^{j2\pi(f-f_0)(t_1-t_2)}\tilde{\mathcal{S}}(f)\right\},
\end{aligned}$$

where we have expressed $\tilde{\mathcal{R}}(\tau)$ by means of its Fourier transform, that is, the power spectral density $\tilde{\mathcal{S}}(f)$ of the process. We have used the simpler notation $\int \mathrm{d}x \int \mathrm{d}y f(x, y)$ instead of $\int \left(\int f(x, y)\,\mathrm{d}y\right)\mathrm{d}x$. Substituting the time integration variables according to $t_1' = t + \tau - t_1$ and $t_2' = t - t_2$ and noting that $t_1 - t_2 = \tau - t_1' + t_2'$, we get the expression

$$\begin{aligned}
&2\,\mathrm{E}\left\{\int_{-\infty}^{\infty} \mathrm{d}f\,\tilde{\mathcal{S}}(f)\int_{-\infty}^{\infty} \mathrm{d}t_1'\,\Phi(t_1')\mathrm{e}^{j2\pi(f-f_0)(\tau-t_1')}\int_{-\infty}^{\infty} \mathrm{d}t_2'\,\Phi(t_2')\mathrm{e}^{+j2\pi(f-f_0)t_2'}\right\} \\
&\quad = 2\,\mathrm{E}\left\{\int_{-\infty}^{\infty} \mathrm{d}f\,\mathrm{e}^{j2\pi(f-f_0)\tau}\tilde{\mathcal{S}}(f)\,\Pi\left(\frac{f-f_0}{B}\right)\right\} \\
&\quad = 2\,\mathrm{E}\left\{\int_{-\infty}^{\infty} \mathrm{d}f\,\mathrm{e}^{j2\pi f\tau}\tilde{\mathcal{S}}(f+f_0)\,\Pi\left(\frac{f}{B}\right)\right\},
\end{aligned}$$

which completes the proof. We note that

$$\mathcal{S}(f) = 2\,\tilde{\mathcal{S}}(f+f_0)\,\Pi\left(\frac{f}{B}\right),$$

the power spectral density of the process in the complex baseband, is the Fourier transform of the complex baseband autocorrelation function

$$\mathcal{R}(\tau) = \Phi(\tau) * \left[2\,\exp\left(-j2\pi f_0\tau\right)\,\tilde{\mathcal{R}}(\tau)\right].$$

The proof of Equation (1.60) is similar. Applying again the convolution in the definition of $z(t)$ leads to

$$\begin{aligned}
&\mathrm{E}\left\{z(t+\tau)z(t)\right\} \\
&\quad = 2\,\mathrm{E}\left\{\int_{-\infty}^{\infty} \mathrm{d}t_1\,\Phi(t+\tau-t_1)\mathrm{e}^{-j2\pi f_0 t_1}\tilde{z}(t_1)\int_{-\infty}^{\infty} \mathrm{d}t_2\,\Phi(t-t_2)\mathrm{e}^{-j2\pi f_0 t_2}\tilde{z}(t_2)\right\} \\
&\quad = 2\,\mathrm{E}\left\{\int_{-\infty}^{\infty} \mathrm{d}t_1 \int_{-\infty}^{\infty} \mathrm{d}t_2\,\Phi(t+\tau-t_1)\Phi(t-t_2)\mathrm{e}^{-j2\pi f_0(t_1+t_2)}\tilde{\mathcal{R}}(t_1-t_2)\right\} \\
&\quad = 2\,\mathrm{E}\left\{\int_{-\infty}^{\infty} \mathrm{d}t_1 \int_{-\infty}^{\infty} \mathrm{d}t_2 \int_{-\infty}^{\infty} \mathrm{d}f\;\Phi(t+\tau-t_1)\Phi(t-t_2)\mathrm{e}^{-j2\pi f_0(t_1+t_2)}\mathrm{e}^{j2\pi f(t_1-t_2)}\tilde{\mathcal{S}}(f)\right\}.
\end{aligned}$$

We now substitute the time integration variables according to $t_1' = t + \tau - t_1$ and $t_2' = t - t_2$ and get the expression

$$2\,\mathrm{E}\left\{\int_{-\infty}^{\infty} \mathrm{d}f\, \tilde{S}(f) \int_{-\infty}^{\infty} \mathrm{d}t_1' \Phi(t_1') \mathrm{e}^{j2\pi(f-f_0)(t+\tau-t_1')} \int_{-\infty}^{\infty} \mathrm{d}t_2' \Phi(t_2') \mathrm{e}^{-j2\pi(f+f_0)(t_2'-t)}\right\}$$

$$= 2\,\mathrm{E}\left\{\int_{-\infty}^{\infty} \mathrm{d}f\, \tilde{S}(f)\, \mathrm{e}^{j2\pi(f-f_0)(t+\tau)}\, \Pi\left(\frac{f-f_0}{B}\right) \mathrm{e}^{j2\pi(f+f_0)t}\, \Pi\left(\frac{f+f_0}{B}\right)\right\}.$$

We note that

$$\Pi\left(\frac{f-f_0}{B}\right) \Pi\left(\frac{f+f_0}{B}\right) = 0,$$

which completes the proof.

1.3.3 The discrete AWGN channel

Consider a complex baseband signal $s(t)$ band limited to $B/2$ to be transmitted at the carrier frequency f_0. The corresponding passband transmit signal $\tilde{s}(t)$ given by Equation (1.34) is corrupted by AWGN, resulting in the receive signal

$$\tilde{r}(t) = \tilde{s}(t) + w(t).$$

The IQ-demodulated complex baseband receive signal is then given by

$$r(t) = s(t) + n(t), \tag{1.61}$$

where $n(t)$ is complex baseband AWGN as introduced in the preceding subsection. Let $\{g_k(t)\}_{k=1}^{K}$ be an orthogonal transmit base, for example, a Nyquist base, and

$$s(t) = \sum_{k=1}^{K} s_k g_k(t). \tag{1.62}$$

Let $n_k = \mathcal{D}_{g_k}[n]$ be the detector outputs of the noise for the detector $g_k(t)$. The detector outputs at the receiver $r_k = \mathcal{D}_{g_k}[r]$ are then given by

$$r_k = s_k + n_k. \tag{1.63}$$

We conclude from Equations (1.55) and (1.56) that n_k is discrete complex AWGN characterized by

$$\mathrm{E}\{n_i n_k^*\} = N_0 \delta_{ik} \tag{1.64}$$

and

$$\mathrm{E}\{n_i n_k\} = 0. \tag{1.65}$$

For $n_k = x_k + j\, y_k$, these two equations are equivalent to

$$\mathrm{E}\{x_i x_k\} = \frac{N_0}{2}\delta_{ik}, \tag{1.66}$$

$$\mathrm{E}\{y_i y_k\} = \frac{N_0}{2}\delta_{ik}, \tag{1.67}$$

and

$$\mathrm{E}\{x_i\, y_k\} = 0. \tag{1.68}$$

The random variables x_k have the joint pdf

$$p(x_1, \ldots, x_K) = \frac{1}{\sqrt{2\pi\sigma^2}^K} \exp\left(-\frac{1}{2\sigma^2}\left(x_1^2 + \cdots + x_K^2\right)\right)$$

for $\sigma^2 = N_0/2$. The random variables y_k have the same pdf. Defining the vectors

$$\mathbf{x} = (x_1, \ldots, x_K)^T,\ \mathbf{y} = (y_1, \ldots, y_K)^T$$

results in

$$p(\mathbf{x}) = \frac{1}{\sqrt{2\pi\sigma^2}^K} \exp\left(-\frac{1}{2\sigma^2}\|\mathbf{x}\|^2\right),$$

$$p(\mathbf{y}) = \frac{1}{\sqrt{2\pi\sigma^2}^K} \exp\left(-\frac{1}{2\sigma^2}\|\mathbf{y}\|^2\right).$$

The joint pdf for the x_k, y_k is the product of both. We define the complex noise vector

$$\mathbf{n} = (n_1, \ldots, n_K)^T,$$

and write, with $p(\mathbf{n}) = p(\mathbf{x}, \mathbf{y}) = p(\mathbf{x})p(\mathbf{y})$,

$$p(\mathbf{n}) = \frac{1}{\sqrt{2\pi\sigma^2}^{2K}} \exp\left(-\frac{1}{2\sigma^2}\|\mathbf{n}\|^2\right).$$

Using the vector notation $\mathbf{s} = (s_1, \ldots, s_K)^T$ and $\mathbf{r} = (r_1, \ldots, r_K)^T$ for the transmit symbols and the detector outputs, we write the discrete AWGN transmission channel (1.63) as

$$\mathbf{r} = \mathbf{s} + \mathbf{n}. \tag{1.69}$$

If the symbols s_k are real numbers, one can depict this as a transmission mission of a vector in the K-dimensional real Euclidean space with the canonical base $\mathbf{e}_1 = (1, 0, 0, \ldots, 0)^T$, $\mathbf{e}_2 = (0, 1, 0, \ldots, 0)^T, \ldots$. For complex s_k, one may think of a $2K$-dimensional real Euclidean space, because it has the same distance structure as a K-dimensional complex space.

We have assumed that the transmit base is orthonormal. This is not a fundamental restriction, because one can always perform a base transform to an orthonormal base. In that case, the symbols s_i are related to the original transmit symbols x_k by a transform $\mathbf{s} = \mathbf{Bx}$, where $\mathbf{B}$ is the matrix that describes the coordinate transform.

1.4 Detection of Signals in Noise

1.4.1 Sufficient statistics

For the sake of simplicity, consider the model of Equations (1.63) and (1.69) for only three real dimensions as illustrated in Figure 1.14. Assume that two real symbols s_1, s_2

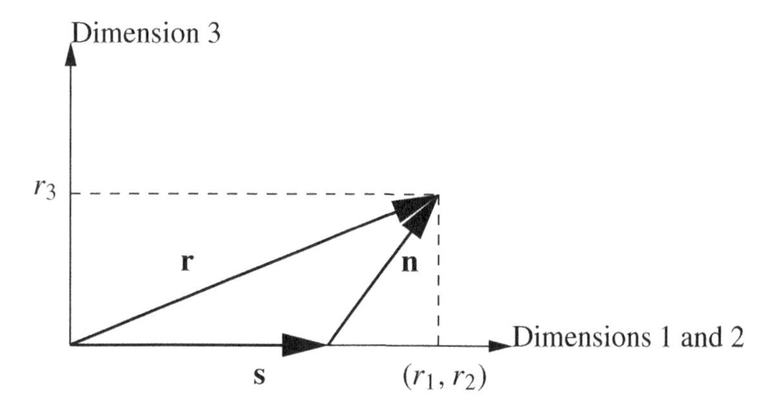

Figure 1.14 The noise in dimension 3 is irrelevant for the decision.

chosen from a finite alphabet are transmitted, while nothing is transmitted ($s_3 = 0$) in the third dimension. At the receiver, the detector outputs r_1, r_2, r_3 for three real dimensions are available. We can assume that the signal and the noise are statistically independent. We know that the Gaussian noise samples n_1, n_2, n_3, as outputs of orthogonal detectors, are statistically independent. It follows that the detector outputs r_1, r_2, r_3 are statistically independent. We argue that only the receiver outputs for those dimensions where a symbol has been transmitted are relevant for the decision and the others can be ignored because they are statistically independent, too. In our example, this means that we can ignore the receiver output r_3. Thus, we expect that

$$P(s_1, s_2|r_1, r_2, r_3) = P(s_1, s_2|r_1, r_2) \tag{1.70}$$

holds, that is, the probability that s_1, s_2 was transmitted conditioned by the observation of r_1, r_2, r_3 is the same as conditioned by the observation of only r_1, r_2. We now show that this equation follows from the independence of the detector outputs. From Bayes rule (Feller 1970), we get

$$P(s_1, s_2|r_1, r_2, r_3) = \frac{p(s_1, s_2, r_1, r_2, r_3)}{p(r_1, r_2, r_3)}, \tag{1.71}$$

where $p(a, b, \ldots)$ denotes the joint pdf for the random variable $a, b, \ldots$. Since r_3 is statistically independent from the other random variables s_1, s_2, r_1, r_2, it follows that

$$P(s_1, s_2|r_1, r_2, r_3) = \frac{p(s_1, s_2, r_1, r_2)p(r_3)}{p(r_1, r_2)p(r_3)}. \tag{1.72}$$

From

$$P(s_1, s_2|r_1, r_2) = \frac{p(s_1, s_2, r_1, r_2)}{p(r_1, r_2)}, \tag{1.73}$$

we obtain the desired property given by Equation (1.70). Note that, even though this property is seemingly intuitively obvious, we have made use of the fact that the noise is Gaussian. White noise outputs of orthogonal detectors are uncorrelated, but the Gaussian property ensures that they are statistically independent, so that their pdfs can be factorized.

The above argument can obviously be generalized to more dimensions. We only need to detect in those dimensions where the signal has been transmitted. The corresponding detector outputs are then called a *set of sufficient statistics*. For a more detailed discussion, see (Benedetto and Biglieri 1999; Blahut 1990; Wozencraft and Jacobs 1965).

1.4.2 Maximum likelihood sequence estimation

Again we consider the discrete-time model of Equations (1.63) and (1.69) and assume a finite alphabet for the transmit symbols s_k , so that there is a finite set of possible transmit vectors $\mathbf{s}$. Given a receive vector $\mathbf{r}$, we ask for the most probable transmit vector $\hat{\mathbf{s}}$, that is, the one for which the conditional probability $P(\mathbf{s}|\mathbf{r})$ that $\mathbf{s}$ was transmitted given that $\mathbf{r}$ has been received becomes maximal. The estimate of the symbol is

$$\hat{\mathbf{s}} = \arg\max_{\mathbf{s}} P(\mathbf{s}|\mathbf{r}). \tag{1.74}$$

From Bayes law, we have

$$P(\mathbf{s}|\mathbf{r})p(\mathbf{r}) = p(\mathbf{r}|\mathbf{s})P(\mathbf{s}), \tag{1.75}$$

where $p(\mathbf{r})$ is the pdf for the receive vector $\mathbf{r}$, $p(\mathbf{r}|\mathbf{s})$ is the pdf for the receive vector $\mathbf{r}$ given a fixed transmit vector $\mathbf{s}$, and $P(\mathbf{s})$ is the *a priori* probability for $\mathbf{s}$. We assume that all transmit sequences have equal *a priori* probability. Then, from

$$p(\mathbf{r}|\mathbf{s}) \propto \exp\left(-\frac{1}{2\sigma^2}\|\mathbf{r}-\mathbf{s}\|^2\right), \tag{1.76}$$

we conclude that

$$\hat{\mathbf{s}} = \arg\min_{\mathbf{s}} \|\mathbf{r}-\mathbf{s}\|^2. \tag{1.77}$$

Thus, the most likely transmit vector minimizes the squared Euclidean distance. From

$$\|\mathbf{r}-\mathbf{s}\|^2 = \|\mathbf{r}\|^2 + \|\mathbf{s}\|^2 - 2\Re\left\{\mathbf{s}^\dagger\mathbf{r}\right\},$$

we obtain the alternative condition

$$\hat{\mathbf{s}} = \arg\max_{\mathbf{s}} \left(\Re\left\{\mathbf{s}^\dagger\mathbf{r}\right\} - \frac{1}{2}\|\mathbf{s}\|^2\right). \tag{1.78}$$

The first (scalar product) term can be interpreted as a cross correlation between the transmit and the receive signal. The second term is half the signal energy. Thus, the most likely transmit signal is the one that maximizes the cross correlation with the receive signal, thereby taking into account a correction term for the energy. If all transmit signals have the same energy, this term can be ignored.

The receiver technique described above, which finds the most likely transmit vector, is called *maximum likelihood sequence estimation* (MLSE). It is of fundamental importance in communication theory, and we will often need it in the following chapters.

A continuous analog to Equation (1.78) can be established. We recall that the continuous transmit signal $s(t)$ and the components s_k of the discrete transmit signal vector $\mathbf{s}$ are related by

$$s(t) = \sum_{k=1}^{K} s_k g_k(t),$$

and the continuous receive signal $r(t)$ and the components r_k of the discrete transmit signal vector $\mathbf{r}$ are related by

$$r_k = \mathcal{D}_{g_k}[r] = \int_{-\infty}^{\infty} g_k^*(t) r(t)\,\mathrm{d}t.$$

From these relations, we easily conclude that

$$\mathbf{s}^\dagger \mathbf{r} = \int_{-\infty}^{\infty} s^*(t) r(t)\,\mathrm{d}t$$

holds. Equation (1.78) is then equivalent to

$$\hat{s} = \arg\max_{s} \left(\Re\{\mathcal{D}_s[r]\} - \frac{1}{2}\|s\|^2 \right) \tag{1.79}$$

for finding the maximum likelihood (ML) transmit signal $\hat{s}(t)$. In the first term of this expression,

$$\mathcal{D}_s[r] = \int_{-\infty}^{\infty} s^*(t) r(t)\,\mathrm{d}t$$

means that the detector outputs (= sampled MF outputs) for all possible transmit signals $s(t)$ must be taken. For all these signals, half of their energy

$$\|s\|^2 = \int_{-\infty}^{\infty} |s(t)|^2\,\mathrm{d}t$$

must be subtracted from the real part of the detector output to obtain the likelihood of each signal.

Example 3 (Walsh Demodulator) *Consider a transmission with four possible transmit vectors* $\mathbf{s}_1$, $\mathbf{s}_2$, $\mathbf{s}_3$ *and* $\mathbf{s}_4$ *given by the columns of the matrix*

$$[\mathbf{s}_1, \mathbf{s}_2, \mathbf{s}_3, \mathbf{s}_4] = \begin{bmatrix} 1 & 1 & 1 & 1 \\ 1 & -1 & 1 & -1 \\ 1 & 1 & -1 & -1 \\ 1 & -1 & -1 & 1 \end{bmatrix},$$

each being transmitted with the same probability. This is just orthogonal Walsh modulation for $M = 4$. *We ask for the most probable transmit vector* $\hat{\mathbf{s}}$ *on the condition that the vector* $\mathbf{r} = (1.5, -0.8, 1.1, -0.2)^T$ *has been received. Since all transmit vectors have equal energy, the most probable transmit vector is the one that maximizes the scalar product with* $\mathbf{r}$. *We calculated the scalar products as*

$$\mathbf{s}_1 \cdot \mathbf{r} = 2.0, \quad \mathbf{s}_2 \cdot \mathbf{r} = 3.2, \quad \mathbf{s}_3 \cdot \mathbf{r} = 0.4, \quad \mathbf{s}_4 \cdot \mathbf{r} = 1.4.$$

We conclude that $\mathbf{s}_2$ *has most probably been transmitted.*

1.4.3 Pairwise error probabilities

Consider again a discrete AWGN channel as given by Equation (1.69). We write

$$\mathbf{r} = \mathbf{s} + \mathbf{n}_c,$$

where $\mathbf{n}_c$ is the complex AWGN vector. For the geometrical interpretation of the following derivation of error probabilities, it is convenient to deal with real vectors instead of complex ones. By defining

$$\mathbf{y} = \begin{bmatrix} \Re\{\mathbf{r}\} \\ \Im\{\mathbf{r}\} \end{bmatrix}, \quad \mathbf{x} = \begin{bmatrix} \Re\{\mathbf{s}\} \\ \Im\{\mathbf{s}\} \end{bmatrix},$$

and

$$\mathbf{n} = \begin{bmatrix} \Re\{\mathbf{n}_c\} \\ \Im\{\mathbf{n}_c\} \end{bmatrix},$$

we can investigate the equivalent discrete real AWGN channel

$$\mathbf{y} = \mathbf{x} + \mathbf{n}. \tag{1.80}$$

Consider the case that $\mathbf{x}$ has been transmitted, but the receiver decides for another symbol $\hat{\mathbf{x}}$. The probability for this event (excluding all other possibilities) is called the *pairwise error probability* (PEP) $P(\mathbf{x} \mapsto \hat{\mathbf{x}})$. Define the decision variable

$$X = \|\mathbf{y} - \mathbf{x}\|^2 - \|\mathbf{y} - \hat{\mathbf{x}}\|^2$$

as the difference of squared Euclidean distances. If $X > 0$, the receiver will take an erroneous decision for $\hat{\mathbf{x}}$. Then, using simple vector algebra (see Problem 7), we obtain

$$X = 2\left[\left(\mathbf{y} - \frac{\mathbf{x} + \hat{\mathbf{x}}}{2}\right) \cdot (\hat{\mathbf{x}} - \mathbf{x})\right].$$

The geometrical interpretation is depicted in Figure 1.15. The decision variable is (up to a factor) the projection of the difference between the receive vector $\mathbf{y}$ and the center point $\frac{1}{2}(\mathbf{x} + \hat{\mathbf{x}})$ between the two possible transmit vectors on the line between them. The decision threshold is a plane perpendicular to that line. Define $\mathbf{d} = \frac{1}{2}(\hat{\mathbf{x}} - \mathbf{x})$ as the difference vector between $\hat{\mathbf{x}}$ and the center point, that is, $d = \|\mathbf{d}\|$ is the distance of the two possible transmit signals from the threshold. Writing $\mathbf{y} = \mathbf{x} + \mathbf{n}$ and using $\mathbf{x} = \frac{1}{2}(\mathbf{x} + \hat{\mathbf{x}}) - \mathbf{d}$, the scaled decision variable $\tilde{X} = \frac{1}{4d}X$ can be written as

$$\tilde{X} = (-\mathbf{d} + \mathbf{n}) \cdot \frac{\mathbf{d}}{d}.$$

It can easily be shown that

$$n = \mathbf{n} \cdot \frac{\mathbf{d}}{d},$$

the projection of the noise onto the relevant dimension, is a Gaussian random variable with zero mean and variance $\sigma^2 = N_0/2$ (see Problem 8). Since $\tilde{X} = -d + n$, the error probability is given by

$$P(\tilde{X}) > 0) = P(n > d).$$

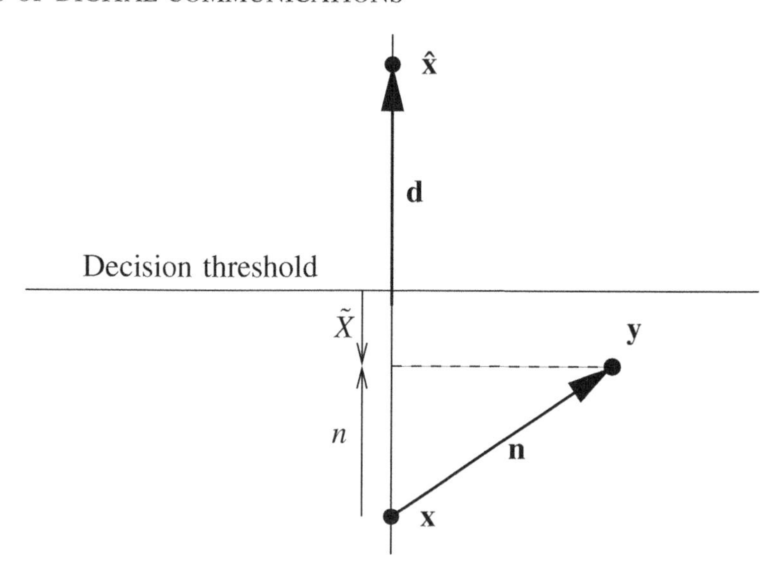

Figure 1.15 Decision threshold.

This equals

$$P(n > d) = Q\left(\frac{d}{\sigma}\right), \tag{1.81}$$

where the Gaussian probability integral is defined by

$$Q(x) = \frac{1}{\sqrt{2\pi}} \int_x^{\infty} e^{-\frac{1}{2}\xi^2}\, \mathrm{d}\xi$$

The Q-function defined above can be expressed by the complementary Gaussian error function $\operatorname{erfc}(x) = 1 - \operatorname{erf}(x)$, where $\operatorname{erf}(x)$ is the Gaussian error function, as

$$Q\,(x) = \frac{1}{2}\operatorname{erfc}\left(\frac{x}{\sqrt{2}}\right). \tag{1.82}$$

The pairwise error probability can then be expressed by

$$P(\mathbf{x} \mapsto \hat{\mathbf{x}}) = \frac{1}{2}\operatorname{erfc}\left(\sqrt{\frac{1}{4N_0}\,\|\mathbf{x} - \hat{\mathbf{x}}\|^2}\right). \tag{1.83}$$

Since the norms of complex vectors and the equivalent real vectors are identical, we can also write

$$P(\mathbf{s} \mapsto \hat{\mathbf{s}}) = \frac{1}{2}\operatorname{erfc}\left(\sqrt{\frac{1}{4N_0}\,\|\mathbf{s} - \hat{\mathbf{s}}\|^2}\right). \tag{1.84}$$

For the continuous signal,

$$s(t) = \sum_{k=1}^{K} s_k g_k(t), \tag{1.85}$$

this is equivalent to

$$P\left(s(t) \mapsto \hat{s}(t)\right) = \frac{1}{2}\mathrm{erfc}\left(\sqrt{\frac{1}{4N_0}\int_{-\infty}^{\infty} |s(t) - \hat{s}(t)|^2 \, \mathrm{d}t}\right). \tag{1.86}$$

It has been pointed out by Simon and Divsalar (Simon and Divsalar 1998) that, for many applications, the following polar representation of the complementary Gaussian error function provides a simpler treatment of many problems, especially for fading channels.

Proposition 1.4.1 (Polar representation of the Gaussian erfc function)

$$\frac{1}{2}\mathrm{erfc}(x) = \frac{1}{\pi}\int_0^{\pi/2} \exp\left(-\frac{x^2}{\sin^2\theta}\right) \mathrm{d}\theta. \tag{1.87}$$

Proof. The idea of the proof is to view the one-dimensional problem of pairwise error probability as two-dimensional and introduce polar coordinates. AWGN is a Gaussian random variable with mean zero and variance $\sigma^2 = 1$. The probability that the random variable exceeds a positive real value, x, is given by the Gaussian probability integral

$$Q(x) = \int_x^{\infty} \frac{1}{\sqrt{2\pi}} \exp\left(-\frac{1}{2}\xi^2\right) \mathrm{d}\xi. \tag{1.88}$$

This probability does not change if noise of the same variance is introduced in the second dimension. The error threshold is now a straight line parallel to the axis of the second dimension, and the probability is given by

$$Q(x) = \int_x^{\infty} \left(\int_{-\infty}^{\infty} \frac{1}{2\pi} \exp\left(-\frac{1}{2}(\xi^2 + \eta^2)\right) \mathrm{d}\eta\right) \mathrm{d}\xi. \tag{1.89}$$

This integral can be written in polar coordinates (r, ϕ) as

$$Q(x) = \int_{-\pi/2}^{\pi/2} \left(\int_{x/\cos\phi}^{\infty} \frac{r}{2\pi} \exp\left(-\frac{1}{2}r^2\right) \mathrm{d}r\right) \mathrm{d}\phi. \tag{1.90}$$

The integral over r can immediately be solved to give

$$Q(x) = \int_{-\pi/2}^{\pi/2} \frac{1}{2\pi} \exp\left(-\frac{1}{2}\frac{x^2}{\cos^2\phi}\right) \mathrm{d}\phi. \tag{1.91}$$

A simple symmetry argument now leads to the desired form of $\frac{1}{2}\mathrm{erfc}(x) = Q(\sqrt{2}x)$.

An upper bound of the erfc function can easily be obtained from this expression by upper bounding the integrand by its maximum value,

$$\frac{1}{2}\mathrm{erfc}(x) \leq \frac{1}{2}\mathrm{e}^{-x^2}. \tag{1.92}$$

Example 4 (PEP for Antipodal Modulation) *Consider the case of only two possible transmit signals $s_1(t)$ and $s_2(t)$ given by*

$$s_{1,2}(t) = \pm\sqrt{E_S}\, g(t),$$

where $g(t)$ is a pulse normalized to $\|g\|^2 = 1$, and E_S is the energy of the transmitted signal. To obtain the PEP, according to Equation (1.86), we calculate the squared Euclidean distance

$$\|s_1 - s_2\|^2 = \int_{-\infty}^{\infty} |s_1(t) - s_2(t)|^2 \, \mathrm{d}t$$

between two possible transmit signals $s_1(t)$ and $s_2(t)$ and obtain

$$\|s_1 - s_2\|^2 = \left\| \sqrt{E_s}\, g - \left(-\sqrt{E_s}\, g\right) \right\|^2 = 4E_S.$$

The PEP is then given by Equation (1.86) as

$$P\left(s_1(t) \mapsto s_2(t)\right) = \frac{1}{2}\operatorname{erfc}\left(\sqrt{\frac{E_S}{N_0}}\right).$$

One can transmit one bit by selecting one of the two possible signals. Therefore, the energy per bit is given by $E_b = E_S$ leading to the PEP

$$P\left(s_1(t) \mapsto s_2(t)\right) = \frac{1}{2}\operatorname{erfc}\left(\sqrt{\frac{E_b}{N_0}}\right).$$

Example 5 (PEP for Orthogonal Modulation) *Consider an orthonormal transmit base $g_k(t)$, $k = 1, \ldots, M$. We may think of the Walsh base or the Fourier base as an example, but any other choice is possible. Assume that one of the M possible signals*

$$s_k(t) = \sqrt{E_S}\, g_k(t)$$

is transmitted, where E_S is again the signal energy. In case of the Walsh base, this is just Walsh modulation. In case of the Fourier base, this is just (orthogonal) FSK (frequency shift keying). To obtain the PEP, we have to calculate the squared Euclidean distance

$$\|s_i - s_k\|^2 = \int_{-\infty}^{\infty} |s_i(t) - s_k(t)|^2 \, \mathrm{d}t$$

between two possible transmit signals $s_i(t)$ and $s_k(t)$ with $i \neq k$. Because the base is orthonormal, we obtain

$$\|s_i - s_k\|^2 = E_S \, \|g_i - g_k\|^2 = 2E_S.$$

The PEP is then given by

$$P\left(s_i(t) \mapsto s_k(t)\right) = \frac{1}{2}\operatorname{erfc}\left(\sqrt{\frac{E_S}{2N_0}}\right).$$

One can transmit $\log_2(M)$ bits by selecting one of M possible signals. Therefore, the energy per bit is given by $E_b = E_S / \log_2(M)$, leading to the PEP

$$P\left(s_i(t) \mapsto s_k(t)\right) = \frac{1}{2}\operatorname{erfc}\left(\sqrt{\log_2(M)\frac{E_b}{2N_0}}\right).$$

Concerning the PEP, we see that for $M = 2$, orthogonal modulation is inferior compared to antipodal modulation, but it is superior if more than two bits per signal are transmitted. The price for that robustness of high-level orthogonal modulation is that the number of the required signal dimensions and thus the required bandwidth increases exponentially with the number of bits.

1.5 Linear Modulation Schemes

Consider some digital information that is given by a finite bit sequence. To transmit this information over a physical channel by a passband signal $\tilde{s}(t) = \Re\left\{s(t)e^{j2\pi f_0 t}\right\}$, we need a mapping rule between the set of bit sequences and the set of possible signals. We call such a mapping rule a *digital modulation scheme*. A *linear* digital modulation scheme is characterized by the complex baseband signal

$$s(t) = \sum_{k=1}^{K} s_k g_k(t),$$

where the information is carried by the complex transmit symbols s_k. The modulation scheme is called *linear*, because this is a *linear mapping* from the vector $\mathbf{s} = (s_1, \ldots, s_K)^T$ of transmit symbols to the continuous transmit signal $s(t)$. In the following subsections, we will briefly discuss the most popular *signal constellations* for the modulation symbols s_k that are used to transmit information by choosing one of M possible points of that constellation. We assume that M is a power of two, so each complex symbol s_k carries $\log_2(M)$ bits of the information. Although it is possible to combine several symbols to a higher-dimensional constellation, the following discussion is restricted to the case where each symbol s_k is modulated separately by a tuple of $m = \log_2(M)$ bits. The rule how this is done is called the *symbol mapping* and the corresponding device is called the symbol mapper. In this section, we always deal with orthonormal base pulses $g_k(t)$. Then, as discussed in the preceding sections, we can restrict ourselves to a discrete-time transmission setup where the complex modulation symbols

$$s_k = x_k + jy_k$$

are corrupted by complex discrete-time white Gaussian noise n_k.

1.5.1 Signal-to-noise ratio and power efficiency

Since we have assumed orthonormal transmit pulses $g_k(t)$, the corresponding detector outputs are given by

$$r_k = s_k + n_k,$$

where n_k is discrete complex AWGN. We note that, because the pulses are normalized according to

$$\int_{-\infty}^{\infty} g_i^*(t) g_k(t)\, dt = \delta_{ik},$$

the detector changes the dimension of the signal; the squared continuous signals have the dimension of a power, but the squared discrete detector output signals have the dimension of an energy.

The average signal energy is given by

$$E = \mathrm{E}\left\{\int_{-\infty}^{\infty} |s(t)|^2 \, dt\right\} = \mathrm{E}\left\{\sum_{k=1}^{K} |s_k|^2\right\} = K\,\mathrm{E}\left\{|s_k|^2\right\},$$

where we have assumed that all the K symbols s_k have identical statistical properties. The energy per symbol $E_S = E/K$ is given by

$$E_S = \mathrm{E}\left\{|s_k|^2\right\}.$$

The energy of the detector output of the noise is

$$E_N = \mathrm{E}\left\{|n_k|^2\right\} = N_0,$$

so the signal-to-noise ratio, *SNR*, defined as the ratio between the signal energy and the relevant noise, results in

$$SNR = \frac{E_S}{N_0}.$$

When thinking of practical receivers, it may be confusing that a detector changes the dimension of the signal, because we have interpreted it as a matched filter together with a sampling device. To avoid this confusion, we may introduce a proper constant. For signaling with the Nyquist base, $g_k(t) = g(t - kT_S)$, one symbol s_k is transmitted in each time interval of length T_S. We then define the matched filter by its impulse response

$$h(t) = \frac{1}{\sqrt{T_S}} g^*(-t)$$

so that the matched filter output $h(t) * r(t)$ has the same dimension as the input signal $r(t)$. The samples of the matched filter output are given by

$$\frac{1}{\sqrt{T_S}} r_k = \frac{1}{\sqrt{T_S}} s_k + \frac{1}{\sqrt{T_S}} n_k.$$

Then, the power of the sampled useful signal is given by

$$P_S = \mathrm{E}\left\{\left|\frac{1}{\sqrt{T_S}} s_k\right|^2\right\} = \frac{E_S}{T_S},$$

and the noise power is

$$P_N = \mathrm{E}\left\{\left|\frac{1}{\sqrt{T_S}} n_k\right|^2\right\} = \frac{N_0}{T_S}.$$

Thus, the SNR may equivalently be defined as

$$SNR = \frac{P_S}{P_N},$$

which is the more natural definition for practical measurements.

The SNR is a physical quantity that can easily be measured, but it does not say anything about the power efficiency. To evaluate the power efficiency, one must know the average energy E_b per useful bit at the receiver that is needed for a reliable recovery of the information. If $\log_2(M)$ useful bits are transmitted by each symbol s_k, the relation

$$E_S = \log_2(M)\, E_b$$

holds, which relates both quantities by

$$SNR = \log_2(M)\, \frac{E_b}{N_0}.$$

We note the important fact that $E_b = P_S/R_b$ is just the average signal power P_S needed per useful bit rate R_b. Therefore, a modulation that needs less E_b/N_0 to achieve a reliable transmission is more *power efficient* .

In the following sections, we discuss the most popular symbol mappings and their properties.

1.5.2 ASK and QAM

For M-ASK (*amplitude-shift keying*), a tuple of $m = \log_2(M)$ bits will be mapped only on the real part x_k of s_k, while the imaginary part y_k will be set to zero. The M points will be placed equidistant and symmetrically about zero. Denoting the distance between two points by $2d$, the signal constellation for 2-ASK is given by $x_l \in \{\pm d\}$, for 4-ASK by $x_l \in \{\pm d, \pm 3d\}$ and for 8-ASK by $x_l \in \{\pm d, \pm 3d, \pm 5d, \pm 7d\}$. We consider *Gray mapping*, that is, two neighboring points differ only in one bit. In Figure 1.16, the M-ASK signal constellations are depicted for $M = 2, 4, 8$.

Assuming the same *a priori* probability for each signal point, we easily calculate the symbol energies as $E_S = \mathrm{E}\left\{|s_k|^2\right\} = d^2,\ 5d^2,\ 21d^2$ for these constellations, leading to the respective energies per bit $E_b = E_S/\log_2(M) = d^2,\ 2.5d^2,\ 7d^2$.

Adjacent points have the distance $2d$, so the distance to the corresponding decision threshold is given by d. If a certain point of the constellation is transmitted, the probability that an error occurs because the discrete noise with variance $\sigma^2 = N_0/2$ (per real dimension) exceeds the distance to the decision threshold with distance d is given by

$$P_{err} = \mathrm{Q}\left(\frac{d}{\sigma}\right) = \frac{1}{2}\,\mathrm{erfc}\left(\sqrt{\frac{d^2}{N_0}}\right), \tag{1.93}$$

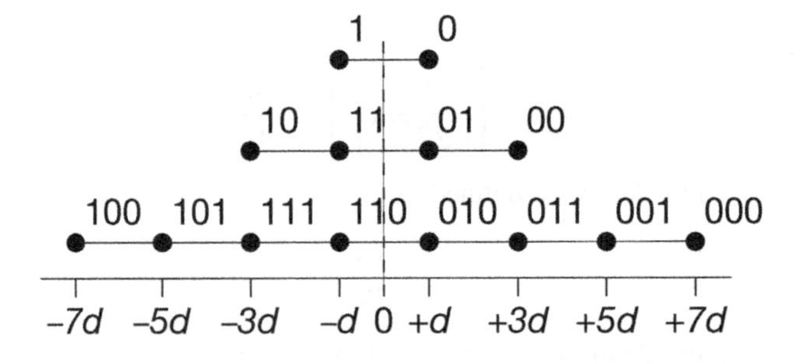

Figure 1.16 M-ASK Constellation for $M = 2, 4, 8$.

see Equation (1.81). For the two outer points of the constellation, this is just the probability that a symbol error occurs. In contrast, for $M > 2$, each inner point has two neighbors, leading to a symbol error probability of $2P_{err}$ for these points. Averaging over the symbol error probabilities for all points of each constellation, we get the symbol error probabilities

$$P_S^{2-ASK} = \mathrm{Q}\left(\frac{d}{\sigma}\right), \quad P_S^{4-ASK} = \frac{3}{2}\mathrm{Q}\left(\frac{d}{\sigma}\right), \quad P_S^{8-ASK} = \frac{7}{4}\mathrm{Q}\left(\frac{d}{\sigma}\right).$$

For Gray mapping, we can make the approximation that each symbol error leads only to one bit error. Thus, we readily obtain the bit error probabilities expressed by the bit energy for $M = 2, 4, 8$ as

$$P_b^{2-ASK} = \frac{1}{2}\operatorname{erfc}\left(\sqrt{\frac{E_b}{N_0}}\right),$$

and

$$P_b^{4-ASK} \approx \frac{3}{8}\operatorname{erfc}\left(\sqrt{\frac{2}{5}\frac{E_b}{N_0}}\right), \tag{1.94}$$

$$P_b^{8-ASK} \approx \frac{7}{24}\operatorname{erfc}\left(\sqrt{\frac{1}{7}\frac{E_b}{N_0}}\right).$$

For ASK constellations, only the I-component, corresponding to the cosine wave, will be modulated, while the sine wave will not be present in the passband signal. Since, in general, every passband signal of a certain bandwidth may have both components, 50% of the bandwidth resources remain unused. A simple way to use these resources is to apply the same ASK modulation for the Q-component too. We thus have complex modulation symbols $s_k = x_k + jy_k$, where both x_k and y_k are taken from an M-ASK constellation. The result is a square constellation of M^2 signal points in the complex plane, as depicted in Figure 1.17 for $M^2 = 64$. We call this an M^2-QAM (quadrature amplitude modulation). The bit error performance of M^2-QAM as a function of E_b/N_0 is the same as for M-ASK, that is,

$$P_b^{4-QAM} = \frac{1}{2}\operatorname{erfc}\left(\sqrt{\frac{E_b}{N_0}}\right),$$

$$P_b^{16-QAM} \approx \frac{3}{8}\operatorname{erfc}\left(\sqrt{\frac{2}{5}\frac{E_b}{N_0}}\right) \tag{1.95}$$

and

$$P_b^{64-QAM} \approx \frac{7}{24}\operatorname{erfc}\left(\sqrt{\frac{1}{7}\frac{E_b}{N_0}}\right).$$

This is because the I- and the Q-component can be regarded as completely independent channels that do not influence each other. Thus, M^2-QAM can be regarded as M-ASK multiplexed to the orthogonal I- and Q-channel. Note that the bit error rates are not identical if they are plotted as a function of the signal-to-noise ratio. The bit error probabilities of Equations (1.95) are depicted in Figure 1.18. For high values of E_b/N_0, 16-QAM shows

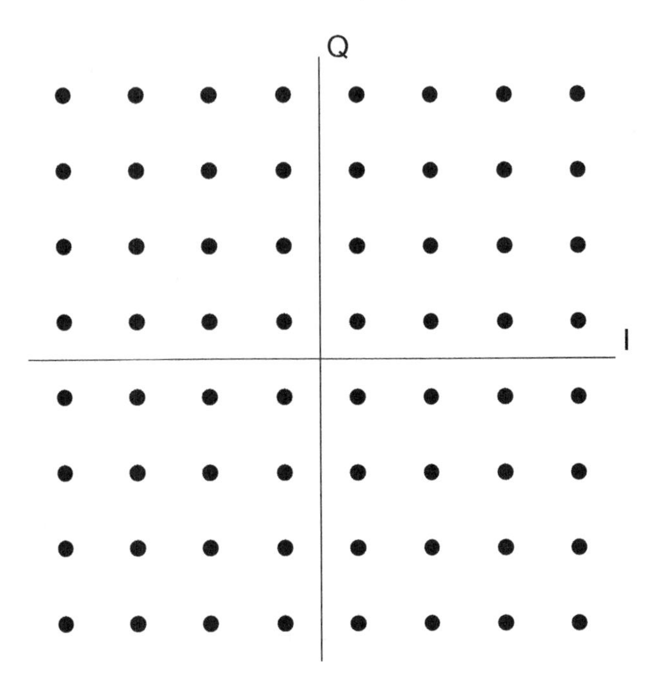

Figure 1.17 The 64-QAM constellation.

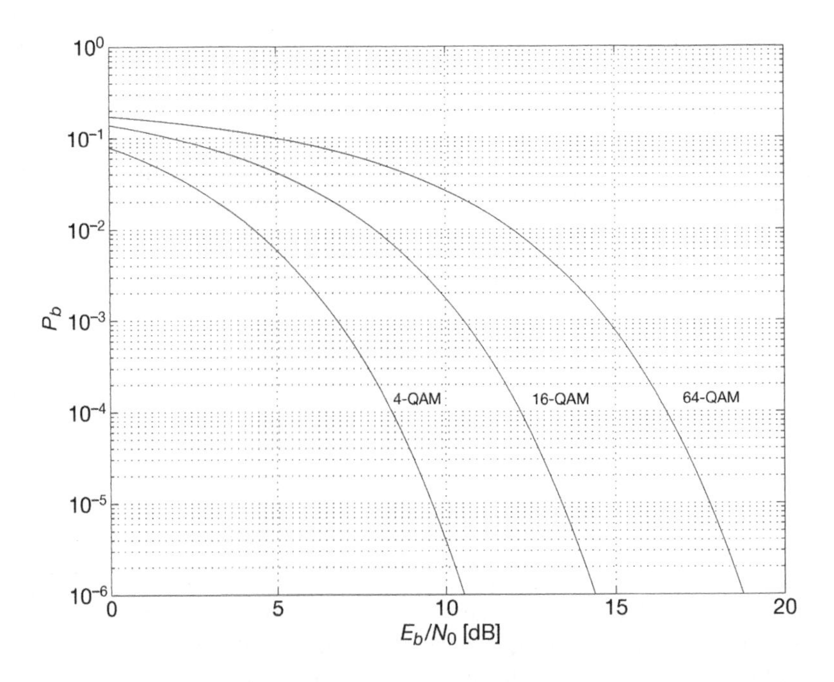

Figure 1.18 Bit error probabilities for 4-QAM, 16-QAM, and 64-QAM.

a performance loss of $10\lg(2.5) \approx 4$ dB compared to 4-QAM, while 64-QAM shows a performance loss of $10\lg(7) \approx 8.5$ dB. This is the price that has to be paid for transmitting twice, respectively three times the data rate in the same bandwidth.

We finally note that nonsquare QAM constellations are also possible like, for example, 8-QAM, 32-QAM and 128-QAM, but we will not discuss these constellations in this text.

1.5.3 PSK

For M-PSK (phase-shift keying), the modulation symbols s_k can be written as

$$s_k = \sqrt{E_S}\, e^{j\phi_k},$$

that is, all the information is contained in the M possible phase values ϕ_k of the symbol. Two adjacent points of the constellation have the phase difference $2\pi/M$. It is a matter of convenience whether $\phi = 0$ is a point of the constellation or not. For 2-PSK – often called *BPSK* (binary PSK) – the phase may take the two values $\phi_k \in \{0, \pi\}$ and thus 2-PSK is just the same as 2-ASK. For 4-PSK – often called QPSK (quaternary PSK) – the phase may take the four values $\phi_k \in \left\{\pm\frac{\pi}{4}, \pm\frac{3\pi}{4}\right\}$ and thus 4-PSK is just the same as 4-QAM. The constellation for 8-PSK with Gray mapping, as an example, is depicted in Figure 1.19.

The approximate error probabilities for M-PSK with Gray mapping can be easily obtained. Let the distance between two adjacent points be $2d$. From elementary geometrical consideration, we get

$$d = \sqrt{E_S}\sin\left(\frac{\pi}{M}\right).$$

For $M > 2$, each constellation point has two nearest neighbors. All the other signal points corresponding to symbol errors lie beyond the two corresponding decision thresholds. By

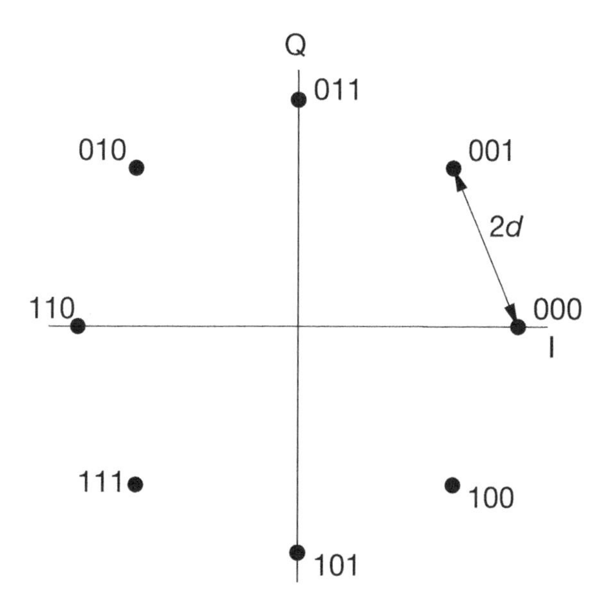

Figure 1.19 Signal constellation for 8-PSK.

a simple union-bound argument, we find that the symbol error probability can be tightly upper bounded by

$$P_S \leq 2\mathrm{Q}\left(\frac{\mathbf{d}}{\sigma}\right) = \mathrm{erfc}\left(\sqrt{\sin^2\left(\frac{\pi}{M}\right)\frac{E_S}{N_0}}\right).$$

By assuming that only one bit error occurs for each symbol error and taking into account the relation $E_S = \log_2(M)\, E_b$, we get the approximate expression

$$P_b \approx \frac{1}{\log_2(M)}\mathrm{erfc}\left(\sqrt{\log_2(M)\,\sin^2\left(\frac{\pi}{M}\right)\frac{E_b}{N_0}}\right) \tag{1.96}$$

for the bit error probability. The bit error probabilities of Equation (1.96) are depicted in Figure 1.20. For high values of E_b/N_0, 8-PSK shows a performance loss of $10\,\mathrm{lg}(3\sin^2(\pi/8)) \approx 3.6$ dB compared to 4-PSK, while 16-PSK shows a performance loss of $10\,\mathrm{lg}(4\sin^2(\pi/16)) \approx 8.2$ dB. Thus, higher-level PSK modulation leads to a considerable loss in power efficiency compared to higher-level QAM at the same spectral efficiency.

1.5.4 DPSK

For DPSK (differential PSK), the phase *difference* between two adjacent transmit symbols carries the information, not the phase of the transmit symbol itself. This means that for a sequence of transmit symbols

$$s_k = \sqrt{E_S}\,\mathrm{e}^{j\phi_k},$$

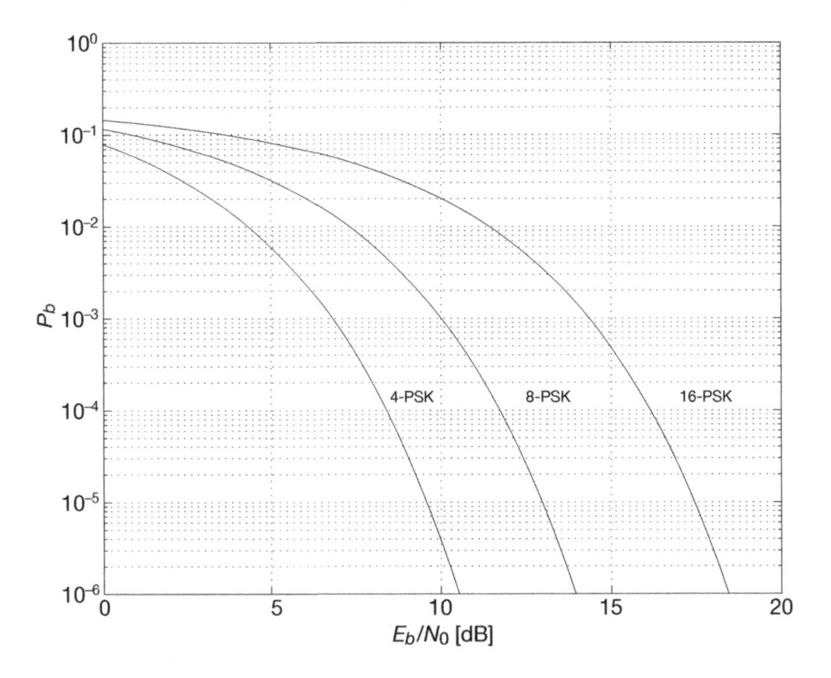

Figure 1.20 Bit error probabilities for 4-PSK, 8-PSK and 16-PSK.

the information is carried by

$$\Delta\phi_k = \phi_k - \phi_{k-1},$$

and

$$z_k = e^{j\Delta\phi_k}$$

is a symbol taken from an M-PSK constellation with energy one. The transmit symbols are then given by the recursion

$$s_k = z_k \cdot s_{k-1}$$

with a start symbol s_0 that may have some arbitrary reference phase ϕ_0. We may set this phase equal to zero and write

$$s_0 = \sqrt{E_S}.$$

Because of this phase reference symbol, the transmit signal

$$s(t) = \sum_{k=0}^{K} s_k g_k(t)$$

carries $K + 1$ transmit symbols s_k, but only K useful PSK symbols z_k. Typically, the phase reference symbol will be transmitted at the beginning of a frame, and the frame length is large enough so that the loss in data rate due to the reference symbol can be neglected.

Again it is a matter of convenience whether the PSK constellation for z_k contains the phase (difference) $\Delta\phi_k = 0$ or not. For the most popular QPSK constellation, $\Delta\phi_k \in \left\{\pm\frac{\pi}{4}, \pm\frac{3\pi}{4}\right\}$ or

$$z_k \in \left\{\frac{1}{\sqrt{2}}(\pm 1 \pm j)\right\}.$$

Obviously, this leads to eight possible values of the transmit symbol s_k, corresponding to the absolute phase values

$$\phi_k \in \left\{0, \pm\frac{\pi}{4}, \pm\frac{\pi}{2}, \pm\frac{3\pi}{4}, \pi\right\},$$

see Figure 1.21, where the possible transitions are marked by arrows.

For even values of k,

$$\phi_k \in \left\{0, \pm\frac{\pi}{2}, \pi\right\}$$

and for odd values of k,

$$\phi_k \in \left\{\pm\frac{\pi}{4}, \pm\frac{3\pi}{4}\right\}.$$

We thus have two different constellations for s_k, which are phase shifted by $\pi/4$. This modulation scheme is therefore called $\pi/4$-*DQPSK*.

Differential PSK is often used because it does not require an absolute phase reference. In practice, the channel introduces an unknown phase θ, that is, the receive signal is

$$r_k = e^{j\theta} s_k + n_k.$$

In a coherent PSK receiver, the phase must be estimated and back- rotated. A differential receiver compares the phase of two adjacent symbols by calculating

$$u_k = r_k r_{k-1}^* = s_k s_{k-1}^* + e^{j\theta} s_k n_{k-1}^* + n_k e^{-j\theta} s_{k-1}^* + n_k n_{k-1}^*.$$

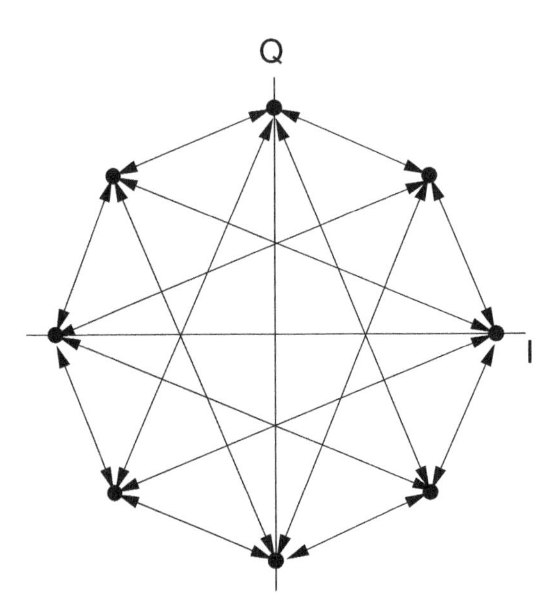

Figure 1.21 Transmit symbols for $\frac{\pi}{4}$-DQPSK.

In the noise-free case, $u_k/\sqrt{E_S} = z_k$ represents original PSK symbols that carry the information. However, we see from the above equation that we have additional noise terms that do not occur for coherent signaling and that degrade the performance. The performance analysis of DPSK is more complicated than for coherent PSK (see e.g. (Proakis 2001)). We will later refer to the results when we need them for the applications.

1.6 Bibliographical Notes

This chapter is intended to give a brief overview of the basics that are needed in the following chapters and to introduce some concepts and notations. A more detailed introduction into digital communication and detection theory can be found in many text books (see e.g. (Benedetto and Biglieri 1999; Blahut 1990; Kammeyer 2004; Lee and Messerschmidt 1994; Proakis 2001; Van Trees 1967; Wozencraft and Jacobs 1965)). We assume that the reader is familiar with Fourier theory and has some basic knowledge of probability and stochastic processes. We will not define these concepts further; one may refer to standard text books (see e.g. (Bracewell 2000; Feller 1970; Papoulis 1991)).

We have emphasized the vector space properties of signals. This allows a geometrical interpretation that makes the solution of many detection problems intuitively obvious. The interpretation of signals as vectors is not new. We refer to the excellent classical text books (Van Trees 1967; Wozencraft and Jacobs 1965).

We have emphasized the concept of a detector as an integral operation that performs a measurement. A Fourier analyzer is such a device that may be interpreted as a set of detectors, one for each frequency. The integral operation is given by a scalar product if the signal is well behaved (i.e. of finite energy). If not, the signal has to be understood as a generalized function (which is called *distribution* or *functional* in mathematical literature (Reed and Simon 1980)), and the detection is the action of this signal on a well-behaved *test function*. It is interesting to note that this is the same situation as in quantum theory, where such a test function is interpreted as a detection device for the quantum state of a physical system. In this context it is worth noting that $\delta(t)$, the most important generalized function in communication theory, has been introduced by one of the quantum theory pioneers, P.A.M. Dirac.

1.7 Problems

1. Let $\{g_k(t)\}_{k=1}^{K}$ be an orthonormal transmit base and

$$\mathbf{s} = (s_1, \ldots, s_K)^T$$

and

$$\mathbf{x} = (x_1, \ldots, x_K)^T$$

two transmit symbol vectors. Let

$$s(t) = \sum_{k=1}^{K} s_k g_k(t)$$

and

$$x(t) = \sum_{k=1}^{K} x_k g_k(t).$$

Show that

$$\langle s, x \rangle = \mathbf{s}^{\dagger}\mathbf{x}.$$

2. Let $S(f)$ denote the Fourier transform of the signal $s(t)$ and define

$$\tilde{s}(t) = \sqrt{2}\,\Re\{s(t)e^{j2\pi f_0 t}\}.$$

Show that the Fourier transform of that signal is given by

$$\tilde{S}(f) = \frac{1}{\sqrt{2}}\left(S\left(f - f_0\right) + S^*\left(-f - f_0\right)\right).$$

3. Let $x(t)$ and $y(t)$ be finite-energy low-pass signals strictly band- limited to $B/2$ and let $f_0 > B/2$. Show that the two signals

$$\tilde{x}(t) = \sqrt{2}\cos\left(2\pi f_0 t\right) x(t)$$

and

$$\tilde{y}(t) = -\sqrt{2}\sin\left(2\pi f_0 t\right) y(t)$$

are orthogonal. Let $u(t)$ and $v(t)$ be two other finite-energy signals strictly band-limited to $B/2$ and define

$$\tilde{u}(t) = \sqrt{2}\cos(2\pi f_0 t)\, u(t)$$

and

$$\tilde{v}(t) = -\sqrt{2}\sin(2\pi f_0 t)\, v(t).$$

Show that

$$\langle \tilde{u}, \tilde{x} \rangle = \langle u, x \rangle$$

and

$$\langle \tilde{v}, \tilde{y} \rangle = \langle v, y \rangle$$

hold. Hint: Transform all the signals into the frequency domain and use Parseval's equation.

4. Show that, from the definition of the time-variant linear systems I and Q, the definitions (given in Subsection 1.2.2) of the time-variant linear systems I^D and Q^D are uniquely determined by

$$\langle \tilde{u}, Iv \rangle = \left\langle I^D \tilde{u}, v \right\rangle$$

and

$$\langle \tilde{u}, Qv \rangle = \left\langle Q^D \tilde{u}, v \right\rangle$$

for any (real-valued) finite-energy signal $\tilde{u}(t)$ and $v(t)$. Mathematically speaking, this means that I^D and Q^D are defined as the *adjoints* of the linear operators I and Q. For the theory of linear operators, see for example (Reed and Simon 1980).

5. Show that the definitions

$$\mathrm{E}\{w(t_1)w(t_2)\} = \frac{N_0}{2}\delta(t_1 - t_2)$$

and

$$\mathrm{E}\left\{\mathcal{D}_{\phi_1}[w]\mathcal{D}_{\phi_2}[w]\right\} = \frac{N_0}{2}\langle \phi_1, \phi_2 \rangle$$

are equivalent conditions for the *whiteness* of the (real-valued) noise $w(t)$.

6. Let $g(t)$ be a transmit pulse and $n(t)$ complex baseband white (not necessarily Gaussian) noise. Let

$$\mathcal{D}_h[r] = \int_{-\infty}^{\infty} h^*(t) r(t)\, dt$$

be a detector for a (finite-energy) pulse $h(t)$ and $r(t) = g(t) + n(t)$ be the transmit pulse corrupted by the noise. Show that the signal-to-noise ratio after the detector defined by

$$SNR = \frac{|\mathcal{D}_h[g]|^2}{\mathrm{E}\left\{|\mathcal{D}_h[n]|^2\right\}}$$

becomes maximal if $h(t)$ is chosen to be proportional to $g(t)$.

7. Show the equality

$$\|\mathbf{y}-\mathbf{x}\|^2 - \|\mathbf{y}-\hat{\mathbf{x}}\|^2 = 2\left[\left(\mathbf{y}-\frac{\mathbf{x}+\hat{\mathbf{x}}}{2}\right)(\hat{\mathbf{x}}-\mathbf{x})\right].$$

8. Let $\mathbf{n} = (n_1, \ldots, n_K)^T$ be a K-dimensional real-valued AWGN with variance $\sigma^2 = N_0/2$ in each dimension and $\mathbf{u} = (u_1, \ldots, u_K)^T$ be a vector of length $|\mathbf{u}| = 1$ in the K-dimensional Euclidean space. Show that $n = \mathbf{n} \cdot \mathbf{u}$ is a Gaussian random variable with mean zero and variance $\sigma^2 = N_0/2$.

9. We consider a digital data transmission from the Moon to the Earth. Assume that the digital modulation scheme (e.g. QPSK) requires $E_b/N_0 = 10$ dB at the receiver for a sufficiently low bit error rate of, for example, $BER = 10^{-5}$. For free-space propagation, the power at the receiver is given by

$$P_r = G_t G_r \frac{\lambda^2}{(4\pi R)^2} P_t.$$

We assume $G_t = G_r = 10$ dB for the antenna gains at the receiver and the transmitter, respectively, and $\lambda = 40$ cm for the wavelength. The distance of the moon is approximately given by $R = 400\,000$ km. The receiver noise (including a noise figure of 4 dB) is given by $N_0 = -170$ dBm/Hz. How much transmit power is necessary for a bit rate of $R_b = 1$ bit/s?

2

Mobile Radio Channels

2.1 Multipath Propagation

Mobile radio reception is severely affected by multipath propagation; the electromagnetic wave is scattered, diffracted and reflected, and reaches the antenna via various ways as an incoherent superposition of many signals with different delay times that are caused by the different path lengths of these signals. This leads to an interference pattern that depends on the frequency and the location or – for a mobile receiver – the time. The mobile receiver moves through an interference pattern that may change within milliseconds and that varies over the transmission bandwidth. One says that the mobile radio channel is characterized by *time variance* and *frequency selectivity*.

The time variance is determined by the relative speed v between receiver and transmitter and the wavelength $\lambda = c/f_0$, where f_0 is the transmit frequency and c is the velocity of light. The relevant physical quantity is the maximum Doppler frequency shift given by

$$\nu_{\max} = \frac{v}{c} f_0 \approx \frac{1}{1080} \frac{f_0}{\text{MHz}} \frac{v}{\text{km/h}} \text{Hz}.$$

Table 2.1 shows some practically relevant figures for $\nu_{\max}$ for speeds from a slowly moving person (2.4 km/h) to a high-speed train or car (192 km/h).

For an angle α between the direction of the received signal and the direction of motion, the Doppler shift ν is given by

$$\nu = \nu_{\max} \cos \alpha.$$

Consider a carrier wave transmitted at frequency f_0. Typically, the received signal is a superposition of many scattered and reflected signals from different directions resulting in a spatial interference pattern. For a vehicle moving through this interference pattern, the received signal amplitude fluctuates in time, which is called *fading*. In the frequency domain, we see a superposition of many Doppler shifts corresponding to different directions resulting in a Doppler spectrum instead of a sharp spectral line located at f_0. Figure 2.1 shows an example of the amplitude fluctuations of the received time signal for $\nu_{\max} = 50$ Hz, corresponding for example, to a transmit signal at 900 MHz for a vehicle

Theory and Applications of OFDM and CDMA Henrik Schulze and Christian Lüders

Table 2.1 Doppler frequencies

Radio frequency	Doppler frequency for a speed of			
	$v = 2.4$ km/h	$v = 48$ km/h	$v = 120$ km/h	$v = 192$ km/h
$f_0 =$ 225 MHz	0.5 Hz	10 Hz	25 Hz	40 Hz
$f_0 =$ 900 MHz	2.0 Hz	40 Hz	100 Hz	160 Hz
$f_0 =$ 2025 MHz	4.5 Hz	90 Hz	225 Hz	360 Hz

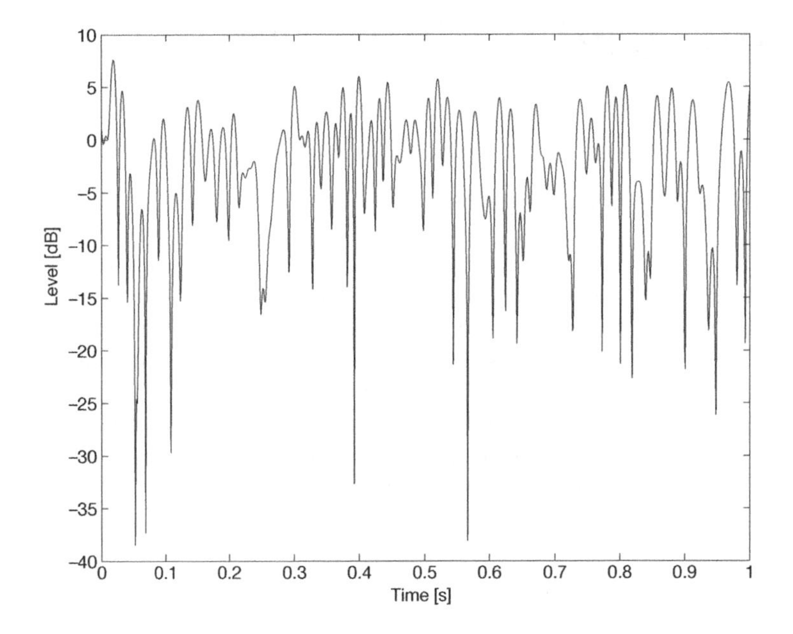

Figure 2.1 Time variance (selectivity) of the fading amplitude for 50 Hz maximum Doppler frequency.

speed $v = 60$ km/h. The figure shows deep amplitude fades up to -40 dB. If a car stands still at the corresponding location (e.g. at a traffic light), the reception breaks down. If the car moves half a wavelength, it may get out of the deep fade.

The superposition of Doppler-shifted carrier waves leads to a fluctuation of the carrier amplitude and phase. This means that the received signal is amplitude and phase modulated by the channel.

Figure 2.2 shows the trace of the phasor in the complex plane for the same channel parameters as above. For digital phase modulation, these rapid phase fluctuations cause severe problems if the carrier phase changes too much during the time T_S that is needed to transmit one digitally modulated symbol. The amplitude and the phase fluctuate randomly. The typical frequency of the variation is of the order of $\nu_{\max}$ corresponding to a timescale of the variations given by

$$t_{\text{corr}} = \nu_{\max}^{-1},$$

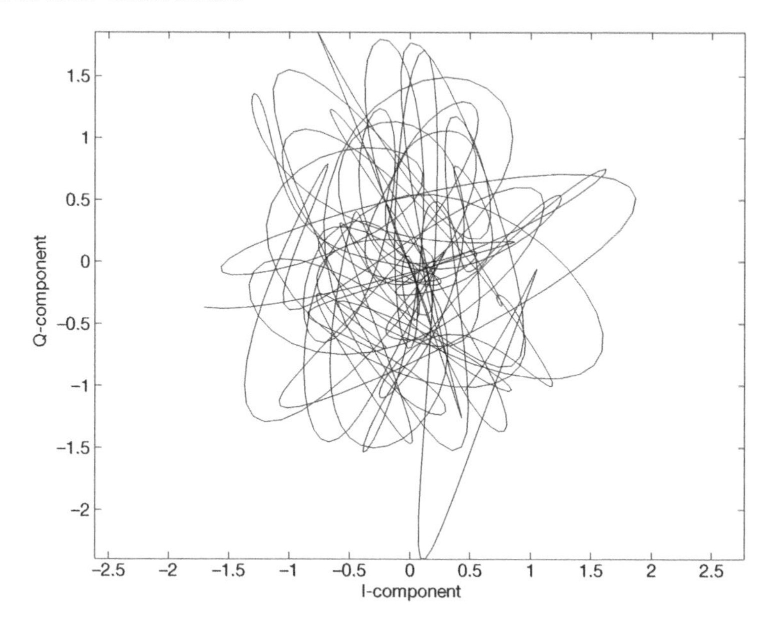

Figure 2.2 Time variance (selectivity) shown above as a curve in the complex plane.

which we call the correlation time. Digital transmission with symbol period T_S is only possible if the channel remains nearly constant during that period, which requires $T_S \ll t_{\mathrm{corr}}$ or, equivalently, the condition

$$\nu_{\max} T_S \ll 1$$

must hold.

The frequency selectivity of the channel is determined by the different delay times of the signals. They can be calculated as the ratio between the traveling distances and the velocity of light. 1 μs delay time difference corresponds to 300 m of path difference. A few microseconds are typical for cellular mobile radio. For a broadcasting system for a large area, echoes up to 100 μs are possible in a hilly or mountainous region. In a so-called *single frequency network* (see Section 4.6), the system must cope with even longer echoes. Longer echoes correspond to more fades within the transmission bandwidth. Figure 2.3 shows an example for a received signal level as a function of the frequency (relative to the center frequency) at a fixed location in a situation with delay time differences of the signals corresponding to a few kilometers. In the time domain, intersymbol interference disturbs the transmission if the delay time differences are not much smaller than the symbol duration T_S. A data rate of 200 kbit/s leads to $T_S = 10$ μs for the QPSK modulation. This is of the same order as the echoes for such a scenario. This means that digital transmission of that data rate is not possible without using more sophisticated methods such as equalizers, spread spectrum techniques, or multicarrier modulation. We define a correlation frequency

$$f_{\mathrm{corr}} = \Delta\tau^{-1},$$

where $\Delta\tau$ is the square root of the variance of the power distribution of the echoes, which we call the *delay spread*. f_{corr} is often called *coherence* (or *coherency*) *bandwidth* because the

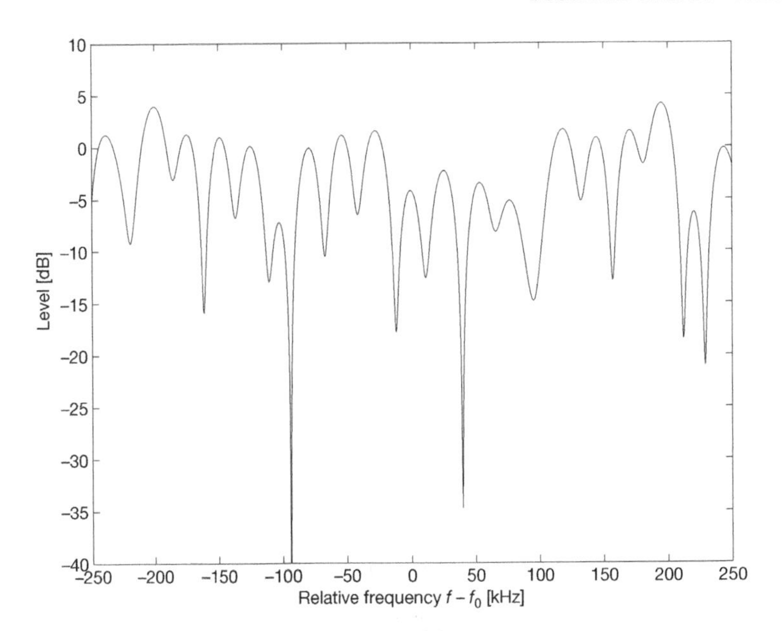

Figure 2.3 Frequency selectivity (variance) of the fading amplitude for a broadcasting channel with long echoes.

channel can be regarded as frequency-nonselective within a bandwidth B with $B \ll f_{\mathrm{corr}}$. If B is in the order of T_s^{-1} as it is the case for signaling with the Nyquist base, this is equivalent to the condition

$$\Delta\tau \ll T_S$$

that intersymbol interference can be neglected.

2.2 Characterization of Fading Channels

2.2.1 Time variance and Doppler spread

Consider a modulated carrier wave

$$\tilde{s}(t) = \sqrt{2}\,\Re\left\{s(t)e^{j2\pi f_0 t}\right\} \tag{2.1}$$

at frequency f_0 that is modulated by a complex baseband signal $s(t)$. For a moving receiver with velocity v and an incoming wave with an angle of incidence α relative to the direction of motion, the carrier frequency will be shifted by the Doppler frequency given by

$$\nu = \nu_{\max}\cos\alpha. \tag{2.2}$$

The same Doppler shift occurs for a fixed receiver and a transmitter moving with velocity v. Because the angle α from the left-hand side causes the same Doppler shift as the angle

$-\alpha$ from the right-hand side, we identify both cases and let the angle run from 0 to π. The Doppler-shifted receive signal is given by

$$\tilde{r}(t) = \sqrt{2}\,\Re\left\{a e^{j\theta} e^{j2\pi\nu t} s(t) e^{j2\pi f_0 t}\right\}, \tag{2.3}$$

where a is an attenuation factor and θ the phase of the carrier wave at the receiver. Here, we have made some reasonable assumptions that simplify the treatment:

- The angle α is constant during the time of consideration. This is true if the distance between transmitter and receiver is sufficiently large and we can assume that many bits are transmitted during a very small change of the angle. This is in contrast to the case of the acoustic Doppler shift of an ambulance car, where the angle runs from 0 to π during the observation time and the listener hears a tone decreasing in frequency from $f_0 + \nu_{\max}$ to $f_0 - \nu_{\max}$.

- The signal is of sufficiently small bandwidth so that the Doppler shift can be assumed to be the same for all spectral components.

Furthermore, we have only taken into account that the delay of the RF signal results in a phase delay, ignoring the group delay of the complex baseband signal $s(t)$. We will study the effect of such a delay in the following subsection. Here, we assume that these delays are so small that they can be ignored. Typically, the received signal is the superposition of several signals, scattered from different obstacles, with attenuation factors a_k, carrier phases θ_k and Doppler shifts $\nu_k = \nu_{\max} \cos \alpha_k$, resulting in

$$\tilde{r}(t) = \sqrt{2} \sum_{k=1}^{N} \Re\left\{a_k e^{j\theta_k} e^{j2\pi\nu_k t} s(t) e^{j2\pi f_0 t}\right\}. \tag{2.4}$$

The complex baseband transmit and receive signals $s(t)$ and $r(t)$ are thus related by

$$r(t) = c(t)s(t), \tag{2.5}$$

where

$$c(t) = \sum_{k=1}^{N} a_k e^{j\theta_k} e^{j2\pi\nu_k t} \tag{2.6}$$

is the time-variant complex fading amplitude of the channel. Typically, this complex fading amplitude looks as shown in Figures 2.1 and 2.2. In the special case of two-path channels ($N = 2$), the fading amplitude shows a more regular behavior. In this case, the time-variant power gain $|c(t)|^2$ of the channel can be calculated as

$$|c(t)|^2 = a_1^2 + a_2^2 + 2a_1a_2 \cos\left(2\pi\left(\nu_1 - \nu_2\right)t + \theta_1 - \theta_2\right).$$

Figure 2.4 shows $|c(t)|^2$ for $a_1 = 0.75$ and $a_2 = \sqrt{7}/4$. The average power is normalized to one, the maximum power is $(a_1 + a_2)^2 \approx 1.99$, the minimum power is $(a_1 - a_2)^2 \approx 0.008$, resulting in level fluctuations of about 24 dB. The fading amplitude is periodic with period $|\nu_1 - \nu_2|^{-1}$. Such a two-path channel can occur in reality, for instance, in the special situation where the received signal is a superposition of a direct signal and

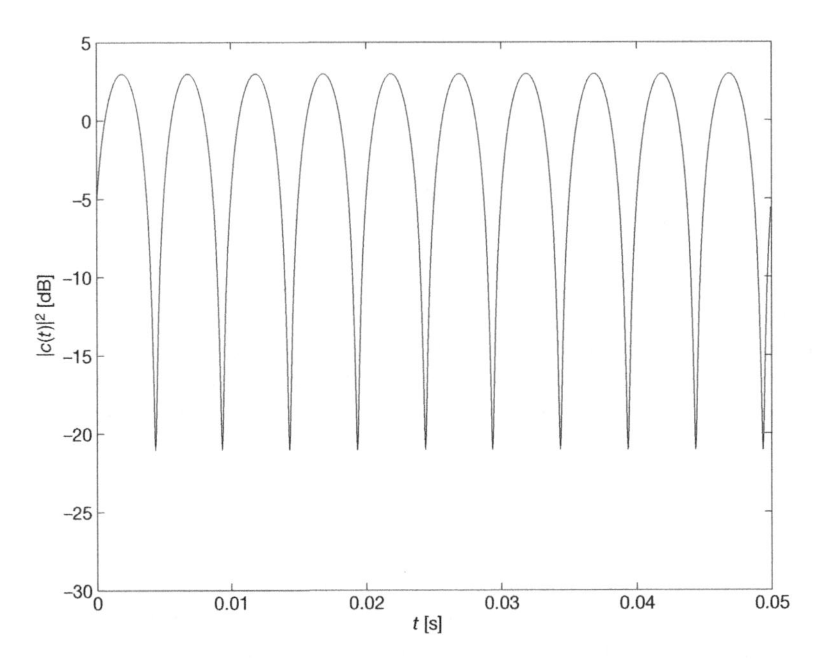

Figure 2.4 Time variance of a two-path channel.

a strong reflection. If, for example, $\nu_1 = \nu_{\max}$ and $\nu_2 = -\nu_{\max}$, that is, one signal from the front and one from the back, then the period is $(2\nu_{\max})^{-1}$. Since $\nu_{\max} = v/\lambda$, the spatial separation of two power maxima (or minima) is $\lambda/2$. This interference pattern is well known in physics as a standing wave. In the example, we have chosen $\nu_1 = 100$ Hz and $\nu_2 = -100$ Hz, which corresponds to 120 km/h at 900 MHz. For that frequency, $\lambda/2 \approx 16.7$ cm.

It is a usual assumption to think of a general complex fading amplitude $c(t)$ given by Equation (2.6) as a *stationary* random signal. This is a convenient assumption, but one should keep in mind that this introduces a simplified mathematical model for the physical reality. We therefore add the following remarks:

- The real physical $c(t)$ is deterministic because a_k, ν_k, and θ_k are deterministic. But at least the phases θ_k are completely unknown. It is reasonable and a common practice in communications engineering to model unknown phases as random variables. Philosophically speaking, statistics is not due to the randomness of nature, but to our lack of knowledge. This is also a common practice in physics, for example, in statistical thermodynamics (Landau and Lifshitz 1958).

- Strictly speaking, stationarity cannot be true because the environment changes. This slow change of the channel is called *long-term fading*, in contrast to the short-term fading considered here. Long-term fading is of primary interest for network planning (see Subsection 5.1.2), but for the performance analysis of communication systems we have to focus on the short-term fading, thereby assuming that the environment is constant during the time period that is necessary to measure, for example, the bit error rate.

- Because mathematical modeling of physical reality is a coarse procedure, we regard it as splitting hairs to distinguish between wide-sense stationary (WSS) and strict-sense stationary (SSS) stochastic processes. For SSS processes, time-shift invariance is given for all statistical properties, while for WSS this is only true up to the second-order properties (covariances). For a theoretical analysis, for example, to identify bit error probabilities with bit error rates, we need to assume even some ergodic properties (which is a stronger assumption than SSS) that are mathematically necessary but physically not given.

The random process $c(t)$ given by Equation (2.6) has a discrete power spectral density (psd) $\mathcal{S}_c(\nu)$ as shown in Figure 2.5(a) for $N = 5$. We call $\mathcal{S}_c(\nu)$ the *Doppler spectrum*. However, in most real situations, the received signal is a continuous rather than a discrete superposition of Doppler-shifted signals, resulting in a continuous psd $\mathcal{S}_c(\nu)$ as shown in Figure 2.5(b).

As a result of Equation (2.2), each Doppler frequency corresponds to an angle $\alpha \in [0, \pi]$. Therefore, the Doppler spectrum is related to the angular power density $\mathcal{S}_{\text{angle}}(\alpha)$ by

$$-\mathcal{S}_c(\nu)\,\mathrm{d}\nu = \mathcal{S}_{\text{angle}}(\alpha)\,\mathrm{d}\alpha.$$

The negative sign is due to the fact that, because the cosine is a decreasing function over the relevant interval, a positive infinitesimal $\mathrm{d}\alpha$ corresponds to a negative infinitesimal $\mathrm{d}\nu$. By Equation (2.2), we get

$$\mathrm{d}\nu = -\nu_{\max}\sin\alpha\,\mathrm{d}\alpha = -\nu_{\max}\sqrt{1-\frac{\nu^2}{\nu_{\max}^2}}\,\mathrm{d}\alpha$$

resulting in

$$\mathcal{S}_c(\nu)\sqrt{\nu_{\max}^2-\nu^2} = \mathcal{S}_{\text{angle}}\left(\arccos\left(\frac{\nu}{\nu_{\max}}\right)\right), \quad -\nu_{\max} < \nu < \nu_{\max}.$$

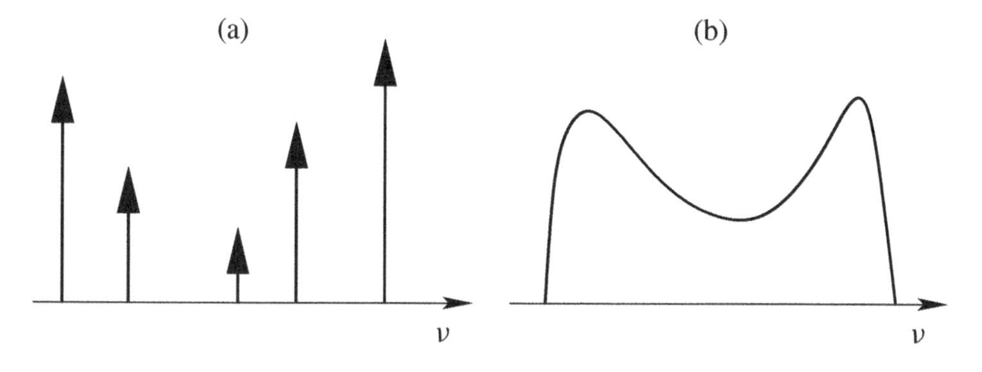

Figure 2.5 Example of a discrete (a) and continuous (b) Doppler spectrum.

A simple model – and also a kind of worst case – is the isotropic angular power distribution $\mathcal{S}_{\text{angle}}(\alpha) = \pi^{-1}$. In that case, we obtain the power spectral density

$$\mathcal{S}_c(\nu) = \frac{1}{\pi \nu_{\max} \sqrt{1 - \frac{\nu^2}{\nu_{\max}^2}}} \tag{2.7}$$

for $-\nu_{\max} < \nu < \nu_{\max}$ and zero outside that interval. This spectral shape is sometimes called the *isotropic* or *Jakes* Doppler spectrum (Jakes 1975). Figure 2.6 shows this (normalized) Doppler spectrum $\nu_{\max}\mathcal{S}_c(\nu)$. Note the singularities at the edges that have their origin in geometry.

We assumed that $c(t)$ is the complex baseband signal corresponding to a (wide-sense) stationary stochastic process, which is the carrier wave affected by the Doppler spread. The autocorrelation function (ACF) of such a process is given by

$$\mathcal{R}_c(t) = \mathrm{E}\left\{c(t_1 + t)c^*(t_1)\right\}$$

(see Proposition 1.3.3). The power spectrum is the Fourier transform of the ACF, that is,

$$\mathcal{S}_c(\nu) = \int_{-\infty}^{\infty} \mathrm{e}^{-j2\pi\nu t} \mathcal{R}_c(t)\, \mathrm{d}t.$$

For the Jakes spectrum, the ACF given by

$$\mathcal{R}_c(t) = \mathrm{J}_0\left(2\pi \nu_{\max} t\right), \tag{2.8}$$

where $\mathrm{J}_0(x)$ is the Bessel function of the first kind of order 0.

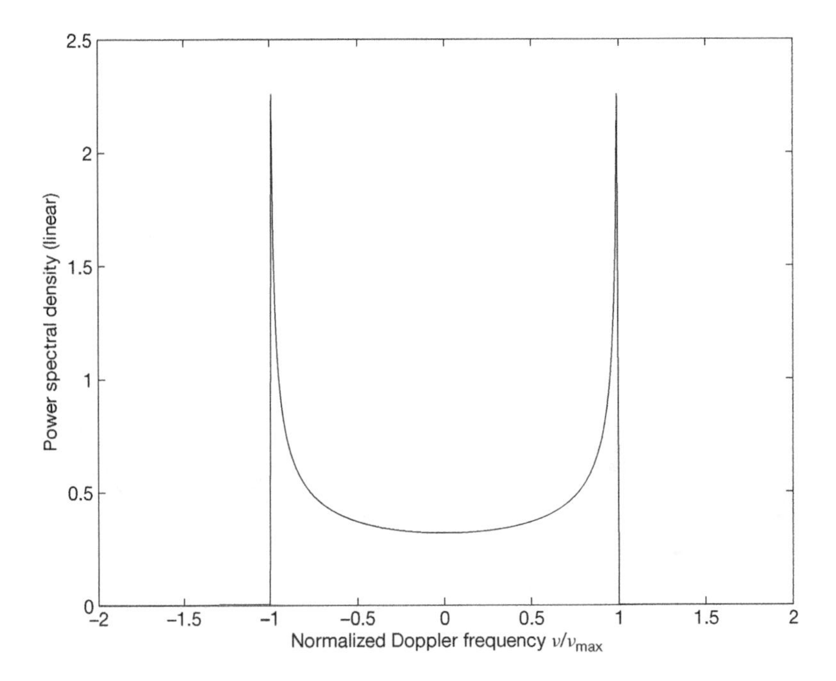

Figure 2.6 The Jakes Doppler spectrum corresponding to the isotropic power distribution.

Because of the coarse modeling of reality, relevant quantities like error probabilities should not strongly depend on the shape of the Doppler spectrum, and in fact, they do not for all known examples. We will illustrate this by the following consideration. As discussed above, digital transmission is only possible if the channel does not vary too fast relative to the symbol duration, which can be expressed by the condition $\nu_{\max} T_S \ll 1$ over relevant times (of the order T_S). The correlation time $t_{\text{corr}} = \nu_{\max}^{-1}$ must be large enough so that the channel samples are highly correlated. These correlations are characterized by the ACF $\mathcal{R}_c(t)$. Thus, only $\mathcal{R}_c(t)$ for small values $|t| \ll t_{\text{corr}}$ is relevant for the performance and we may approximate $\mathcal{R}_c(t)$ by a Taylor series. We note that

$$\left.\frac{\mathrm{d}^n}{\mathrm{d}t^n} \mathcal{R}_c(t)\right|_{t=0} = (2\pi j)^n \mu_n \{\mathcal{S}_c(\nu)\},$$

where

$$\mu_n \{\mathcal{S}_c(\nu)\} = \int_{-\infty}^{\infty} \nu^n \mathcal{S}_c(\nu)\, \mathrm{d}\nu$$

is the nth moment of the power spectral density. $\mathcal{R}_c(t)$ can thus be expanded into the Taylor series

$$\mathcal{R}_c(t) = \sum_{n=0}^{\infty} \frac{1}{n!} (2\pi j)^n \mu_n \{\mathcal{S}_c(\nu)\}\, t^n.$$

Note that $\mu_0 \{\mathcal{S}_c(\nu)\} = 1$ due to energy normalization and $\mu_1 \{\mathcal{S}_c(\nu)\} = 0$ can always be achieved by a frequency shift. Since $\mu_n \{\mathcal{S}_c(\nu)\} \leq (2\nu_{\max})^n$, the absolute value of the nth term in the Taylor series is bounded by

$$\frac{1}{n!} |4\pi\, \nu_{\max} t|^n,$$

which is very small for $|t| \ll t_{\text{corr}} = \nu_{\max}^{-1}$. We thus approximate $\mathcal{R}_c(t)$ by the lowest non-trivial order of the Taylor series

$$\mathcal{R}_c(t) \approx 1 - \frac{1}{2} (2\pi)^2 \mu_2 \{\mathcal{S}_c(\nu)\}\, t^2.$$

Thus, one should expect that only the second moment of the Doppler spectrum – rather than the exact shape – is relevant for the performance of the system. We will see later in Subsection 2.4.6 that the bit error rate of differential QPSK (DQPSK) in a time-variant fading channel depends on the time variance through $\mathcal{R}_c(T_S)$.

Thus, it is therefore not important to use the real shape of the Doppler spectrum. We may use, for example, the Jakes spectrum as a coarse reflection of reality. We note that the second moment typically becomes smaller if the angles of receive signals are not isotropically distributed.

Furthermore, we should always keep in mind that the assumed stationarity (or wide-sense stationarity) is, strictly speaking, not true. We can only say that $c(t)$ cannot be distinguished from a stationary process when observed over a relatively short time of, say, a few seconds. This makes it reasonable to treat it like a stationary process, because this is mathematically convenient.

Finally, we state that for the Jakes spectrum the second moment can easily be calculated as

$$\mu_2\left\{\mathcal{S}_c(\nu)\right\} = \frac{\nu_{\max}^2}{2}$$

(see Problem 2).

The inverse of the variance

$$\Delta\nu = \sqrt{\mu_2\left\{\mathcal{S}_c(\nu)\right\}}$$

also appears to be a reasonable choice for the definition of the correlation time t_{corr}. However, $v_{\max}$ is easy to obtain from the carrier frequency and the vehicle speed and is therefore a better choice for the practice. For typical shapes as the Jakes spectrum, both quantities are of the same order.

2.2.2 Frequency selectivity and delay spread

Consider again a transmit signal as given by Equation (2.1). We now assume that both transmitter and receiver are at rest (or the time variance is so slow that it can be neglected for the time period under consideration) and we can ignore any Doppler shifts. But, unlike in the above treatment, we do not ignore the delays $\tau_k = l_k/c$ of the complex baseband signal $s(t) \mapsto s(t-\tau_k)$ for the different propagation paths of length l_k. Instead of Equation (2.4), we now get for the received signal

$$\tilde{r}(t) = \sqrt{2}\sum_{k=1}^{N} \Re\left\{a_k e^{j\theta_k} s(t-\tau_k) e^{j2\pi f_0 t}\right\}. \tag{2.9}$$

The delays of the carrier are already included in the phases θ_k. The complex baseband transmit and receive signals $s(t)$ and $r(t)$ are related by

$$r(t) = h(t) * s(t), \tag{2.10}$$

where

$$h(t) = \sum_{k=1}^{N} a_k e^{j\theta_k} \delta(t-\tau_k) \tag{2.11}$$

is the impulse response of the channel. The corresponding channel transfer function is given by

$$H(f) = \sum_{k=1}^{N} a_k e^{j\theta_k} e^{-j2\pi f \tau_k}. \tag{2.12}$$

We note the strong similarity to Equation (2.6). Typically, the frequency response looks as shown in Figure 2.3. In the special case of two-path channels ($N = 2$), the transfer function shows a more regular behavior. In this case, the power gain $|H(f)|^2$ of the channel can be calculated as

$$|H(f)|^2 = a_1^2 + a_2^2 + 2a_1a_2\cos\left(2\pi f\,(\tau_1-\tau_2) + \theta_2 - \theta_1\right).$$

The picture is similar to the one depicted in Figure 2.4, where time is replaced by frequency. The transfer function is periodic with period $|\tau_1 - \tau_2|^{-1}$.

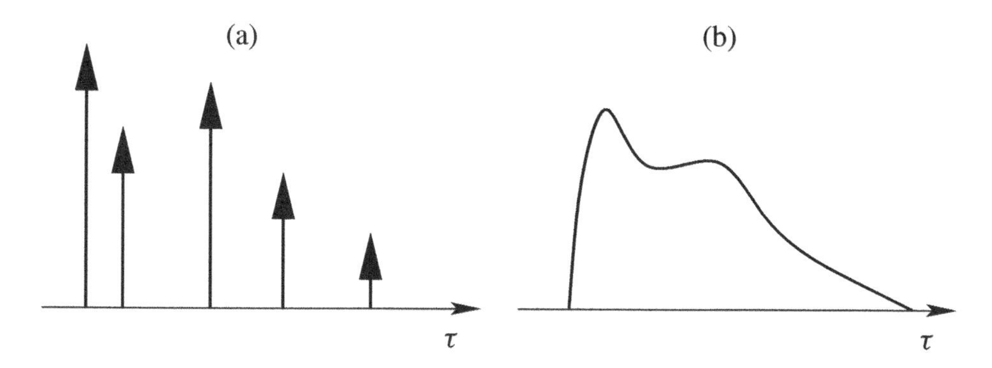

Figure 2.7 Example of a discrete (a) and continuous (b) delay power spectrum.

With the same arguments as for the treatment of the time-variant fading amplitude $c(t)$, we may regard $H(f)$ as a random transfer function, or – more formally – as a stochastic process in a frequency variable. Frequency-shift invariance (corresponding to stationarity for the time variable) can only be an approximation. The same remarks about modeling the reality as made in the preceding subsection apply here. Since the variable for this process is a frequency, there is a power density distribution as a function of a time variable τ that can be identified as the delay time. Figure 2.7(a) illustrates such a *delay power spectrum* $\mathcal{S}_H(\tau)$ corresponding to the process given by the Equations (2.11) and (2.12). However, in most real situations, the received signal is a continuous rather than a discrete superposition of delayed signal components, resulting in a continuous delay power spectrum $\mathcal{S}_H(\tau)$ as shown in Figure 2.7(b). Note that the delay power spectrum reflects the distribution of path length. We identify the delay spread as

$$\Delta\tau = \sqrt{\mu_2\{\mathcal{S}_H(\tau)\} - \mu_1^2\{\mathcal{S}_H(\tau)\}}.$$

Here, in contrast to the Doppler spectrum, we do not set the first moment to zero because this would lead to negative delay times τ. However, with similar arguments as for the Doppler spectrum, we expect that typically the performance of a communication system in a frequency selective fading channel will not depend on the shape of $\mathcal{S}_H(\tau)$, but only on the second moment.

One popular model for $\mathcal{S}_H(\tau)$ is an exponential distribution

$$\mathcal{S}_H(\tau) = \frac{1}{\tau_m}\mathrm{e}^{-\tau/\tau_m}$$

for $\tau > 0$ and zero elsewhere. The mean value τ_m of this distribution equals the delay spread $\Delta\tau$. The exponential power delay spectrum reflects the idea that the power of the paths decreases strongly with their delay. This is of course a very rough model, but it can be refined by adding components due to significant distant reflectors.

We have assumed frequency-shift invariance (which is similar to what is called *wide-sense stationarity* for the time variable). The (frequency) ACF is given by

$$\mathcal{R}_H(f) = \mathrm{E}\left\{H(f_1 + f)H^*(f_1)\right\}.$$

The delay power spectrum is the inverse Fourier transform of the ACF, that is,

$$\mathcal{S}_H(\tau) = \int_{-\infty}^{\infty} \mathrm{e}^{j2\pi f\tau} \mathcal{R}_H(f)\,\mathrm{d}f.$$

For the exponential delay power spectrum, the ACF given by

$$\mathcal{R}_H(f) = \frac{1}{1 + j2\pi f \tau_m}.$$

2.2.3 Time- and frequency-variant channels

We now consider a channel that is both time and frequency selective. We combine the effects of the Equations (2.4) and (2.9) and obtain a receive signal given by

$$\tilde{r}(t) = \sqrt{2} \sum_{k=1}^{N} \Re \left\{ a_k \mathrm{e}^{j\theta_k} \mathrm{e}^{j2\pi \nu_k t} s(t - \tau_k) \mathrm{e}^{j2\pi f_0 t} \right\}. \tag{2.13}$$

The complex baseband transmit and receive signals $s(t)$ and $r(t)$ are related by

$$r(t) = \int_{-\infty}^{\infty} h(\tau, t) s(t - \tau)\,\mathrm{d}\tau, \tag{2.14}$$

where

$$h(\tau, t) = \sum_{k=1}^{N} a_k \mathrm{e}^{j\theta_k} \mathrm{e}^{j2\pi \nu_k t} \delta(t - \tau_k) \tag{2.15}$$

is the time-variant impulse response of the channel. Note that Equation (2.14) contains Equations (2.5) and (2.10) as special cases by setting either $h(\tau, t) = c(t)\delta(\tau)$ or $h(\tau, t) = h(\tau)$. We note that $h(\tau, t)$ is the channel response of an impulse with travel time τ received at time t, that is, transmitted at time $t - \tau$.

The time-variant channel transfer function will now be defined as the Fourier transform in the delay variable τ of the time-variant impulse response

$$H(f, t) = \int_{-\infty}^{\infty} \mathrm{e}^{-j2\pi f\tau} h(\tau, t)\,\mathrm{d}\tau$$

which equals

$$H(f, t) = \sum_{k=1}^{N} a_k \mathrm{e}^{j\theta_k} \mathrm{e}^{j2\pi \nu_k t} \mathrm{e}^{-j2\pi f \tau_k}$$

in our special case. The receive signal $r(t)$ is related to the Fourier transform $S(f)$ of the transmit signal $s(t)$ by

$$r(t) = \int_{-\infty}^{\infty} \mathrm{e}^{-j2\pi f\tau} H(f, t) S(f)\,\mathrm{d}\tau.$$

The power density with respect to Doppler and delay is now given by a common power density function $\mathcal{S}(\tau, \nu)$ called *scattering function* with the properties

$$\mathcal{S}_c(\nu) = \int_{-\infty}^{\infty} \mathcal{S}(\tau, \nu)\,\mathrm{d}\tau$$

and

$$\mathcal{S}_H(\tau) = \int_{-\infty}^{\infty} \mathcal{S}(\tau, \nu)\,\mathrm{d}\nu.$$

Since the Doppler frequency is related to the angle of the incoming wave by $\nu = \nu_{\max} \cos \alpha$, and the delay is related to the echo path length l by $\tau = l/c$, the scattering function reflects the geometrical distribution of the scatterers and their corresponding power contributions. Like the Doppler and the delay power spectrum, the scattering function is typically a continuous rather than a discrete superposition of Doppler-shifted and delayed components.

The ACF of the two-dimensional random process is defined by

$$\mathcal{R}(f, t) = \mathrm{E}\left\{H(f_1 + f, t_1 + t)H^*(f_1, t_1)\right\}.$$

It is related to the scattering function by a two-dimensional Fourier (back) transform

$$\mathcal{S}(\tau, \nu) = \int_{-\infty}^{\infty} \mathrm{d}f \int_{-\infty}^{\infty} \mathrm{d}t\, \mathrm{e}^{j2\pi f\tau} \mathrm{e}^{-j2\pi\nu t} \mathcal{R}(f, t).$$

Here, we have used the simpler notation $\int \mathrm{d}x \int \mathrm{d}y f(x, y)$ instead of $\int \left(\int f(x, y)\,\mathrm{d}y\right) \mathrm{d}x$. We note that

$$\mathcal{R}(0, t) = \mathcal{R}_c(t)$$

and

$$\mathcal{R}(f, 0) = \mathcal{R}_H(f)$$

hold.

Up to now, the treatment was mainly heuristic, which seems to be adequate for a channel that can only be modeled roughly. However, formally we have to deal with time-variant random systems. In the following section, we characterize such systems and give more formal definitions. However, these are formal mathematical concepts and their relation to reality must be argued as done above.

2.2.4 Time-variant random systems: the WSSUS model

Consider a linear, but typically not time-invariant system. The output $r(t)$ for an input $s(t)$ of such a system can formally be written as

$$r(t) = \int_{-\infty}^{\infty} k(t, t')s(t')\,\mathrm{d}t', \tag{2.16}$$

where $k(t, t')$ is the so-called *integral kernel* of the system. This means that the output at time t is a continuous superposition of the input signal taken at time instants t' multiplied by a weight factor $k(t, t')$. We add the following remarks:

- For (finite-dimensional) vectors, every linear mapping is given by a matrix multiplication. Equation (2.16) is just a natural generalization of a matrix multiplication to the continuous case.

- In the case in which the kernel depends only on the difference argument $t - t'$, that is $k(t, t') = \tilde{k}(t - t')$, the system is time invariant, and Equation (2.16) reduces to a convolution.

- Equation (2.16) is a formal notation. The integral kernel $k(t, t')$ may contain singularities, for example, terms like δ-functions. But this is not really a problem in practice.

We now substitute the integration variable by $\tau = t - t'$, which can be interpreted as the delay between the input time t' and the output time t, resulting in

$$r(t) = \int_{-\infty}^{\infty} k(t, t - \tau)s(t - \tau)\, d\tau.$$

Furthermore, we define the time-variant impulse response as

$$h(\tau, t) = k(t, t - \tau)$$

and obtain

$$r(t) = \int_{-\infty}^{\infty} h(\tau, t)s(t - \tau)\, d\tau.$$

The time-variant channel transfer function is defined as the Fourier transform in the delay variable τ of the time-variant impulse response

$$H(f, t) = \int_{-\infty}^{\infty} e^{-j2\pi f\tau} h(\tau, t)\, d\tau.$$

The receive signal $r(t)$ is related to the Fourier transform $S(f)$ of the transmit signal $s(t)$ by

$$r(t) = \int_{-\infty}^{\infty} e^{j2\pi f\tau} H(f, t)S(f)\tau,$$

where $H(f, t)$ is the time-variant transfer function of the channel.

We now assume that the time-variant transfer function $H(f, t)$ is a two-dimensional random process, and, without losing generality, that the mean value is zero[1]. We further assume that the two-dimensional autocorrelation is time- and frequency-shift invariant, that is,

$$\mathrm{E}\left\{H(f_1 + f, t_1 + t)H^*(f_1, t_1)\right\} = \mathrm{E}\left\{H(f_2 + f, t_2 + t)H^*(f_2, t_2)\right\}. \tag{2.17}$$

Such a process is called a *wide-sense stationary uncorrelated scattering* (WSSUS) process. Wide-sense stationarity means time-shift invariance only up to second-order statistical expectation values. This is weaker than SSS, where all statistical expectation values must be time-shift invariant. Frequency-shift invariance up to the second order is named uncorrelated scattering for reasons that will become obvious soon. The two-dimensional ACF of the two-dimensional random process is defined by

$$\mathcal{R}(f, t) = \mathrm{E}\left\{H(f_1 + f, t_1 + t)H^*(f_1, t_1)\right\} \tag{2.18}$$

and the scattering function by the two-dimensional (2-D) Fourier (back) transform

$$\mathcal{S}(\tau, \nu) = \int_{-\infty}^{\infty} df \int_{-\infty}^{\infty} dt\, e^{j2\pi f\tau} e^{-j2\pi\nu t} \mathcal{R}(f, t). \tag{2.19}$$

[1] If the mean value does not equal zero, we may consider the stochastic process of the difference to the mean value.

Because $H(f, t)$ is the complex baseband process for a WSSUS process, we have the property

$$\mathrm{E}\{H(f_1 + f, t_1 + t)H(f_1, t_1)\} = 0,$$

which is the generalization of the result of Proposition 1.3.3. to two-dimensional stochastic processes.

We define the 2-D Fourier transform of the time-variant transfer function as

$$G(\tau, \nu) = \int_{-\infty}^{\infty} \mathrm{d}f \int_{-\infty}^{\infty} \mathrm{d}t\, \mathrm{e}^{j2\pi f\tau} \mathrm{e}^{-j2\pi\nu t} H(f, t). \tag{2.20}$$

The inverse is given by

$$H(f, t) = \int_{-\infty}^{\infty} \mathrm{d}\tau \int_{-\infty}^{\infty} \mathrm{d}\nu\, \mathrm{e}^{-j2\pi f\tau} \mathrm{e}^{j2\pi\nu t} G(\tau, \nu). \tag{2.21}$$

The term *uncorrelated scattering* stems from the following.

Proposition 2.2.1 (Uncorrelated scattering) *The condition (2.17) is equivalent to the condition*

$$\mathrm{E}\left\{G(\tau_1, \nu_1)G^*(\tau_2, \nu_2)\right\} = \delta(\tau_1 - \tau_2)\delta(\nu_1 - \nu_2)\mathcal{S}(\tau_1, \nu_2)$$

with $\mathcal{S}(\tau, \nu)$ defined by Equations (2.19) and (2.18).

Proof. From Equations (2.20) and (2.18), we conclude that the left-hand side equals the fourfold integral

$$\int_{-\infty}^{\infty} \mathrm{d}f_1\, \mathrm{d}t_1\, \mathrm{d}f_2\, \mathrm{d}t_2\, \mathrm{e}^{j2\pi f_1\tau_1} \mathrm{e}^{-j2\pi\nu_1 t_1} \mathrm{e}^{-j2\pi f_2\tau_2} \mathrm{e}^{j2\pi\nu_2 t_2} \cdot \mathrm{E}\left\{H(f_1, t_1)H^*(f_2, t_2)\right\}$$

$$= \int_{-\infty}^{\infty} \mathrm{d}f_1 \int_{-\infty}^{\infty} \mathrm{d}t_1 \int_{-\infty}^{\infty} \mathrm{d}f_2 \int_{-\infty}^{\infty} \mathrm{d}t_2\, \mathrm{e}^{j2\pi(f_1\tau_1 - f_2\tau_2)} \mathrm{e}^{-j2\pi(\nu_1 t_1 - \nu_2 t_2)} \mathcal{R}(f_1 - f_2, t_1 - t_2).$$

We change the order of integration and substitute $f = f_1 - f_2$ for f_1 and $t = t_1 - t_2$ for t_1 to obtain

$$\int_{-\infty}^{\infty} \mathrm{d}f_2 \int_{-\infty}^{\infty} \mathrm{d}t_2 \int_{-\infty}^{\infty} \mathrm{d}f \int_{-\infty}^{\infty} \mathrm{d}t\, \mathrm{e}^{j2\pi(f+f_2)\tau_1} \mathrm{e}^{-j2\pi\nu_1(t+t_2)} \mathrm{e}^{-j2\pi f_2\tau_2} \mathrm{e}^{j2\pi\nu_2 t_2} \mathcal{R}(f, t)$$

$$= \int_{-\infty}^{\infty} \mathrm{d}f_2 \mathrm{e}^{j2\pi f_2(\tau_1 - \tau_2)} \int_{-\infty}^{\infty} \mathrm{d}t_2 \mathrm{e}^{-j2\pi(\nu_1 - \nu_2)t_2} \int_{-\infty}^{\infty} \mathrm{d}f \int_{-\infty}^{\infty} \mathrm{d}t\, \mathrm{e}^{j2\pi f\tau_1} \mathrm{e}^{-j2\pi\nu_1 t} \mathcal{R}(f, t).$$

The first integral equals $\delta(\tau_1 - \tau_2)$, the second equals $\delta(\nu_1 - \nu_2)$ and the third (twofold) integral equals $\mathcal{S}(\tau_1, \nu_2)$.

We add the following remarks:

- This is a generalization of a known property for WSS processes to two dimensions: The Fourier transform $X(f)$ of a WSS process $x(t)$ has the property that its values $X(f_1)$ and $X(f_2)$ for different frequencies f_1 and f_2 are uncorrelated (see e.g. (Papoulis 1991)). The proposition says that the values of $G(\tau, \nu)$ for different Doppler frequencies and for different delays are uncorrelated.

- We must note that in a real transmission setup, uncorrelated scattering will no longer be given because any receive filter will introduce a correlation between delays τ_1 and τ_2.

2.2.5 Rayleigh and Ricean channels

To apply the WSSUS model to concrete problems in digital communications (e.g. to calculate or simulate bit error rates), the statistics of the two-dimensional fading process $H(f,t)$ has to be specified. In many typical situations without line of sight (LOS), the received signal is a superposition of many scattered components. Owing to the central limit theorem (Feller 1970; Papoulis 1991; van Kampen 1981), it is reasonable in this case to model $H(f,t)$ by a Gaussian process with zero mean. If a LOS component is present, this can be taken into account simply by adding a constant mean value. A WSSUS process that is Gaussian is sometimes called a *GWSSUS* process. We note that a wide-sense stationary Gaussian process is already a strict-sense stationary (see e.g. (Feller 1970; Papoulis 1991)). Gaussian processes are completely characterized by their properties up to second order, that is, the mean and the autocorrelation (one can take this as a definition of a Gaussian process). For a GWSSUS process with zero mean, every sample $H(f_1,t_1)$ for fixed frequency and time is a complex Gaussian random variable that we write as

$$H(f_1,t_1) = X + jY.$$

From the property

$$\mathrm{E}\{H(f_1+f,t_1+t)H(f_1,t_1)\} = 0$$

we conclude that X and Y are uncorrelated and have the same variance (see Problem 1). Because these random variables are Gaussian, they are even statistically independent, identically distributed real Gaussian random variables . We normalize the average power gain of the channel to one so that

$$\mathrm{E}\left\{X^2\right\} = \mathrm{E}\left\{Y^2\right\} = \frac{1}{2}.$$

The probability density functions of X and Y are then given by

$$p_X(x) = \frac{1}{\sqrt{\pi}}\mathrm{e}^{-x^2},\ p_Y(y) = \frac{1}{\sqrt{\pi}}\mathrm{e}^{-y^2}.$$

We introduce polar coordinates $X = A\cos\Phi$, $Y = A\sin\Phi$ to calculate the joint pdf $p_{A,\Phi}(a,\phi)$ of amplitude and phase. From the condition

$$p_{A,\Phi}(a,\phi)\,\mathrm{d}a\,\mathrm{d}\phi = p_X(x)p_Y(y)\,\mathrm{d}x\,\mathrm{d}y$$

and $\mathrm{d}x\,\mathrm{d}y = a\,\mathrm{d}a\,\mathrm{d}\phi$, we obtain

$$p_{A,\Phi}(a,\phi) = \frac{1}{2\pi}2a\mathrm{e}^{-a^2}.$$

Thus, $p_{A,\Phi}(a,\phi) = p_A(a)p_\Phi(\phi)$ with

$$p_\Phi(\phi) = \frac{1}{2\pi}$$

and

$$p_A(a) = 2a\mathrm{e}^{-a^2}. \tag{2.22}$$

This probability density function for the amplitude is called *Rayleigh distribution.* We therefore call this mean-zero GWSSUS channel a *Rayleigh fading channel.* It is interesting to ask for the pdf $p_{\mathrm{power}}(\gamma)$ of the power A^2. From the condition

$$p_{\mathrm{power}}(\gamma)\,\mathrm{d}\gamma = p_A(a)\,\mathrm{d}a$$

we easily obtain

$$p_{\mathrm{power}}(\gamma) = \mathrm{e}^{-\gamma},$$

that is, the power is exponentially distributed for a Rayleigh fading channel. The probability $\mathrm{P}(A^2 < \gamma)$ that the power falls below a certain value γ is given by

$$\mathrm{P}(A^2 < \gamma) = \int_0^{\gamma} \mathrm{e}^{-x}\,\mathrm{d}x = 1 - \mathrm{e}^{-\gamma} = \gamma - \frac{1}{2}\gamma^2 + \cdots.$$

For small values of γ, the approximation $\mathrm{P}(A^2 < \gamma) \approx \gamma$ is allowed. We can now easily conclude that, in a Rayleigh fading channel, the probability for a deep fade of -20 dB (relative to the mean power) is approximately 1%, the probability for a deep fade below -30 dB is approximately 0.1% and so on.

For a LOS channel, we assume a GWSSUS channel with a constant mean value, that is, the Doppler shift of the LOS is assumed to be zero. This can always be achieved by a proper choice of the carrier frequency f_0, that is, each Doppler shift is defined as the shift relative to the LOS. The probability density function for the amplitude is now a Gaussian distribution with nonzero mean. It can be shown to be (Proakis 2001)

$$p_A(a) = 2a\,(1+K)\,\mathrm{e}^{-\left(K + a^2(1+K)\right)} \cdot \mathrm{I}_0\left(2a\sqrt{K\,(1+K)}\right),$$

where $\mathrm{I}_0\,(x)$ is the modified Bessel function of the first kind and order zero. This pdf is called the *Rice distribution* and the GWSSUS channel is called a *Ricean channel.* The parameter K called *Rice factor* is the power ratio between the LOS component and the scattering components. The special case $K = 0$ corresponds to the absence of a LOS and leads to a Rayleigh channel. The special case $K \to \infty$ corresponds to the absence of scattering and leads to an AWGN channel. Figure 2.8 shows the Ricean pdf for different values of K. For large K, it approaches a δ function. The pdf for the power can be obtained as

$$p_{\mathrm{power}}(\gamma) = (1+K)\,\mathrm{e}^{-(K+\gamma(1+K))} \cdot \mathrm{I}_0\left(2\sqrt{\gamma K\,(1+K)}\right).$$

The probability $\mathrm{P}(A^2 < \gamma)$ that the power falls below a certain value γ must be calculated by numerical integration. Figure 2.9 shows this quantity for different values of K.

2.3 Channel Simulation

To evaluate the performance of a digital communication system in a mobile radio channel by means of computer simulations, we need a simulation method that can be implemented in a computer program and that reflects the relevant statistical properties of the channels discussed above. In this section, we introduce a practical simulation method that is quite simple to implement and has been adopted by many authors because of its computational efficiency.

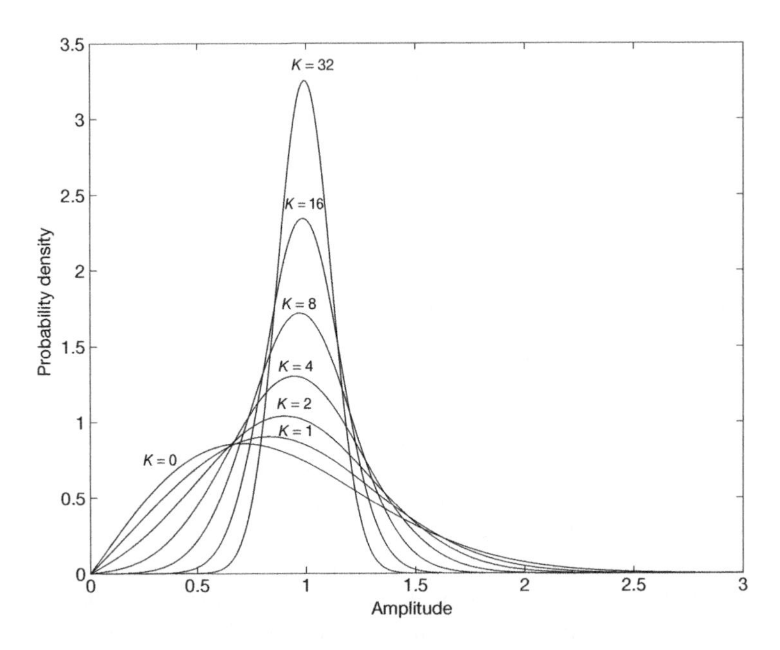

Figure 2.8 The Ricean pdf for $K = 0$ (Rayleigh), and $K = 1, 2, 4, 8, 16, 32$.

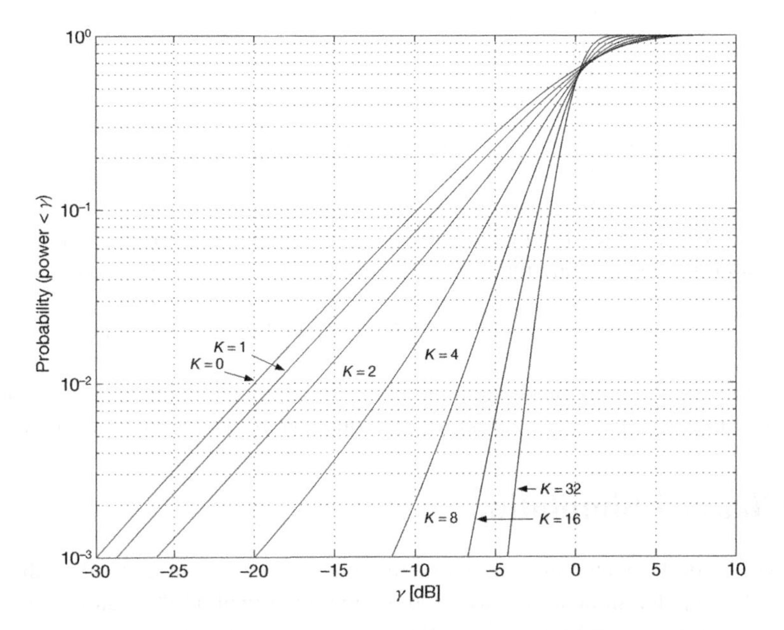

Figure 2.9 The probability $\mathrm{P}(A^2 < \gamma)$ that the power below a certain value γ for the Ricean channel with $K = 0$ (Rayleigh), and $K = 1, 2, 4, 8, 16, 32$.

The main idea of this simulation model is to reverse the line of thought that leads to the statistical Gaussian WSSUS model to reflect the physical reality. The central limit theorem gives the justification to apply the mathematical model of a Gaussian process for a physical signal that is the superposition of many unknown, that is, (pseudo-) random components. On the other hand, owing to the central limit theorem, a superposition of sufficiently many independent random signal components should be a good approximation for a Gaussian process and should thus be a good model for computer simulations. The remaining task is to find out what statistical properties of the components are needed to achieve the appropriate statistical properties of the composed process. For a zero mean GWSSUS channel, we make the following *ansatz* for the process of the simulation model

$$H_N(f,t) = \frac{1}{\sqrt{N}} \sum_{k=1}^{N} \mathrm{e}^{j\theta_k} \mathrm{e}^{j2\pi \nu_k t} \mathrm{e}^{-j2\pi f \tau_k}, \tag{2.23}$$

where θ_k, ν_k, τ_k are random variables that are statistically independent and identically distributed for different values of k. To be more specific, the random phase θ_k is assumed to be independent of ν_k and τ_k and it is uniformly distributed over the unit circle. The random variables ν_k and τ_k have a joint pdf $p(\tau,\nu)$ that needs to be adjusted to the statistical properties of the mathematical GWSSUS channel, which is completely characterized by the autocorrelation $\mathcal{R}(f,t)$ or, alternatively, by the scattering function $\mathcal{S}(\tau,\nu)$. The factor $1/\sqrt{N}$ in Equation (2.23) has been introduced to normalize the average power to one. Note that we did not introduce different amplitudes for the different fading paths. It turns out that this is not necessary. Because of the central limit theorem, in the limit $N \to \infty$, $H_N(f,t)$ approaches a Gaussian process. The following theorem states that $H_N(f,t)$ is WSSUS for any finite value of N.

Theorem 2.3.1 *For any value of N,*

$$\mathrm{E}\left\{H_N(f_1+f,t_1+t)H_N^*(f_1,t_1)\right\} = \mathrm{E}\left\{H_N(f_2+f,t_2+t)H_N^*(f_2,t_2)\right\}$$

is independent of f_1 *and* t_1 *and*

$$\mathrm{E}\left\{H_N(f_1+f,t_1+t)H_N^*(f_1,t_1)\right\} = \int_{-\infty}^{\infty} \mathrm{d}\tau \int_{-\infty}^{\infty} \mathrm{d}\nu\, \mathrm{e}^{-j2\pi f\tau} \mathrm{e}^{j2\pi\nu t} p(\tau,\nu)$$

holds. Furthermore

$$\mathrm{E}\left\{H_N(f,t)\right\} = 0.$$

Proof. We insert Equation (2.23) into the left-hand side of the above equation and obtain the expression

$$\frac{1}{N} \sum_{k=1}^{N} \sum_{l=1}^{N} \mathrm{E}\left\{\mathrm{e}^{j(\theta_k-\theta_l)} \mathrm{e}^{j2\pi(\nu_k(t_1+t)-\nu_l t_1)} \mathrm{e}^{-j2\pi((f_1+f)\tau_k - f_1\tau_l)}\right\}$$

$$= \frac{1}{N} \sum_{k=1}^{N} \sum_{l=1}^{N} \mathrm{E}\left\{\mathrm{e}^{j(\theta_k-\theta_l)}\right\} \mathrm{E}\left\{\mathrm{e}^{j2\pi(\nu_k(t_1+t)-\nu_l t_1)} \mathrm{e}^{-j2\pi((f_1+f)\tau_k - f_1\tau_l)}\right\}.$$

The last equality holds because θ_k is independent of all other random variables. For $k \neq l$, $\mathrm{E}\left\{e^{j(\theta_k-\theta_l)}\right\} = \mathrm{E}\left\{e^{j\theta_k}\right\}\mathrm{E}\left\{e^{j\theta_l}\right\} = 0$. Therefore, all terms with $k \neq l$ in the double sum vanish and we obtain the single sum

$$\frac{1}{N}\sum_{k=1}^{N}\mathrm{E}\left\{e^{j2\pi\nu_k t}e^{-j2\pi f\tau_k}\right\},$$

which is obviously independent of f_1 and t_1. Because the random variables ν_k and τ_k have the same statistics for all values of k with joint pdf $p(\tau, \nu)$, we eventually obtain for the sum

$$\mathrm{E}\left\{e^{j2\pi\nu_k t}e^{-j2\pi f\tau_k}\right\} = \int_{-\infty}^{\infty} d\tau \int_{-\infty}^{\infty} d\nu\, e^{-j2\pi f\tau}e^{j2\pi\nu t}p(\tau, \nu).$$

The second property follows from $\mathrm{E}\left\{e^{j\theta_l}\right\} = 0$.

The theorem states that $H_N(f, t)$ is a WSSUS process with an ACF

$$\mathcal{R}_{\text{model}}(f, t) = \mathrm{E}\left\{H_N(f_1 + f, t_1 + t)H_N^*(f_1, t_1)\right\}$$

and a scattering function given by

$$\mathcal{S}_{\text{model}}(\tau, \nu) = \int_{-\infty}^{\infty} df \int_{-\infty}^{\infty} dt\, e^{j2\pi f\tau}e^{-j2\pi\nu t}\mathcal{R}_{\text{model}}(f, t).$$

The inverse relation is

$$\mathcal{R}_{\text{model}}(f, t) = \int_{-\infty}^{\infty} d\tau \int_{-\infty}^{\infty} d\nu\, e^{-j2\pi f\tau}e^{j2\pi\nu t}\mathcal{S}_{\text{model}}(\tau, \nu).$$

Comparing this equation with the statement

$$\mathcal{R}_{\text{model}}(f, t) = \int_{-\infty}^{\infty} d\tau \int_{-\infty}^{\infty} d\nu\, e^{-j2\pi f\tau}e^{j2\pi\nu t}p(\tau, \nu)$$

of the theorem, we find that

$$\mathcal{S}_{\text{model}}(\tau, \nu) = p(\tau, \nu).$$

We thus have found that for the model channel, the scattering function can be interpreted as a probability density function of the delays and Doppler frequencies.

We add the following remarks:

- The statement of the theorem and the interpretation is true for any N, even for $N = 1$. However, the model is a reasonable approximation only if N is sufficiently large.
- For finite values of N, the model process is not ergodic.
- The model process is Gaussian only in the limit $N \to \infty$. Thus, it only approximates the GWSSUS process, and it does so in a similar manner, as the GWSSUS process is only a model of the physical reality.

- For bit error simulations, it is not necessary to choose N extremely large. For small amplitudes, only a moderately large number of superposed random variables already approximates a Gaussian distribution quite well. Bit errors in a Rayleigh fading channel occur mainly during deep fades, that is, at small channel amplitudes. For many practical simulations, $N = 100$ has proven to be a good choice.

Channel simulations will be done in the time domain. The fading channel model is then given by the receive signal

$$r(t) = \frac{1}{\sqrt{N}} \sum_{k=1}^{N} e^{i\theta_k} e^{i2\pi \nu_k t} s(t - \tau_k) + n(t), \tag{2.24}$$

where $s(t)$ is the transmit signal and $n(t)$ is AWGN. The random variables ν_k and τ_k will be generated by a random number generator matched to the scattering function. Approximations are allowed. As discussed above, the exact structure of the scattering function is not important, and it can be expected that the model is quite good if the second moments agree.

An additional direct component to simulate Ricean fading can easily be included into the model by writing

$$r(t) = \frac{K}{1+K} \cdot s(t) + \frac{1}{1+K} \frac{1}{\sqrt{N}} \sum_{k=1}^{N} e^{i\theta_k} e^{i2\pi \nu_k t} s(t - \tau_k) + n(t) \tag{2.25}$$

for a direct component with Doppler shift zero. K is the Rice factor.

The tapped delay line model

Another kind of model, which is widely used for simulations of mobile communication systems, is called the *tapped delay line model*. This model may be interpreted as a superposition of a certain number M of discrete fading paths (taps) corresponding to propagation delay values τ_m, $m = 1, 2, \ldots, M$. To fix such a model, the Doppler spectrum of each tap, the delay values (relative to the first tap) and the corresponding mean relative amplitudes or power values of each tap have to be specified. The channel according to such a model may be generated using Equation (2.23) by inserting the fixed delay values τ_m of the taps instead of choosing random delay values. The number of summands corresponding to one tap has to be selected proportional to the relative amplitude of that tap.

Some important tapped delay line models used for Global System for Mobile Communication (GSM) and Universal Mobile Telecommunication System (UMTS) simulations are illustrated in Figure 2.10: the GSM models for a typical urban (TU) and a hilly terrain (HT) as well as the International Telecommunication Union (ITU) channel models for the so-called *vehicular* (V) and *indoor-to-outdoor-pedestrian* (IOP) environment. For each environment, two channel models have been specified by the ITU (see e.g. (ETSI TR 101 112 1998)) – one with a low delay spread (channel A) and one with a higher delay spread (channel B). For the evaluation of the system proposals of UMTS, mainly the low delay spread models (channel A) have been used.

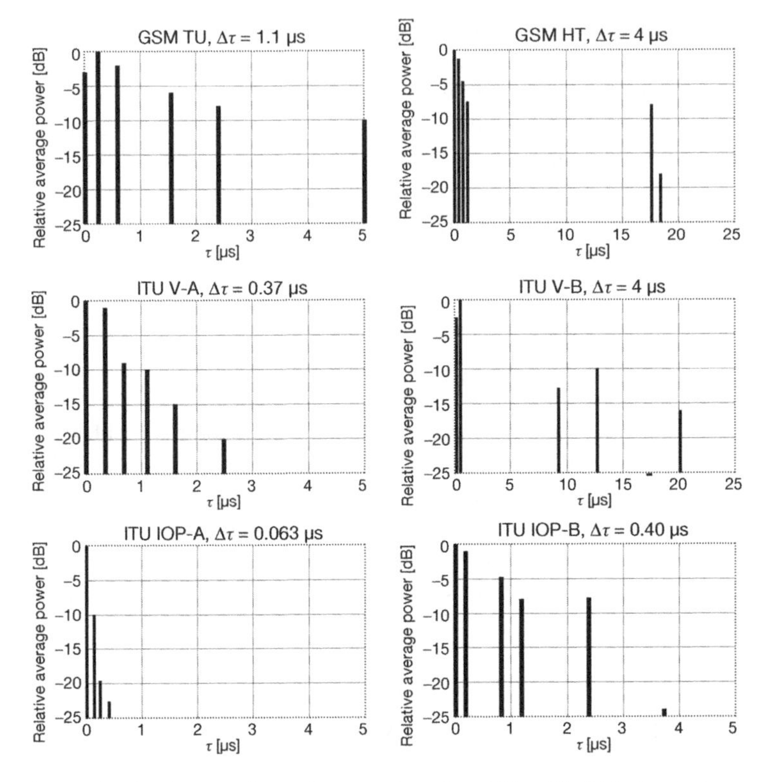

Figure 2.10 Tapped delay line models for GSM, UMTS simulations.

2.4 Digital Transmission over Fading Channels

2.4.1 The MLSE receiver for frequency nonselective and slowly fading channels

Consider a time variant, but frequency nonselective fading channel as discussed in Subsection 2.2.1 with a complex baseband receive signal given by

$$r(t) = c(t)s(t) + n(t),$$

where $c(t)$ is the complex fading amplitude of the channel, $n(t)$ is complex baseband AWGN and the transmit signal $s(t)$ is given by

$$s(t) = \sum_{k=1}^{K} s_k g_k(t)$$

with an orthonormal transmit base $g_k(t)$ and transmit symbols s_k. The assumption that the fading is frequency nonselective – resulting only in a multiplicative fading amplitude – is justified if the symbol duration T_S is much longer than the delay spread $\Delta\tau$ of the channel. The fading channel transforms the transmit base to a new base with pulses given by $h_k(t) =$

$c(t)g_k(t)$ that must be taken as detector pulses at the receiver. In general, these pulses are not orthogonal and the treatment becomes more involved. However, it can be simplified for the important special case that the fading amplitude varies slowly compared to the duration of the pulse $g_k(t)$ (which is typically of the order T_S) and can thus be treated as a constant c_k for the transmission of s_k. This means that the pulse $g_k(t)$ is concentrated in time compared to the variation of the channel so that the approximation $h_k(t) = c(t)g_k(t) \approx c_k g_k(t)$ can be made and

$$\int_{-\infty}^{\infty} g_i^*(t)c(t)g_k(t)\,dt \approx \int_{-\infty}^{\infty} g_i^*(t)c_k g_k(t)\,dt = c_k\delta_{ik}$$

holds. Because the channel only changes the transmit base vectors by a multiplicative factor, the outputs r_k of the detectors for the base $g_k(t)$ are a set of sufficient statistics and the treatment is similar to the one described in Section 1.4 with s_k replaced by $c_k s_k$, resulting in discrete-time channel model with receive symbols given by

$$r_k = c_k s_k + n_k,\ k = 1, \ldots, K. \tag{2.26}$$

We can write this in vector notation as

$$\mathbf{r} = \mathbf{Cs} + \mathbf{n}, \tag{2.27}$$

where $\mathbf{C} = \mathrm{diag}(c_1, \ldots, c_K)$ is the diagonal matrix of the complex fading amplitudes.

We note that the discrete channel model also applies to frequency selective channels as long as the channel is frequency-flat for one transmission pulse. Such a situation may occur if the transmit pulses $g_k(t)$ are located at different frequencies, for example, for multicarrier transmission (OFDM), frequency hopping or frequency diversity, when the same information is repeated at another frequency. Let $G_k(f)$ be the Fourier transform of $g_k(t)$. The fading channel transforms the transmit base to a new base with pulses given by

$$h_k(t) = \int_{-\infty}^{\infty} e^{j2\pi ft} H(f, t)G_k(f)\,df$$

that have to be taken as detector pulses at the receiver. We first assume that $G_k(f)$ is concentrated at the frequency f_k so that $H(f, t)$ can be replaced by $H(f_k, t)$ under the integral. Then, we can write $h_k(t) \approx H(f_k, t)g_k(t)$. We then assume that $g_k(t)$ is concentrated at the frequency t_k so that $H(f_k, t) \approx H(f_k, t_k)$ during the pulse duration. We can then write

$$h_k(t) \approx c_k g_k(t)$$

with $c_k = H(f_k, t_k)$.

Given a receive vector $\mathbf{r}$, we now ask for the maximum likelihood transmit vector $\hat{\mathbf{s}}$. If we assume that the complex fading amplitudes are known at the receiver, we can apply the analysis of Subsection 1.4.2, only replacing the vector $\mathbf{s}$ by $\mathbf{Cs}$. Modifying Equation (1.77) in this way, we obtain the most likely transmit vector $\hat{\mathbf{s}}$ as

$$\hat{\mathbf{s}} = \arg\min_{\mathbf{s}} \|\mathbf{r} - \mathbf{Cs}\|^2, \tag{2.28}$$

where

$$\|\mathbf{r} - \mathbf{Cs}\|^2 = \sum_{k=1}^{K} |c_k|^2 \left|c_k^{-1} r_k - s_k\right|^2. \tag{2.29}$$

This equation allows an interesting interpretation of the optimum receiver. First, the receive symbols r_k are multiplied by the inverse of the complex channel coefficient $c_k = a_k e^{j\varphi_k}$. This means that, by multiplying with c_k^{-1}, the channel phase shift φ_k is back rotated, and the receive symbol is divided by the channel amplitude a_k to adjust the symbols to their original size. We may regard this as an *equalizer*. Each properly equalized receive symbol will be compared with the possible transmit symbol s_k by means of the squared Euclidean distance. These individual decision variables for each index k must be summed up with weighting factors given by $|c_k|^2$, the squared channel amplitude. Without these weighting factors, the receiver would inflate the noise for the very unreliable receive symbols. If a deep fade occurs at the index k, the channel transmit power $|c_k|^2$ may be much less than the power of the noise. The receive symbol r_k is nearly absolutely unreliable and provides us with nearly no useful information about the most likely transmit vector $\hat{\mathbf{s}}$. It would thus be much better to ignore that very noisy receive symbol instead of amplifying it and using it like the more reliable ones. The factor $|c_k|^2$ just takes care of the weighting with the individual reliabilities.

As in Subsection 1.4.2, we may use another form of the maximum likelihood condition. Replacing the vector $\mathbf{s}$ by $\mathbf{Cs}$ in Equation (1.78), we obtain

$$\hat{\mathbf{s}} = \arg\max_{\mathbf{s}} \left(\Re\left\{\mathbf{s}^\dagger \mathbf{C}^\dagger \mathbf{r}\right\} - \frac{1}{2} \|\mathbf{Cs}\|^2 \right). \tag{2.30}$$

There is one difference to the AWGN case: in the first term, before cross-correlating with all possible transmit vectors $\mathbf{s}$, the receive vector $\mathbf{r}$ will first be processed by multiplication with the matrix $\mathbf{C}^\dagger$. This operation performs a back rotation of the channel phase shift φ_k for each receive symbol r_k and a weighting with the channel amplitude a_k. The resulting vector

$$\mathbf{C}^\dagger \mathbf{r} = \begin{bmatrix} c_1^* r_1 \\ \vdots \\ c_K^* r_K \end{bmatrix}.$$

must be cross-correlated with all possible transmit vectors. The second term takes the different energies of the transmit vectors $\mathbf{Cs}$ into account, including the multiplicative fading channel. If all transmit symbols s_k have the same constant energy $E_S = |s_k|^2$ as it is the case for PSK signaling,

$$\|\mathbf{Cs}\|^2 = \sum_{k=1}^{K} |c_k|^2 |s_k|^2 = E_S \sum_{k=1}^{K} |c_k|^2$$

is the same for all transmit vectors $\mathbf{s}$ and can therefore be ignored for the decision.

2.4.2 Real-valued discrete-time fading channels

Even though complex notation is a common and familiar tool in communication theory, there are some items where it is more convenient to work with real-valued quantities. If Euclidean distances between vectors have to be considered – as it is the case in the derivation of estimators and in the evaluation of error probabilities – things often become simpler if one recalls that a K-dimensional complex vector space has equivalent distances as a $2K$-dimensional real vector space. We have already made use of this fact in Subsection 1.4.3, where pairwise error probabilities for the AWGN channel were derived. For a discrete fading channel, things become slightly more involved because of the multiplication of the complex transmit symbols s_k by the complex fading coefficients c_k. In the corresponding two-dimensional real vector space, this corresponds to a multiplication by a rotation matrix together with an attenuation factor. Surely, one prefers the simpler complex multiplication by $c_k = a_k e^{j\varphi_k}$, where a_k and φ_k are the amplitude and the phase of the channel coefficient. At the receiver, the phase will be back rotated by means of a complex multiplication with $e^{j\varphi_k}$ corresponding to multiplication by the inverse rotation matrix in the real vector space. Obviously, no information is lost by this back rotation, and we still have a set of sufficient statistics. We may thus work with a discrete channel model that includes the back rotation and where the fading channel is described by a multiplicative real fading amplitude.

To proceed as described above, we rewrite Equation (2.27) as

$$\mathbf{r} = \mathbf{C}\mathbf{s} + \mathbf{n}_c.$$

Here $\mathbf{C}$ is the diagonal matrix of complex fading amplitudes $c_k = a_k e^{j\varphi_k}$ and $\mathbf{n}_c$ is complex AWGN. We may write

$$\mathbf{C} = \mathbf{D}\mathbf{A}$$

with

$$\mathbf{A} = \mathrm{diag}(a_1, \ldots, a_K)$$

is the diagonal matrix of real fading amplitudes and

$$\mathbf{D} = \mathrm{diag}(e^{j\varphi_1}, \ldots, e^{j\varphi_K})$$

is the diagonal matrix of phase rotations. We note that $\mathbf{D}$ is a unitary matrix, that is, $\mathbf{D}^{-1} = \mathbf{D}^{\dagger}$. The discrete channel can be written as

$$\mathbf{r} = \mathbf{D}\mathbf{A}\mathbf{s} + \mathbf{n}_c.$$

We apply the back rotation of the phase and get

$$\mathbf{D}^{\dagger}\mathbf{r} = \mathbf{A}\mathbf{s} + \mathbf{n}_c.$$

Note that a phase rotation does not change the statistical properties of the Gaussian white noise, so that we can write $\mathbf{n}_c$ instead of $\mathbf{D}^{\dagger}\mathbf{n}_c$. We now decompose the complex vectors into their real and imaginary parts as

$$\mathbf{s} = \mathbf{x}_1 + j\mathbf{x}_2,$$

$$\mathbf{D}^{\dagger}\mathbf{r} = \mathbf{y}_1 + j\mathbf{y}_2$$

and

$$\mathbf{n}_c = \mathbf{n}_1 + j\mathbf{n}_2.$$

Then the complex discrete channel can be written as two real channels in K dimensions given by

$$\mathbf{y}_1 = \mathbf{A}\mathbf{x}_1 + \mathbf{n}_1$$

and

$$\mathbf{y}_2 = \mathbf{A}\mathbf{x}_2 + \mathbf{n}_2$$

corresponding to the inphase and the quadrature component, respectively. Depending on the situation, one may consider each K-dimensional component separately, as in the case of square QAM constellations and then drop the index. Or one may multiplex both together to a $2K$-dimensional vector, as in the case of PSK constellations. One must keep in mind that each multiplicative fading amplitude occurs twice because of the two components. In any case, we may write

$$\mathbf{y} = \mathbf{A}\mathbf{x} + \mathbf{n} \tag{2.31}$$

for the channel with an appropriately redefined matrix $\mathbf{A}$. We finally mention that Equation (2.30) has its equivalent in this real model as

$$\hat{\mathbf{x}} = \arg\max_{\mathbf{x}} \left(\mathbf{x} \cdot \mathbf{A}\mathbf{y} - \frac{1}{2} \|\mathbf{A}\mathbf{s}\|^2 \right).$$

2.4.3 Pairwise error probabilities for fading channels

In this subsection, we consider the case that the fading amplitude is even constant during the whole transmission of a complete transmit vector, that is, the channel of Equation (2.31) reduces to

$$\mathbf{y} = a\mathbf{x} + \mathbf{n}$$

with a constant real fading amplitude a. A special case is, of course, a symbol by symbol transmission where only one symbol is be considered. If that symbol is real, the vector $\mathbf{x}$ reduces to a scalar. If the symbol is complex, $\mathbf{x}$ is a two-dimensional vector.

Let the amplitude a be a random variable with pdf $p(a)$. For a fixed amplitude value a, we can apply the results of Subsection 1.4.3 with $\mathbf{x}$ replaced by $a\mathbf{x}$. Then Equation (1.83) leads to the conditioned pairwise error probability

$$P(\mathbf{x} \mapsto \hat{\mathbf{x}}|a) = Q\left(\frac{ad}{\sigma}\right) = \frac{1}{2}\mathrm{erfc}\left(\sqrt{\frac{a^2}{4N_0}\|\mathbf{x} - \hat{\mathbf{x}}\|^2}\right)$$

with $\sigma^2 = N_0/2$ and

$$d = \frac{1}{2}\|\mathbf{x} - \hat{\mathbf{x}}\|.$$

The overall pairwise error probability

$$P(\mathbf{x} \mapsto \hat{\mathbf{x}}) = \int_0^\infty P(\mathbf{x} \mapsto \hat{\mathbf{x}}|a)p(a)\,\mathrm{d}a$$

is obtained by averaging over the fading amplitude a.

We first consider the Rayleigh fading channel and insert the integral expression for $Q(x)$ to obtain

$$P(\mathbf{x} \mapsto \hat{\mathbf{x}}) = \int_0^\infty \mathrm{d}a\, 2ae^{-a^2} \frac{1}{\sqrt{2\pi\sigma^2/d^2}} \int_a^\infty \mathrm{d}t e^{-\frac{d^2t^2}{2\sigma^2}}.$$

We change the order of integration resulting in

$$P(\mathbf{x} \mapsto \hat{\mathbf{x}}) = \frac{1}{\sqrt{2\pi\sigma^2/d^2}} \int_0^\infty \mathrm{d}t\, \mathrm{e}^{-\frac{d^2t^2}{2\sigma^2}} \int_0^t \mathrm{d}a\, 2a\mathrm{e}^{-a^2}.$$

The second integral is $1 - \mathrm{e}^{-t^2}$ so that

$$P(\mathbf{x} \mapsto \hat{\mathbf{x}}) = \frac{1}{\sqrt{2\pi\sigma^2/d^2}} \int_0^\infty \left(\mathrm{e}^{-\frac{d^2t^2}{2\sigma^2}} - \mathrm{e}^{-\frac{d^2+2\sigma^2}{2\sigma^2}t^2} \right) \mathrm{d}t,$$

which can be solved resulting in

$$P(\mathbf{x} \mapsto \hat{\mathbf{x}}) = \frac{1}{2} \left[1 - \sqrt{\frac{d^2/2\sigma^2}{1 + d^2/2\sigma^2}} \right]$$

or

$$P(\mathbf{x} \mapsto \hat{\mathbf{x}}) = \frac{1}{2} \left[1 - \sqrt{\frac{\frac{1}{4N_0} \left\| \mathbf{x} - \hat{\mathbf{x}} \right\|^2}{1 + \frac{1}{4N_0} \left\| \mathbf{x} - \hat{\mathbf{x}} \right\|^2}} \right].$$

For BPSK and QPSK transmission, each bit error corresponds to an error for one real symbol x, that is, $P_b = P(x \mapsto \hat{x})$ with $\hat{x} = -x$ and

$$\frac{1}{4N_0} \left\| x - \hat{x} \right\|^2 = \frac{E_b}{N_0}$$

holds. Thus,

$$P_b = \frac{1}{2} \left[1 - \sqrt{\frac{\frac{E_b}{N_0}}{1 + \frac{E_b}{N_0}}} \right].$$

To discuss the asymptotic behavior for large E_b/N_0 of this expression, we observe that $\sqrt{1+x} \approx 1 + x/2$ for small values of $x = N_0/E_b$ and find the approximation

$$P_b \approx \frac{1}{2} \frac{1}{1 + 2\frac{E_b}{N_0}} \approx \left(4 \frac{E_b}{N_0} \right)^{-1}$$

for large SNRs. For other modulation schemes than BPSK or QPSK,

$$P(\mathbf{x} \mapsto \hat{\mathbf{x}}) \approx \frac{1}{2} \frac{1}{1 + \frac{1}{2N_0} \left\| \mathbf{x} - \hat{\mathbf{x}} \right\|^2} \approx \left(\frac{1}{N_0} \left\| \mathbf{x} - \hat{\mathbf{x}} \right\|^2 \right)^{-1}$$

holds. There is always the proportionality

$$\frac{1}{4N_0} \left\| \mathbf{x} - \hat{\mathbf{x}} \right\|^2 \propto \mathit{SNR} \propto \frac{E_b}{N_0}.$$

As a consequence, the error probabilities always decrease asymptotically as SNR^{-1} or $(E_b/N_0)^{-1}$.

2.4.4 Diversity for fading channels

In a Rayleigh fading channel, the error probabilities P_{error} decrease asymptotically as slow as $P_{\text{error}} \propto SNR^{-1}$. To lower P_{error} by a factor of 10, the signal power must be increased by a factor of 10. This is related to the fact that, for an average receive signal power γ_m, the probability $P\left(A^2 < \gamma\right)$ that the signal power A^2 falls below a value γ is given by

$$P\left(A^2 < \gamma\right) = 1 - \mathrm{e}^{-\frac{\gamma}{\gamma_m}}$$

which decreases as

$$P\left(A^2 < \gamma\right) \approx \frac{\gamma}{\gamma_m} \propto SNR^{-1}$$

for high SNRs.

The errors occur during the deep fades, and thus the error probability is proportional to the probability of deep fades. A simple remedy against this is twofold (or L-fold) *diversity reception*: if two (or L) replicas of the same information reach the transmitter via two (or L) channels with statistically independent fading amplitudes, the probability that the whole received information is affected by a deep fade will be (asymptotically) decrease as SNR^{-2} (or SNR^{-L}). The same power law will then be expected for the probability of error. L is referred to as the *diversity degree* or the number of *diversity branches*. The following diversity techniques are commonly used:

- *Receive antenna diversity* can be implemented by using two (or L) receive antennas that are sufficiently separated in space. To guarantee statistical independence, the antenna separation Δx should be much larger than the wavelength λ. For a mobile receiver, $\Delta x \approx \lambda/2$ is often regarded as sufficient (without guarantee). For the base station receiver, this is certainly not sufficient.

- *Transmit antenna diversity* techniques were developed only a few years ago. Since then, these methods have evolved in a widespread area of research. We will discuss the basic concept later in a separate subsection.

- *Time diversity* reception can be implemented by transmitting the same information at two (or L) sufficiently separated time slots. To guarantee statistical independence, the time difference Δt should be much larger than the correlation time $t_{\text{corr}} = \nu_{\max}^{-1}$.

- *Frequency diversity* reception can be implemented by transmitting the same information at two (or L) sufficiently separated frequencies. To guarantee statistical independence, the frequency separation Δf should be much larger than $f_{\text{corr}} = \Delta\tau^{-1}$, that is, the correlation frequency (coherency bandwidth) of the channel.

It is obvious that L-fold time or frequency diversity increases the bandwidth requirement for a given data rate by a factor of L. Antenna diversity does not increase the required bandwidth, but increases the hardware expense. Furthermore, it increases the required space, which is a critical item for mobile reception.

The replicas of the information that have been received via several and (hopefully) statistical independent fading channels can be combined by different methods:

- *Selection diversity combining* simply takes the strongest of the L signals and ignores the rest. This method is quite crude, but it is easy to implement. It needs a selector, but only one receiver is required.

- *Equal gain combining* (EGC) needs L receivers. The receiver outputs are summed as they are (i.e. with *equal gain*), thereby ignoring the different reliabilities of the L signals.
- *Maximum ratio combining* (MRC) also needs L receivers. But in contrast to EGC, the receiver outputs are properly weighted by the fading amplitudes, which must be known at the receiver. The MRC is just a special case of the maximum likelihood receiver that has been derived in Subsection 2.4.1. The name *maximum ratio* stems from the fact that the maximum likelihood condition always minimizes the noise (i.e. *maximizes* the signal-to-noise *ratio*) (see Problem 3).

Let E_b be the total energy per data bit available at the receiver and let $E_S = \mathrm{E}\{|s_i|^2\}$ be the average energy per complex transmit symbol s_i. We assume M-ary modulation, so each symbol carries $\log_2(M)$ data bits. We normalize the average power gain of the channel to one, that is, $\mathrm{E}\left\{A^2\right\} = 1$. Thus, for L-fold diversity, the energy E_S is available L times at the receiver. Therefore, the total energy per data bit E_b and the symbol energy are related by

$$LE_S = \log_2(M)E_b. \tag{2.32}$$

As discussed in Section 1.5, for linear modulations schemes $SNR = E_S/N_0$ holds, that is,

$$SNR = \frac{\log_2(M)}{L}\frac{E_b}{N_0}. \tag{2.33}$$

Because the diversity degree L is a multiplicative factor between SNR and E_b/N_0, it is very important to distinguish between both quantities when speaking about *diversity gain*. A fair comparison of the power efficiency must be based on how much energy per bit, E_b, is necessary at the receiver to achieve a reliable reception. If the power has a fixed value and we transmit the same signal via L diversity branches, for example, L different frequencies, each of them must reduce the power by a factor of L to be compared with a system without diversity. This is also true for receive antenna diversity: L receive antennas have L times the area of one antenna. But this is an antenna gain, not a diversity gain. We must therefore compare, for example, a setup with L antenna dishes of 1 m^2 with a setup with one dish of $L\,\mathrm{m}^2$. We state that there is no diversity gain in an AWGN channel. Consider for example, BPSK with transmit symbols $x_k = \pm\sqrt{E_S}$. For L-fold diversity, there are only two possible transmit sequences. The pairwise error probability then equals the bit error probability

$$P_b = P(\mathbf{x} \mapsto \hat{\mathbf{x}}) = \frac{1}{2}\mathrm{erfc}\left(\sqrt{\frac{1}{4N_0}\,\|\mathbf{x} - \hat{\mathbf{x}}\|^2}\right).$$

With $\mathbf{x} = -\hat{\mathbf{x}}$ and $\|\mathbf{x}\|^2 = LE_S$ we obtain

$$P_b = \frac{1}{2}\mathrm{erfc}\left(\sqrt{L \cdot SNR}\right)$$

for P_b as a function of the SNR but

$$P_b = \frac{1}{2}\mathrm{erfc}\left(\sqrt{\frac{E_b}{N_0}}\right)$$

for P_b as a function of E_b/N_0. Thus, for time or frequency diversity, we have wasted bandwidth by a factor of L without any gain in power efficiency.

2.4.5 The MRC receiver

We will now analyze the MRC receiver in some more detail. For L-fold diversity, L replicas of the same information reach the transmitter via L statistically independent fading amplitudes. In the simplest case, this information consists only of one complex PSK or QAM symbol, but in general, it may be any sequence of symbols, for example, of chips in the case of orthogonal modulation with Walsh vectors. The general case is already included in the treatment of Subsection 2.4.1. Here we will discuss the special case of repeating only one symbol in more detail.

Consider a single complex PSK or QAM symbol $s \equiv s_1$ and repeat it L times over different channels. The diversity receive vector can be described by Equation (2.26) by setting $s_1 = \cdots = s_L$ with K replaced by L. The maximum likelihood transmit symbol $\hat{s}$ is given by Equations (2.28) and (2.29), which simplifies to

$$\hat{s} = \arg\min_{s} \sum_{i=1}^{L} |c_i|^2 \left| c_i^{-1} r_i - s \right|^2 ,$$

that is, the receive symbols r_i are equalized, and next the squared Euclidean distances to the transmit symbol are summed up with the weights given by the powers of the fading amplitudes.

We may write Equation (2.27) in a simpler form as

$$\mathbf{r} = s\mathbf{c} + \mathbf{n}_c, \tag{2.34}$$

with the channel vector $\mathbf{c}$ given by

$$\mathbf{c} = (c_1, \ldots, c_L)^T$$

and complex AWGN $\mathbf{n}_c$. The vector $\mathbf{c}$ defines a (complex) one-dimensional transmission base, and sufficient statistics is given by calculating the scalar product $\mathbf{c}^{\dagger}\mathbf{r}$ at the receiver. The complex number $\mathbf{c}^{\dagger}\mathbf{r}$ is the output of the maximum ratio combiner, which, for each receive symbol r_i back rotates the phase φ_i, weights each with the individual channel amplitude $a_i = |c_i|$, and forms the sum of all these L signals.

Here we note that EGC cannot be optimal because at the receiver, the scalar product $\left(e^{-j\varphi_1}, \ldots, e^{-j\varphi_L}\right)\mathbf{r}$ is calculated, and this is not a set of sufficient statistics because $\left(e^{j\varphi_1}, \ldots, e^{j\varphi_L}\right)^T$ does not span the transmit space.

Minimizing the squared Euclidean distance yields

$$\hat{s} = \arg\min_{s} \|\mathbf{r} - s\mathbf{c}\|^2$$

or

$$\hat{s} = \arg\max_{s} \left(\Re\left\{ s^* \mathbf{c}^{\dagger}\mathbf{r} \right\} - \frac{1}{2}|s|^2 \|\mathbf{c}\|^2 \right), \tag{2.35}$$

which is a special case of Equation (2.30).

The block diagram for the MRC receiver is depicted in Figure 2.11. First, the combiner calculates the quantity

$$v = \mathbf{c}^{\dagger}\mathbf{r} = \sum_{k=1}^{L} c_k^* r_k = \sum_{k=1}^{L} a_k e^{-j\varphi_k} r_k,$$

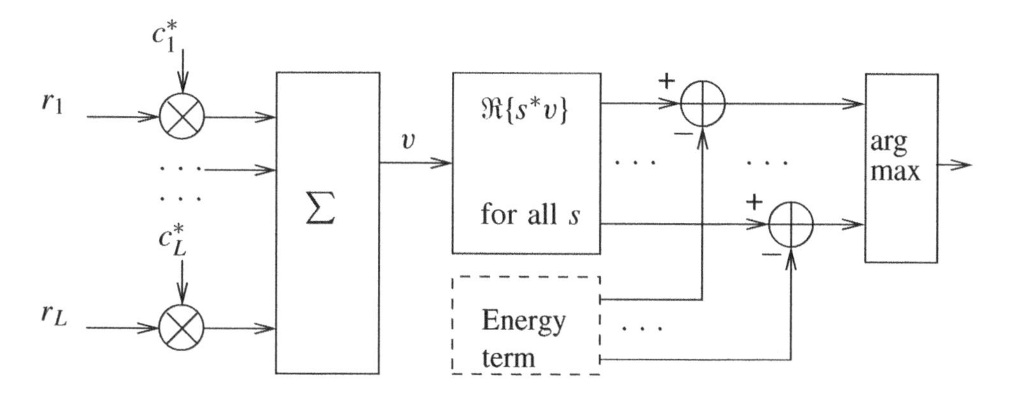

Figure 2.11 Block diagram for the MRC diversity receiver.

that is, it back rotates the phase for each receive symbol r_k and then sums them up (*combines* them) with a weight given by the channel amplitude $a_k = |c_k|$. The first term in Equation (2.35) is the correlation between the MRC output $v = \mathbf{c}^\dagger \mathbf{r}$ and the possible transmit symbols s. For general signal constellations, second (energy) term in Equation (2.35) has to be subtracted from the combiner output before the final decision. For PSK signaling, it is independent of s and can thus be ignored. For BPSK, the bit decision is given by the sign of $\Re\{v\}$. For QPSK, the two bit decisions are obtained from the signs of $\Re\{v\}$ and $\Im\{v\}$.

For the theoretical analysis, it is convenient to consider the transmission channel including the combiner. We define the composed real fading amplitude

$$a = \sqrt{\sum_{i=1}^{L} a_i^2}$$

and normalize the combiner output by

$$u = a^{-1}\mathbf{c}^\dagger \mathbf{r}.$$

We multiply Equation (2.34) by $a^{-1}\mathbf{c}^\dagger$ and obtain the one-dimensional scalar transmission model

$$u = as + n_c,$$

where $n_c = a^{-1}\mathbf{c}^\dagger \mathbf{n}_c$ can easily be proven to be one-dimensional discrete complex AWGN with variance $\sigma^2 = N_0$. A two-dimensional equivalent real-valued vector model

$$\mathbf{y} = a\mathbf{x} + \mathbf{n}, \tag{2.36}$$

can be obtained by defining real transmit and receive vectors

$$\mathbf{x} = \begin{bmatrix} \Re\{s\} \\ \Im\{s\} \end{bmatrix}, \quad \mathbf{y} = \begin{bmatrix} \Re\{u\} \\ \Im\{u\} \end{bmatrix}.$$

Here, $\mathbf{n}$ is two-dimensional real AWGN. Minimizing the squared Euclidean distance in the real vector space yields

$$\hat{\mathbf{x}} = \arg\min_{s} \|\mathbf{y} - a\mathbf{x}\|^2$$

or

$$\hat{\mathbf{x}} = \arg\max_{\mathbf{x}} \left(a\mathbf{y} \cdot \mathbf{x} - \frac{1}{2} |a^2 \, \|\mathbf{x}\|^2 \right). \tag{2.37}$$

The first term is the correlation (scalar product) of the combiner output $a\mathbf{y}$ and the transmit symbol vector $\mathbf{x}$, and the second is the energy term. For PSK signaling, this term is independent of $\mathbf{x}$ and can thus be ignored. In that case, the maximum likelihood transmit symbol vector $\mathbf{x}$ is the one with the smallest angle to the MRC output. For QPSK with Gray mapping, the two dimensions of $\mathbf{x}$ are independently modulated, and thus the signs of the components of $\mathbf{y}$ lead directly to bit decisions.

2.4.6 Error probabilities for fading channels with diversity

Consider again the frequency nonselective, slowly fading channel with the receive vector

$$\mathbf{r} = \mathbf{C}\mathbf{s} + \mathbf{n}$$

as discussed in Subsection 2.4.1. Assume that the diagonal matrix of complex fading amplitudes $\mathbf{C} = \mathrm{diag}\,(c_1, \ldots, c_K)$ is fixed and known at the receiver. We ask for the conditional pairwise error probability $P(\mathbf{s} \mapsto \hat{\mathbf{s}}|\mathbf{C})$ that the receiver erroneously decides for $\hat{\mathbf{s}}$ instead of $\mathbf{s}$ for that given channel. Since $P(\mathbf{s} \mapsto \hat{\mathbf{s}}|\mathbf{C}) = P(\mathbf{C}\mathbf{s} \mapsto \mathbf{C}\hat{\mathbf{s}})$, we can apply the results of Subsection 1.4.3 by replacing $\mathbf{s}$ with $\mathbf{C}\mathbf{s}$ and $\hat{\mathbf{s}}$ with $\mathbf{C}\hat{\mathbf{s}}$ and get

$$P(\mathbf{s} \mapsto \hat{\mathbf{s}}|\mathbf{C}) = \frac{1}{2}\mathrm{erfc}\left(\sqrt{\frac{1}{4N_0} \|\mathbf{C}\mathbf{s} - \mathbf{C}\hat{\mathbf{s}}\|^2} \right).$$

Let $\mathbf{s} = (s_1, \ldots, s_K)$ and $\hat{\mathbf{s}} = (\hat{s}_1, \ldots, \hat{s}_K)$ differ exactly in $L \leq K$ positions. Without losing generality we assume that these are the first ones. This leads to the expression

$$P(\mathbf{s} \mapsto \hat{\mathbf{s}}|\mathbf{C}) = \frac{1}{2}\mathrm{erfc}\left(\sqrt{\frac{1}{4N_0} \sum_{i=1}^{L} |c_i|^2 \, |s_i - \hat{s}_i|^2} \right).$$

The pairwise error probability is the average $\mathrm{E}_{\mathbf{C}}\{\cdot\}$ over all fading amplitudes, that is,

$$P(\mathbf{s} \mapsto \hat{\mathbf{s}}) = \mathrm{E}_{\mathbf{C}} \left\{ \frac{1}{2}\mathrm{erfc}\left(\sqrt{\frac{1}{4N_0} \sum_{i=1}^{L} |c_i|^2 \, |s_i - \hat{s}_i|^2} \right) \right\}. \tag{2.38}$$

For the following treatment, we use the polar representation

$$\frac{1}{2}\mathrm{erfc}(x) = \frac{1}{\pi} \int_0^{\pi/2} \exp\left(-\frac{x^2}{\sin^2\theta} \right) d\theta$$

of the complementary error integral (see Subsection 1.4.3) and obtain the expression

$$P(\mathbf{s} \mapsto \hat{\mathbf{s}}) = \frac{1}{\pi} \int_0^{\pi/2} \mathrm{E}_{\mathbf{C}} \left\{ \exp\left(-\frac{1}{4N_0 \sin^2\theta} \sum_{i=1}^{L} |c_i|^2 \, |s_i - \hat{s}_i|^2 \right) \right\} d\theta. \tag{2.39}$$

This method proposed by Simon and Alouini (2000); Simon and Divsalar (1998) is very flexible because the expectation of the exponential is just the moment generating function of the pdf of the power, which is usually known. The remaining finite integral over θ is easy to calculate by simple numerical methods. Let us assume that the fading amplitudes are statistically independent. Then the exponential factorizes as

$$\mathrm{E}_{\mathbf{C}}\left\{\exp\left(-\frac{1}{4N_0\sin^2\theta}\sum_{i=1}^{L}|c_i|^2|s_i-\hat{s}_i|^2\right)\right\}=\prod_{i=1}^{L}\mathrm{E}_{a_i}\left\{\exp\left(-\frac{a_i^2|s_i-\hat{s}_i|^2}{4N_0\sin^2\theta}\right)\right\},$$

where $\mathrm{E}_{a_i}\{\cdot\}$ is the expectation over the fading amplitude $a_i=|c_i|$. We note that with this expression, it will not cause additional problems if the L fading amplitudes have different average powers or even have different types of probability distribution. If they are identically distributed, the expression further simplifies to

$$\mathrm{E}_{\mathbf{C}}\left\{\exp\left(-\frac{1}{4N_0\sin^2\theta}\sum_{i=1}^{L}|c_i|^2|s_i-\hat{s}_i|^2\right)\right\}=\prod_{i=1}^{L}\mathrm{E}_{a}\left\{\exp\left(-\frac{a^2\Delta_i^2}{N_0\sin^2\theta}\right)\right\},$$

where $\Delta_i=\frac{1}{2}|s_i-\hat{s}_i|$ and $\mathrm{E}_a\{\cdot\}$ is the expectation over the fading amplitude $a=a_i$. For Rayleigh fading, the moment generating function of the squared amplitude can easily be calculated as

$$\mathrm{E}_a\left\{\mathrm{e}^{-xa^2}\right\}=\int_0^\infty 2a\mathrm{e}^{-a^2}\mathrm{e}^{-xa^2}\,da$$

resulting in

$$\mathrm{E}_a\left\{\mathrm{e}^{-xa^2}\right\}=\frac{1}{1+x}.$$

With this expression, Equation (2.39) now simplifies to

$$P(\mathbf{s}\mapsto\hat{\mathbf{s}})=\frac{1}{\pi}\int_0^{\pi/2}\prod_{i=1}^{L}\frac{1}{1+\frac{\Delta_i^2}{N_0\sin^2\theta}}\,d\theta. \tag{2.40}$$

We note that an upper bound can easily be obtained by upper bounding the integrand by its maximum at $\theta=\pi/2$, leading to

$$P(\mathbf{s}\mapsto\hat{\mathbf{s}})\leq\frac{1}{2}\prod_{i=1}^{L}\frac{1}{1+\frac{\Delta_i^2}{N_0}}. \tag{2.41}$$

Obviously, this quantity decreases asymptotically as SNR^{-L}. We note that bounds of this type – but without the factor $1/2$ in front – are commonly obtained by Chernoff bound techniques (Jamali and Le-Ngoc 1994). A method described by Viterbi (Viterbi 1995) improved those bounds by a factor of two and yields (2.41).

A similar bound that is tighter for high SNRs but worse for low SNRs can be obtained by using the inequality

$$\frac{1}{1+\frac{1}{\sin^2\theta}\frac{\Delta_i^2}{N_0}}\leq\sin^2\theta\left(\frac{\Delta_i^2}{N_0}\right)^{-1}$$

to upper bound the integrand. The integral can then be solved resulting in the asymptotically tight upper bound

$$P(\mathbf{s} \mapsto \hat{\mathbf{s}}) \leq \frac{1}{2} \frac{1}{4^L} \binom{2L}{L} \prod_{i=1}^{L} \frac{N_0}{\Delta_i^2}, \tag{2.42}$$

which again shows that the error probability decreases with the power L of the inverse SNR for L-fold repetition diversity.

Consider BPSK as an example. Here, $\Delta_i^2 = E_S$ for all values of i and, by Equation (2.32) with $M = 2$,

$$E_S = \frac{1}{L} E_b$$

holds. Thus, by Equation (2.40), the expression for the bit error rate $P_b = P(\mathbf{s} \mapsto \hat{\mathbf{s}})$ becomes

$$P_b = \frac{1}{\pi} \int_0^{\pi/2} \left(\frac{1}{1 + \frac{1}{L} \frac{E_b}{N_0 \sin^2 \theta}} \right)^L \mathrm{d}\theta \tag{2.43}$$

which may be upper bounded by

$$P(\mathbf{s} \mapsto \hat{\mathbf{s}}) \leq \frac{1}{2} \left(\frac{1}{1 + \frac{1}{L} \frac{E_b}{N_0}} \right)^L.$$

It is interesting to note that we can see from Equation (2.43) that in the limit that the diversity degrees approach infinity, we reach the performance of an AWGN channel. Using the formula

$$\lim_{n \to \infty} \left(1 + \frac{1}{n} x \right)^n = \mathrm{e}^x$$

we obtain

$$\lim_{L \to \infty} \frac{1}{\pi} \int_0^{\pi/2} \left(\frac{1}{1 + \frac{1}{L} \frac{E_b}{N_0 \sin^2 \theta}} \right)^L \mathrm{d}\theta = \frac{1}{\pi} \int_0^{\pi/2} \exp\left(-\frac{E_b}{N_0 \sin^2 \theta} \right) \mathrm{d}\theta.$$

By the polar representation of the error integral, the r.h.s. equals

$$P_b = \frac{1}{2} \mathrm{erfc} \left(\sqrt{\frac{E_b}{N_0}} \right),$$

which is the BER for BPSK in the AWGN channel. Figure 2.12 shows these curves for $L = 1, 2, 4, 8, 16, 32, 64$ compared to the AWGN limit.

For Ricean fading with Rice factor K, the characteristic function can be calculated as well, resulting in the expression for the error probability

$$P(\mathbf{s} \mapsto \hat{\mathbf{s}}) = \frac{1}{\pi} \int_0^{\pi/2} \prod_{i=1}^{L} R_K \left(\frac{\Delta_i^2}{N_0 \sin^2 \theta} \right) \mathrm{d}\theta \tag{2.44}$$

with the abbreviation

$$R_K(x) = \frac{1 + K}{1 + K + x} \exp\left(-\frac{Kx}{1 + K + x} \right)$$

(see (Benedetto and Biglieri 1999; Jamali and Le-Ngoc 1994)).

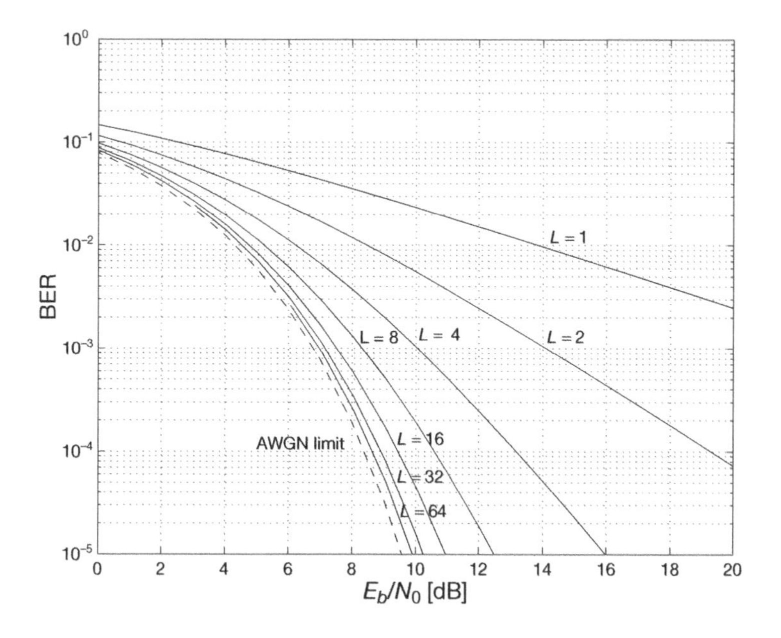

Figure 2.12 Error rates for BPSK with L-fold diversity for different values of L in a Rayleigh fading channel.

Some alternative expressions for error probabilities with diversity reception

The above method that utilizes the polar representation of the complementary Gaussian error integral is quite flexible because the fading coefficients may or may not have equal power. We will see later that it is also suited very well to investigate error probabilities for codes QAM. One drawback of this method is that it does not apply to differential modulation.

For the sake of completeness, we will now present some formulas that are valid for differential and coherent BPSK ($M = 2$) and QPSK ($M = 4$). These formulas can be found in (Hagenauer 1982; Hagenauer *et al.* 1990; Proakis 2001). We define the SNR

$$\gamma_S = E_S/N_0 = \log_2(M)E_b/N_0.$$

For coherent modulation, we define a parameter

$$\xi = \begin{cases} \sqrt{\frac{\gamma_s}{1+\gamma_s}} : & M = 2 \\ \sqrt{\frac{\gamma_s}{2+\gamma_s}} : & M = 4 \end{cases}.$$

Note that

$$\xi = \sqrt{\frac{E_b/N_0}{1 + E_b/N_0}}$$

for both cases. For differential modulation, we define

$$\xi = \begin{cases} \frac{\gamma_s R_1}{1+\gamma_s} & : \quad M = 2 \\ \frac{\gamma_s R_1}{\sqrt{2(1+\gamma_s)^2 - \gamma_s^2 R_1^2}} & : \quad M = 4 \end{cases},$$

where

$$R_1 = \mathcal{R}_c(T_S)$$

is the value of the time autocorrelation $\mathcal{R}_c(t)$ of the channel, taken at the symbol duration $t = T_S$. For the Jakes Doppler spectrum, we obtain from Equation (2.8)

$$\mathcal{R}_c(T_S) = \mathrm{J}_0\left(2\pi\nu_{\max}T_S\right),$$

which is obviously a function of the product $\nu_{\max}T_S$. As discussed in Subsection 2.2.1, for $\nu_{\max}T_S \ll 1$, we can approximate $\mathcal{R}_c(T_S)$ by the second order of the Taylor series as

$$\mathcal{R}_c(T_S) \approx 1 - (\pi\nu_{\max}T_S)^2 .$$

The bit error probabilities P_L for L-fold diversity can then be expressed by either of the three following equivalent expressions:

$$P_L = \frac{1}{2}\left[1 - \xi\sum_{k=0}^{L-1}\binom{2k}{k}\left(\frac{1-\xi^2}{4}\right)^k\right]$$

$$P_L = \left(\frac{1-\xi}{2}\right)^L\sum_{k=0}^{L-1}\binom{L-1+k}{k}\left(\frac{1+\xi}{2}\right)^k$$

$$P_L = \left(\frac{1-\xi}{2}\right)^{2L-1}\sum_{k=0}^{L-1}\binom{2L-1}{k}\left(\frac{1+\xi}{1-\xi}\right)^k$$

We note that – as expected – these expressions are identical for coherent BPSK and QPSK if they are written as functions of E_b/N_0. DBPSK (differential BPSK) and DQPSK are numerically very close together.

2.4.7 Transmit antenna diversity

In many wireless communications scenarios, the practical boundary conditions at the transmitter and at the receiver are asymmetric. Consider as an example the situation of mobile radio. The mobile unit must be as small as possible and it is practically not feasible to apply receive antenna diversity and mount two antennas with a distance larger than half the wavelength. At the base station site, this is obviously not a problem. Transmit antenna diversity is thus very desirable because multiple antennas at the base station can then be used for the uplink and for the downlink as well. However, if one transmits signals from two or more antennas via the same physical medium, the signals will interfere and the receiver will be faced with the difficult task of disentangling the information from the superposition of signals. It is apparent that this cannot be done without loss in any case. Only a few years ago, Alamouti (Alamouti 1998) found a quite simple setup for two antennas for which it is possible to disentangle the information without loss. The scheme can even be proven to be in some sense equivalent to a setup with two receive antennas. For the geometrical interpretation of this fact and the situation for more than two antennas, we refer to the discussion in (Schulze 2003a).

The Alamouti scheme

This scheme uses two transmit antennas to transmit a pair of two complex symbols (s_1, s_2) during two time slots. We assume that for both antennas the channel can be described by the discrete-time fading model with complex fading coefficients c_1 and c_2, respectively. We assume that the time variance of the channel is slow enough so that we can assume that these coefficients do not change from one time slot to the other.

The pair of complex transmit symbols (s_1, s_2) is then processed in the following way: at time slot 1, the symbol s_1 is transmitted from antenna 1 and s_2 is transmitted from antenna 2. The received signal (without noise) at time slot 1 is then given by $c_1 s_1 + c_2 s_2$. At time slot 2, the symbol s_2^* is transmitted from antenna 1 and $-s_1^*$ is transmitted from antenna 2. The received signal at time slot 2 is then given by $-c_2 s_1^* + c_1 s_2^*$. It is convenient for the analysis to take the complex conjugate of the received symbol in the second time slot before any other further processing at the receiver. We can therefore say that, at time slot 2, s_1 and s_2 have been transmitted over channel branches with fading coefficients $-c_2^*$ and c_1^*, respectively. The received symbols with additive white Gaussian noise at time slots 1 and 2 are given by

$$r_1 = c_1 s_1 + c_2 s_2 + n_{c1}$$

and

$$r_2 = -c_2^* s_1 + c_1^* s_2 + n_{c2},$$

respectively, where n_{c1} and n_{c2} are the independent complex Gaussian noise components with variance $\sigma^2 = N_0/2$ in each real dimension. We can write this in vector notation as

$$\mathbf{r} = \mathbf{C}\mathbf{s} + \mathbf{n}_c \tag{2.45}$$

with the vectors $\mathbf{s} = (s_1, s_2)^T$, $\mathbf{r} = (r_1, r_2)^T$, $\mathbf{n}_c = (n_{c1}, n_{c2})^T$ and the *channel matrix*

$$\mathbf{C} = \begin{bmatrix} c_1 & c_2 \\ -c_2^* & c_1^* \end{bmatrix}. \tag{2.46}$$

We observe that the channel matrix has the property

$$\mathbf{C}^\dagger \mathbf{C} = \mathbf{C}\mathbf{C}^\dagger = \left(|c_1|^2 + |c_2|^2\right) \mathbf{I}_2, \tag{2.47}$$

where $\mathbf{I}_2$ is the 2×2 identity matrix. Equation (2.47) means that the channel matrix can be written as

$$\mathbf{C} = \sqrt{|c_1|^2 + |c_2|^2}\, \mathbf{U},$$

where $\mathbf{U}$ is a unitary matrix, that is, a matrix with the property $\mathbf{U}^\dagger \mathbf{U} = \mathbf{U}\mathbf{U}^\dagger = \mathbf{I}_2$. Unitary matrices (like orthogonal matrices for real vector spaces) are invertible matrices that leave

Euclidean distances invariant (see e.g. (Horn and Johnson 1985)). They can be visualized as rotations (possibly combined with a reflection) in an Euclidean space. This means that the transmission channel given by Equation (2.45) can be separated into three parts:

1. A rotation of the vector $\mathbf{s}$ in two complex (= four real) dimensions.

2. An attenuation by the composed fading amplitude $\sqrt{|c_1|^2 + |c_2|^2}$.

3. An AWGN channel.

Keeping in mind that multiplicative fading is just a phase rotation together with an attenuation by a real fading amplitude, we can now interpret this transmission according to Equation (2.45) with a matrix given by Equation (2.46) as a generalization of the familiar multiplicative fading from one to two complex dimensions, or, if this is easier to visualize, from two to four real dimensions. The two-dimensional rotation by the channel phase is replaced by a four-dimensional rotation, and the (real-valued) channel fading amplitude has to be replaced by the composed fading amplitude

$$a = \sqrt{|c_1|^2 + |c_2|^2}.$$

This geometrical view shows that the receiver must back rotate the receive signal $\mathbf{r}$, and then estimate the transmit vector $\mathbf{s}$ in the familiar way as known for the AWGN channel, thereby taking into account the amplitude factor a.

For the formal derivation of the diversity combiner, we proceed as in Subsection 2.4.1. The channel given by Equation (2.45) looks formally the same as the channel considered there, only the matrix $\mathbf{C}$ has a different structure. The maximum likelihood transmit vector $\hat{\mathbf{s}}$ is the one that minimizes the squared Euclidean distance, that is,

$$\hat{\mathbf{s}} = \arg\min_{\mathbf{s}} \|\mathbf{r} - \mathbf{C}\mathbf{s}\|^2 .$$

Equivalently, we may write

$$\hat{\mathbf{s}} = \arg\max_{\mathbf{s}} \left(\Re\left\{\mathbf{s}^\dagger \mathbf{C}^\dagger \mathbf{r}\right\} - \frac{1}{2} \|\mathbf{C}\mathbf{s}\|^2 \right).$$

Using Equation (2.47), we can evaluate the energy term and obtain the expression

$$\hat{\mathbf{s}} = \arg\max_{\mathbf{s}} \left(\Re\left\{\mathbf{s}^\dagger \mathbf{C}^\dagger \mathbf{r}\right\} - \frac{1}{2} \left(|c_1|^2 + |c_2|^2\right) \|\mathbf{s}\|^2 \right).$$

We note that the energy term can be discarded if all signal vectors $\mathbf{s}$ have the same energy. This is obviously the case if the (two-dimensional) symbols s_i have always equal energy as for PSK signaling, but this is not necessary. It remains true if the symbol energy differs, but all vectors of a four-dimensional signal constellation lie on a four-dimensional sphere.

The diversity combiner processes the receive vector $\mathbf{r}$ to the vector

$$\mathbf{C}^\dagger \mathbf{r} = \begin{bmatrix} c_1^* r_1 - c_2 r_2 \\ c_2^* r_1 + c_1 r_2 \end{bmatrix}$$

and correlates it with all possible transmit vectors, thereby – if necessary – taking into account their different energies. We note that for QPSK with Gray mapping, the signs of the real and imaginary parts of $\mathbf{C}^{\dagger}\mathbf{r}$ provide us directly with the bit decisions.

The strong formal similarity to two-antenna receive antenna diversity becomes evident in the equivalent real-valued model. We multiply Equation (2.45) by the back-rotation matrix $\mathbf{U}^{\dagger} = |\mathbf{c}|^{-1}\mathbf{C}^{\dagger}$ and obtain

$$\mathbf{U}^{\dagger}\mathbf{r} = a\mathbf{s} + \mathbf{n}_c,$$

where we made use of the fact that $\mathbf{U}^{\dagger}\mathbf{n}_c$ has the same statistical properties as $\mathbf{n}_c$. We define the four-dimensional real transmit and receive vectors

$$\mathbf{x} = \begin{bmatrix} \Re\{\mathbf{s}\} \\ \Im\{\mathbf{s}\} \end{bmatrix}, \quad \mathbf{y} = \begin{bmatrix} \Re\{\mathbf{U}^{\dagger}\mathbf{r}\} \\ \Im\{\mathbf{U}^{\dagger}\mathbf{r}\} \end{bmatrix}$$

and obtain the real-valued time-discrete vector model

$$\mathbf{y} = a\mathbf{x} + \mathbf{n}, \tag{2.48}$$

where $\mathbf{n}$ is four-dimensional real AWGN. This is formally the same as the real-valued model for the MRC combiner given by Equation (2.36). The only difference is the extension from two to four dimensions. As for the MRC combiner, minimizing the squared Euclidean distance in the real vector space yields

$$\hat{\mathbf{x}} = \arg\min_{s} \|\mathbf{y} - a\mathbf{x}\|^2$$

or

$$\hat{\mathbf{x}} = \arg\max_{\mathbf{x}} \left(a\mathbf{y} \cdot \mathbf{x} - \frac{1}{2}|a^2 \|\mathbf{x}\|^2 \right).$$

The first term is the correlation (scalar product) of the combiner output $a\mathbf{y}$ and the transmit symbol vector $\mathbf{x}$, and the second is the energy term. For PSK signaling, this term is independent of $\mathbf{x}$ and can thus be ignored. In that case, the maximum likelihood transmit symbol vector $\mathbf{x}$ is the one with the smallest angle to the MRC output. For QPSK with Gray mapping, the two dimensions of $\mathbf{x}$ are independently modulated, and thus the signs of the components of $\mathbf{y}$ lead directly to bit decisions.

The conditional pairwise error probability given a fixed channel matrix $\mathbf{C}$ will be obtained similar to that of conventional diversity discussed in Subsection 2.4.6 as

$$P(\mathbf{s} \mapsto \hat{\mathbf{s}}|\mathbf{C}) = \frac{1}{2}\mathrm{erfc}\left(\sqrt{\frac{1}{4N_0}\|\mathbf{C}\mathbf{s} - \mathbf{C}\hat{\mathbf{s}}\|^2}\right).$$

In Subsection 2.4.6, the matrix $\mathbf{C}$ is diagonal, but here it has the property given by Equations (2.47). Thus the squared Euclidean distance can be simplified according to $\|\mathbf{C}\mathbf{s} - \mathbf{C}\hat{\mathbf{s}}\|^2 = \left(|c_1|^2 + |c_2|^2\right)\|\mathbf{s} - \hat{\mathbf{s}}\|^2$ and we find the expression for the pairwise error probability

$$P(\mathbf{s} \mapsto \hat{\mathbf{s}}) = \mathrm{E}_{\mathbf{C}}\left\{\frac{1}{2}\mathrm{erfc}\left(\sqrt{\frac{\|\mathbf{s} - \hat{\mathbf{s}}\|^2}{4N_0}\left(|c_1|^2 + |c_2|^2\right)}\right)\right\},$$

where $\mathrm{E}_{\mathbf{C}}$ means averaging over the channel. This equation is a special case of Equation (2.38), and it can be analyzed using the same methods.

Now let E_b be the total energy per data bit available at the receiver and E_S the energy per complex transmit symbol s_1 or s_2. We assume M-ary modulation, so each of them carries $\log_2(M)$ data bits. Both symbols are assumed to be of equal (average) energy, which means that $E_S = \mathrm{E}\{|s_1|^2\} = \mathrm{E}\{|s_2|^2\} = \mathrm{E}\{\|\mathbf{s}\|^2/2\}$. We normalize the average power gain for each antenna channel coefficient to one, that is, $\mathrm{E}\left\{|c_1|^2\right\} = \mathrm{E}\left\{|c_2|^2\right\} = 1$. Then, for each time slot, the total energy $2E_S$ is transmitted at both antennas together and the same (average) energy is available at each of the receive antenna. Therefore, the total energy available at the receiving site for that time slot is

$$2E_S = \log_2(M)E_b.$$

For only one transmit antenna, $SNR = E_S/N_0$ is the SNR at the receive antenna for linear modulation. For two transmit antennas, we have $SNR = 2E_S/N_0$. For uncoded BPSK or QPSK transmission, the value of each data bit affects only one real dimension. The event of an erroneous bit decision corresponds to the squared Euclidean distance

$$\|\mathbf{s} - \hat{\mathbf{s}}\|^2 = 4E_S/\log_2(M) = 2E_b,$$

which means that both BPSK ($M = 2$) and QPSK ($M = 4$) have the bit error probability

$$P_b = \mathrm{E}_{\mathbf{C}}\left\{\frac{1}{2}\mathrm{erfc}\left(\sqrt{\frac{E_b}{2N_0}\left(|c_1|^2 + |c_2|^2\right)}\right)\right\}$$

as a function of E_b/N_0. This is exactly the same as for twofold receive antenna diversity, which is a special case of the results in Subsection 2.4.6 and we can apply the formulas given there for independent Rayleigh or Ricean fading.

We note that Alamouti's twofold transmit antenna diversity has the same performance as the twofold receive antenna diversity only if we write P_b as a function of E_b/N_0 because $SNR = \log_2 M \cdot E_b/N_0$ holds for the first case and $SNR = \frac{1}{2}\log_2 M \cdot E_b/N_0$ holds for the latter.

2.5 Bibliographical Notes

The classical textbook about mobile radio channels is (Jakes 1975). However, fading channels are treated in many modern textbooks about digital communication techniques (see e.g. (Benedetto and Biglieri 1999; Kammeyer 2004; Proakis 2001)). We also recommend the introductory chapter of (Jamali and Le-Ngoc 1994).

The system theory of WSSUS processes goes back to the classical paper of (Bello 1963). The practical simulation method described in this chapter has been developed by one of the authors (Schulze 1988) and has later been refined and extended by Hoeher (1992). We would like to point out that the line of thought for this model was mainly inspired by the way physicists looked at statistical mechanics (see e.g. (Hill 1956; Landau and Lifshitz 1958)). All measurements are time averages, while the statistical theory deals with statistical (so-called *ensemble*) averages. This replacement (the so-called *ergodic hypothesis*) is mathematically nontrivial, but is usually heuristically justified. Systems in statistical physics that are too complex to be studied analytically are often investigated by the so-called *Monte-Carlo* Simulations. The initial conditions (locations and velocities) of the particles (e.g. molecules) are generated as (pseudo) random variables by a computer, and the dynamics of the system has to be calculated, for example, by the numerical solution of differential equations. From these solutions, time averages of physical quantities are calculated. For a mobile radio system, we generate phases, Doppler shifts and delay (*as initial conditions* of the system) and calculate the system dynamics from these quantities. Finally, time averages, for example, for the bit error rate are calculated.

2.6 Problems

1. Let $z(t) = x(t) + jy(t)$ be a stochastic process with the property

$$\mathrm{E}\{z(t+\tau)z(t)\} = 0.$$

Show the properties

$$\mathrm{E}\{x(t+\tau)x(t)\} = \mathrm{E}\{y(t+\tau)y(t)\}$$

and

$$\mathrm{E}\{x(t+\tau)y(t)\} = -\mathrm{E}\{y(t+\tau)x(t)\}.$$

2. Show that for the Jakes Doppler spectrum $\mathcal{S}_c(\nu)$, the second moment is given by

$$\mu_2\{\mathcal{S}_c(\nu)\} = \frac{\nu_{\max}^2}{2}.$$

3. An L-branch diversity channel is given by the receive signal vector

$$\mathbf{r} = s\mathbf{c} + \mathbf{n},$$

with the channel vector $\mathbf{c}$ given by

$$\mathbf{c} = (c_1, \ldots, c_L)^T,$$

where s is the transmit symbol and $\mathbf{n}$ is L-dimensional (not necessarily Gaussian) complex white noise. A linear combiner is given by the operation $u = \mathbf{v}^\dagger\mathbf{r}$ with a given vector $\mathbf{v}$. Show that the SNR for the combiner output is maximized for $\mathbf{v} = \mathbf{c}$.

3

Channel Coding

3.1 General Principles

3.1.1 The concept of channel coding

Channel coding is a common strategy to make digital transmission more reliable, or, equivalently, to achieve the same required reliability for a given data rate at a lower power level at the receiver. This gain in power efficiency is called *coding gain*. For mobile communication systems, channel coding is often indispensable. As discussed in the preceding chapter, the bit error rate in a Rayleigh fading channel decreases as $P_b \sim (E_b/N_0)^{-1}$, which would require an unacceptable high transmit power to achieve a sufficiently low bit error rate. We have seen that one possible solution is diversity. We will see in the following sections that channel coding can achieve the same gain as diversity with less redundancy.

This chapter gives a brief but self-contained overview over the channel coding techniques that are commonly applied in OFDM and CDMA systems. For a more detailed discussion, we refer to standard text books cited in the Bibliographical Notes.

Figure 3.1 shows the classical channel coding setup for a digital transmission system. The channel encoder adds redundancy to digital data b_i from a data source. For simplicity, we will often speak of data bits b_i and channel encoder output bits c_i, keeping in mind that other data symbol alphabets than binary ones are possible and the same discussion applies to that case. We briefly review some basic concepts and definitions.

- The output of the encoder is called a *code word*. The set of all possible code words is the *code*. The encoder itself is a mapping rule from the set of possible data words into the code. We remark that a code (which is a set) may have many different encoders (i.e. different mappings with that same set as the image).

- *Block codes*: If the channel encoder always takes a data block $\mathbf{b} = (b_1, \ldots, b_K)^T$ of a certain length K and encodes it to a code word $\mathbf{c} = (c_1, \ldots, c_N)^T$ of a certain length N, we speak of an (N, K) *block code*. For other codes than block codes, for example, convolutional codes, it is often convenient to work with code words of finite length, but it is not necessary, and the length is not determined by the code.

Theory and Applications of OFDM and CDMA Henrik Schulze and Christian Lüders

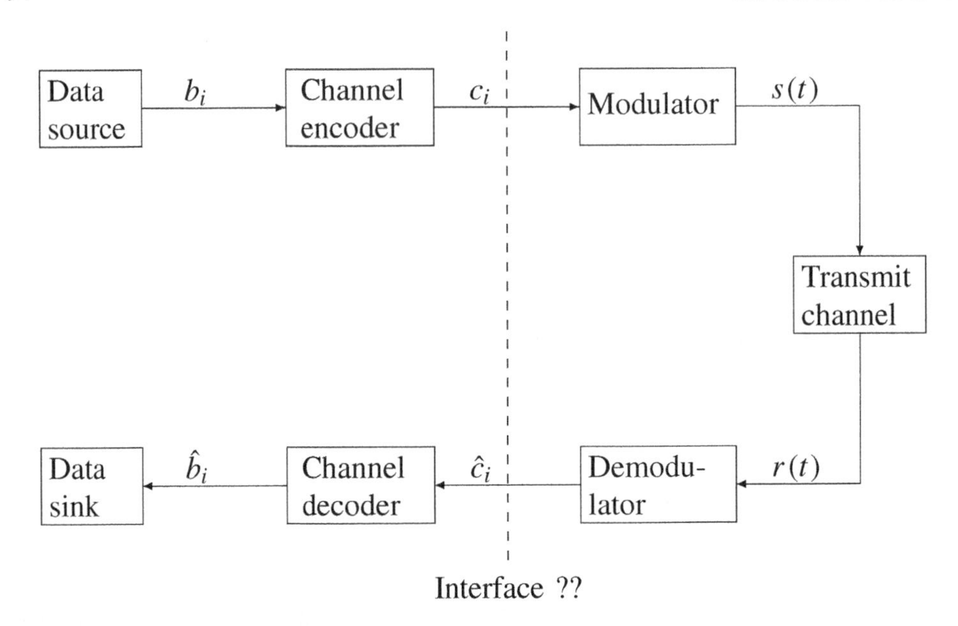

Figure 3.1 Block diagram for a digital transmission setup with channel coding.

- If the encoder maps $\mathbf{b} = (b_1, \ldots, b_K)^T$ to the code word $\mathbf{c} = (c_1, \ldots, c_N)^T$, the ratio $R_c = K/N$ is called the *code rate*.

- If two code words differ in d positions, then d is called the *Hamming distance* between the two code words. The minimum Hamming distance between any two code words is called the *Hamming distance of the code* and is usually denoted by d_H. For an (N, K) block code, we write the triple (N, K, d_H) to characterize the code.

- If the vector sum of any two code words is always a code word, the code is called *linear*.

- The Hamming distance of a linear code equals the minimum number of nonzero elements in a code word, which is called the *weight* of the code. A code can correct up to t errors if $2t + 1 \leq d_H$ holds.

- An encoder is called *systematic* if the data symbols are a subset of the code word. Obviously, it is convenient but not necessary that these systematic (i.e. data) bits (or symbols) are positioned at the beginning of the code word. In that case the encoder maps $\mathbf{b} = (b_1, \ldots, b_K)^T$ to the code word $\mathbf{c} = (c_1, \ldots, c_N)^T$ and $b_i = c_i$ for $i = 1, \ldots, K$. The nonsystematic symbols of the code word are called *parity check* (PC) *symbols.*

The channel decoder outputs c_i are the inputs of the modulator. Depending on these data, the modulator transmits one out of a set of possible signals $s(t)$. For binary codes, there are 2^K code words and thus there are 2^K possible signals $s(t)$. For a linear modulation scheme, such a signal can be written as

$$s(t) = \sum_{i=1}^{L} s_i g_i(t)$$

with an orthonormal transmit base $\{g_i(t)\}_{i=1}^{L}$. Each possible signal can then uniquely be characterized by the corresponding transmit symbol vector $\mathbf{s} = (s_1, \ldots, s_L)^T$. Generally, there is a one-to-one correspondence between the code word vectors $\mathbf{c} = (c_1, \ldots, c_N)^T$ and the transmit symbol vector $\mathbf{s} = (s_1, \ldots, s_L)^T$. Because each code word $\mathbf{c}$ is uniquely determined by the corresponding data word $\mathbf{b}$, the mapping $\mathbf{b} \mapsto \mathbf{c}$ is uniquely defined and one may regard the channel encoder and the modulator as one single device that corresponds to the mapping $\mathbf{b} \mapsto \mathbf{s}$. This natural, but modern concept is called *coded modulation* and it has the advantage that it is quite natural to look for a joint optimization of channel coding and modulation (see e.g. (Biglieri *et al.* 1991; Jamali and Le-Ngoc 1994; Ungerboeck 1982)).

The more traditional concept as depicted in Figure 3.1 keeps both parts separated, and the modulator is one of the classical concatenation schemes discussed in Section 1.4. We may have a simple M-ary PSK or QAM symbol mapping for the s_i. Then, a subset of $m = \log_2(M)$ bits of one code word will be mapped on each symbol s_i. Or, alternatively, $m = \log_2(M)$ bits taken from m different code words will be mapped on each symbol s_i. The latter turns out to be a better choice for fading channels to avoid two bits of the same code word being affected by the same fading amplitude. We note that for M-ary signal constellations and a code rate R_c, only $R_c \log_2(M)$ useful bits are transmitted per complex symbol s_i. Thus, the energy E_b per useful bit is related to the average symbol energy $E_S = \mathrm{E}\left\{|s_k|^2\right\}$ by the equation

$$E_S = R_c \log_2(M) \cdot E_b. \tag{3.1}$$

Following Figure 3.1, the modulated signal is corrupted by the noisy transmission channel. Thus, some of the demodulator output bits $\hat{c}_i$ will be erroneously decided from the received signal $r(t)$. The channel decoder then uses the redundancy of the code to correct the errors and delivers (hopefully) correct data $\hat{b}_i$ identical to the source data b_i.

The crucial point of this traditional conception of correcting bit errors is the interface between CoDec (Coder/Decoder) and the MoDem (Modulator/Demodulator). We have seen that there is no fundamental reason to distinguish between these blocks at the transmitter. And, as we know from the detection theory as discussed in Section 1.3, the MLSE will find the most probable transmit vector $\hat{\mathbf{s}}$ that corresponds uniquely to a data vector $\hat{\mathbf{b}}$. The only reason to separate demodulator and decoder is a practical one: often, especially for long block codes, the MLSE requires quite an exhaustive search that cannot be implemented. But there are often algebraic decoding techniques for such codes that require binary inputs[1]. There may be good reasons to proceed this way but one should always keep in mind that any hard bit decision before the decoder causes loss of information. For convolutional codes, the MLSE can be easily implemented by the Viterbi decoder (see the following text). Hard decisions would cause a needless loss of approximately 2 dB in performance for convolutional codes.

The following example shows how the same transmission setup can be interpreted according to different points of view.

[1] Or, for nonbinary codes, discrete valued inputs are taken from some other finite signal alphabet.

Example 6 (Walsh–Hadamard codes) *Consider an encoder map* $\mathbf{b} \mapsto \mathbf{c}$ *for a linear* (8, 3, 4) *block code given by*

$$\begin{bmatrix} 0 & 1 & 0 & 1 & 0 & 1 & 0 & 1 \\ 0 & 0 & 1 & 1 & 0 & 0 & 1 & 1 \\ 0 & 0 & 0 & 0 & 1 & 1 & 1 & 1 \end{bmatrix} \mapsto \begin{bmatrix} 0 & 0 & 0 & 0 & 0 & 0 & 0 & 0 \\ 0 & 1 & 0 & 1 & 0 & 1 & 0 & 1 \\ 0 & 0 & 1 & 1 & 0 & 0 & 1 & 1 \\ 0 & 1 & 1 & 0 & 0 & 1 & 1 & 0 \\ 0 & 0 & 0 & 0 & 1 & 1 & 1 & 1 \\ 0 & 1 & 0 & 1 & 1 & 0 & 1 & 0 \\ 0 & 0 & 1 & 1 & 1 & 1 & 0 & 0 \\ 0 & 1 & 1 & 0 & 1 & 0 & 0 & 1 \end{bmatrix},$$

where the eight possible data words $\mathbf{b}$ *are the columns of the left matrix, and the corresponding code words* $\mathbf{c}$ *are the columns of the right matrix. One can easily verify that this code is linear (see Proposition 3.1.1) and that the Hamming distance is given by* $d_H = 4$*. Thus, it can correct one bit error. We now use a BPSK modulator with the symbol mapping* $c_i \mapsto s_i = (-1)^{c_i}$*. The composed mapping* $\mathbf{b} \mapsto \mathbf{s}$ *is then given by*

$$\begin{bmatrix} 0 & 1 & 0 & 1 & 0 & 1 & 0 & 1 \\ 0 & 0 & 1 & 1 & 0 & 0 & 1 & 1 \\ 0 & 0 & 0 & 0 & 1 & 1 & 1 & 1 \end{bmatrix} \mapsto \begin{bmatrix} 1 & 1 & 1 & 1 & 1 & 1 & 1 & 1 \\ 1 & -1 & 1 & -1 & 1 & -1 & 1 & -1 \\ 1 & 1 & -1 & -1 & 1 & 1 & -1 & -1 \\ 1 & -1 & -1 & 1 & 1 & -1 & -1 & 1 \\ 1 & 1 & 1 & 1 & -1 & -1 & -1 & -1 \\ 1 & -1 & 1 & -1 & -1 & 1 & -1 & 1 \\ 1 & 1 & -1 & -1 & -1 & -1 & 1 & 1 \\ 1 & -1 & -1 & 1 & -1 & 1 & 1 & -1 \end{bmatrix}.$$

The resulting matrix is the Hadamard matrix for $M = 8$*, and it is obvious that this scheme is nothing but orthogonal Walsh modulation as discussed in Subsection 1.1.4. The columns of the matrix just represent the signs of the Walsh functions of Figure 1.7. These codes corresponding to Hadamard matrices are called* Walsh–Hadamard (WH) *codes. The example here is the WH(8, 3, 4) code.*

We now consider as an example detector outputs given by the receive vector $\mathbf{r} = (1.5, 1.2, 0.9, 0.4, 0.8, \text{-}0.2, 1.2, -0.3)^T$*. The hard decision BPSK demodulator produces the output* $\hat{\mathbf{c}} = (0, 0, 0, 0, 0, 1, 0, 1)^T$*. This vector differs in two positions from the first and in two positions from the second code word. Thus, (at least) two errors have occurred that cannot be corrected. However, the MLSE for orthogonal Walsh modulation calculates the eight scalar products* $\mathbf{r} \cdot \mathbf{s}$ *that are the elements of the row vector* $(5.4, 3.2, 1.2, 1.2, -0.6, 2.6, 1.6, 0.2)$ *and decides in favor of the first vector of the Hadamard matrix with the maximal scalar product* $\mathbf{r} \cdot \mathbf{s} = 5.4$*.*

This example shows that there is often an ambiguity in what we call channel coding and what we call modulation. For example, we may regard any linear modulation scheme as a code. The symbol mapper is an encoder that produces real or complex outputs. Here, we prefer real symbols, that is, we identify the complex plane with the two-dimensional (real) space. An M-PSK or M-QAM symbol is a code word of length $N = 2$ with M code words labeled by $K = \log_2 M$ data bit. This is a (N, K) block code. The alphabet of the coded symbols is a finite set of real numbers. For 8-PSK, for example, this is given by $\{0, \pm\frac{1}{2}\sqrt{2}, \pm 1\}$.

A code word is given by a vector

$$\mathbf{x} = (x_1, \ldots, x_N)^T$$

of real-valued modulation symbols. To interpret the orthogonal Walsh modulation of Subsection 1.1.4 as channel coding, we set $N = M$, and the vectors $\mathbf{x}$ are the columns of the $M \times M$ Hadamard matrix. The transmit signal is given by

$$s(t) = \sum_{m=1}^{M} x_m g_m(t).$$

Note that now the base pulses $g_m(t)$ of this linear modulation correspond to what has been called *chip pulses* in that subsection.

As discussed in Subsection 2.4.1, we may consider the real-valued fading channel with AWGN disturbance given by

$$y_i = a_i x_i + n_i$$

with real-valued receive symbols y_i, fading amplitudes a_i, transmit symbols x_i and noise samples n_i with variance $\sigma^2 = N_0/2$ in each dimension. Using vector notation, we may write

$$\mathbf{y} = \mathbf{A}\mathbf{x} + \mathbf{n}.$$

The MLSE receiver calculates the most probable transmit vector $\mathbf{x}$ from the receive vector $\mathbf{y}$. Because the input of the MLSE, that is, the components y_i of the receive vector, are real numbers, the MLSE is called a *soft decision* receiver, in contrast to a *hard decision* receiver, where (*hard*) bit decisions are taken from the y_i and these bits are passed to the decoder (see Figure 3.1). However, the output of the MLSE receiver is the bit sequence that labels the most likely transmit vector. Thus, the MLSE is a *hard output* receiver. Receivers with soft output (i.e. with reliability information about the decisions) are desirable for *concatenated coding* schemes as described in Subsection 3.1.4. A receiver with soft inputs and soft outputs is called a *SISO*(*soft-in, soft-out*) receiver.

3.1.2 Error probabilities

In this subsection, we discuss error rates for binary codes with antipodal signaling, that is, BPSK. The same formulas apply for QPSK, which can be separated into antipodal signaling for both the in-phase and quadrature component.

Error probabilities for the MLSE receiver and the AWGN channel

For the MLSE receiver, general expressions for the pairwise error probabilities in the AWGN channel were derived in Subsection 1.4.3. The probability that the receiver erroneously decides for the transmit vector $\mathbf{s}$ instead of the transmitted vector $\hat{\mathbf{s}}$ is given by Equation (1.84), that is, the expression

$$P(\mathbf{s} \mapsto \hat{\mathbf{s}}) = \frac{1}{2}\mathrm{erfc}\left(\sqrt{\frac{1}{4N_0}\|\mathbf{s} - \hat{\mathbf{s}}\|^2}\right).$$

For BPSK transmission, we have $\mathbf{s} = (s_1, \ldots, s_L)^T$ with $s_i = \pm\sqrt{E_S}$, where E_S is the symbol energy. We assume a binary code with rate R_c and Hamming distance d_H. Assume that $\mathbf{s}$ corresponds to a code word $\mathbf{c}$ and the receiver decides for a code word $\hat{\mathbf{c}}$ corresponding to the signal vector $\hat{\mathbf{s}}$, and the code words $\mathbf{c}$ and $\hat{\mathbf{c}}$ have the Hamming distance d. Then,

$$\|\mathbf{s} - \hat{\mathbf{s}}\|^2 = 4dE_S.$$

For each transmitted symbol, only R_c useful bits are transmitted. Thus, $E_S = R_c E_b$, and the error event probability P_d for an erroneous decision corresponding to a Hamming distance d is given by

$$P_d = \frac{1}{2}\operatorname{erfc}\left(\sqrt{dR_c\frac{E_b}{N_0}}\right). \tag{3.2}$$

For high values of E_b/N_0, the total error probability is dominated by the most probable error event corresponding to $d = d_H$. Asymptotically, the number of such events (that leads to a factor in front of the complementary error function) can be ignored and we may say that we obtain an *asymptotic coding gain* of

$$G_a = 10\log_{10}(d_H R_c)\,\mathrm{dB}$$

compared to uncoded BPSK. We note that the expression (3.2) for P_d – written as a function of E_b/N_0 – also holds for QPSK with Gray mapping and the coding gain is also the same. For higher level modulation schemes, the analysis is more complicated because the Euclidean distances between the symbols are different.

Error probabilities for the MLSE receiver and the Rayleigh fading channel

We assume a discrete-time fading channel with receive symbols given by

$$r_i = a_i s_i + n_i,\ i = 1, \ldots, N$$

with discrete AWGN n_i and real fading amplitudes a_i, that is, the phase has already been back rotated. We consider BPSK transmission with $s_i = \pm\sqrt{E_S}$ and a binary code of rate R_c and Hamming distance d_H. Assume that $\mathbf{s}$ corresponds to a code word $\mathbf{c}$ of length N and the receiver decides for a code word $\hat{\mathbf{c}}$ corresponding to the signal vector $\hat{\mathbf{s}}$, and the code words $\mathbf{c}$ and $\hat{\mathbf{c}}$ have the Hamming distance d. We ask for the probability P_d that the code word $\mathbf{c}$ has been transmitted and the receiver erroneously decides for another code word $\hat{\mathbf{c}}$. To keep the notation simple, we assume a renumbering of the indices in such a way that the different positions are those with $i = 1, \ldots, d$ and write $\mathbf{s} = (s_1, \ldots, s_d)^T$ for the symbol vector corresponding to the first d positions of $\mathbf{c}$ and $\hat{\mathbf{s}} = (\hat{s}_1, \ldots, \hat{s}_d)^T$ for the symbol vector corresponding to the first d positions of $\hat{\mathbf{c}}$. The last $N - d$ symbols are irrelevant for the decision. This is just the same as the problem of d-fold diversity that was treated in Subsection 2.4.6. We can apply Equation (2.38) with $|s_i - \hat{s}_i|^2 = 4E_S$ and obtain the expression

$$P_d = \mathrm{E}_{a_i}\left\{\frac{1}{2}\operatorname{erfc}\left(\sqrt{\frac{E_S}{N_0}\sum_{i=1}^{d}|a_i|^2}\right)\right\}, \tag{3.3}$$

where $\mathrm{E}_{a_i}\{\cdot\}$ is the average over all fading amplitudes. For a Rayleigh channel with identically distributed and independent fading amplitudes of average power one, the average can be easily performed as shown in Subsection 2.4.5 resulting in

$$P_d = \frac{1}{\pi} \int_0^{\pi/2} \left(\frac{1}{1 + R_c \frac{E_b}{N_0 \sin^2 \theta}} \right)^d \mathrm{d}\theta, \tag{3.4}$$

where we have used $E_S = R_c E_b$. As an alternative to this polar representation expression, the formulas given at the end of Subsection 2.4.6 can be applied. We thus have seen that the Hamming distance can be interpreted as diversity that is provided by the code. Using Equation (2.42), P_d can be tightly upper bounded by

$$P_d \leq \frac{1}{2} \frac{1}{4^L} \binom{2d}{d} \left(R_c \frac{E_b}{N_0} \right)^{-d}.$$

Thus, for a code with Hamming distance d_H, the error rates asymptotically decay as SNR^{-d_H} in an independently fading Rayleigh channel. These expressions for P_d – written as a function of E_b/N_0 – also hold for QPSK if the two bits corresponding to a QPSK symbol s_i belong to different code words. Otherwise, two bits of the same code word will be affected by the same fading amplitude, which would result in a loss of diversity.

Residual bit error rates for hard decision decoding of block codes

We consider an (N, K, d_H) binary block code with hard decision error correction capability of t bit errors. For an odd Hamming distance d_H, we have $d_H = 2t + 1$, and for even d_H, we have $d_H = 2t + 2$. Let p be the channel bit error probability and assume that $i \geq t + 1$ channel bit errors occurs inside the code word. The probability for a certain error pattern with i bit errors is given by

$$p^i (1 - p)^{N-i}.$$

There are $\binom{N}{i}$ such possible patterns for a code word of length N. Thus, the block error probability, that is, the probability for a wrong decoding decision for the code word is given by

$$P_{\mathrm{Block}} = \sum_{i=t+1}^{N} \binom{N}{i} p^i (1 - p)^{N-i}. \tag{3.5}$$

Often, one is interested not only in the block error probability, but in the bit error probability. If the error correction capability is t, the decoder may change at most t bits inside the code word when trying to correct the error. If $i > t$ errors have occurred, this will result in a wrong decision and the decoder will erroneously change at most t additional bits, resulting in at most $i + t$ errors. Thus, if we consider only error events corresponding to exactly i errors ($i > t$), the bit error probability for such an event is bounded by

$$\frac{t + i}{N} \binom{N}{i} p^i (1 - p)^{N-i}.$$

Keeping in mind that not more than N errors may occur, the bound can be slightly improved if we replace $t + i$ in the numerator by $\min(t + i, N)$. Summing up over all possible numbers of errors, we get the bound for the bit error probability

$$P_b \leq \sum_{i=t+1}^{N} \frac{\min(t+i, N)}{N} \binom{N}{i} p^i (1-p)^{N-i}. \tag{3.6}$$

Although it is more popular to talk about bit error probabilities rather than block error probabilities, one should keep in mind that the block error probability is often more relevant for an application than the bit error probability. For the application, the average time between two error events is often the most relevant figure. In case of a decoding error, at least d_H bit errors occur at the same time. The error event corresponding to d_H errors is the most probable one, and it is dominant if the channel is not too bad. Looking only at P_b without knowledge of the code may give rise to wrong interpretations. A residual bit error rate of $P_b = 10^{-6}$ for a rate of 10 kbit/s does not mean that in average one bit error occurs every 100 seconds, but approximately d_H errors occur every 100 d_H seconds. For large values of d_H, this makes a great difference for the application.

3.1.3 Some simple linear binary block codes

In this subsection, we will present some facts about linear binary block codes and give some examples. We will not go into further details and refer to the text books about channel coding cited in the Bibliographical Notes.

Let $\mathcal{C}$ be a linear binary (N, K) block code. We write all the $M = 2^K$ code words of length N as binary columns $\mathbf{c}_m \in \mathcal{C}$, $m = 1, \ldots, M$ and join them together to a matrix

$$\mathbf{C} = [\mathbf{c}_1, \ldots, \mathbf{c}_M].$$

There are M bit tuples of length K. We write them as column vectors $\mathbf{b}_m$, $m = 1, \ldots, M$ with the LSB (Least Significant Bit) in the upper position. We join them together to a binary tuple matrix

$$\mathbf{B} = [\mathbf{b}_1, \ldots, \mathbf{b}_M].$$

For $M = 8$, for example, this matrix is given by

$$\mathbf{B} = \begin{pmatrix} 0 & 1 & 0 & 1 & 0 & 1 & 0 & 1 \\ 0 & 0 & 1 & 1 & 0 & 0 & 1 & 1 \\ 0 & 0 & 0 & 0 & 1 & 1 & 1 & 1 \end{pmatrix}.$$

We write $\mathcal{B}$ for the set of all these binary vectors. The encoder can then be written as a linear mapping

$$\mathbf{G} : \mathcal{B} \to \mathcal{C},\ \mathbf{b}_m \mapsto \mathbf{c}_m = \mathbf{G}\mathbf{b}_m$$

between the vector spaces $\mathcal{B}$ and $\mathcal{C}$. $\mathbf{G}$ is called the *generator matrix*. Using matrix notation, we may also write

$$\mathbf{C} = \mathbf{G}\mathbf{B}.$$

From linear algebra we know that $\mathbf{G}$ is given by an $N \times K$ matrix, with K columns given by the images of the K canonical base vectors, that is, those vectors in $\mathbf{B}$ with only one 1 and all other entries equal to 0. From the structure of $\mathbf{B}$, we see that these are the vectors $\mathbf{c}_m$ with $m = 2^k + 1$, $k = 0, 1, \ldots, K-1$, that is,

$$\mathbf{G} = [\mathbf{c}_2, \mathbf{c}_3, \mathbf{c}_5, \mathbf{c}_9, \ldots, \mathbf{c}_{M/2+1}].$$

The dual code $\mathcal{C}^{\perp}$ is just the orthogonal complement of $\mathcal{C}$ in the N-dimensional vector space over the binary numbers. Its $(N-K) \times N$ generator matrix $\mathbf{H}$ is related to $\mathbf{G}$ by

$$\mathbf{H}^T \mathbf{G} = 0.$$

Because $\mathbf{H}^T \mathbf{c} = 0$ for each $\mathbf{c} \in \mathcal{C}$, $\mathbf{H}$ is called the *parity check matrix* of the code $\mathcal{C}$.

Repetition (RP) codes

A very naive idea for coding is a simple repetition of the bits. An RP$(N, 1, N)$ code has $d_H = N$ and $R_c = 1/N$ and, thus, the coding gain is zero. Obviously, RP coding is just another word for diversity, and, in a fading channel, it has a diversity gain if the fading amplitudes of the received coded symbols are sufficiently independent. The generator matrix of this code is the all-one column vector of length N.

Single parity check (SPC) codes

The matrix of code words $\mathbf{C}$ for the SPC$(K+1, K, 2)$ code is obtained from the binary tuple matrix $\mathbf{B}$ by appending one row in such a way that the (modulo 2) sum over each column equals zero, that is, all code words must have even parity. For the SPC(4, 3, 2) code, for example, we have

$$\mathbf{C} = \begin{bmatrix} 0 & 1 & 0 & 1 & 0 & 1 & 0 & 1 \\ 0 & 0 & 1 & 1 & 0 & 0 & 1 & 1 \\ 0 & 0 & 0 & 0 & 1 & 1 & 1 & 1 \\ 0 & 1 & 1 & 0 & 1 & 0 & 0 & 1 \end{bmatrix}$$

and

$$\mathbf{G} = \begin{bmatrix} 1 & 0 & 0 \\ 0 & 1 & 0 \\ 0 & 0 & 1 \\ 1 & 1 & 1 \end{bmatrix}.$$

An SPC$(K+1, K, 2)$ code has Hamming distance $d_H = 2$ and rate $R_c = K/(K+1)$. The most popular application for SPC (single parity check) codes is error detection by checking the parity. Obviously, the detection is not very reliable because only one error can be detected, not two. It is generally believed that SPC codes can only detect errors, but cannot correct them. This is only the case for hard decision. When using an MLSE receiver, SPC codes have an asymptotic coding gain of

$$G_a = 10 \log_{10} \frac{2K}{K+1} \text{ dB},$$

which approximately approaches a 3 dB gain for high values of N.

Walsh–Hadamard (WH) codes

As already discussed in the example in the last subsection, Walsh–Hadamard codes will be obtained from the Hadamard matrices by replacing each $+1$ by a 0 and -1 by a 1. The code words of the WH$(M, \log_2 M, M/2)$ code of length $M = 2^K$ and Hamming distance $M/2$

are the columns of the resulting $M \times M$ matrix. Transmitting these code words with BPSK modulation brings us back to orthogonal signaling with Walsh functions. The asymptotic coding gain

$$G_a = 10 \log_{10} \frac{K}{2} \text{ dB}$$

has already been obtained in Subsection 1.4.3, where the pairwise error probabilities of orthogonal modulation has been derived.

Proposition 3.1.1 *The WH codes are linear codes.*

Proof. The proof is by induction over K, where $M = 2^K$. The statement is trivially true for $K = 0$, that is, $M = 1$. We will show that if it is true for any $M/2$, it is true for M. Let $\mathbf{C} = [\mathbf{c}_1, \ldots, \mathbf{c}_M]$ be the $M \times M$ matrix obtained from the $M \times M$ Hadamard matrix as described above. The column vectors are the code words. Then, by construction of the Hadamard matrices, the code words $\mathbf{c}_m$ are exactly those vectors that have either the structure

$$\mathbf{c}_m = \begin{bmatrix} \mathbf{c}' \\ \mathbf{c}' \end{bmatrix} \text{ for } m \in \{1, \ldots, M/2\}$$

or

$$\mathbf{c}_m = \begin{bmatrix} \mathbf{c}' \\ \mathbf{c}' \oplus 1 \end{bmatrix} \text{ for } m \in \{M/2 + 1, \ldots, M\},$$

where $\mathbf{c}'$ is a code word of the WH($M/2, \log_2 M - 1, M/4$) code, which, by assumption, has already been proven to be a linear code. From the above decomposition it can easily be seen that the sum of any such vectors has again this structure and thus is a code word of WH($M, \log_2 M, M/2$).

From the recursive construction of the Hadamard matrices, we observe that the column $\mathbf{c}_{M/2+1}$ equals the last row of $\mathbf{B}$ (the MSB (Most Significant Bit) row), the column $\mathbf{c}_{M/4+1}$ equals the second last row of $\mathbf{B}$, and so on. Thus, we find that for the WH code, the interesting property

$$\mathbf{G} = \mathbf{B}^T$$

holds, and the matrix of code words is given by

$$\mathbf{C} = \mathbf{B}^T \mathbf{B}.$$

Since $\mathbf{B}$ has the canonical base as column numbers $m = 2^k + 1$, $k = 0, 1, \ldots, K - 1$, the generator matrix $\mathbf{G}$ provides a systematic encoder with the systematic bit number $k + 1$ at the position $m = 2^k + 1$, $k = 0, 1, \ldots, K - 1$ in the code word. For $M = 8$, for example, the three systematic bit can be found in the positions 2, 3 and 5 of the code word.

Simplex (SPL) codes

An SPL($M - 1, \log_2 M, M/2$) is obtained by omitting the first bit of every WH ($M, \log_2 M, M/2$) code word. This can be done without any loss in performance because this bit is always zero. Because of the higher code rate, the performance is even better, especially for small values of M. For $M = 8$, the matrix of code words is given by

$$\mathbf{C} = \begin{bmatrix} 0 & 1 & 0 & 1 & 0 & 1 & 0 & 1 \\ 0 & 0 & 1 & 1 & 0 & 0 & 1 & 1 \\ 0 & 1 & 1 & 0 & 0 & 1 & 1 & 0 \\ 0 & 0 & 0 & 0 & 1 & 1 & 1 & 1 \\ 0 & 1 & 0 & 1 & 1 & 0 & 1 & 0 \\ 0 & 0 & 1 & 1 & 1 & 1 & 0 & 0 \\ 0 & 1 & 1 & 0 & 1 & 0 & 0 & 1 \end{bmatrix},$$

and the generator matrix is

$$\mathbf{G} = \begin{bmatrix} 1 & 0 & 0 \\ 0 & 1 & 0 \\ 1 & 1 & 0 \\ 0 & 0 & 1 \\ 1 & 0 & 1 \\ 0 & 1 & 1 \\ 1 & 1 & 1 \end{bmatrix}.$$

Hamming codes

Hamming codes are the dual codes of simplex codes. A $(2^K - 1, 2^K - 1 - K, 3)$ Hamming code can correct one error. Hamming codes are simple and weak codes, but they are popular to explain the concepts of algebraic coding. We do not discuss them further because they are exhaustively treated in most text books.

3.1.4 Concatenated coding

If an application requires very low bit error rates, concatenated coding is often the most efficient method to reach this goal. In such a setup, two codes are combined to a stronger overall concatenated code (see Figure 3.2). At the transmitter, the source data will first be encoded by the *outer code*. The code words of this code will then serve as the input data for the *inner code*[2]. Between the two encoders, the order of the symbols inside the stream of code words may be changed by a device that is called *interleaver*. Interleaver structures will be discussed later in Subsection 4.4.2. The code words of the inner code are transmitted over the channel and then decoded by the inner decoder. The inner decoder has to be matched to the channel. At its output, the error rate will be low, but the errors are not uniformly distributed. The inner decoder will typically produce error bursts, that is, connected sequences of unreliable symbols between long sequences of reliable symbols. For block codes, a burst error corresponds to an erroneously decided code word. For convolutional codes, an error burst corresponds to the sequence of states in the trellis, where the correct path and the maximum likelihood path are different (see Subsection 3.2.2). The deinterleaver inverts the interleaver. The interleaving scheme breaks up the error bursts and has to be matched to the error correction capabilities of the outer code, that is, to its Hamming distance and code word length. For a properly designed concatenated coding scheme, the output of the outer decoder will be nearly error free. Thus, one can

[2]The naming *inner* and *outer* code stems from the fact that the inner code is closer to the channel than the outer code.

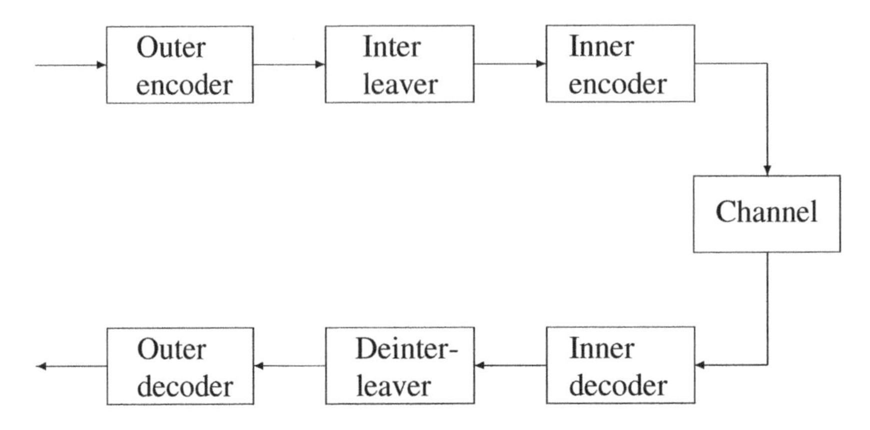

Figure 3.2 Block diagram for a concatenated coding setup.

visualize the inner decoder as a device that is suited for coarsely cleaning up the errors that are produced by a severely corrupted transmission channel. The outer decoder will then clean up the residual errors left by the outer decoder.

We note that in some cases the inner decoder may pass soft decision values rather than hard decision bits to the outer decoder. This is, for example, an important item for a concatenated coding scheme with two convolutional codes. Turbo codes are a setup of two parallel concatenated codes (see Subsection 3.2.5). The classical concatenation described above is sometimes called *serial* concatenation to distinguish from such setups. For serially concatenated convolutional codes, we refer to (Benedetto and Biglieri 1999).

If we regard QAM or PSK modulation as (nonbinary) coding schemes (see the discussion at the end of Subsection 3.1.1), QAM or PSK with additional (convolutional) coding is a concatenated coding scheme. The outer QAM or PSK decoder will typically pass *soft bits* to the outer (convolutional) decoder. The optimal soft bits are the LLRs (log-likelihood ratios) calculated by the MAP receiver (see Subsection 3.1.5).

Probably the most popular concatenated coding scheme is an inner convolutional code with an outer Reed–Solomon (RS) code. Both types of codes are discussed in the following sections. Convolutional codes with soft decision decoding are well suited for channels that are severely corrupted by a high noise level and/or by multipath fading. MLSE can be easily implemented by the Viterbi decoder. However, because of their typically quite low Hamming distance, the BER curves show a poor decay. Thus, a high SNR is needed if very low BERs are required. The convolutional decoder produces bursts of erroneously decided bits. RS codes can be designed as strong codes with high Hamming distances. They are based on byte arithmetics rather than bit arithmetics, that is, they correct byte errors rather than bit errors. The decoder works with hard decision input bytes. The favorable input for the decoder is a data stream with bit errors that are grouped together into bytes, but the byte error structure should not be bursty. Thus, a convolutional decoder together with a byte interleaving produces the favorite input for the RS decoder.

Deep space communication was one of the first applications of such a scheme with a convolutional code and an RS code (see (McEliece and Swanson 2001)). Because of the power limitation, the physical channel is a very noisy AWGN channel that makes convolutional codes the best choice. If, for example, data compressed pictures have to be transmitted, the bit error rate has to be very low. Therefore an outer RS will give a considerable gain (see (McEliece and Swanson 2001)). For this application, the outer codes give an additional coding gain of 2.5 dB at BER $= 10^{-6}$ or a gain in data rate of 78%.

Another concatenated coding scheme with a convolutional code and an RS code is applied in the DVB-T system, which will be described in Subsection 4.6.2. DVB-T is an OFDM system with QAM modulation. Because QAM with convolutional coding can already be interpreted as code concatenation, we have a twofold concatenation in this system. The outer RS code has been chosen because the goal has been to reach the extremely low residual bit error rate 10^{-11} in the video data stream. This cannot be reached with convolutional codes at a reasonable effort, especially in a fading channel.

We note that in a concatenated decoding setup as depicted in Figure 3.2, the decoding will be done stepwise and the two decoders are separate devices. This is in general suboptimal compared to a MLSE or MAP receiver for the complete concatenated code. Improvements can be obtained if the decoders can help each other by means of iterative decoding. Turbo codes may be the most famous application (see Subsection 3.2.5). The idea of iterative decoding had been applied earlier in deep space communications (Hagenauer *et al.* 2001). Another application is multistage coding, which is implemented in the OFDM system DRM (Digital Radio Mondiale). The iterative decoding of a convolutionally coded QAM system with OFDM will be discussed in Subsection 4.5.2.

3.1.5 Log-likelihood ratios and the MAP receiver

Until now, we have discussed only the maximum likelihood sequence estimator at the receiver. For a channel with additive white Gaussian noise, and a given receive signal vector $\mathbf{r}$, this receiver estimates the most probable transmit signal vector $\hat{\mathbf{s}}$ out of a finite set of possible transmit vectors. This most probable transmit vector is the one for which the squared Euclidean distance $\|\mathbf{r} - \hat{\mathbf{s}}\|^2$ is minimized, or, for transmit vectors of equal energy, the correlation given by $\Re\{\mathbf{r} \cdot \hat{\mathbf{s}}\}$ is maximized. The MLSE estimates transmit vectors (*sequences*), not bits. Since there is a one-to-one correspondence to the bit sequences $\mathbf{b}$, we get an estimate for each bit. Thus, MLSE does not provide us directly with an estimate for the most probable value of a bit. Both may sometimes be different [3] (see Problem 2).

In this subsection, we introduce a receiver that gives an estimate for the bit together with a reliability information of the decision. It does not need the assumption of Gaussian statistics, and is able to provide a reliability measure for each decision, even for single bit decisions. Furthermore, we can incorporate *a priori* probabilities of the bits. This is of high value for any kind of iterative decoding algorithms, as applied in the turbo decoding and multistage decoding. The basic tool is the log-likelihood ratio, which is defined by

Definition 3.1.2 (Log-likelihood ratio) *Let $P(A)$ be the probability of an event A. Then the LLR of this event is defined as*

$$L(A) = \log \frac{P(A)}{1 - P(A)}. \tag{3.7}$$

We add the following remarks.

[3] Even though in practice, for reasonably high SNRs, both will be the same with extremely high probability.

- The LLR is the ratio between the probability that something is true and that it is not true, viewed on a logarithmic scale. This is extremely useful if the probabilities are very high or very small. This is quite similar to the familiar decibel calculus for signal power levels. Even though the logarithm in Equation (3.7) is usually understood as a natural logarithm, one can in principle also use $10\log_{10}$ there and then, for example, regard the probability $P(A) = 0.999$ as an LLR of approximately 30 dB or the probability $P(A) = 0.001$ as an LLR of approximately −30 dB.
- One easily finds that the probability can be expressed by the LLR as

$$P(A) = \frac{e^{\frac{1}{2}L(A)}}{e^{\frac{1}{2}L(A)} + e^{-\frac{1}{2}L(A)}}. \tag{3.8}$$

- The LLR of the complementary event $\bar{A}$ is given by

$$L(\bar{A}) = -L(A).$$

Soft bits

We often deal with the probability that a random bit takes the value 0 or 1. We will therefore derive some simple properties of the corresponding random LLRs. To do this, it is very convenient to replace bits by signs, that is, write $+1$ for the bit value 0 and -1 for the bit value 1. This can be interpreted as antipodal (BPSK) signaling with energy 1 given by the mapping

$$s = (-1)^b = 1 - 2b$$

between the bit b and the sign (BPSK symbol) s. In the following text, we will speak of bits and their corresponding signs synonymously. To avoid confusion, we will (at this place) carefully distinguish, in our notation, between random variables (written as capital letters) and their values. That is, we deal with a random variable S, which is a random sign that takes values $s \in \{+1, -1\}$. The LLR for the event $S = s$ is then given by

$$L(S = s) = \log \frac{P(S = s)}{P(S = -s)}.$$

From $L(S = s) = -L(S = -s)$, we easily see that $L(S = s) = sL(S = +1)$. If no confusion may arise, we simply write

$$L = L(S = +1) \tag{3.9}$$

We call this the L-value of the random bit S. We note that

$$L(S = s) = sL. \tag{3.10}$$

The L-value of a random bit S (written as a sign) has a very natural interpretation as a *soft* bit. The sign of L says which of the two possible events $S = +1$ or $S = -1$ is more probable, that is, the sign gives a *hard* bit decision. The absolute value of L is a logarithmic measure for the reliability of this decision. For those who like to express everything in decibels, we can write $10\log_{10}$ for the logarithm in the definition of the LLR. Then, for example, an LLR value of $+30$ dB means that the random bit equals zero with a probability 0.999 (approximately), and an LLR value of −30 dB means that the random bit equals one with a probability 0.999 (approximately).

Using Equation (3.8), the probability for the random bit can be expressed by the LLR as

$$P(S = s) = \frac{e^{\frac{1}{2}sL}}{e^{\frac{1}{2}L} + e^{-\frac{1}{2}L}}. \tag{3.11}$$

We note that

$$P(S = s) = C\, e^{\frac{1}{2}sL},$$

where the constant $C = \left(e^{\frac{1}{2}L} + e^{-\frac{1}{2}L}\right)^{-1}$ does not depend on the value of s. Using the definition of the LLR, we find

$$L(S = s) = \log\left(\frac{P(S = s)}{\sqrt{P(S = -s)P(S = s)}}\right)^2$$

and

$$P(S = s) = \sqrt{P(S = +1)P(S = -1)}\, e^{\frac{1}{2}sL}, \tag{3.12}$$

which is an alternative representation with the constant C expressed by the probabilities.

Sequences of random signs

Now consider a vector $\mathbf{S} = (S_1, \ldots, S_K)^T$ of statistically independent random signs. Then, the probability for the event that $\mathbf{S}$ takes a certain value $\mathbf{s} = (s_1, \ldots, s_K)^T$ factorizes into the probabilities of the single signs, that is,

$$P(\mathbf{S} = \mathbf{s}) = \prod_{i=1}^{K} P(S_i = s_i).$$

Using Equation (3.12), we may write this as

$$P(\mathbf{S} = \mathbf{s}) = \prod_{i=1}^{K} \sqrt{P(S_i = +1)P(S_i = -1)}\, e^{\frac{1}{2}s_i L_i},$$

where $L_i = \log(P(S_i = +1)/P(S_i = -1))$. This takes the more compact form

$$P(\mathbf{S} = \mathbf{s}) = C\, \exp\left(\frac{1}{2}\sum_{i=1}^{K} s_i L_i\right) = C\, \exp\left(\frac{1}{2}\mathbf{s} \cdot \mathbf{L}\right),$$

where we have defined a log-likelihood vector $\mathbf{L} = (L_1, \ldots, L_K)$. The constant C is now given by

$$C = \prod_{i=1}^{K} \sqrt{P(S_i = +1)P(S_i = -1)}.$$

We see that as in the case of the MLSE estimator for the AWGN channel, we find the most probable sequence (which is here directly a bit sequence) by maximizing a correlation that is given by the scalar product $\mathbf{s} \cdot \mathbf{L}$. Here we have derived the general case, and the LLR expressions must be evaluated for any special statistics.

Products of random signs

Consider a random sign S. The expectation value of S is given by

$$\mathrm{E}\{S\} = (+1) \cdot \Pr(S = +1) + (-1) \cdot \Pr(S = -1).$$

Using Equation (3.11), we easily obtain

$$\mathrm{E}\{S\} = \tanh\left(\frac{1}{2}L\right),$$

where again we have used the shorthand notation $L = L(S = +1)$. Now let S_1 and S_2 be two independent random signs and define $S = S_1 \cdot S_2$ and write $L_1,\ L_2,\ L$ for the corresponding LLRs for the positive sign. From

$$\mathrm{E}\{S\} = \mathrm{E}\{S_1\}\,\mathrm{E}\{S_2\},$$

we conclude

$$\tanh\left(\frac{1}{2}L\right) = \tanh\left(\frac{1}{2}L_1\right)\tanh\left(\frac{1}{2}L_2\right)$$

or

$$L = 2\,\mathrm{artanh}\left(\tanh\left(\frac{1}{2}L_1\right)\tanh\left(\frac{1}{2}L_2\right)\right). \tag{3.13}$$

Using the relations

$$2\,\mathrm{artanh}\,(x) = \log\frac{1+x}{1-x}$$

and

$$\tanh(x) = \frac{\mathrm{e}^x - 1}{\mathrm{e}^x + 1},$$

one can show that this equals

$$L = \log\frac{1 + \mathrm{e}^{L_1}\mathrm{e}^{L_2}}{\mathrm{e}^{L_1} + \mathrm{e}^{L_2}}.$$

This can be approximated as

$$L \approx \mathrm{sign}\,(L_1)\,\mathrm{sign}\,(L_2)\min\left(|L_1|\,, |L_2|\right)$$

(see Problem 3). We can interpret this expression as follows: multiplication of random signs corresponds to the modulo 2 addition of the corresponding random bits. Thus, the modulo 2 addition of two random bits can (approximately) be realized as follows: the hard decision value of the result is obtained as modulo 2 addition of the hard decision values of the two bits. Its reliability given by $|L|$ equals the reliability of the least reliable of the two bits, similar to the saying that a chain is as strong as its weakest link. This becomes especially obvious if we generalize the approximation to the modulo 2 addition of more than two random bits and write

$$L \approx \prod_{i=1}^{K} \mathrm{sign}\,(L_i) \cdot \min\left(|L_1|\,, \ldots, |L_K|\right). \tag{3.14}$$

LLRs for BPSK and additive white Gaussian noise

Until now, no assumption has been made about the special statistics. It is instructive to exploit the concept of LLRs for the special case of antipodal transmission in an additive white Gaussian noise channel. In the context of channel coding, it is often convenient to deal with quantities that have no physical dimension. We rewrite BPSK transmission in an AWGN channel as

$$y_i = x_i + n_i,$$

where $x_i \in \{\pm 1\}$ and n_i is real and normalized discrete AWGN with variance

$$\sigma^2 = \frac{N_0}{2E_b}.$$

The transmit symbols x_i and the receive symbols y_i are regarded as random variables[4]. According to Bayes' law, the probability $P(x_i|y_i)$ that x_i has been transmitted under the condition that y_i has been received is given by

$$P(x_i|y_i) = \frac{p(y_i|x_i)P(x_i)}{p(y_i)}.$$

Here $P(x_i)$ is the *a priori* probability that x_i was transmitted, $p(y_i)$ is the probability density for the receive symbol y_i and $p(y_i|x_i)$ is the probability density for the receive symbol y_i under the condition that x_i was transmitted. This is just the Gaussian probability density with mean x_i and variance σ^2, that is,

$$p(y_i|x_i = \pm 1) = \frac{1}{\sqrt{2\pi\sigma^2}} \exp\left(-\frac{1}{2\sigma^2} |y_i \mp 1|^2\right).$$

We write

$$L_i = \log \frac{P(x_i = +1|y_i)}{P(x_i = -1|y_i)}$$

and get

$$L_i = \log \frac{\exp\left(-\frac{1}{2\sigma^2} |y_i - 1|^2\right) P(x_i = +1)}{\exp\left(-\frac{1}{2\sigma^2} |y_i + 1|^2\right) P(x_i = -1)},$$

which can eventually be written as

$$L_i = L_i^c + L_i^a, \tag{3.15}$$

where

$$L_i^c = \log \frac{p(y_i|x_i = +1)}{P(y_i|x_i = -1)} = \frac{2}{\sigma^2} y_i$$

is the *channel* LLR obtained from the symbol transmitted over the channel, and

$$L_i^a = \log \frac{P(x_i = +1)}{P(x_i = -1)} \tag{3.16}$$

is the *a priori* LLR corresponding to the *a priori* probability of the bit.

[4]Here we fall back into the lax usage of notation that is common in engineering not to distinguish between random variables and their values.

For a multiplicative fading channel given by

$$y_i = a_i x_i + n_i,$$

where a_i is a real fading amplitude, one easily sees that the channel LLR turns out to be

$$L_i^c = \frac{2}{\sigma^2} a_i y_i, \tag{3.17}$$

that is, the receive symbol must be weighted by the fading amplitude. We note that a one bit hard BPSK decision will be based on the sign of

$$L_i = \frac{2}{\sigma^2}\left(a_i y_i + \frac{\sigma^2}{2} L_i^a\right).$$

A positive sign corresponds to a 0, a negative to 1. The second term corresponds to a shift of the decision threshold that is proportional to the *a priori* LLR and to the noise.

The conditional probability for the BPSK symbol x_i under the condition that y_i was received is given by

$$P(x_i|y_i) = \sqrt{P(x_i = +1|y_i)P(x_i = -1|y_i)}\, \mathrm{e}^{\frac{1}{2}x_i L_i}.$$

For uncoded transmission, the transmit symbols are statistically independent, and thus the conditional probability $P(\mathbf{x}|\mathbf{y})$ for the whole BPSK vector $\mathbf{x} = (x_1, \ldots, x_K)^T$ given that $\mathbf{y}$ was received factorizes to

$$P(\mathbf{x}|\mathbf{y}) = \prod_{i=1}^{K} P(x_i|y_i) = \prod_{i=1}^{K} \sqrt{P(x_i = +1|y_i)P(x_i = -1|y_i)}\, \mathrm{e}^{\frac{1}{2}x_i L_i}.$$

General signal constellation and coded transmission

We consider a set $\mathcal{C}$ of $M = 2^K$ real transmit signal vectors $\mathbf{x} = (x_1, \ldots, x_N)^T$ labeled by bit tuples of length K written as binary vectors $\mathbf{b} = (b_1, \ldots, b_K)^T$. We write $\mathcal{B}$ for the set of these tuples. As discussed at the end of Subsection 3.1.1, the mapping

$$X : \mathcal{B} \to \mathcal{C}, \quad \mathbf{b} \mapsto \mathbf{x}$$

can be interpreted as a joint channel encoder and symbol mapper. When using LLR values, it is convenient to replace a binary vector $\mathbf{b}$ by its equivalent sign vector $\mathbf{u} = (u_1, \ldots, u_K)^T$ defined by $u_k = (-1)^{b_k}$. The set of these vectors will be denoted by $\mathcal{U}$.

First, we ask for the conditional probability $P(\mathbf{x}|\mathbf{y})$ for a transmitted signal vector $\mathbf{x}$ given that the vector $\mathbf{y}$ has been received. According to Bayes' law, we have

$$P(\mathbf{x}|\mathbf{y}) = \frac{p(\mathbf{y}|\mathbf{x})P(\mathbf{x})}{p(\mathbf{y})}.$$

Here, $P(\mathbf{x})$ is the *a priori* probability that $\mathbf{x}$ was transmitted, $p(\mathbf{y})$ is the probability density function for the receive symbol vector $\mathbf{y}$ and $p(\mathbf{y}|\mathbf{x})$ is the probability density function for the receive symbol $\mathbf{y}$ under the condition that $\mathbf{x}$ has been transmitted. We have fallen back into the loose notation for random variables, that is, we use $\mathbf{x}$ for the random variable and for the value of the random variable as well. We may replace $P(\mathbf{x})$ by $P(\mathbf{b})$, which is the *a priori* probability for the bit tuple $\mathbf{b}$ that corresponds to $\mathbf{x}$ or by the *a priori* probability $P(\mathbf{u})$ of the sign vector $\mathbf{u}$. Assuming that the *a priori* probabilities of the bits are statistically independent and that $p(\mathbf{y}|\mathbf{x})$ factorizes as well because of statistical independence, we may write

$$P(\mathbf{x}|\mathbf{y}) = \frac{1}{p(\mathbf{y})} \prod_{i=1}^{N} p(y_i|x_i) \prod_{k=1}^{K} p(u_k).$$

To find the most probable transmit vector for a general signal constellation, this expression must be maximized.

For the special case of a binary code with antipodal signaling with $x_i \in \{\pm 1\}$, we define

$$L_i^c = \log \frac{p(y_i|x_i = +1)}{P(y_i|x_i = -1)}, \quad L_k^a = \log \frac{P(u_k = +1)}{P(u_k = -1)}.$$

It is convenient to define channel and *a priori* LLR vectors as $\mathbf{L}^c = (L_1^c, \ldots, L_K^c)$ and $\mathbf{L}^a = (L_1^a, \ldots, L_N^a)$. The channel LLR vector can be written as

$$\mathbf{L}^c = \frac{2}{\sigma^2} \mathbf{A}\mathbf{y}$$

with the diagonal matrix of fading amplitudes $\mathbf{A} = \mathrm{diag}(a_1, \ldots, a_K)$. The conditional probability that a certain sequence has been transmitted is then given by

$$P(\mathbf{x}|\mathbf{y}) = C \exp\left(\frac{1}{\sigma^2} \mu(\mathbf{x})\right) \tag{3.18}$$

with a constant C that does not depend on the transmitted sequence and a *metric* defined by

$$\mu(\mathbf{x}) = \frac{\sigma^2}{2} \left(\mathbf{x} \cdot \mathbf{L}^c + \mathbf{u} \cdot \mathbf{L}^a\right). \tag{3.19}$$

To find the most probable sequence $\mathbf{x}$, this correlation metric $\mu(\mathbf{x})$ has to be maximized. We have chosen this normalization, because in the case of BPSK without *a priori* information, Equation (3.17) yields $\mu(\mathbf{x}) = \mathbf{A}\mathbf{y} \cdot \mathbf{x}$, which does not depend on σ.

For the special case of a systematic code, the vector $\mathbf{x}$ consists of a systematic part $\mathbf{x}_s = \mathbf{u}$ and a parity check part $\mathbf{x}_p$. In that case, the correlation metric becomes

$$\mu(\mathbf{x}) = \mathbf{x}_s \cdot \left(\mathbf{L}^a + \mathbf{L}_s^c\right) + \mathbf{x}_p \cdot \mathbf{L}_p^c. \tag{3.20}$$

Thus, the channel LLR can be split up to a part $\mathbf{L}_s^c$ corresponding to the systematic symbol vector and a part $\mathbf{L}_p^c$ corresponding to the parity check symbol vector. The *a priori* LLR will be added to the channel LLR for the systematic part.

For an arbitrary signal constellation and a (real-valued) discrete fading channel with AWGN of variance σ^2, we have

$$P(\mathbf{x}|\mathbf{y}) = C_1 \exp\left(\frac{1}{2\sigma^2} \|\mathbf{y} - \mathbf{A}\mathbf{x}\|^2\right) \cdot \exp\left(\frac{1}{2}\mathbf{u} \cdot \mathbf{L}^a\right),$$

which can be written as

$$P(\mathbf{x}|\mathbf{y}) = C_2 \exp\left(\frac{1}{\sigma^2}\left(\mathbf{x} \cdot \mathbf{A}\mathbf{y} - \frac{1}{2} \|\mathbf{A}\mathbf{x}\|^2\right)\right) \cdot \exp\left(\frac{1}{2}\mathbf{u} \cdot \mathbf{L}^a\right),$$

with properly defined constant C_1 and C_2. This can be written as Equation (3.18) with the metric expression of Equation (3.19) replaced by

$$\mu(\mathbf{x}) = \frac{\sigma^2}{2}\left(\mathbf{x} \cdot \mathbf{L}^c + \mathbf{u} \cdot \mathbf{L}^a\right) - \frac{1}{2}\left\|\mathbf{A}\mathbf{x}\right\|^2, \tag{3.21}$$

that is, an energy term must be taken into account.

The bitwise MAP receiver

We now want to analyze decisions about single bits instead of whole sequences. We ask for the probability $P(b_k = 0|\mathbf{y})$ that the kth bit b_k in the tuple $\mathbf{b}$ has the value zero under the condition that the vector $\mathbf{y}$ was received. The corresponding LLR is given by

$$L(b_k = 0|\mathbf{y}) = \log \frac{P(b_k = 0|\mathbf{y})}{P(b_k = 1|\mathbf{y})}.$$

We write $\mathcal{B}_k^{(0)}$ for the set of those vectors $\mathbf{b} \in \mathcal{B}$ for which $b_k = 0$ and $\mathcal{B}_k^{(1)}$ for the set of those for which $b_k = 1$. Then,

$$P(b_k = 0|\mathbf{y}) = \sum_{\mathbf{b}\in\mathcal{B}_k^{(0)}} P(\mathbf{b}|\mathbf{y})$$

and

$$P(b_k = 1|\mathbf{y}) = \sum_{\mathbf{b}\in\mathcal{B}_k^{(1)}} P(\mathbf{b}|\mathbf{y})$$

hold and the LLR can be written as

$$L(b_k = 0|\mathbf{y}) = \log\left(\frac{\sum_{\mathbf{b}\in\mathcal{B}_k^{(0)}} P(\mathbf{b}|\mathbf{y})}{\sum_{\mathbf{b}\in\mathcal{B}_k^{(1)}} P(\mathbf{b}|\mathbf{y})}\right). \tag{3.22}$$

Applying Bayes' law for $P(\mathbf{b}|\mathbf{y})$, we get

$$L(b_k = 0|\mathbf{y}) = \log\left(\frac{\sum_{\mathbf{b}\in\mathcal{B}_k^{(0)}} p(\mathbf{y}|\mathbf{b})P(\mathbf{b})}{\sum_{\mathbf{b}\in\mathcal{B}_k^{(1)}} p(\mathbf{y}|\mathbf{b})P(\mathbf{b})}\right), \tag{3.23}$$

where $P(\mathbf{b})$ is the *a priori* probability of vector $\mathbf{b}$, and $p(\mathbf{y}|\mathbf{b})$ is the conditional probability density for the receive vector $\mathbf{y}$ given that the signal vector $\mathbf{x}$ corresponding to $\mathbf{b}$ was transmitted. If $L(b_k = 0|\mathbf{y}) > 0$, the receiver decides for $b_k = 0$ and otherwise for $b_k = 1$. Furthermore, this receiver provides information about the reliability of the decision, which is given by the absolute value of $L(b_k = 0|\mathbf{y})$. We call such a receiver a *(bitwise) maximum a posteriori probability* (MAP) receiver. As we have already seen above, we may write

$$p(\mathbf{y}|\mathbf{b})P(\mathbf{b}) = C \exp\left(\frac{1}{\sigma^2}\mu\left(\mathbf{x}\right)\right) \tag{3.24}$$

with a metric $\mu(\mathbf{x})$ that depends on the code and the signal constellation. The constant C is the same in numerator and denominator.

The SISO decoder for systematic codes

For the special case of a binary systematic code, the metric $\mu(\mathbf{x})$ is given by Equation (3.20). We insert that expression into Equation (3.23) and get the LLR expression

$$L(b_k = 0|\mathbf{y}) = \log\left(\frac{\sum_{\mathbf{b}\in\mathcal{B}_k^{(0)}} \exp\left(\frac{1}{2}\left(\mathbf{x}_s\cdot\left(\mathbf{L}^a+\mathbf{L}_s^c\right)+\mathbf{x}_p\cdot\mathbf{L}_p^c\right)\right)}{\sum_{\mathbf{b}\in\mathcal{B}_k^{(1)}} \exp\left(\frac{1}{2}\left(\mathbf{x}_s\cdot\left(\mathbf{L}^a+\mathbf{L}_s^c\right)+\mathbf{x}_p\cdot\mathbf{L}_p^c\right)\right)}\right) \tag{3.25}$$

for the MAP receiver. Here, the sum has to be understood in such a way that, for each binary vector $\mathbf{b}$, there are uniquely determined corresponding vectors $\mathbf{x}_s$ and $\mathbf{x}_p$. We may now split up this equation into three terms that have an intuitively obvious interpretation. To do this, we keep the index k fixed and split up the exponents in the numerator and the denominator as

$$\mathbf{x}_s\cdot\left(\mathbf{L}^a+\mathbf{L}_s^c\right)+\mathbf{x}_p\cdot\mathbf{L}_p^c = x_{sk}(L_k^a+L_{sk}^c)+\sum_{i\neq k} x_{si}(L_i^a+L_{si}^c)+\mathbf{x}_p\cdot\mathbf{L}_p^c.$$

In the numerator, the first term is constantly $+(L_k^a+L_{sk}^c)$, and in the denominator, it is constantly $-(L_k^a+L_{sk}^c)$. We can thus extract this term from the sum. $L_k = L(b_k=0|\mathbf{y})$ in Equation (3.25) can then be written as

$$L_k = L_k^a + L_{sk}^c + L_k^e \tag{3.26}$$

with the *extrinsic* LLR defined by

$$L_k^e = \log\left(\frac{\sum_{\mathbf{b}\in\mathcal{B}_k^{(0)}} \exp\left(\frac{1}{2}\left(\sum_{i\neq k} x_{si}(L_i^a+L_{si}^c)+\mathbf{x}_p\cdot\mathbf{L}_p^c\right)\right)}{\sum_{\mathbf{b}\in\mathcal{B}_k^{(1)}} \exp\left(\frac{1}{2}\left(\sum_{i\neq k} x_{si}(L_i^a+L_{si}^c)+\mathbf{x}_p\cdot\mathbf{L}_p^c\right)\right)}\right).$$

Comparing this with Equation (3.15) for uncoded transmission, we see that this extrinsic LLR corresponds to the additional information gain due to channel coding. It depends on all other received symbols – systematic symbols and parity check symbols – except the one corresponding to b_k. This LLR adds to the LLRs L_k^a and L_{sk}^c that already occur in case of uncoded transmission.

We now write Equation (3.26) in vector notation as

$$\mathbf{L} = \mathbf{L}^a + \mathbf{L}^c + \mathbf{L}^e$$

and interpret the MAP as a SISO receiver that has soft LLR vectors as input and output as depicted in Figure 3.3. The input vectors of the SISO are $\mathbf{L}^a$ and $\mathbf{L}^c$, where $\mathbf{L}^c$ consists of the two parts $\mathbf{L}_s^c$ and $\mathbf{L}_p^c$. The systematic MAP of Equation (3.25) has the two inputs $\mathbf{L}^a+\mathbf{L}_s^c$ and $\mathbf{L}_p^c$. The output vector is $\mathbf{L}$. The extrinsic LLR vector $\mathbf{L}^e$ of that vector is obtained from the output $\mathbf{L}$ by

$$\mathbf{L}^e = \mathbf{L} - \mathbf{L}^a - \mathbf{L}_s^c.$$

The SISO may be regarded as a device that calculates extrinsic LLR information due to the code. The total LLR is the sum of this extrinsic LLR and the (*a priori* and channel) LLR information already available before decoding.

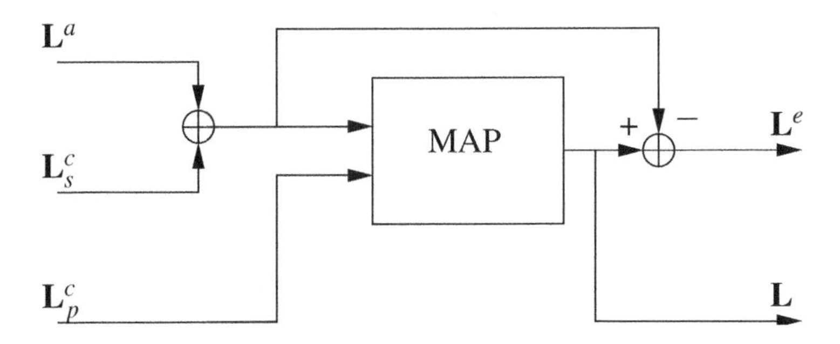

Figure 3.3 The soft-input/soft-output (SISO) decoder.

The maxlog MAP approximation

The expression (3.23) for the MAP receiver contains transcendental functions and thus it may sometimes be too costly to implement. The *maxlog* MAP approximation is often used in practice because it is quite accurate if the SNR is not extremely low. In that case, the sums in the numerator and denominator are dominated by their respective largest term and Equation (3.23) can be approximated by

$$L(b_k = 0|\mathbf{y}) \approx \max_{\mathbf{b}\in\mathcal{B}_k^{(0)}} \log\left(p(\mathbf{y}|\mathbf{b})P(\mathbf{b})\right) - \max_{\mathbf{b}\in\mathcal{B}_k^{(1)}} \log\left(p(\mathbf{y}|\mathbf{b})P(\mathbf{b})\right). \tag{3.27}$$

We note that the hard bit decisions obtained by this maxlog MAP receiver are always identical to those obtained from the MLSE receiver because the latter searches for $\max_{\mathbf{b}\in\mathcal{B}} \log\left(p(\mathbf{y}|\mathbf{b})P(\mathbf{b})\right)$ and decides for the corresponding sequence $\mathbf{b} = \hat{\mathbf{b}}$ and thus $L(b_k = 0|\mathbf{y})$ has always the same sign as the sign obtained from the MLSE.

If we insert Equation (3.24) for the argument of the logarithm, we can write this as

$$L(b_k = 0|\mathbf{y}) \approx \frac{1}{\sigma^2}\left(\max_{\mathbf{b}\in\mathcal{B}_k^{(0)}} \mu\left(\mathbf{x}\right) - \max_{\mathbf{b}\in\mathcal{B}_k^{(1)}} \mu\left(\mathbf{x}\right)\right).$$

For a binary code with antipodal signaling, the metric is given by

$$\mu(\mathbf{x}) = \mathbf{A}\mathbf{y}\cdot\mathbf{x} + \frac{\sigma^2}{2}\mathbf{L}^a\mathbf{u}.$$

If no *a priori* information is available, both the input LLR $\mathbf{L}^c = \frac{1}{\sigma^2}\mathbf{A}\mathbf{y}\cdot\mathbf{x}$ and the output LLR $\mathbf{L}$ are linear in the SNR value σ^{-2}. This linear scale factor can be omitted without any loss and is only needed for the calculation of absolute probabilities. This is in contrast to the exact MAP, where the output is a nonlinear function in the SNR and even the hard decision value may depend on its value.

3.2 Convolutional Codes

3.2.1 General structure and encoder

In contrast to block codes, convolutional codes do not have a defined block structure. A continuously flowing data stream will be encoded into a continuously flowing code word.

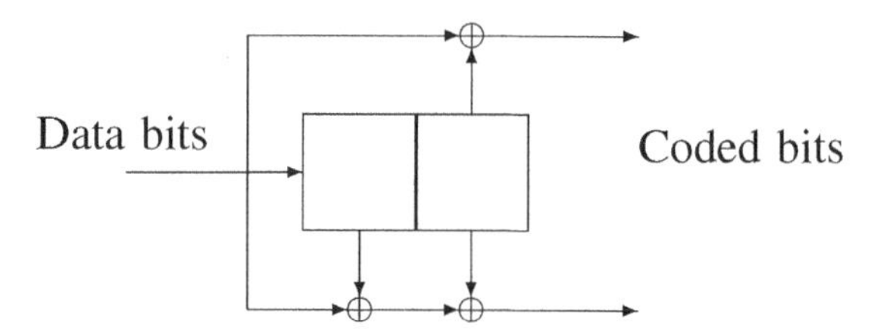

Figure 3.4 Block diagram for a simple convolutional encoder.

Even though it is reasonable for practical and theoretical reasons to work with finite sets of data and coded bits, the length of a code word is given by other requirements than the structure of the code. Convolutional encoders are linear and time-invariant systems given by the convolution of a binary data stream with generator sequences. They can be implemented by shift registers. Figure 3.4 shows a simple example of such an encoder with rate $R_c = 1/2$ and memory $m = 2$.

Given an input bit stream $\{b_i\}_{i=0}^{\infty}$, a convolutional encoder of code rate $R_c = 1/n$ produces n parallel data streams $\{c_{\nu,i}\}_{i=0}^{\infty}$, $\nu = 1, \ldots, n$ that may be multiplexed to one serial code word before transmission. It can be written as

$$c_{\nu,i} = \sum_{k=0}^{m} g_{\nu,k} b_{i-k},$$

where we have set $b_i = 0$ für $i < 0$. The sum has to be understood as *modulo* 2 sum. Here, $g_{\nu,k}$, $(\nu = 1, \ldots, n;\ k = 0, \ldots, m)$ are the generators that can also be written as generator polynomials

$$g_\nu(D) = \sum_{k=0}^{m} g_{\nu,k} D^k,$$

where D is a formal variable that can be interpreted as *delay*. In the example of Figure 3.4, we have

$$g_1(D) = 1 + D^2 \equiv (101) \equiv 5_{\text{oct}},$$
$$g_2(D) = 1 + D + D^2 \equiv (111) \equiv 7_{\text{oct}},$$

where we have introduced the binary vector notation and the octal notation for the generators.

It is often convenient to work with formal power series instead of sequences. This is similar to the formalism in signal processing, where we may switch to the frequency domain to replace convolutions by multiplications. We define the power series

$$b(D) = \sum_{k=0}^{\infty} b_k D^k$$

for the data word and

$$c_\nu(D) = \sum_{k=0}^{\infty} c_{\nu k} D^k,\ \nu = 1, \ldots, n$$

for the code word. Then the encoder can be described by

$$c_\nu(D) = b(D)g_\nu(D) \tag{3.28}$$

or, in vector notation, as

$$\mathbf{c}(D) = b(D)\mathbf{g}(D), \tag{3.29}$$

that is,

$$\begin{bmatrix} c_1(D) \\ \vdots \\ c_n(D) \end{bmatrix} = b(D) \begin{bmatrix} g_1(D) \\ \vdots \\ g_n(D) \end{bmatrix}.$$

Convolutional codes are linear codes. Thus, the Hamming distance of the code is the minimum weight. This is called the *free distance* and will be denoted by d_{free}.

Trellis diagrams

For any time instant i, one can characterize the encoding step by the actual state $\mathbf{s} = (s_1, \ldots, s_m)$, that is, the content of the shift register (s_1 is the most recent bit that has been shifted into the register) and the actual input bit b_i. This uniquely defines the next state $\mathbf{s}' = (b_i, s_1, \ldots, s_{m-1})$ and the encoded output bits $c_{1i}, c_{2i}, \ldots, c_{ni}$. For the code given by Figure 3.4, there are four possible states $(s_1 s_2) \in \{(00), (10), (01), (11)\}$ and eight possible transitions from one stage to the following, as depicted in Figure 3.5. For each transition, there is an input bit b_i and a pair of output bits (c_{1i}, c_{2i}) denoted by $b_i/c_{1i}c_{2i}$ at each transition line in the figure. Now consider a certain number of such transitions. They can

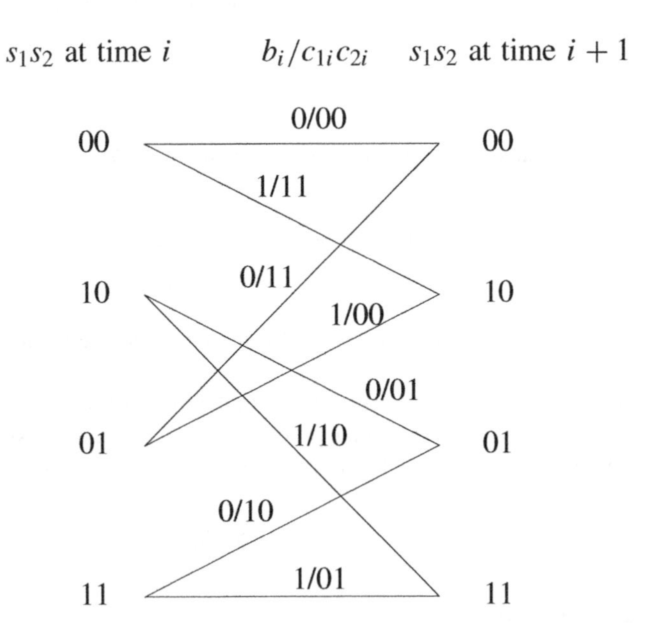

Figure 3.5 State transitions for the $(1 + D^2, 1 + D + D^2)$ convolutional code.

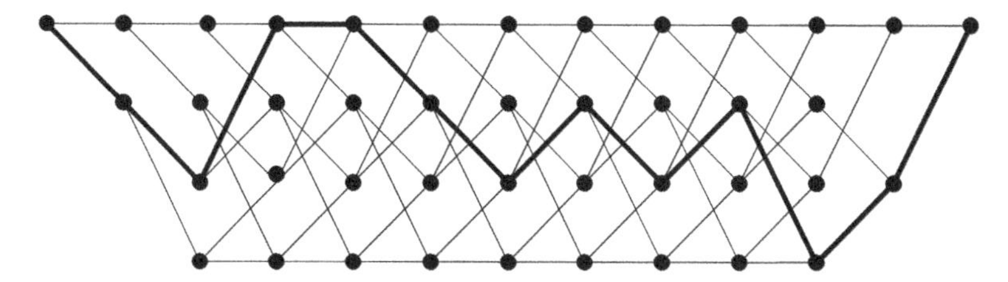

Figure 3.6 Trellis diagram for the $(1 + D^2, 1 + D + D^2)$ convolutional code.

be depicted by successively appending such transitions as shown in Figure 3.6. This is called a *trellis diagram*. Given a defined initial state of the shift register (usually the all-zero state), each code word is characterized by sequence of certain transitions. We call this a *path* in the trellis. In Figure 3.6, the path corresponding to the data word 1000 0111 0100 and the code word 11 01 11 00 00 11 10 01 10 00 01 11 is depicted by bold lines for the transitions in the trellis. In this example, the last $m = 2$ bits are zero and, as a consequence, the final state in the trellis is the all-zero state. It is common practice to start and to stop with the all-zero state because it helps the decoder. This can easily be achieved by appending m zeros – the so-called *tail bits* – to the useful bit stream.

State diagrams

One can also characterize the encoder by states and inputs and their corresponding transitions as depicted in part (a) of Figure 3.7 for the code under consideration. This is known as a *Mealy automat*. To evaluate the free distance of a code, it is convenient to cut open the automat diagram as depicted in part (b) of Figure 3.7. Each path (code word) that starts in the all-zero state and comes back to that state can be visualized by a sequence of states that starts at the all-zero state on the left and ends at the all-zero state on the right. We look at the coded bits in the labeling $b_i/c_{1i}c_{2i}$ and count the bits that have the value one. This is just the Hamming distance between the code word corresponding to that sequence and the all-zero code word. From the diagram, one can easily obtain the smallest distance d_{free} to the all-zero code word. For the code of our example, the minimum distance corresponds to the sequence of transitions $00 \to 10 \to 01 \to 00$ and turns out to be $d_{\text{free}} = 5$. The alternative sequence $00 \to 10 \to 11 \to 01 \to 00$ has the distance $d = 6$. All other sequences include loops that produce higher distances.

From the state diagram, we may also find the so-called *error coefficient* c_d. These error coefficients are multiplicative coefficients that relate the probability P_d of an error event of distance d to the corresponding bit error probability. To obtain c_d, we have to count all the nonzero data bits of all error paths of distance d to the all-zero code word. Using $P(A_1 \cup A_2) \leq P(A_1) + P(A_2)$, we obtain the union bound

$$P_b \leq \sum_{d=d_{\text{free}}}^{\infty} c_d P_d$$

for the bit error probability. The coefficients c_d for most relevant codes can be found in text books. The error event probability P_d, for example, for antipodal signaling is given

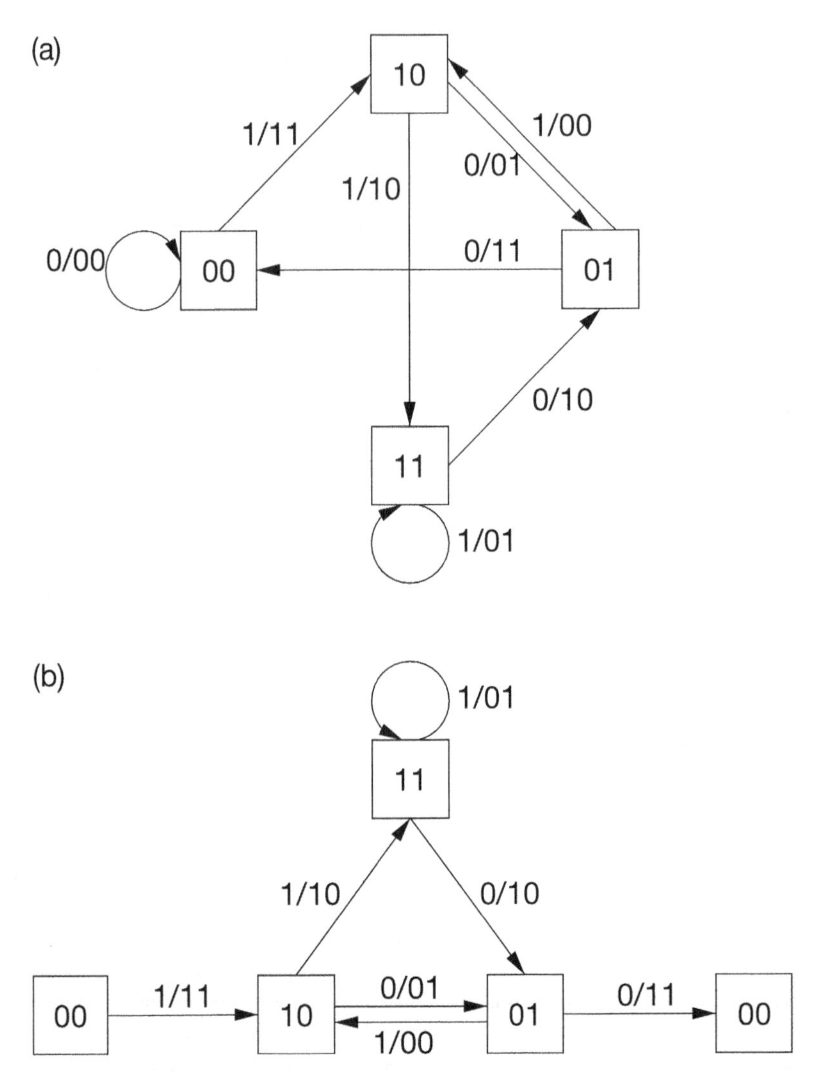

Figure 3.7 State diagram (Mealy automat) for the $(1 + D^2, 1 + D + D^2)$ convolutional code.

by Equation (3.2) for the AWGN channel and by Equation (3.4) for the Rayleigh fading channel.

Catastrophic codes

The state diagram also enables us to find a class of encoders called *catastrophic encoders* that must be excluded because they have the undesirable property of error propagation: if there is a closed loop in the state diagram where all the coded bits $c_{1i}c_{2i}$ are equal to zero, but at least one data bit b_i equals one, then there exists a path of infinite length with an infinite number of ones in the data, but with only a finite number of ones in the code word. As a

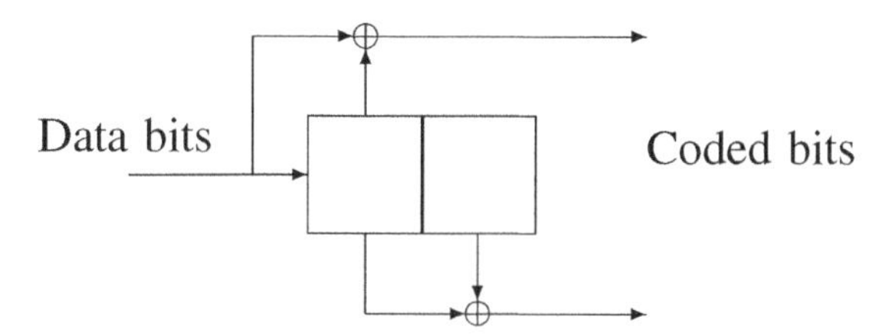

Figure 3.8 Example of a catastrophic convolutional encoder.

consequence, a finite number of channel bit errors may lead to an infinite number of errors in the data, which is certainly a very undesirable property. An example for a catastrophic encoder is the one characterized by the generators $(3,\ 6)_{\text{oct}} = (1 + D, D + D^2)$, which is depicted in Figure 3.8. Once in the state 11, the all-one input sequence will be encoded to the all-zero code word.

Punctured convolutional codes

Up to now, we have only considered convolutional codes of rate $R_c = 1/n$. There are two possibilities to obtain $R_c = k/n$ codes. The classical one is to use k parallel shift registers and combine their outputs. This, however, makes the implementation more complicated. A simpler and more flexible method called *puncturing* is usually preferred in practical communication systems. We explain it by means of the example of an $R_c = 1/2$ code that can be punctured to obtain an $R_c = 2/3$ code. The encoder produces two parallel encoded data streams $\{c_{1,i}\}_{i=0}^{\infty}$ and $\{c_{2,i}\}_{i=0}^{\infty}$. The first data stream will be left unchanged. From the other data stream every second bit will be discarded, that is, only the bits with even time index i will be multiplexed to the serial code word and then transmitted. Instead of the original code word

$$\left(c_{10} \quad c_{20} \quad c_{11} \quad c_{21} \quad c_{12} \quad c_{22} \quad c_{13} \quad c_{23} \quad c_{14} \quad \cdots \right)$$

the punctured code word

$$\left(c_{10} \quad c_{20} \quad c_{11} \quad \bigcirc \quad c_{12} \quad c_{22} \quad c_{13} \quad \bigcirc \quad c_{14} \quad \cdots \right)$$

will be transmitted. Here we have indicated the punctured bits by $\bigcirc$. At the receiver, the puncturing positions must be known. A soft decision (e.g. MLSE) receiver has metric values $\mu_{\nu i}$ as inputs that correspond to the encoded bits $c_{\nu i}$. The absolute value of $\mu_{\nu i}$ is an indicator for the reliability of the bit. Punctured bits can be regarded as bits with reliability zero. Thus, the receiver has to add dummy receive bits at the punctured positions of the code word and assign them the metric values $\mu_{\nu i} = 0$.

Recursive systematic convolutional encoders

Recursive systematic convolutional (RSC) encoders have become popular in the context of parallel concatenated codes and turbo decoding (see below). For every nonsystematic convolutional (NSC) $R_c = 1/n$ encoder, one can find an equivalent RSC encoder that

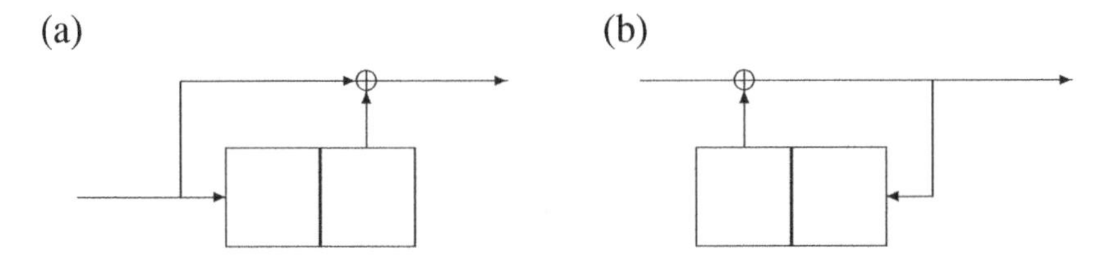

Figure 3.9 Inversion circuit for the generator polynomial $1 + D^2$.

produces the same code (i.e. the same set of code words) with a different relation between the data word and the code word. It can be constructed in such a way that the first of the n parallel encoded bit stream of the code word is systematic, that is, it is identical to the data word.

As an example, we consider the $R_c = 1/2$ convolutional code of Figure 3.4 that can be written in compact power series notation as

$$\begin{bmatrix} c_1(D) \\ c_2(D) \end{bmatrix} = b(D) \begin{bmatrix} 1 + D^2 \\ 1 + D + D^2 \end{bmatrix}.$$

The upper branch corresponding to the generator polynomial $g_1(D) = 1 + D^2$ of the shift register circuit depicted in part (a) of Figure 3.9 defines a one-to-one map from the set of all data words to itself. One can easily check that the inverse is given by the recursive shift register circuit depicted in part (b) of Figure 3.9. This can be described by the formal power series

$$g_1^{-1}(D) = \left(1 + D^2\right)^{-1} = 1 + D^2 + D^4 + D^6 + \cdots$$

This power series description of feedback shift registers is formally the same as the description of linear systems in digital signal processing[5], where the delay is usually denoted by $e^{-j\omega}$ instead of D. The shift register circuits of Figure 3.9 invert each other. Thus, $g_1^{-1}(D)$ is a one-to-one mapping between bit sequences. As a consequence, combining the convolutional encoder with that recursive shift register circuit as depicted in part (a) of Figure 3.10 leads to the same set of code words. This circuit is equivalent to the one depicted in part (b) of Figure 3.10. This RSC encoder with generator polynomials $(5,\ 7)_{\text{oct}}$ can formally be written as

$$\begin{bmatrix} c_1(D) \\ c_2(D) \end{bmatrix} = \tilde{b}(D) \begin{bmatrix} 1 \\ \frac{1+D+D^2}{1+D^2} \end{bmatrix},$$

where the bit sequences are related by $\tilde{b}(D) = (1 + D^2)\, b(D)$.

For a general $R_c = 1/n$ convolutional code, we have the NSC encoder given by the generator polynomial vector

$$\mathbf{g}(D) = \begin{bmatrix} g_1(D) \\ \vdots \\ g_n(D) \end{bmatrix}.$$

[5]In signal processing, we have an interpretation of ω as a (normalized) frequency, which has no meaning for convolutional codes. Furthermore, here all additions are modulo 2. However, all formal power series operations are the same.

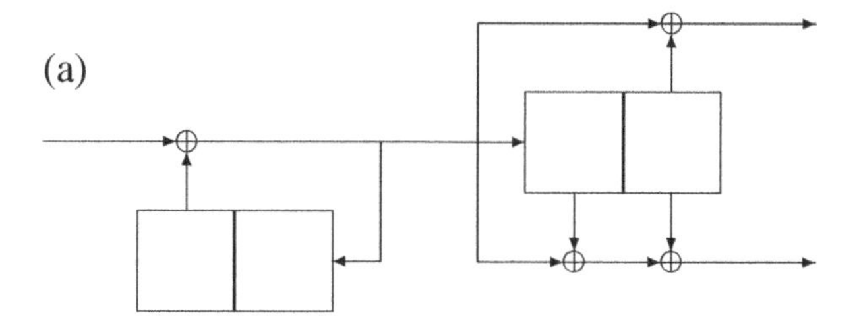

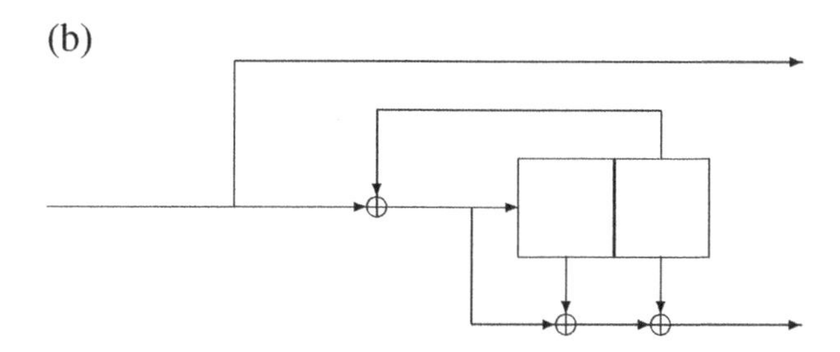

Figure 3.10 Recursive convolutional encoder.

The equivalent RSC encoder is given by the generator vector

$$\tilde{\mathbf{g}}(D) = \begin{bmatrix} 1 \\ g_2(D)/g_1(D) \\ \vdots \\ g_n(D)/g_1(D) \end{bmatrix}.$$

The bits sequence $b(D)$ encoded by $\mathbf{g}(D)$ results in the same code word as the bit sequence $\tilde{b}(D) = g_1(D)b(D)$ encoded by $\tilde{\mathbf{g}}(D) = g_1^{-1}(D)\mathbf{g}(D)$, that is,

$$\mathbf{c}(D) = b(D)\mathbf{g}(D) = \tilde{b}(D)\tilde{\mathbf{g}}(D).$$

An MLSE decoder will find the most likely code word that is uniquely related to a data word corresponding to an NSC encoder and another data word corresponding to an RSC encoder. As a consequence, one may use the same decoder for both and then relate the sequences as described above. But note that this is true only for a decoder that makes decisions about sequences. This is not true for a decoder that makes bitwise decisions like the MAP decoder.

3.2.2 MLSE for convolutional codes: the Viterbi algorithm

Let us consider a convolutional code with memory m and a finite sequence of K input data bits $\{b_k\}_{k=1}^{K}$. We denote the coded bits as c_i. We assume that the corresponding trellis starts and ends in the all-zero state. In our notation, the tail bits are included in $\{b_k\}_{k=1}^{K}$, that is, there are only $K - m$ bits that really carry information.

Although the following discussion is not restricted to that case, we first consider the concrete case of antipodal (BPSK) signaling, that is, transmit symbols $x_i = (-1)^{c_i} \in \{\pm 1\}$ written as a vector $\mathbf{x}$ and a real discrete AWGN channel given by

$$\mathbf{y} = \mathbf{x} + \mathbf{n},$$

where $\mathbf{y}$ is the vector of receive symbols and $\mathbf{n}$ is the real AWGN vector with components n_i of variance

$$\sigma^2 = \mathrm{E}\left\{n_i^2\right\} = \frac{N_0}{2E_S}.$$

Here, we have normalized the noise by the symbol energy E_S. We know from the discussion in Subsection 1.3.2 that, given a fixed receive vector $\mathbf{y}$, the most probable transmit sequence $\mathbf{x}$ for this case is the one that maximizes the *correlation metric* given by the scalar product

$$\mu(\mathbf{x}) = \mathbf{y} \cdot \mathbf{x}. \tag{3.30}$$

For an $R_c = 1/n$ convolutional code, the code word consists of nK encoded bits, and the metric can be written as a sum

$$\mu(\mathbf{x}) = \sum_{k=1}^{K} \mu_k \tag{3.31}$$

of metric increments

$$\mu_k = \mathbf{y}_k \cdot \mathbf{x}_k$$

corresponding to the K time steps $k = 1, \ldots, K$ of the trellis. Here $\mathbf{x}_k$ is the vector of the n symbols x_i that correspond to encoded bits for the time step number k where the bit b_k is encoded, and $\mathbf{y}_k$ is the vector of the corresponding receive vector.

The task now is to find the vector $\mathbf{x}$ that maximizes the metric given by Equation (3.31), thereby exploiting the special trellis structure of a convolutional code. We note that the following treatment is quite general and it is by no means restricted to the special case of the AWGN metric given by Equation (3.30). For instance, any metric that is given by expressions like Equations (3.19–3.21) can be written as Equation (3.31). Thus, *a priori* information about the bits also can be included in a straightforward manner by the expressions presented in Subsection 3.1.5, see also (Hagenauer 1995).

For a reasonable sequence length K, it is not possible to find the vector $\mathbf{x}$ by exhaustive search because this would require a computational effort that is proportional to 2^K. But, owing to the trellis structure of convolutional codes, this is not necessary. We consider two code words $\mathbf{x}$ and $\hat{\mathbf{x}}$ with corresponding paths merging at a certain time step k in a common state $\mathbf{s}_k$ (see Figure 3.11). Assume that for both paths the *accumulated metrics*, that is, the sum of all metric increments up to that time step

$$\Sigma_k = \sum_{i=1}^{k} \mu_i$$

for $\mathbf{x}$ and

$$\hat{\Sigma}_k = \sum_{i=1}^{k} \hat{\mu}_i$$

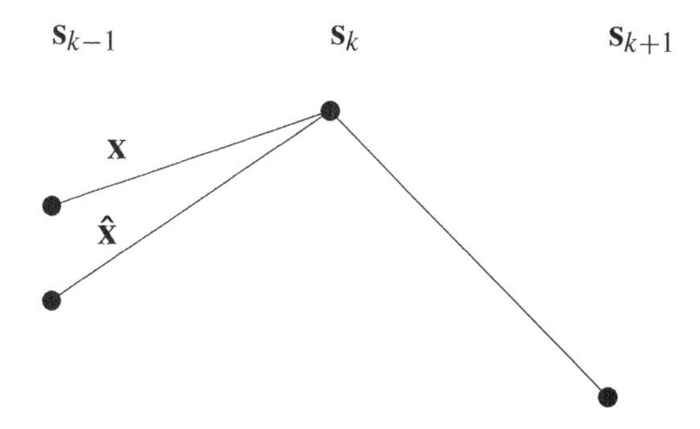

Figure 3.11 Transition where the paths $\mathbf{x}$ and $\hat{\mathbf{x}}$ merge.

for $\hat{\mathbf{x}}$ have been calculated. Because the two paths merge at time step k and will be identical for the whole future,

$$\mu(\hat{\mathbf{x}}) - \mu(\mathbf{x}) = \hat{\Sigma}_k - \Sigma_k$$

holds and we can already make a decision between both paths. Assume $\mu(\hat{\mathbf{x}}) - \mu(\mathbf{x}) > 0$. Then, $\hat{\mathbf{x}}$ is more likely than $\mathbf{x}$, and we can discard $\mathbf{x}$ from any further considerations. This fact allows us to sort out unlikely paths before the final decision and thus an effort that is exponentially growing with the time can be avoided.

The algorithm that does this is the Viterbi algorithm and it works as follows: starting from the initial state, the metric increments μ_k for all transitions between all the state $\mathbf{s}_{k-1}$ and $\mathbf{s}_k$ are calculated recursively and added up to the time accumulated metrics Σ_{k-1}. Then, for the two transitions with the same new state $\mathbf{s}_k$, the values of $\Sigma_{k-1} + \mu_k$ are compared. The larger value will serve as the new accumulated metric $\Sigma_k = \Sigma_{k-1} + \mu_k$, and the other one will be discarded. Furthermore, a pointer will be stored, which points from $\mathbf{s}_k$ to the preceding state corresponding to the larger metric value. Thus, going from the left to the right in the trellis diagram, for each time instant k and for all possible states, the algorithm executes the following steps:

1. Calculate the metric increments μ_k for all the $2 \cdot 2^m$ transitions between all the 2^m states $\mathbf{s}_{k-1}$ and all the 2^m states $\mathbf{s}_k$ and add them to the to the 2^m accumulated metric values Σ_{k-1} corresponding to the states $\mathbf{s}_{k-1}$.

2. For all states $\mathbf{s}_k$ compare the values of $\Sigma_{k-1} + \mu_k$ for the two transitions ending at $\mathbf{s}_k$ and select the one that is the maximum and then set $\Sigma_k = \Sigma_{k-1} + \mu_k$, which is the accumulated metric of that state.

3. Place a pointer to the state $\mathbf{s}_{k-1}$ that is the most likely preceding state for that transition.

Then, when all these calculations and assignments have been done, we start at the end of the trellis and trace back the pointers that indicate the most likely preceding states. This procedure finally leads us to the most likely path in the trellis.

3.2.3 The soft-output Viterbi algorithm (SOVA)

The soft-output Viterbi algorithm (SOVA) is a relatively simple modification of the Viterbi algorithm that allows to obtain an additional *soft* reliability information for the hard decision bits provided by the MLSE.

By construction, the Viterbi algorithm is a sequence estimator, not a bit estimator. Thus, it does not provide reliability information about the bits corresponding to the sequence. However, it can provide us with information about the reliability of the decision between two sequences. Let $\mathbf{x}$ and $\hat{\mathbf{x}}$ be two possible transmit sequences. Then, according to Equation (3.18), the conditional probability that this sequence has been transmitted given that $\mathbf{y}$ has been received is

$$P(\mathbf{x}|\mathbf{y}) = C\ \exp\left(\frac{1}{\sigma^2}\mu(\mathbf{x})\right)$$

for $\mathbf{x}$ and

$$P(\hat{\mathbf{x}}|\mathbf{y}) = C\ \exp\left(\frac{1}{\sigma^2}\mu(\hat{\mathbf{x}})\right)$$

for $\hat{\mathbf{x}}$. Now assume that $\hat{\mathbf{x}}$ is the maximum likelihood sequence obtained by the Viterbi algorithm. If one could be sure that one of the two sequences $\mathbf{x}$ or $\hat{\mathbf{x}}$ is the correct one (and not any other one), then $\Pr(\hat{\mathbf{x}}|\mathbf{y}) = 1 - \Pr(\mathbf{x}|\mathbf{y})$ and the LLR for a correct decision would be given by

$$L(\hat{\mathbf{x}}) = \log\frac{P(\hat{\mathbf{x}}|\mathbf{y})}{P(\mathbf{x}|\mathbf{y})} = \frac{1}{\sigma^2}\left(\mu(\hat{\mathbf{x}}) - \mu(\mathbf{x})\right), \tag{3.32}$$

that is, the metric difference is a measure for the reliability of the decision between the two sequences. We note that this LLR is conditioned by the event that one of both paths is the correct one.

We now consider a data bit $\hat{b}_k$ at a certain position in the bit stream corresponding to the ML sequence $\hat{\mathbf{x}}$ estimated by the Viterbi Algorithm[6]. The goal now is to gain information about the reliability of this bit by looking at the reliability of the decisions between $\hat{\mathbf{x}}$ and other sequences $\mathbf{x}^{(\beta)}$ whose paths merge with the ML path at some state $\mathbf{s}_k$. Any decision in favor of $\hat{\mathbf{x}}$ instead of the alternative sequence $\mathbf{x}^{(\beta)}$ with a bit $b_k^{(\beta)}$ is only relevant for that bit decision if $b_k^{(\beta)} \neq b_k$. Thus, we can restrict our consideration to the relevant sequences $\mathbf{x}^{(\beta)}$. Each of the relevant alternative paths labeled by the index β is the source of a possible erroneous decision in favor of $\hat{b}_k$ instead of $b_k^{(\beta)}$. We define a random error bit $e_k^{(\beta)}$ that takes the value $e_k^{(\beta)} = 1$ for an erroneous decision in favor of $\hat{b}_k$ instead of $b_k^{(\beta)}$ and $e_k^{(\beta)} = 0$ otherwise. We write $L_k^{(\beta)} = L\left(e_k^{(\beta)} = 0\right)$ for the L-values of the error bits. By construction, it is given by

$$L_k^{(\beta)} = \frac{1}{\sigma^2}\left(\mu(\hat{\mathbf{x}}) - \mu(\mathbf{x}^{\beta})\right).$$

Note that $L_k^{(\beta)} > 0$ holds because b_k belongs to the maximum likelihood path that is *per definitionem* more likely than any other.

It is important to note that all the corresponding probabilities are *conditional* probabilities because in any case it is assumed that one of the two sequences $\hat{\mathbf{x}}$ or $\mathbf{x}^{(\beta)}$ is the correct

[6]The same arguments apply if we consider a symbol $\hat{x}_i$ of the transmit sequence.

one. Furthermore, we only consider paths that merge directly with the ML path. Therefore, all paths that are discarded after comparing them with another path than the ML path are not considered. It is possible (but not very likely in most cases) that the correct path is among these discarded paths. This rare event has been excluded in our approximation. We further assume that the random error bits $e_k^{(\beta)}$ are statistically independent. All the random error bits $e_k^{(\beta)}$ together result in an error bit e_k that is assumed to be given by the modulo 2 sum

$$e_k = \sum_{\text{relevant } \beta} \oplus e_k^{(\beta)}.$$

We further write $L_k = L\,(e_k = 0)$ for the L-value of the resulting error bit. Using Equation (3.14), the L-value for this resulting error bit is approximately given by

$$L_k \approx \min_{\text{relevant } \beta} \left(L_k^{(\beta)} \right) = \min_{\text{relevant } \beta} \left(\frac{1}{\sigma^2} \left(\mu(\hat{\mathbf{x}}) - \mu(\mathbf{x}^{(\beta)}) \right) \right),$$

where we have used Equation (3.32). It is intuitively simple to understand that this is a reasonable reliability information about the bit b_k. We consider all the sequence decisions that are relevant for the decision of this bit. Then, according to the intuitively obvious rule that a chain is as strong as its weakest link, we assign the smallest of those sequence reliabilities as the bit reliability.

Now, in the Viterbi algorithm, the reliability information about the merging paths have to be stored for each state in addition to the accumulated metric and the pointer to the most likely preceding state. Then the reliability of the bits of the ML path will be calculated. First, they will all be initialized with $+\infty$, that is, practically speaking, with a very large number. Then, for each relevant decision between two paths, this value will be updated, that is, the old reliability will be replaced by the reliability of the path decision if the latter is smaller. To do this, every path corresponding to any sequence $\mathbf{x}^{(\beta)}$ that has been discarded in favor of the ML sequence $\hat{\mathbf{x}}$ has to be traced back to a point where both paths merge.

We finally note that the reliability information can be assigned to the transmit symbols $x_i \in \{\pm 1\}$ (i.e. the signs corresponding to the bits of the code word) as well as to the data bit itself.

3.2.4 MAP decoding for convolutional codes: the BCJR algorithm

To obtain LLR information about bits rather than about sequences, the bitwise MAP receiver of Equation (3.23) has to be applied instead of a MLSE. This equation cannot be applied directly because it would require an exhaustive search through all code words. For a convolutional code, the exhaustive search for the MLSE can be avoided in the Viterbi algorithm by making use of the trellis structure. For the MAP receiver, the exhaustive search can be avoided in the BCJR (Bahl, Cocke, Jelinek, Raviv) algorithm (Bahl *et al.* 1974). In contrast to the SOVA, it provides us with the exact LLR value for a bit, not just an approximate one. The price for this exact information is the higher complexity. The BCJR algorithm has been known for a long time, but it became very popular not before its widespread application in turbo decoding.

We consider a vector of data bits $\mathbf{b} = (b_1, \ldots, b_K)^T$ encoded to a code word $\mathbf{c}$ and transmitted with symbols $\mathbf{x}_k$. Given a receive symbol sequence $\mathbf{y} = (y_1, \ldots, y_N)^T$, we

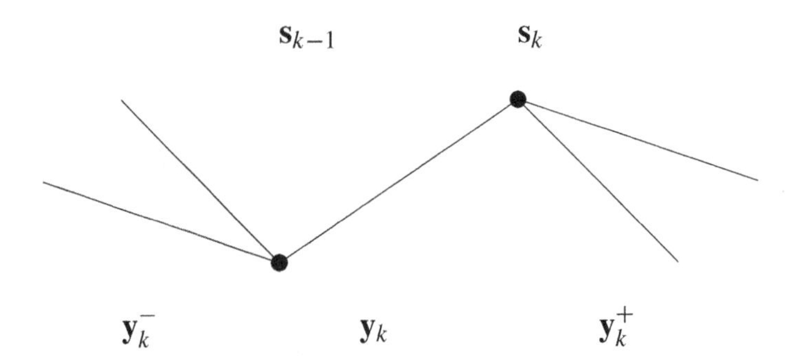

Figure 3.12 Transition.

want to calculate the LLR for a data bit b_k given as

$$L(b_k = 0|\mathbf{y}) = \log \frac{\sum_{\mathbf{b}\in\mathcal{B}_k^{(0)}} P(\mathbf{b}|\mathbf{y})}{\sum_{\mathbf{b}\in\mathcal{B}_k^{(1)}} P(\mathbf{b}|\mathbf{y})}. \tag{3.33}$$

Here, $\mathcal{B}_k^{(0)}$ is the set of those vectors $\mathbf{b} \in \mathcal{B}$ for which $b_k = 0$ and $\mathcal{B}_k^{(1)}$ is the set of those for which $b_k = 1$. We assume that the bit b_k is encoded during the transition between the states $\mathbf{s}_{k-1}$ and $\mathbf{s}_k$ of a trellis. For each time instant k, there are 2^m such transitions corresponding to $b_k = 0$ and 2^m transitions corresponding to $b_k = 1$. Each probability term $P(\mathbf{b}|\mathbf{y})$ in the numerator or denominator of Equation (3.33) can be written as the conditional probability $P(\mathbf{s}_k\mathbf{s}_{k-1}|\mathbf{y})$ for the transition between two states $\mathbf{s}_{k-1}$ and $\mathbf{s}_k$. Since the denominator in

$$P(\mathbf{s}_k\mathbf{s}_{k-1}|\mathbf{y}) = \frac{p(\mathbf{y}, \mathbf{s}_k\mathbf{s}_{k-1})}{p(\mathbf{y})}$$

cancels out in Equation (3.33), we can consider the joint probability density function $p(\mathbf{y}, \mathbf{s}_k\mathbf{s}_{k-1})$ instead of the conditional probability $P(\mathbf{s}_k\mathbf{s}_{k-1}|\mathbf{y})$. We now decompose the receive symbol vector into three parts: we write $\mathbf{y}_k^-$ for those receive symbols corresponding to time instants earlier than the transition between the states $\mathbf{s}_{k-1}$ and $\mathbf{s}_k$. We write $\mathbf{y}_k$ for those receives symbols corresponding to time instants at the transition between the states $\mathbf{s}_{k-1}$ and $\mathbf{s}_k$. And we write $\mathbf{y}_k^+$ for those receive symbols corresponding to time instants later than the transition between the states $\mathbf{s}_{k-1}$ and $\mathbf{s}_k$. Thus, the receive vector may be written as

$$\mathbf{y} = \begin{bmatrix} \mathbf{y}_k^- \\ \mathbf{y}_k \\ \mathbf{y}_k^+ \end{bmatrix}$$

(see Figure 3.12), and the probability density may be written as

$$p(\mathbf{y}, \mathbf{s}_k\mathbf{s}_{k-1}) = p(\mathbf{y}_k^+\mathbf{y}_k\mathbf{y}_k^-\mathbf{s}_k\mathbf{s}_{k-1}).$$

If no confusion arises, we dispense with the commas between vectors. Using the definition of conditional probability, we modify the right-hand side and get

$$p(\mathbf{y}, \mathbf{s}_k\mathbf{s}_{k-1}) = p(\mathbf{y}_k^+|\mathbf{y}_k\mathbf{y}_k^-\mathbf{s}_k\mathbf{s}_{k-1})p(\mathbf{y}_k\mathbf{y}_k^-\mathbf{s}_k\mathbf{s}_{k-1}),$$

and, in another step,

$$p(\mathbf{y}, \mathbf{s}_k \mathbf{s}_{k-1}) = p(\mathbf{y}_k^+ | \mathbf{y}_k \mathbf{y}_k^- \mathbf{s}_k \mathbf{s}_{k-1}) p(\mathbf{y}_k \mathbf{s}_k | \mathbf{y}_k^- \mathbf{s}_{k-1}) p(\mathbf{y}_k^- \mathbf{s}_{k-1}).$$

We now make the assumptions

$$p(\mathbf{y}_k^+ | \mathbf{y}_k \mathbf{y}_k^- \mathbf{s}_k \mathbf{s}_{k-1}) = p(\mathbf{y}_k^+ | \mathbf{s}_k)$$

and

$$p(\mathbf{y}_k \mathbf{s}_k | \mathbf{y}_k^- \mathbf{s}_{k-1}) = p(\mathbf{y}_k \mathbf{s}_k | \mathbf{s}_{k-1}),$$

which are quite similar to the properties of a Markov chain. The first equation means that we assume that the random variable $\mathbf{y}_k^+$ corresponding to the receive symbols after state $\mathbf{s}_k$ depends on that state, but is independent of the earlier state $\mathbf{s}_{k-1}$ and any earlier receive symbols corresponding to $\mathbf{y}$ and $\mathbf{y}_k^-$. The second equation means that we assume that the random variable $\mathbf{y}_k$ corresponding to the receive symbols for the transition from the state $\mathbf{s}_{k-1}$ to $\mathbf{s}_k$ does not depend on earlier receive symbols corresponding to $\mathbf{y}_k^-$. For a given fixed receive sequence $\mathbf{y}$, we define

$$\alpha_{k-1}(\mathbf{s}_{k-1}) = p(\mathbf{y}_k^- \mathbf{s}_{k-1}), \quad \beta_k(\mathbf{s}_k) = p(\mathbf{y}_k^+ | \mathbf{s}_k), \quad \gamma_k(\mathbf{s}_k | \mathbf{s}_{k-1}) = p(\mathbf{y}_k \mathbf{s}_k | \mathbf{s}_{k-1}) \tag{3.34}$$

and write

$$p(\mathbf{y}, \mathbf{s}_k \mathbf{s}_{k-1}) = \beta_k(\mathbf{s}_k) \gamma_k(\mathbf{s}_k | \mathbf{s}_{k-1}) \alpha_{k-1}(\mathbf{s}_{k-1}).$$

The probability densities $\gamma_k(\mathbf{s}_k | \mathbf{s}_{k-1})$ for the transition from the state $\mathbf{s}_{k-1}$ to $\mathbf{s}_k$ can be obtained from the metric value μ_k calculated from $\mathbf{y}_k$. As shown in Section 3.1.5, for the AWGN channel with normalized noise variance σ^2 and bipolar transmission, we have simply

$$\gamma_k(\mathbf{s}_k | \mathbf{s}_{k-1}) = C \exp\left(\frac{1}{\sigma^2} \mathbf{x}_k \cdot \mathbf{y}_k\right) \cdot \Pr(\mathbf{x}_k),$$

where $\mathbf{x}_k$ is the transmit symbol and $P(\mathbf{x}_k)$ is the *a priori* probability corresponding to that transition. The α_k and β_k values have to be calculated using recursive relations. We state the following proposition.

Proposition 3.2.1 (Forward-backward recursions) *For α_k, β_k, γ_k as defined by Equation (3.34), the following two recursive relations*

$$\alpha_k(\mathbf{s}_k) = \sum_{\mathbf{s}_{k-1}} \gamma_k(\mathbf{s}_k | \mathbf{s}_{k-1}) \alpha_{k-1}(\mathbf{s}_{k-1}) \tag{3.35}$$

and

$$\beta_{k-1}(\mathbf{s}_{k-1}) = \sum_{\mathbf{s}_k} \beta_k(\mathbf{s}_k) \gamma_k(\mathbf{s}_k | \mathbf{s}_{k-1}) \tag{3.36}$$

hold.

Proof. Forward recursion:

$$\begin{aligned}
\alpha_k(\mathbf{s}_k) &= p(\mathbf{y}_{k+1}^- \mathbf{s}_k) = p(\mathbf{y}_k \mathbf{y}_k^- \mathbf{s}_k) \\
&= \textstyle\sum_{\mathbf{s}_{k-1}} p(\mathbf{y}_k \mathbf{y}_k^- \mathbf{s}_k \mathbf{s}_{k-1}) \\
&= \textstyle\sum_{\mathbf{s}_{k-1}} p(\mathbf{y}_k \mathbf{s}_k | \mathbf{y}_k^- \mathbf{s}_{k-1}) p(\mathbf{y}_k^- \mathbf{s}_{k-1})
\end{aligned}$$

Using the Markov property $p(\mathbf{y}_k \mathbf{s}_k | \mathbf{y}_k^- \mathbf{s}_{k-1}) = p(\mathbf{y}_k \mathbf{s}_k | \mathbf{s}_{k-1})$, we obtain Equation (3.35).

Backward recursion:

$$\begin{aligned}\beta_{k-1}(\mathbf{s}_{k-1}) &= p(\mathbf{y}_{k-1}^{+}|\mathbf{s}_{k-1}) = p(\mathbf{y}_k^{+}\mathbf{y}_k|\mathbf{s}_{k-1}) = p(\mathbf{y}_k^{+}\mathbf{y}_k\mathbf{s}_{k-1})/\Pr(\mathbf{s}_{k-1})\\ &= \textstyle\sum_{\mathbf{s}_k} p(\mathbf{y}_k^{+}\mathbf{y}_k\mathbf{s}_k\mathbf{s}_{k-1})/\Pr(\mathbf{s}_{k-1})\\ &= \textstyle\sum_{\mathbf{s}_k} p(\mathbf{y}_k^{+}|\mathbf{y}_k\mathbf{s}_k\mathbf{s}_{k-1})p(\mathbf{y}_k\mathbf{s}_k\mathbf{s}_{k-1})/\Pr(\mathbf{s}_{k-1}).\end{aligned}$$

Using $p(\mathbf{y}_k\mathbf{s}_k\mathbf{s}_{k-1})/\Pr(\mathbf{s}_{k-1}) = p(\mathbf{y}_k\mathbf{s}_k|\mathbf{s}_{k-1})$ and the Markov property $p(\mathbf{y}_k^{+}|\mathbf{y}_k\mathbf{s}_k\mathbf{s}_{k-1}) = p(\mathbf{y}_k^{+}|\mathbf{s}_k)$, we obtain Equation (3.36).

The BCJR algorithm now proceeds as follows: initialize the initial and the final state of the trellis as $\alpha_0 = 1$ and $\beta_K = 1$ and calculate the α_k values according to the forward recursion of Equation (3.35) from the left to the right in the trellis and then calculate the β_k according to the backward recursion Equation (3.36) from the right to the left in the trellis. Then the LLRs for each transition can be calculated as

$$L(b_k = 0|\mathbf{y}) = \log \frac{\sum_{\mathbf{b}\in\mathcal{B}_k^{(0)}} p(\mathbf{y}, \mathbf{s}_k\mathbf{s}_{k-1})}{\sum_{\mathbf{b}\in\mathcal{B}_k^{(1)}} p(\mathbf{y}, \mathbf{s}_k\mathbf{s}_{k-1})}$$

that is,

$$L(b_k = 0|\mathbf{y}) = \log \frac{\sum_{\mathbf{b}\in\mathcal{B}_k^{(0)}} \alpha_{k-1}(\mathbf{s}_{k-1})\gamma_k(\mathbf{s}_k|\mathbf{s}_{k-1})\beta_k(\mathbf{s}_k)}{\sum_{\mathbf{b}\in\mathcal{B}_k^{(1)}} \alpha_{k-1}(\mathbf{s}_{k-1})\gamma_k(\mathbf{s}_k|\mathbf{s}_{k-1})\beta_k(\mathbf{s}_k)}.$$

In this notation, we understand the sum over all $\mathbf{b} \in \mathcal{B}_k^{(0)}$ as the sum over all transitions from $\mathbf{s}_{k-1}$ to $\mathbf{s}_k$ with $b_k = 0$ and sum over all $\mathbf{b} \in \mathcal{B}_k^{(1)}$ as the sum over all transitions from $\mathbf{s}_{k-1}$ to $\mathbf{s}_k$ with $b_k = 1$.

3.2.5 Parallel concatenated convolutional codes and turbo decoding

During the last decade, great success has been achieved in closely approaching the theoretical limit of channel coding. The codes that have been used for that are often called *turbo codes*. More precisely, one should carefully distinguish between the code and the decoding method. The first turbo code was a parallel concatenated convolutional code (PCCC). Parallel concatenation can be done with block codes as well. Also serial concatenation is possible. The novel decoding method that has been applied to all theses codes deserves the name *turbo decoder* because there is an iterative exchange of extrinsic and *a priori* information between the decoders of the component codes.

To explain the method, we consider the classical scheme with a parallel concatenation of two RSC codes of rate $R_c = 1/2$ as depicted in Figure 3.13. The data bit stream is encoded in parallel by two RSC encoders (that may be identical). The common systematic part $\mathbf{x}_s$ of both codes will be transmitted only once. Thus, the output code word consists of three parallel vectors: the systematic symbol vector $\mathbf{x}_s$ and the two nonsystematic PC symbol vectors $\mathbf{x}_{p1}$ and $\mathbf{x}_{p2}$. The input for the second RSC parity check encoder (RSC-PC2) is interleaved by a pseudo-random permutation Π before encoding. The resulting $R_c = 1/3$ code word may be punctured in the nonsystematic symbols to achieve higher code rates. Lower code rates can be achieved by additional RSC-PCs, together with interleavers. This setup may be regarded as well as a parallel concatenation of the first RSC code of rate $R_c = 1/2$ with an $R_c = 1$ recursive nonsystematic code that produces $\mathbf{x}_{p2}$. However, here

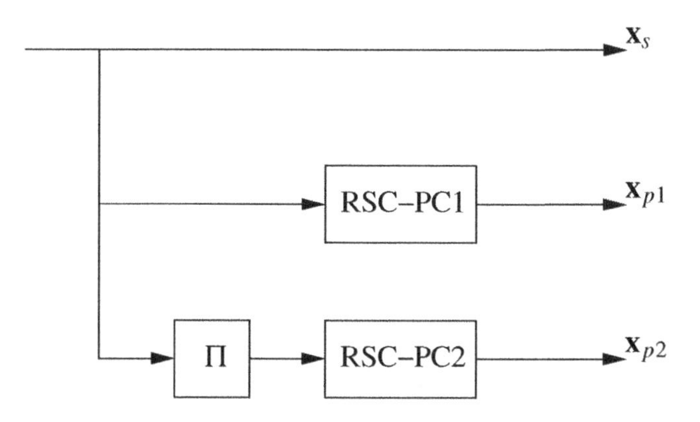

Figure 3.13 PCCC encoder.

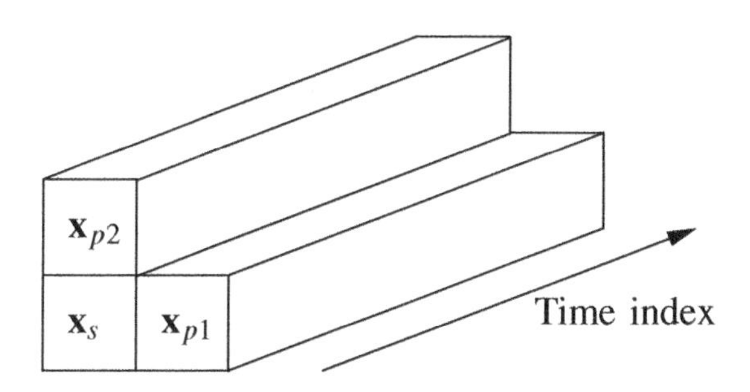

Figure 3.14 PCCC code word.

we prefer the point of view of two equal rate RSC codes with a common systematic symbol stream.

The code word consisting of three parallel symbol streams can be visualized as depicted in Figure 3.14. The vector $\mathbf{x}_{p1}$ can be regarded as a horizontal parity check, the vector $\mathbf{x}_{p2}$ as a vertical parity check. The time index is the third dimension. At the decoder, the corresponding receive vectors are denoted by $\mathbf{y}_s$, $\mathbf{y}_{p1}$ and $\mathbf{y}_{p2}$. With a diagonal matrix of fading amplitudes $\mathbf{A}$, the channel LLRs are

$$\mathbf{L}_s^c = \frac{2}{\sigma^2}\mathbf{A}\mathbf{y}_s, \quad \mathbf{L}_{p1}^c = \frac{2}{\sigma^2}\mathbf{A}\mathbf{y}_{p1}, \quad \mathbf{L}_{p2}^c = \frac{2}{\sigma^2}\mathbf{A}\mathbf{y}_{p2},$$

where σ^{-2} is the channel SNR. We write $\mathbf{L}_1^c = \left(\mathbf{L}_s^c, \mathbf{L}_{p1}^c\right)$ and $\mathbf{L}_2^c = \left(\mathbf{L}_s^c, \mathbf{L}_{p2}^c\right)$ for the respective channel LLRs. In the decoding process, independent extrinsic information $\mathbf{L}_1^e$ and $\mathbf{L}_2^e$ about the systematic part can be obtained from the horizontal and from the vertical decoding, respectively. Thus, the horizontal extrinsic information can be used as *a priori* information for vertical decoding and vice versa.

The turbo decoder setup is depicted in Figure 3.15. It consists of two SISO decoders, SISO1 and SISO2, for the decoding of RSC1 and RSC2, as depicted in Figure 3.3. To

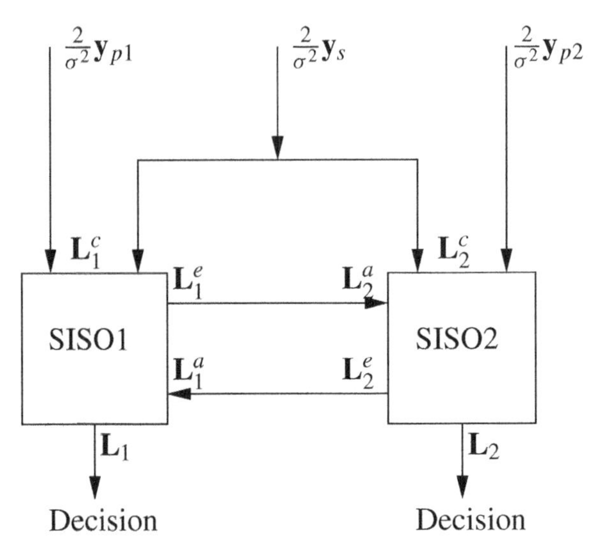

Figure 3.15 Turbo decoder.

simplify the figure, the necessary de–interleaver Π^{-1} at the input for $\mathbf{L}_s^c$ and $\mathbf{L}_2^a$ and the interleaver Π at the output for $\mathbf{L}_2^e$ and $\mathbf{L}_2$ of RSC2 are included inside the SISO2. The MAP decoder for convolutional codes will be implemented by the BCJR algorithm. In the iterative decoding process, the extrinsic output of one SISO decoder serves as the *a priori* input for the other. At all decoding steps, the channel LLR values are available at both SISOs. In the first decoding step, only the channel information, but no *a priori* LLR value is available at SISO1. Then SISO1 calculates the extrinsic LLR value $\mathbf{L}_1^e$ from the horizontal decoding. This serves as the *a priori* input LLR value $\mathbf{L}_2^a$ for SISO2. The extrinsic output $\mathbf{L}_2^e$ then serves as the *a priori* input for SISO1 in the second iteration. These iterative steps will be repeated until a break, and then a final decision can be obtained from the SISO total LLR output value $\mathbf{L}_2$ (or $\mathbf{L}_1$).

We note that the *a priori* input is not really an independent information at the second iteration step or later. This is because all the information of the code has already been used to obtain it. However, the dependencies are small enough so that the information can be successfully used to improve the reliability of the decision by further iterations. On the other hand, it is essential that there will be no feedback of LLR information from the output to the input. Such a feedback would be accumulated at the inputs and finally dominate the decision. Therefore, the extrinsic LLR must be used, where the SISO inputs have been subtracted from the LLR.

We add the following remarks:

- In the ideal case, the SISO is implemented by a BCJR MAP receiver. In practice, the maxlog MAP approximation may be used, which results only in a small loss in performance. This loss is due to the fact that the reliability of the very unreliable symbols is slightly overestimated. The SOVA may also be used, but the performance loss is higher.

- The exact MAP needs the knowledge of the SNR value σ^{-2}, which is normally not available. Thus, a rough estimate must be used. Using the maxlog MAP or SOVA, the SNR is not needed. This is due to the fact that in the first decoding step no *a priori* LLR is used, and, as a consequence, the SNR appears only as a common linear scale factor in all further calculated LLR outputs.

3.3 Reed–Solomon Codes

Reed–Solomon (RS) codes may be regarded as the most important block codes because of their extremely high relevance for many practical applications. These include deep space communications, digital storage media and, last but not least, the digital video broadcasting system (DVB). However, these most useful codes are based on quite sophisticated theoretical concepts that seem to be much closer to mathematics than to electrical engineering. The theory of RS codes can be found in many text books (Blahut 1983; Bossert 1999; Clark and Cain 1988; Lin and Costello 1983; Wicker 1995). In this section about RS codes, we restrict ourselves to some important facts that are necessary to understand the coding scheme of the DVB-T system discussed in Subsection 4.6.2. We will first discuss the basic properties of RS codes as far as they are important for the practical application. Then, we will give a short introduction to the theoretical background. For a deeper understanding of that background, we refer to the text books cited above.

3.3.1 Basic properties

Reed–Solomon codes are based on *byte arithmetics*[7] rather than on bit arithmetics. Thus, RS codes correct *byte errors* instead of bit errors. As a consequence, RS codes are favorable for channels with bursts of bit errors as long as these bursts do not affect too many subsequent bytes. This can be avoided by a proper interleaving scheme. Such bursty channels occur in digital recording. As another example, for a concatenated coding scheme with an inner convolutional code, the Viterbi decoder produces burst errors. An inner convolutional code concatenated with an outer RS code is therefore a favorable setup. It is used in deep space communications and for DVB-T.

Let $N = 2^m - 1$ with an integer number m. For the practically most important RS codes, we have $m = 8$ and $N = 255$. In that case, the symbols of the code word are bytes. For simplicity, in the following text, we will therefore speak of bytes for those symbols. For an RS(N, K, D) code, K data bytes are encoded to a code word of N bytes. The Hamming distance is given by $D = N - K + 1$ bytes. For odd values of D, the code can correct up to t byte errors with $D = 2t + 1$. For even values of D, the code can correct up to t byte errors with $D = 2t + 2$. RS codes are linear codes. For a linear code, any nonsystematic encoder can be transformed into a linear encoder by a linear transform. Figure 3.16 shows the structure of a systematic RS code word with an odd Hamming distance and an even number $N - K = D - 1 = 2t$ of redundancy bytes called *parity check* (PC) *bytes*. In that example, the parity check bytes are placed at the end of the code word. Other choices are possible. RS codes based on byte arithmetics have always the code word length $N = 2^8 - 1 = 255$.

[7]RS codes can be constructed for more general arithmetic structures, but only those based on byte arithmetics are of practical relevance.

K data bytes	$2t$ PC bytes

Figure 3.16 A systematic RS code word.

Table 3.1 Some RS code parameters

RS(255, 253, 3)	$t = 1$
RS(255, 251, 5)	$t = 2$
RS(255, 249, 7)	$t = 3$
...	
RS(255, 239, 17)	$t = 8$
...	

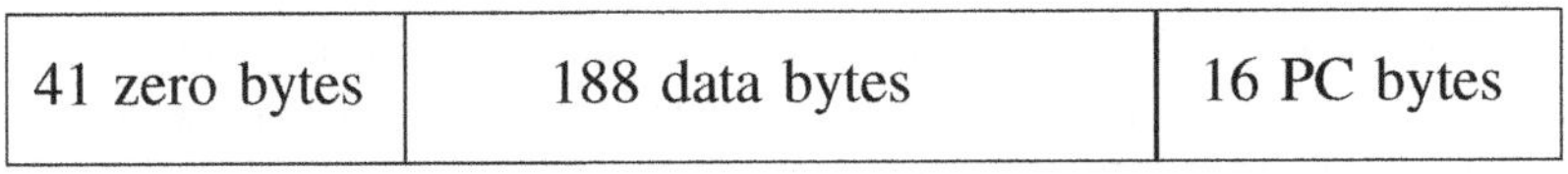

41 zero bytes	188 data bytes	16 PC bytes

Figure 3.17 A shortened RS code word.

They can be constructed for any value of $D \leq N$. Table 3.1 shows some examples for odd values of D.

Shortened RS codes

In practice, the fixed code word length $N = 255$ is an undesirable restriction. One can get more flexibility by using a simple trick. For an RS(N, K, D) code with $N = 255$, we want to encode only $K_1 < K$ data byte and set the first $K - K_1$ bytes of the data word to zero. We then encode the K bytes (including the zeros) with the RS(N, K, D) systematic encoder to obtain a code word of length N whose first $K - K_1$ code words are equal to zero. These bytes contain no information and need not to be transmitted. By this method we have obtained a *shortened* RS(N_1, K_1, D) code word with $N_1 = N - (K - K_1)$. Figure 3.17 shows the code word of a shortened RS(204, 188, 17) code obtained from an RS(255, 239, 17) code. Before decoding, at the receiver, the $K - K_1$ zero bytes must be appended at the beginning of the code word and a RS(255, 239, 17) decoder will be used. This shortened RS code is used as the outer code for the DVB-T system.

Decoding failure

It may happen that the decoder detects errors that cannot be corrected. In the case of decoding failure, an error flag can be set to indicate that the data are in error. The application may then take benefit from this information.

Erasure decoding

If it is known that some received bytes are very unreliable (e.g. from an inner decoder that provides such reliability information), the decoder can make use of this fact in the decoding procedure. These bytes are called *erasures.*

3.3.2 Galois field arithmetics

Reed–Solomon codes are based on the arithmetics of *finite fields* that are usually called *Galois fields*. The mathematical concept of a *field* stands for a system of numbers, where addition and multiplication and the corresponding inverses are defined and which is commutative. The existence of an (multiplicative) inverse is crucial: for any field element a, there must exist a field element a^{-1} with the property $a^{-1}a = 1$. The rational numbers and the real numbers with their familiar arithmetics are fields. The integer numbers are not, because the (multiplicative) inverse of an integer is not an integer (except for the one).

A *Galois field* $GF(q)$ is a field with a finite number q of elements. One can very easily construct a Galois field $GF(q)$ with $q = p$, where p is a prime number. The $GF(p)$ arithmetics is then given by taking the remainder modulo p. For example, $GF(7)$ with the elements 0, 1, 2, 3, 4, 5, 6 is defined by the addition table

+	1	2	3	4	5	6
1	2	3	4	5	6	0
2	3	4	5	6	0	1
3	4	5	6	0	1	2
4	5	6	0	1	2	3
5	6	0	1	2	3	4
6	0	1	2	3	4	5

and the multiplication table

1	2	3	4	5	6
2	4	6	1	3	5
3	6	2	5	1	4
4	1	5	2	6	3
5	3	1	6	4	2
6	5	4	3	2	1

Note that every field element must occur exactly once in each column or row of the multiplication table to ensure the existence of a multiplicative inverse.

A Galois field has at least one *primitive element* α with the property that any nonzero field element can be uniquely written as a power of α. By using the multiplication table of $GF(7)$, we easily see that $\alpha = 5$ is such a primitive element and the nonzero field elements can be written as powers of α in the following way

$$\alpha^0 = 1,\ \alpha^1 = 5,\ \alpha^2 = 4,\ \alpha^3 = 6,\ \alpha^4 = 2,\ \alpha^5 = 3.$$

We note that since $\alpha^6 = \alpha^0 = 1$, negative powers of α like $\alpha^{-2} = \alpha^4$ are defined as well. We can easily visualize the multiplicative structure of $GF(7)$ as depicted in Figure 3.18. Each nonzero element is represented by the edge of a hexagon or the corresponding angle. α^0 has the angle zero, α^1 has the angle $\pi/3$, α^2 has the angle $2\pi/3$, and so on. Obviously,

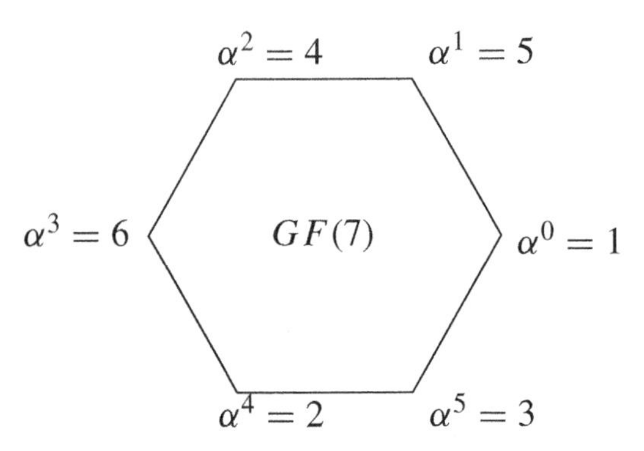

Figure 3.18 $GF(7)$.

the multiplication of field elements is represented by the addition of the corresponding angles. This is the same multiplicative group structure as if we would identify α with the complex phasor $\exp(j2\pi/N)$ with $N = q - 1$. This structure leads directly to a very natural definition of the discrete Fourier transform (DFT) for Galois fields (see below).

The primitive of $GF(q)$ element has the property $\alpha^N = 1$ and thus $\alpha^{iN} = 1$ for $i = 0, 1, \ldots, N-1$. It follows that each element α^i of $GF(q)$ is a root of the polynomial $x^N - 1$, and we may write

$$x^N - 1 = \prod_{i=0}^{N-1} \left(x - \alpha^i\right).$$

Similarly,

$$x^N - 1 = \prod_{i=0}^{N-1} \left(x - \alpha^{-i}\right)$$

holds.

The prime number Galois fields $GF(p)$ are of some tutorial value. Of practical relevance are the *extension fields* $GF(2^m)$, where m is a positive integer. We state that a Galois field $GF(q)$ exists for every $q = p^m$, where p is prime[8]. Almost all practically relevant RS codes are based on $GF(2^8)$ because the field element can be represented as bytes. We will use the smaller field $GF(2^3)$ to explain the arithmetics of the extension fields.

The elements of an extension field $GF(p^m)$ can be represented as polynomials of degree $m-1$ over $GF(p)$. Without going into mathematical details, we state that the primitive element α is defined as the root of a primitive polynomial. The arithmetic is then *modulo* that polynomial. Note that addition and subtraction is the same in $GF(2^m)$.

We explain the arithmetic for the example $GF(2^3)$. The primitive polynomial is given by $p(x) = x^3 + x + 1$. The primitive element α is the root of that polynomial, that is, we can set

$$\alpha^3 + \alpha + 1 \equiv 0.$$

We then write down all powers of alpha and reduce them to modulo $\alpha^3 + \alpha + 1$. For example, we may identify $\alpha^3 \equiv \alpha + 1$. Each element is thus given by a polynomial of

[8]For a proof, we refer to the text books mentioned above.

Table 3.2 Representation of $GF(2^3)$

dec	bin	poly	α^i
0	000	0	*
1	001	1	α^0
2	010	α	α^1
3	011	$\alpha+1$	α^3
4	100	α^2	α^2
5	101	α^2+1	α^6
6	110	$\alpha^2+\alpha$	α^4
7	111	$\alpha^2+\alpha+1$	α^5

degree 2 over the dual number system $GF(2)$ and can therefore be represented by a bit triple or a decimal number. Table 3.2 shows the equivalent representations of the elements of $GF(2^3)$. We note that for a Galois field $GF(2^m)$, the decimal representation of the primitive element is always given by the number 2.

The addition is simply defined as the addition of polynomials, which is equivalent to the vector addition of the bit tuples. Multiplication is defined as the multiplication of polynomials and reduction modulo $\alpha^3+\alpha+1$. The addition table is then given by

+	1	2	3	4	5	6	7
1	0	3	2	5	4	7	6
2	3	0	1	6	7	4	5
3	2	1	0	7	6	5	4
4	5	6	7	0	1	2	3
5	4	7	6	1	0	3	2
6	7	4	5	2	3	0	1
7	6	5	4	3	2	1	0

and the multiplication table by

1	2	3	4	5	6	7
2	4	6	3	1	7	5
3	6	5	7	4	1	2
4	3	7	6	2	5	1
5	1	4	2	7	3	6
6	7	1	5	3	2	4
7	5	2	1	6	4	3

We can visualize the multiplicative structure of $GF(8)$ as depicted in Figure 3.19. This will lead us directly to the discrete Fourier transform that will be defined in the following subsection.

3.3.3 Construction of Reed–Solomon codes

From the communications engineering point of view, the most natural way to introduce Reed–Solomon codes is via the DFT and general properties of polynomials.

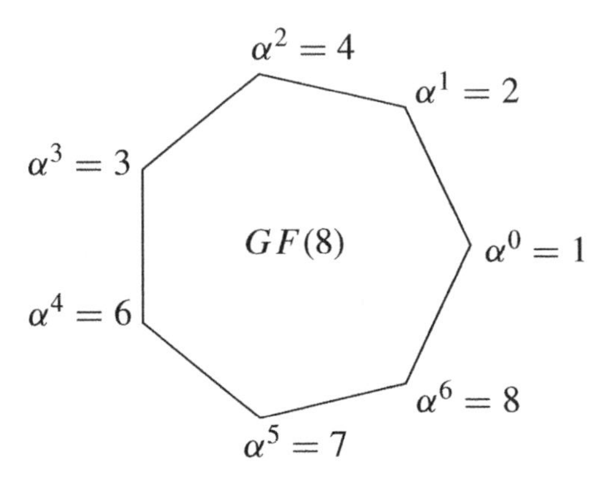

Figure 3.19 $GF(8)$.

The discrete Fourier transforms for Galois fields

Let

$$A = (A_0, A_1, \ldots, A_{N-1})^T$$

be a vector of length $N = q - 1$ with elements $A_i \in GF(q)$. We define the vector

$$a = (a_0, a_1, \ldots, a_{N-1})^T$$

of the DFT by the operation

$$a_j = \sum_{i=0}^{N-1} A_i \alpha^{ij}.$$

We note that the Fourier transform can be described by the multiplication of the vector A by the DFT matrix

$$\mathbf{F} = \begin{pmatrix} 1 & 1 & 1 & \cdots & 1 \\ 1 & \alpha & \alpha^2 & \cdots & \alpha^{N-1} \\ 1 & \alpha^2 & \alpha^4 & \cdots & \alpha^{2N-2} \\ \vdots & \vdots & \vdots & \ddots & \vdots \\ 1 & \alpha^{N-1} & \alpha^{2N-2} & \cdots & \alpha^{(N-1)(N-1)} \end{pmatrix}.$$

As mentioned above, the primitive element α of $GF(q)$ has the same multiplicative properties as $\exp(j2\pi/N)$ with $N = q - 1$. Thus, this is the natural definition of the DFT for Galois fields. We say that A is the *frequency domain vector* and a is the *time domain vector*. The *inverse discrete Fourier transform* (IDFT) in $GF(2^m)$ is given by

$$A_i = \sum_{j=0}^{N-1} a_j \alpha^{-ij}.$$

The proof is the same as for complex numbers, but we must use the fact that

$$\sum_{j=0}^{N-1} \alpha^0 = 1$$

in $GF(2^m)$. For other Galois fields, a normalization factor would occur for the inverse transform.

Any vector can be represented by a formal polynomial. For the frequency domain vector, we may write this formal polynomial as

$$A(x) = A_0 + A_1 x + \cdots + A_{N-1} x^{N-1}.$$

We note that x is only a dummy variable. We add two polynomials $A(x)$ and $B(x)$ by adding their coefficients. If we multiply two polynomials $A(x)$ and $B(x)$ and take the remainder modulo $x^N - 1$, the result is the polynomial that corresponds to the cyclic convolution of the vectors A and B. We write

$$A(x)B(x) \equiv A * B(x) \quad \mathrm{mod}(x^N - 1).$$

The DFT can now simply be defined by

$$a_j = A(\alpha^j),$$

that is, the ith component a_j of the time domain vector a can be obtained by evaluating the frequency domain polynomial $A(x)$ for $x = \alpha^j$. We write the polynomial corresponding to the time domain vector a as

$$a(y) = a_0 + a_1 y + \cdots + a_{N-1} y^{N-1}.$$

Here, y is again a formal variable[9]. The IDFT is then given by

$$A_i = a(\alpha^{-i}).$$

As for the usual DFT, cyclic convolution in the time domain corresponds to elementwise multiplication in the frequency domain and vice versa. We may write this as

$$A * B \longleftrightarrow a \circ b$$

$$A \circ B \longleftrightarrow a * b$$

in $GF(2^m)$. Here we have written $a \circ b$ and $A \circ B$ for the *Hadamard* product, that is, the componentwise multiplication of vectors. We may define it formally as

$$A \circ B(x) = A_0 B_0 + A_1 B_1 x + \cdots + A_{N-1} B_{N-1} x^{N-1}.$$

[9]We may call it x as well.

Frequency domain encoding

We are now ready to define Reed–Solomon in the frequency domain. As an example, we start with the construction of the RS(7, 5, 3) code over $GF(8)$. We want to encode $K = 5$ useful data symbols A_i, $i = 0, 1, 2, 3, 4$ to a code word of length $N = 7$. The polynomial

$$A(x) = A_0 + A_1x + A_2x^2 + A_3x^3 + A_4x^4$$

of degree 4 cannot have more than four zeros. Thus, $a_j = A(\alpha^j)$ cannot be zero for more than four values of j. Then the time domain vector

$$a = (a_0, a_1, a_2, a_3, a_4, a_5, a_6)^T$$

has at least three nonzero components, that is, the *weight* of the vector is at least 3. The vector a is the RS code word. The Hamming distance of that code is then given by (at least) $D = 3$. Figure 3.20 shows this frequency domain encoding. The useful data are given by the data word (2, 7, 4, 3, 6) in decimal notation, where each symbol represents a bit triple according to Table 3.2. The code word in the frequency domain is given by

$$A = (2, 7, 4, 3, 6, 0, 0)^T.$$

Redundancy has been introduced by setting two frequencies equal to zero. This guarantees a minimum weight of three for the time domain code word, which is given by

$$a = (4, 6, 5, 1, 2, 3, 5)^T.$$

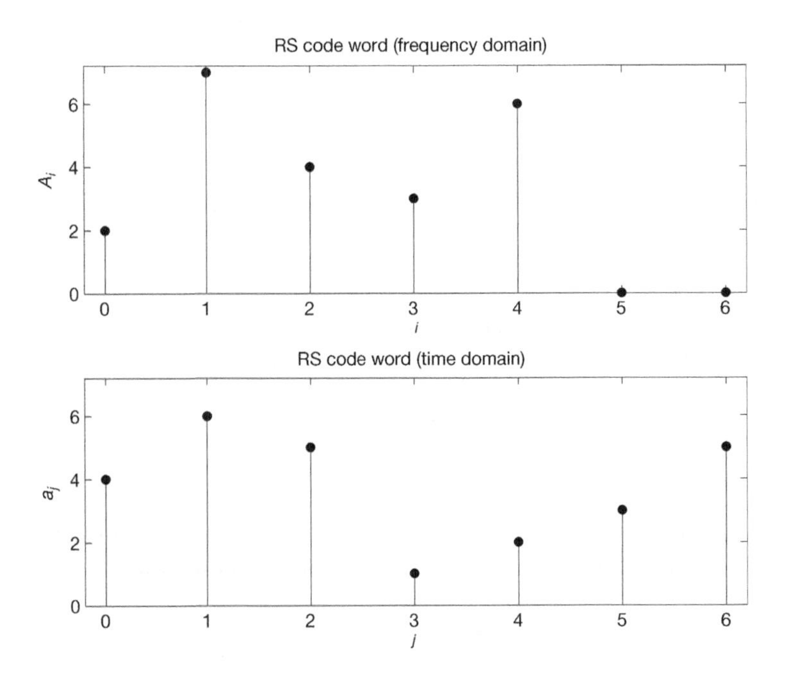

Figure 3.20 The RS(7, 5, 3) code word in the frequency domain.

A general RS(N, K, D) code over $GF(q)$ with $N = q - 1$ can be constructed in the following way. We consider polynomials

$$A(x) = A_0 + A_1 x + \cdots + A_{K-1} x^{K-1}$$

with K useful data symbols, that is, the last $N - K$ components A_i of the frequency domain vector A will be set equal to zero. We perform a DFT of length N. Since $A(x)$ has at most $K - 1$ zeros, $a_j = A(\alpha^j)$ can have at most $K - 1$ zeros. In other words, there are at least $D = N - K + 1$ nonzero components in the time domain code word

$$a = (a_0, a_1, \ldots, a_{N-1})^T.$$

Encoder and parity check

The encoder can be described by the matrix operation

$$\begin{pmatrix} a_0 \\ a_1 \\ a_2 \\ \vdots \\ a_{N-1} \end{pmatrix} = \begin{pmatrix} 1 & 1 & 1 & \cdots & 1 \\ 1 & \alpha & \alpha^2 & \cdots & \alpha^{K-1} \\ 1 & \alpha^2 & \alpha^4 & \cdots & \alpha^{2K-2} \\ \vdots & \vdots & \vdots & \ddots & \vdots \\ 1 & \alpha^{N-1} & \alpha^{2N-2} & \cdots & \alpha^{(K-1)(N-1)} \end{pmatrix} \begin{pmatrix} A_0 \\ A_1 \\ A_2 \\ \vdots \\ A_{K-1} \end{pmatrix},$$

that is, the generator matrix is the matrix of the first K columns of the DFT matrix. The condition that the last $N - K$ components A_i of the frequency domain vector A are equal to zero can be written as

$$\begin{pmatrix} 1 & \alpha^{-K} & \alpha^{-2K} & \cdots & \alpha^{-(N-1)K} \\ 1 & \alpha^{-(K+1)} & \alpha^{-2(K+1)} & \cdots & \alpha^{-(N-1)(K+1)} \\ \vdots & \vdots & \vdots & \ddots & \vdots \\ 1 & \alpha^{-(N-1)} & \alpha^{-2(N-1)} & \cdots & \alpha^{-(N-1)(N-K-1)} \end{pmatrix} \begin{pmatrix} a_0 \\ a_1 \\ a_2 \\ \vdots \\ a_{N-1} \end{pmatrix} = 0,$$

that is, the parity check matrix is the matrix of the last $N - K$ rows of the IDFT matrix.

The condition $A_i = a(\alpha^{-i}) = 0$ for $i = K, \ldots, N - 1$ means that the polynomial $a(x)$ has zeros for $x = \alpha^{-K}, \ldots, \alpha^{-(N-1)}$. We may thus factorize $a(x)$ as

$$a(x) = q(x) \prod_{i=K}^{N-1} \left(x - \alpha^{-i}\right)$$

with some quotient polynomial $q(x)$. We define the *generator polynomial*

$$g(x) = \prod_{i=K}^{N-1} \left(x - \alpha^{-i}\right).$$

The code (i.e. the set of code words) can thus equivalently be defined as those polynomials $a(x)$ that can be written as $a(x) = q(x)g(x)$.

We define the parity check polynomial

$$h(x) = \prod_{i=0}^{K-1} \left(x - \alpha^{-i}\right).$$

Obviously,

$$g(x)h(x) \equiv 0 \mod(x^N - 1)$$

and the code words $a(x)$ must fulfill the *parity check condition*

$$a(x)h(x) \equiv 0 \mod(x^N - 1).$$

3.3.4 Decoding of Reed–Solomon codes

Consider an RS(N, K, D) code with odd Hamming distance $D = 2t + 1$. We assume that a code word a has been transmitted, but another vector $r = (r_0, \ldots, r_{N-1})^T$ with elements $r_j \in GF(q)$ has been received. We write

$$r = a + e,$$

where $e = (e_0, \ldots, e_{N-1})^T$ with elements $e_j \in GF(q)$ has is the *error vector*. We write $E = (E_1, \ldots, E_{N-1})^T$ for the error in the frequency domain and $E(x)$ for the corresponding polynomial. We multiply the above equation by the parity check matrix. The result is the *syndrome vector*

$$\begin{pmatrix} S_1 \\ S_2 \\ S_3 \\ \vdots \\ S_{2t} \end{pmatrix} = \begin{pmatrix} 1 & \alpha^{-K} & \alpha^{-2K} & \cdots & \alpha^{-(N-1)K} \\ 1 & \alpha^{-(K+1)} & \alpha^{-2(K+1)} & \cdots & \alpha^{-(N-1)(K+1)} \\ \vdots & \vdots & \vdots & \ddots & \vdots \\ 1 & \alpha^{-(N-1)} & \alpha^{-2(N-1)} & \cdots & \alpha^{-(N-1)(N-K-1)} \end{pmatrix} \begin{pmatrix} e_0 \\ e_1 \\ e_2 \\ \vdots \\ e_{N-1} \end{pmatrix}.$$

If the syndrome is not equal to zero, then an error has occurred. We note that the syndrome is the vector of the last $N - K = 2t$ components of E, that is, $S_1 = E_K$, $S_2 = E_{K+1}$, $S_{2t} = E_{N-1}$. The task now is to calculate the error vector from the syndrome.

Error locations

First, we must find the error positions, that is, the set of indices

$$\sigma = \{j \mid e_j = E(\alpha^j) \neq 0\}$$

corresponding to the nonzero elements of the error vector. The complement of σ is given by

$$\rho = \{j \mid e_j = E(\alpha^j) = 0\}.$$

We define the *error location polynomial*

$$C(x) = \prod_{j \in \sigma} (x - \alpha^j)$$

and the polynomial of error-free positions

$$D(x) = \prod_{j \in \rho} (x - \alpha^j).$$

By construction,

$$C(x)D(x) \equiv 0 \mod(x^N - 1)$$

holds. Since ρ corresponds to the zeros of $E(x)$, it can be factorized as $E(x) = T(x)D(x)$ with some polynomial $T(x)$. It follows that

$$C(x)E(x) \equiv 0 \mod(x^N - 1).$$

Assume that exactly t errors have occurred. The zeros of $C(x)$ are then given by α^{j_l}, $l = 1, \ldots, t$. We write

$$X_l = \alpha^{-j_l}$$

for their inverses. The error positions are given by

$$j_l = -\log_\alpha X_l.$$

We now renormalize the error location polynomial in such a way that the first coefficient equals one, that is, we define

$$\Lambda(x) = \prod_{l=1}^{t}(1 - \alpha^{-j_l}x) = \prod_{l=1}^{t}(1 - X_l x) = \Lambda_0 + \Lambda_1 x + \Lambda_2 x^2 + \cdots + \Lambda_t x^t$$

with $\Lambda_0 = 1$. Obviously, $C(x)$ and $\Lambda(x)$ have the same zeros and

$$\Lambda(x)E(x) \equiv 0 \mod(x^N - 1)$$

holds, which means $\Lambda * E = 0$ for the cyclic convolution of the vectors. We may write this componentwise as

$$\sum_{i+j=k\,(\mathrm{mod}\,N)} E_i \Lambda_j = 0 \quad \forall k \in \{0, 1, 2, \ldots, N-1\}.$$

We write down the last t of these N linear equations and obtain

$$\begin{array}{cccccccc}
S_1\Lambda_t + & S_2\Lambda_{t-1} + & \ldots & +S_t\Lambda_1 & +S_t\Lambda_0 & = & 0 \\
S_2\Lambda_t + & S_3\Lambda_{t-1} + & \ldots & +S_{t+1}\Lambda_1 & +S_{t+2}\Lambda_0 & = & 0 \\
\vdots & \vdots & \vdots & \vdots & \vdots & \vdots & \vdots \\
S_t\Lambda_t + & S_{t+1}\Lambda_{t-1} + & \ldots & +S_{2t-1}\Lambda_1 & +S_{2t}\Lambda_0 & = & 0
\end{array}.$$

From $\Lambda_0 = 1$, we obtain

$$\begin{pmatrix}
S_1 & S_2 & S_3 & \cdots & S_t \\
S_2 & S_3 & S_5 & \cdots & S_{t+1} \\
S_3 & S_4 & S_5 & \cdots & S_{t+2} \\
\vdots & \vdots & \vdots & \ddots & \vdots \\
S_t & S_{t+1} & S_{t+2} & \cdots & S_{2t-1}
\end{pmatrix}
\begin{pmatrix}
\Lambda_t \\ \Lambda_{t-1} \\ \Lambda_{t-2} \\ \vdots \\ \Lambda_1
\end{pmatrix}
= -
\begin{pmatrix}
S_{t+1} \\ S_{t+2} \\ S_{t+3} \\ \vdots \\ S_{2t}
\end{pmatrix}.$$

This system of linear equations can be solved by matrix inversion. If less than t errors have occurred, the matrix will be singular. In that case, the polynomial $\Lambda(x)$ will be of degree $t-1$ or less. Thus, we delete the first row and first column of the matrix and proceed this way until the remaining matrix is nonsingular. If the last equation $S_{2t-1}\Lambda_1 = -S_{2t}$ is still singular (i.e. $S_{2t-1} = 0$, but the syndrome is not equal to zero), then a decoding failure has occurred.

Once the coefficients of polynomial $\Lambda(x)$ have been found, we have to find the zeros X_l^{-1} of the polynomial. This will be simply done by evaluating the polynomial for all the N nonzero elements of $GF(q)$. This procedure is called *Chien search.* If less zeros than the degree of $\Lambda(x)$ or multiple zeros are found, then a decoding failure has occurred.

Error values

We are now ready to determine the values of the nonzero components of e. For simplicity, we consider the case that t errors have occurred. The treatment of less errors is similar. In the sum

$$E_i = \sum_{j=0}^{N-1} \alpha^{-ij} e_j$$

only those coefficients with $j \in \sigma$ occur, that is,

$$E_i = \sum_{l=1}^{t} \alpha^{-ij_l} e_{j_l} = \sum_{l=1}^{t} X_l^i e_{j_l}.$$

The syndrome coefficients $S_1 = E_K$, $S_2 = E_{K+1}$, $S_{2t} = E_{N-1}$ are known. We thus have $2t$ equations with t unknowns. We take only the first t of them, which leads to the system of linear equations given by

$$\begin{pmatrix} S_1 \\ S_2 \\ S_3 \\ \vdots \\ S_t \end{pmatrix} = \begin{pmatrix} X_1^K & X_2^K & X_3^K & \cdots & X_l^K \\ X_1^{K+1} & X_2^{K+1} & X_3^{K+1} & \cdots & X_l^{K+1} \\ X_1^{K+2} & X_2^{K+2} & X_3^{K+2} & \cdots & X_l^{K+2} \\ \vdots & \vdots & \vdots & \ddots & \vdots \\ X_1^{K+t-1} & X_2^{K+t-1} & X_3^{K+t-1} & \cdots & X_l^{K+t-1} \end{pmatrix} \begin{pmatrix} e_{j_1} \\ e_{j_2} \\ e_{j_3} \\ \vdots \\ e_{j_t} \end{pmatrix}.$$

The Vandermonde matrix is nonsingular and can thus be inverted, which provides us with the error vector e. The corrected code word will then be obtained as $a = r - e$.

3.4 Bibliographical Notes

For a more detailed treatment of channel coding, we refer to the text books (Blahut 1983; Bossert 1999; Clark and Cain 1988; Lin and Costello 1983; Wicker 1995). A delightful introduction into the conceptional ideas of channel coding can be found in the paper (Massey 1984). For a conceptional understanding of concatenated coding, we refer to the classical paper (Forney 1966). For a conceptional understanding of convolutional codes, we refer to (Forney 1970). It is interesting to note that this paper already described RSC encoders before they fell into oblivion for more than 30 years until turbo codes were discovered. The Viterbi algorithm, which is the MLSE for convolutional codes, has been developed by Viterbi (1967), even though the author did not point out that it is really the *optimum* MLSE receiver (note the word *asymptotically optimum* in the title of that paper). The conceptual understanding of the Viterbi algorithm as a MLSE receiver has been established by Forney (1973). Punctured convolutional codes can be found in classical text books (see e.g. (Clark and Cain

1988)). For their application, see (Hagenauer 1988). The concept of log-likelihood ratios is well established in probability theory. However, their usefulness for channel coding and their intuitively amazing visualization has been established in great parts by Hagenauer and coworkers (Hagenauer 1988, 1995; Hagenauer and Hoeher 1989; Hagenauer *et al.* 1996). It is interesting to note that the concept of soft-output decoding introduced as BCJR algorithm (Bahl *et al.* 1974) is nearly as old as the Viterbi algorithm, but seemingly it was too early for an application at that time. A much more popular soft-output decoder was established 25 years later (Hagenauer and Hoeher 1989) as the SOVA. The first turbo code simulations were done with the SOVA, but the BCJR algorithm turned out to be more efficient for that application. Turbo codes were introduced a decade ago (Berrou *et al.* 1993), a great step to their conceptional understanding is the paper (Hagenauer *et al.* 1996). RS codes have been developed by Reed and Solomon (1960). They are treated extensively in the text books cited above. Trellis coded modulation goes back to (Ungerboeck 1982). It is treated in the text books cited above (see also (Biglieri *et al.* 1991)). An interesting overview about their applications can be found in (Wicker and Bhargava 2001). Very enlightening and recommendable overviews about the application of channel coding for deep space communication can be found in (Massey 1992) and (McEliece and Swanson 2001). A concatenated coding scheme with an inner convolutional code and an outer RS code has been standardized for deep space communications (CCSDS 1987). Improvements for the corresponding decoder by using iterative decoding with reliability information are described in (Hagenaucr *et al.* 2001).

3.5 Problems

1. Prove the identity

$$\log\left(e^x + e^y\right) = \max(x, y) + \log\left(1 + e^{-|x-y|}\right).$$

2. Consider a transmission setup with four possible signals given by the columns of the matrix

$$\mathbf{X} = \begin{pmatrix} 1 & 1 & -1 & -1 \\ 1 & -1 & 1 & -1 \\ 1 & -1 & -1 & 1 \end{pmatrix},$$

which is an SPC(3, 2, 2) code with BPSK modulation. Find an example for which the ML receiver takes a different decision for the first bit than the MAP receiver.

3. Let $f(x, y)$ be a function of two real variables x and y defined by

$$f(x, y) = \log\left(\frac{1 + e^x e^y}{e^x + e^y}\right).$$

Show that

$$f(x, y) \approx \operatorname{sign}(x)\operatorname{sign}(y) \cdot \min(|x|, |y|)$$

if either $|x| \ll |y|$ or $|x| \gg |y|$.

4

OFDM

4.1 General Principles

4.1.1 The concept of multicarrier transmission

Let us consider a digital transmission scheme with linear carrier modulation (e.g. M-PSK or M-QAM) and a symbol duration denoted by T_S. Let B be the occupied bandwidth. Typically, B is of the order of T_S^{-1}, for example, $B = (1 + \alpha)T_S^{-1}$ for raised-cosine pulses with rolloff factor α. For a transmission channel with a delay spread τ_m, a reception free of intersymbol interference (ISI) is only possible if the condition

$$\tau_m \ll T_S$$

is fulfilled. As a consequence, the possible bit rate $R_b = \log_2(M)T_S^{-1}$ for a given single carrier modulation scheme is limited by the delay spread of the channel.

The simple idea of multicarrier transmission to overcome this limitation is to split the data stream into K substreams of lower data rate and to transmit these data substreams on adjacent *subcarriers*, as depicted in Figure 4.1 for $K = 8$. This can be regarded as a transmission parallel in the frequency domain, and it does not affect the total bandwidth that is needed. Each subcarrier has a bandwidth B/K, while the symbol duration T_S is increased by a factor of K, which allows for a K times higher data rate for a given delay spread. The factor K, however, cannot be increased arbitrarily, because too long symbol durations make the transmission too sensitive against the time incoherence of the channel that is related to the maximum Doppler frequency $\nu_{\max}$ (see the discussion in Section 2.2). There, we state that the condition

$$\nu_{\max} T_S \ll 1$$

must be fulfilled. Both conditions can only be valid simultaneously if the coherency factor $\kappa = \nu_{\max}\tau_m$ fulfills the condition $\kappa \ll 1$. For a given and sufficiently small factor κ, one should expect that there exists a symbol duration T_S that satisfies both requirements together to give the best possible transmission conditions for that channel. We may then choose this optimal symbol duration that is matched to the channel and parallelize the given data stream in an appropriate way.

Theory and Applications of OFDM and CDMA Henrik Schulze and Christian Lüders

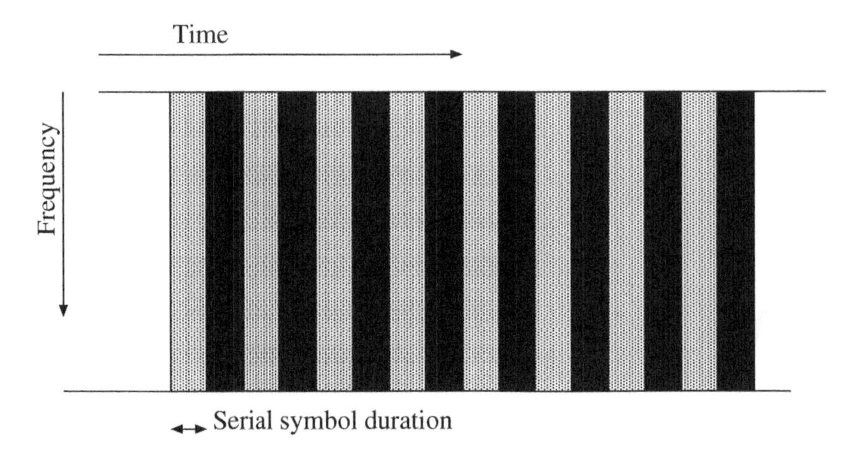

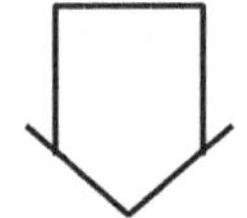

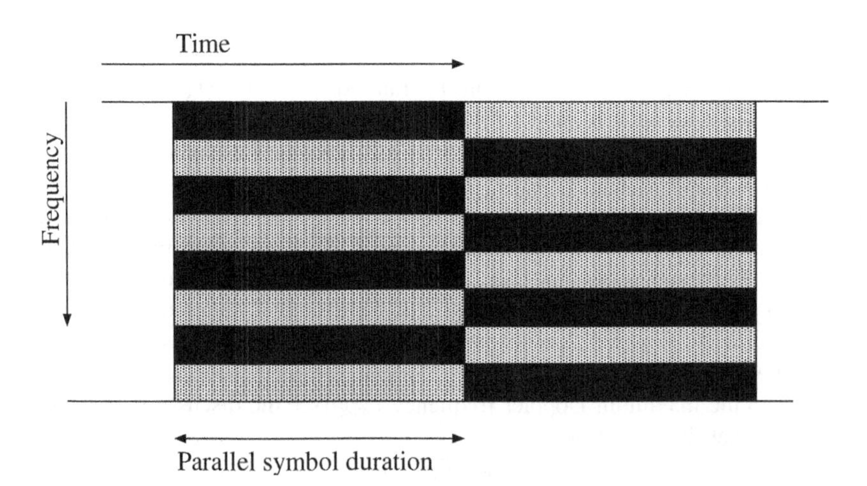

Figure 4.1 The multicarrier concept.

There are two possible ways to look at (and to implement) this idea of multicarrier transmission. Both are equivalent with respect to their transmission properties. Even though mathematically closely related, they differ slightly from the conceptual point of view. The first one emphasizes the multicarrier concept by having K individual carriers that are modulated independently. This concept is the favorite textbook point of view. The second one is based on a filter bank of K adjacent bandpass filters that are excited by a parallel data stream, leading to a transmission parallel in frequency. This concept is usually implemented in practical systems.

The first concept keeps the subcarrier frequency fixed and considers the modulation in time direction for each subcarrier. The second one keeps a time slot of length T_S fixed and considers modulation in frequency direction for each time slot.

In the first setup, the data stream is split up into K parallel substreams, and each one is modulated on its own subcarrier at frequency f_k in the complex baseband, described by the complex harmonic wave $\exp(j2\pi f_k t)$. We denote the complex (e.g. PSK or QAM) modulation symbols by s_{kl}, where k is the frequency index and l is the time index. With a baseband transmission pulse $g(t)$, this setup can be visualized by Figure 4.2: The parallel data stream excites replicas of the same pulse-shaping filter $g(t)$, and the filtered signals are modulated on the different carriers and summed up before transmission. The complex baseband signal is then given by the expression

$$s(t) = \sum_k e^{j2\pi f_k t} \sum_l s_{kl} g(t - lT_S), \tag{4.1}$$

where T_S is the parallel symbol duration. To keep the notation flexible, we do not specify the domain of the summation indices. If it is convenient, the time index l may run from zero or minus infinity to infinity. Since every real transmission starts and stops at some time instant, it is more realistic to let l run from 0 to $L-1$, where L is an integer. The frequency index may only run over a limited domain of, say, K different frequencies. From the mathematical point of view, we may choose $k = 0, 1, \ldots, K-1$. The engineer, however, would prefer to have f_0 in the middle, corresponding to DC in the complex baseband and to the center frequency f_c in the passband, with negative k for the lower sideband and positive k for the upper sideband. For reasons of symmetry, we may then choose the number of carriers to

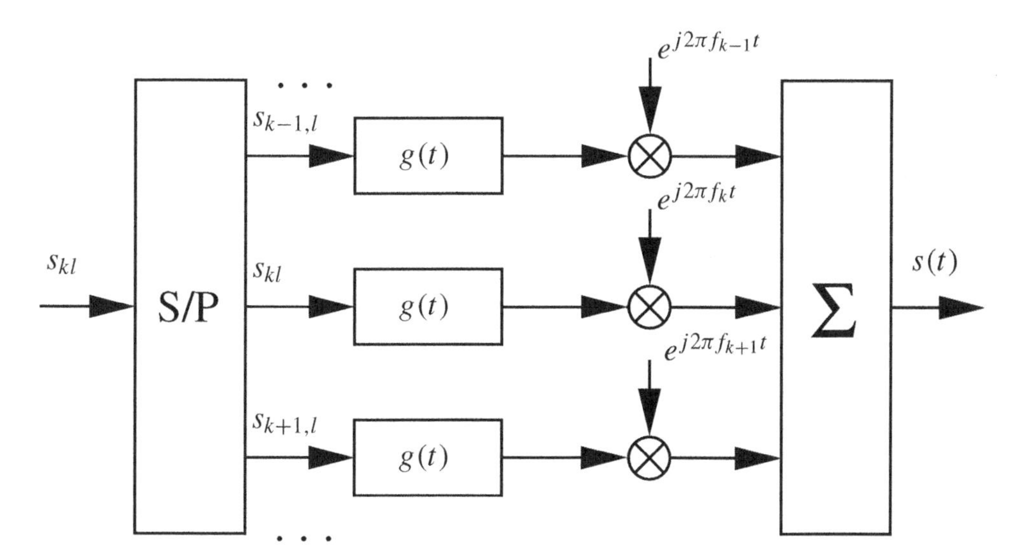

Figure 4.2 Block diagram for multicarrier transmission: Version 1.

be $K+1$, where K is an even integer, and let $k = 0, \pm 1, \pm 2, \ldots, \pm K/2$. The passband signal is then given by

$$\tilde{s}(t) = \Re\left\{\sqrt{2}\,\mathrm{e}^{j2\pi f_c t} s(t)\right\} = \Re\left\{\sqrt{2}\sum_k \mathrm{e}^{j2\pi(f_c+f_k)t} \sum_l s_{kl}\, g(t - lT_S)\right\}.$$

For reasons due to implementation, in practical systems, the DC component will sometimes be left empty, that is, only the subcarriers at $k = \pm 1, \pm 2, \ldots, \pm K/2$ are used.

In the second setup, we start with a base transmit pulse $g(t)$. We obtain frequency-shifted replicas of this pulse as

$$g_k(t) = \mathrm{e}^{j2\pi f_k t} g(t),$$

that is, if $g(t) = g_0(t)$ is located at the frequency $f = 0$, then $g_k(t)$ is located at $f = f_k$. In contrast to the first scheme, for each time instant l, the set of K (or $K+1$) modulation symbols is transmitted by using *different* pulse shapes $g_k(t)$: the parallel data stream excites a filter bank of K (or $K+1$) different bandpass filters. The filter outputs are then summed up before transmission. This setup is depicted in Figure 4.3. The transmit signal in the complex baseband representation is given by

$$s(t) = \sum_l \sum_k s_{kl}\, g_k(t - lT_S).$$

For the domain of the summation indices k and l, the same remarks apply as for the discussion of the first setup. We define

$$g_{kl}(t) = g_k(t - lT_S) = \mathrm{e}^{j2\pi f_k (t - lT_S)} g(t - lT_S)$$

to get the compact expression

$$s(t) = \sum_{kl} s_{kl}\, g_{kl}(t). \tag{4.2}$$

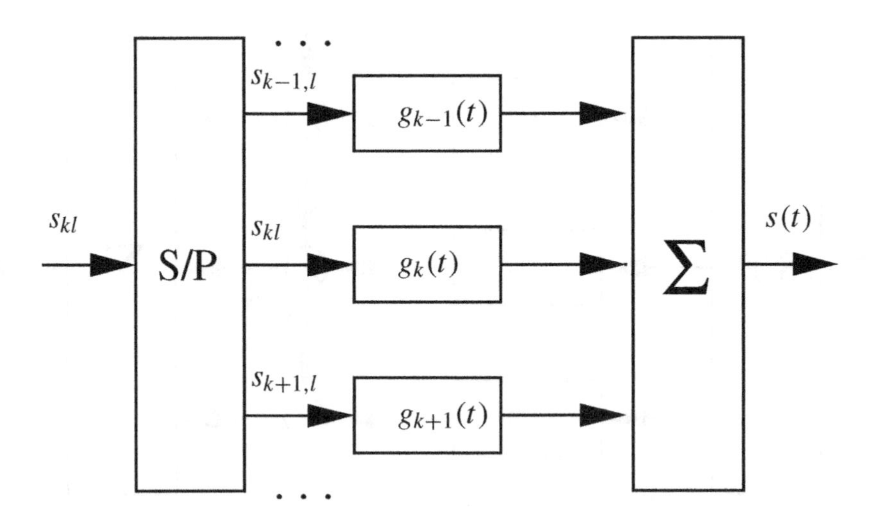

Figure 4.3 Block diagram for multicarrier transmission: Version 2.

It is obvious that we come back to the first setup if we replace the modulation symbols s_{kl} by $s_{kl}e^{-j2\pi f_k l T_S}$ in Equation (4.1). Such a time-frequency-dependent phase rotation does not change the performance, so both methods can be regarded as equivalent. However, the second – the filter bank – point of view is closer to implementation, especially for the case of OFDM, where the filter bank is just an FFT, as it will be later. In the following discussion, we will refer to the second point of view.

4.1.2 OFDM as multicarrier transmission

So far, nothing has been said about the shape of the base transmission pulse $g(t)$. In Chapter 1, we have seen that it is very convenient to have an orthogonal transmit base. It therefore seems to be quite natural to choose the $g_{kl}(t)$ of Equation (4.2) in such a way that they are orthogonal in time and frequency, that is, we require

$$\langle g_{kl}, g_{k'l'} \rangle = \delta_{kk'}\delta_{ll'}. \tag{4.3}$$

Nonorthogonal bases (e.g. Gaussian) are possible and may have interesting properties, see for example, (Kammeyer *et al.* 1992). We will restrict ourselves on pulses with the property (4.3). As discussed in depth in Chapter 1, orthogonality ensures that the modulation symbol can be recovered from the transmit signal without ISI, that is, the detector $\mathcal{D}_{kl}$ for $g_{kl}(t)$ has just the modulation symbol s_{kl} as its output:

$$\mathcal{D}_{kl}[s] = \langle g_{kl}, s \rangle = s_{kl}.$$

In principle, there are two obvious approaches to satisfy the orthogonality condition for multicarrier transmission. We recall that two pulses are always orthogonal if they do not overlap either in time or in frequency domain, and that a pulse cannot be strictly band limited *and* time limited. Thus, we must decide on one of these two options.

The first approach is seemingly the most straightforward one to implement the idea of multicarrier modulation. We choose band-limited pulses that are orthogonal in time. In Subsection 1.1.2, we defined time-orthogonal Nyquist bases. The most important examples for a strictly band-limited Nyquist base are the (square root) raised-cosine pulses. The bandwidth B is related to the rolloff factor α by $BT_S = 1 + \alpha$. Let $g(t)$ be such a pulse that is concentrated in the frequency domain around $f = 0$, so that we may write $g(t) = g_0(t)$, that is, this is the pulse corresponding to the frequency index $k = 0$. The pulses $g_{0l}(t) = g_0(t - lT_S)$ with $l \in \{0, \pm 1, \pm 2, \ldots.\}$ are a Nyquist base, that is, they satisfy the orthogonality condition

$$\langle g_{0l}, g_{0l'} \rangle = \delta_{ll'}$$

in the time domain. With

$$f_k = k\frac{1+\alpha}{T_S},$$

we define

$$g_k(t) = e^{j2\pi f_k t} g_0(t)$$

and

$$g_{kl}(t) = g_k(t - lT_S).$$

Since these pulses are strictly separated in frequency for different k, it is obvious that the condition (4.3) is fulfilled. This multicarrier modulation setup is depicted in Figure 4.4

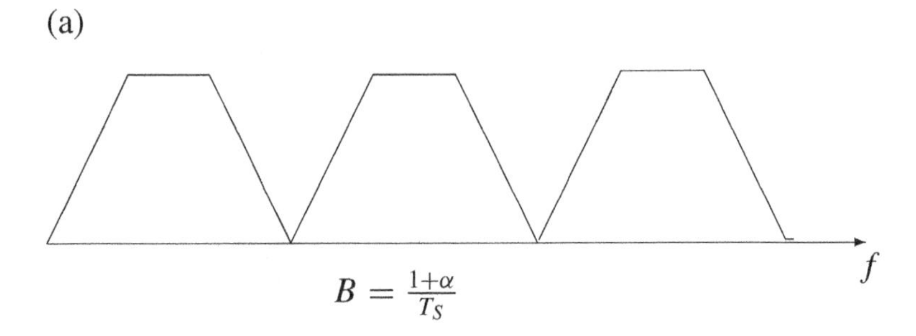

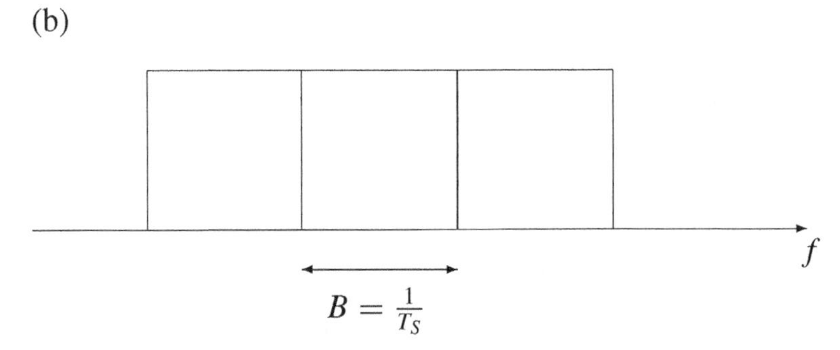

Figure 4.4 Multicarrier spectrum.

for $\alpha = 0.5$ and $\alpha = 0$. In this figure, we have replaced the raised-cosine spectrum by a trapezoidal one, which also corresponds to a Nyquist base. The case $\alpha = 0$ corresponding to an ideal rectangular spectral shape and sinc shaping in time domain and is spectrally most efficient, but not possible to be implemented in practice.

The second approach is to choose time-limited pulses that are orthogonal in frequency. In the example in Subsection 1.1.3, we have already seen that the time-limited complex exponentials of the Fourier series are such a base of time-limited orthogonal pulses. These are just the base pulses for OFDM transmission. However, there are more choices for time-limited orthogonal base pulses. We recall that the Nyquist base discussed in Subsection 1.1.3 fulfills just the condition that one base pulse and its periodically time-shifted replicas are orthogonal. Since the time and the frequency domain are mathematically equivalent, we may state the same orthogonality condition in the frequency domain. Doing this, we obtain strictly time-limited pulses $g_{kl}(t)$ that are orthogonal in frequency by the following construction: choose $g(t)$ to be a pulse that is strictly limited to the time interval[1] $[-T_S/2, T_S/2]$ of duration T_S in such a way that $|g(t)|^2$ has a raised-cosine shape with rolloff factor α. Let $G(f)$ be the pulse in the frequency domain. We define

$$f_k = k\frac{1+\alpha}{T_S}$$

[1] If this is more convenient, we may use the interval $[0, T_S]$ as well.

and the frequency-shifted pulse

$$g_k(t) = e^{j2\pi f_k t} g(t)$$

written in the frequency domain as

$$G_k(f) = G(f - f_k).$$

From the discussion of the Nyquist pulses (with time and frequency domain interchanged), it follows immediately that

$$\langle G_k, G_{k'} \rangle = \langle g_k, g_{k'} \rangle = \delta_{kk'}.$$

We define

$$g_{kl}(t) = g_k(t - lT_S).$$

Using the fact that these pulses are strictly separated in time for different l, it can easily be verified that the condition (4.3) is fulfilled. This multicarrier modulation setup is depicted in Figure 4.5 for $\alpha = 0$. This corresponds to the Fourier bases discussed above. Note that there is always a spectral overlap of subcarriers, but the carriers can be separated due to their orthogonality.

In contrast to the method discussed earlier, $\alpha = 0$ is possible to be implemented with a reasonable accuracy in practical systems. Even though every orthogonal multicarrier pulse transmission as discussed above deserves to be called OFDM (*orthogonal frequency division*

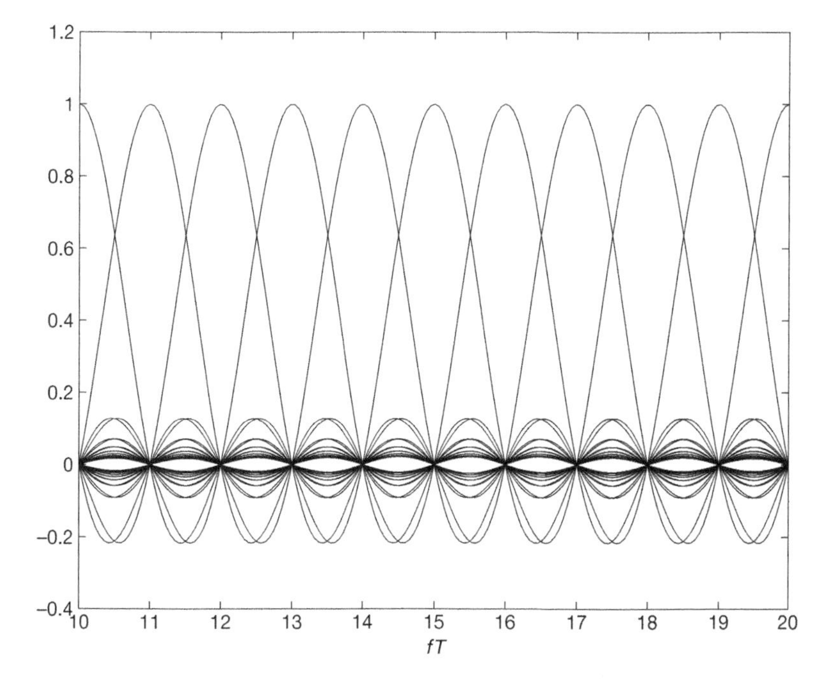

Figure 4.5 Orthogonal overlapping spectral shapes for OFDM.

multiplexing), it is indeed this case with $\alpha = 0$ which is the (narrow-sense) OFDM, because it is usually implemented. In that case, $g(t)$ is just a rectangle over an interval of length T_S, which we choose as $[0, T_S]$ for convenience. Then $f_k = k/T_S$, and the frequency-shifted pulses are just the Fourier base functions

$$g_k(t) = \sqrt{\frac{1}{T}} \exp\left(j2\pi \frac{k}{T} t\right) \Pi\left(\frac{t}{T} - \frac{1}{2}\right) \tag{4.4}$$

over the Fourier period of length $T = T_S$. OFDM transmission is therefore just a Fourier synthesis for every time interval, where the information is contained in the Fourier coefficients s_{kl}. For a receive signal $r(t)$, the detector outputs $\mathcal{D}_k[r] = \mathcal{D}_{g_k}[r]$ at frequency number k for $l = 0$ are just the results of the Fourier analysis given by

$$\mathcal{D}_k[r] = \langle g_k, r\rangle = \sqrt{\frac{1}{T}} \int_0^T \exp\left(-j2\pi \frac{k}{T} t\right) r(t)\,dt, \tag{4.5}$$

which exactly recovers s_{k0} for the ideal transmission channel with $r(t) = s(t)$. For any general l, $\mathcal{D}_{kl}[r] = \langle g_{kl}, r\rangle$ is the Fourier analyzer output for the frequency number k at the time interval shifted by lT_S. We note that for this narrow-sense OFDM the two concepts of Figures 4.2 and 4.3 are equivalent because $f_k = 1/T = 1/T_S$ holds. This property will be lost when a guard interval is introduced (see Subsection 4.1.4).

The power density spectrum of an OFDM signal for $K + 1 = 97$ subcarriers is depicted in Figure 4.6. On the linear scale, it looks very similar to a rectangular spectrum. However, the linear scale is indeed a very flattering presentation of the OFDM spectrum. Note that

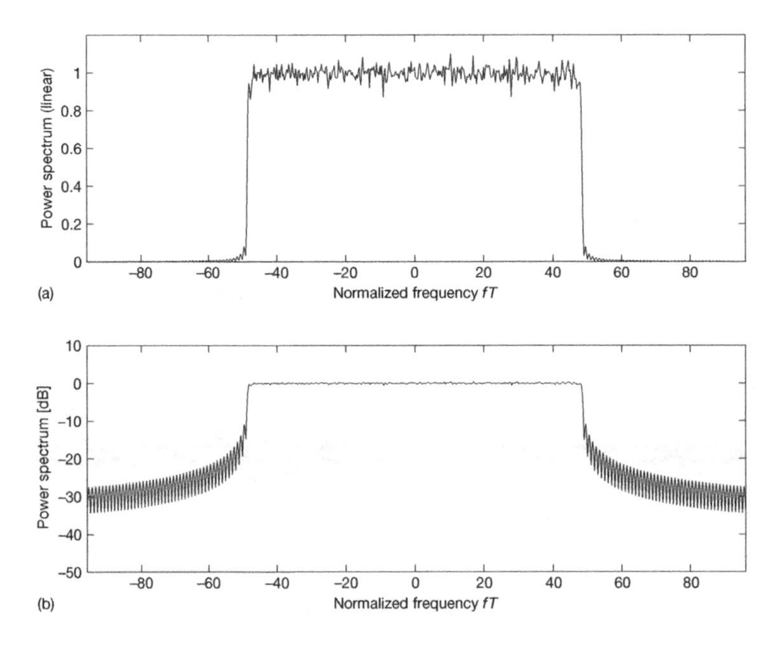

Figure 4.6 The power density spectrum of an OFDM signal on a linear scale (a) and on a logarithmic scale (b).

because of the rectangular pulse shape, at every subcarrier frequency $f_k = k/T$, the spectral shape is given by a sinc function. This can be seen very well at the edges of the spectrum. These edge effects are indeed a severe item and much care must be taken for them in practice. One usually reduces them by applying some filtering or smoothing (see Subsection 4.2.1). The relative size (compared to the bandwidth) of the edge effect region becomes smaller as the number of carriers K (or $K+1$) increases. Therefore, the bandwidth needed to transmit K complex symbols in a time slot of length T grows linearly as $B \approx K/T + C$ with some constant C that is due to the edge effects. Thus, ideal narrow-sense OFDM in the limit $K \to \infty$ has the same spectral efficiency as the ideal sinc pulse shaping in time domain.

4.1.3 Implementation by FFT

The narrow-sense OFDM with the Fourier base is very simple to implement. If we consider one time interval (e.g. that for $l = 0$), the transmit signal is given by

$$s(t) = \frac{1}{\sqrt{T}} \sum_{k=-K/2}^{K/2} s_k \exp\left(j2\pi \frac{k}{T} t\right) \Pi\left(\frac{t}{T} - \frac{1}{2}\right).$$

This means that, for each time interval of length T, OFDM is just a Fourier synthesis for that period. The perfectly synchronized receiver just performs a Fourier analysis to recover the data symbols s_k from the signal:

$$s_k = \langle g_k, s \rangle = \frac{1}{\sqrt{T}} \int_0^T \exp\left(-j2\pi \frac{k}{T} t\right) s(t)\, dt.$$

A Fourier analysis is preferably implemented by means of a fast Fourier transform (FFT), a synthesis by the inverse fast Fourier transform (IFFT), leading to a setup as depicted in Figure 4.7. The stream of digitally modulated symbols s_{kl} is divided into blocks of length K (or $K+1$), discretely Fourier transformed by the IFFT, digital–analog converted and then transmitted. The FFT length N_{FFT} must be chosen to be significantly larger than K to ensure that the edge effects are neglectible at half the sampling frequency and to ensure that the shape of the reconstruction filter of the DAC (digital-to-analog converter) does not affect the significant part of the spectrum. Furthermore, the alias spectra must be suppressed. To give a concrete example, in the European DAB (Digital Audio Broadcasting) and in the DVB-T (Digital Video Broadcasting-Terrestrial) system (EN300401 2001a; EN300744 2001b; Hoeg and Lauterbach 2003), an FFT with $N_{\mathrm{FFT}} = 2048$ is used (among other FFT modes), and the number of modulated carriers is of the order $K \approx 1500$ and $K \approx 1700$, respectively. The $N_{\mathrm{FFT}} - K$ remaining spectral coefficients outside the transmission band are set to zero. At the receiver, the baseband signal will be analog-to-digital converted. Then, for each block of N_{FFT} samples, an FFT of that length is performed, and the K useful coefficients will be extracted from the N_{FFT} spectral coefficients.

This picture is very suggestive from a practical point of view and one feels easily inclined to believe that it should work, because every block on the transmit site has its corresponding inverse on the receive site, so all the data should be perfectly recovered if every block works perfectly. Without explaining anything about orthogonality, the picture is also suited to convince a practical engineer that the concept of OFDM should work.

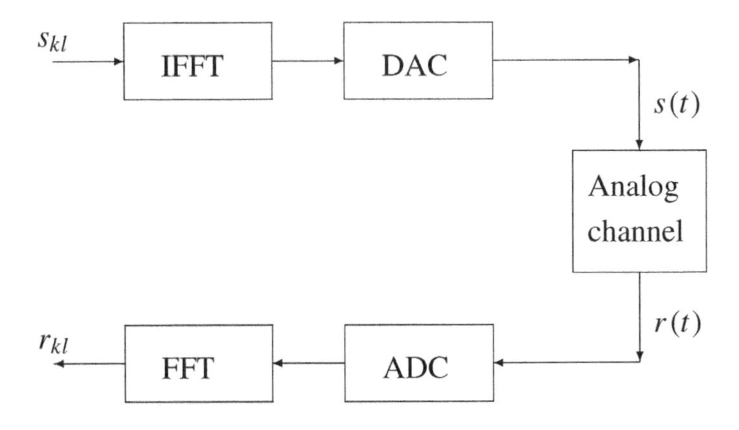

Figure 4.7 OFDM implementation by FFT.

However, the concept of orthogonality is only hidden in this picture under the cover of Fourier transform theory.

One should keep in mind that, for principal reasons, the block diagram of Figure 4.7 can never perfectly reflect the setup given by the theory. This is because the ideal OFDM signal is not strictly band limited due to the sinc shapes in the spectrum, while an analog signal can only be perfectly represented by its samples if it is strictly band limited. However, the problem of aliasing is a familiar one that occurs in many communications systems. For OFDM transmission, special care must be taken because of the poor spectral decay (see Subsection 4.2.1).

4.1.4 OFDM with guard interval

So far, we have always assumed perfect synchronization between transmitter and receiver. In a frequency-selective multipath fading channel, synchronization mismatches are typically of significant order, because every echo component of the signal is a poorly synchronized signal. As a consequence, the base pulses of the original OFDM signal and the delayed version of the signal are no longer orthogonal. This leads to severe intersymbol interference (ISI) in time and frequency as well because the detector output $\mathcal{D}_{kl}[s_\tau] = \langle g_{kl}, s_\tau \rangle$ at frequency number k and time slot l of the delayed signal $s_\tau(t) = s(t - \tau)$ with $0 < \tau < T$ has ISI contributions from pulses at all subcarrier frequencies at time slot l and $l - 1$. This property, which is a consequence of the loss of orthogonality due to the overlap of spectral components, would seemingly disqualify narrow-sense ODFM as a useful technique in a multipath channel. There is, however, a simple trick that modifies the transmit signal in such a way that the orthogonality is preserved in a certain manner in the presence of multipath signal components.

The idea is to introduce a *guard interval* (sometimes called *cyclic prefix*). By doing this, the symbol will be cyclically extended from the original harmonic wave of the Fourier period T by a guard interval of length Δ to become a harmonic of the same frequency and phase, but of duration $T_S = T + \Delta$. As depicted in Figure 4.8, this means that we copy a piece of length Δ from the end of the symbol and paste it in front of the signal. To express

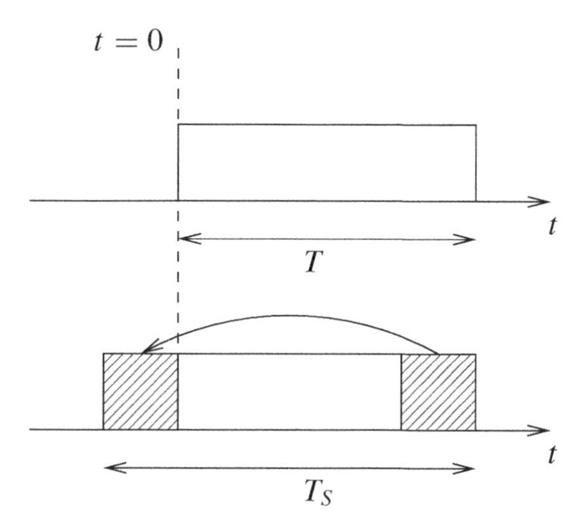

Figure 4.8 Introducing a guard interval.

this more formally, we replace the base pulse $g_k(t)$ as given by Equation (4.4) by a new base pulse defined by

$$g'_k(t) = \sqrt{\frac{1}{T_S}} \exp\left(j2\pi \frac{k}{T} t\right) \Pi\left(\frac{t+\Delta}{T_S} - \frac{1}{2}\right). \tag{4.6}$$

Note that the complex exponential remains exactly the same. The frequency is still $f_k = k/T$, and the phase is the same. Only the interval where the pulse does not vanish has been extended from $t \in [0, T)$ to $t \in [-\Delta, T)$. For convenience, we have chosen the factor in front in such a way that the energy of the pulse remains normalized to one. The transmit signal is then given by

$$s(t) = \sum_{kl} s_{kl} g'_{kl}(t) \tag{4.7}$$

with

$$g'_{kl}(t) = g'_k(t - lT_S). \tag{4.8}$$

We first note that these *transmit* pulses $g'_{kl}(t)$ by themselves are *not* pairwise orthogonal to each others. However, at the receiver, we work with a set of orthonormal *detector* pulses given by

$$g_{kl}(t) = g_k(t - lT_S),$$

where the $g_k(t)$ are still the Fourier base functions for the interval of length T as defined in Equation (4.4). This means that the *Fourier analysis* at the receiver works with the same analysis window of length T, but it will be performed once during the time period T_S instead of once during the time period T. As depicted in Figure 4.9, there is now a gap (or relaxation time) of length Δ between two adjacent analysis windows. We will see in the following text that it is just this gap together with the cyclically extended transmit pulse that allows a synchronization mismatch (and therefore, also echoes) of maximal duration

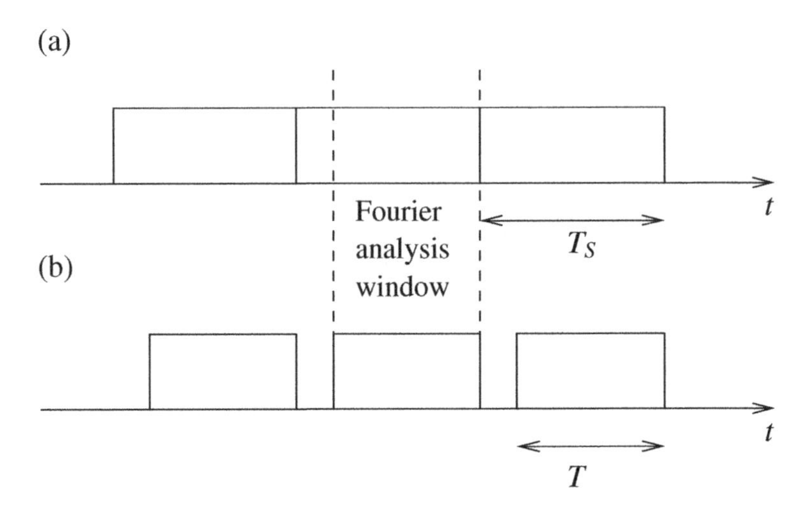

Figure 4.9 OFDM with guard interval: (a) Transmit pulses (b) Detector pulses.

$\tau = \Delta$, corresponding to a shift to the right of the signal in part (a) of Figure 4.9. The output of the detector for $g_{kl}(t)$, given that the pulse $g'_{k'l'}(t)$ has been transmitted, is

$$\langle g_{kl}, g'_{k'l'} \rangle = \sqrt{\frac{T}{T_S}}\, \delta_{kk'}\delta_{ll'}.$$

This means that the transmit base pulses $g_{kl}(t)$ and the detector base pulses $g'_{k'l'}(t)$ are orthogonal unless both the time and the frequency index are identical. Note that if they are identical the output does not take the value 1 but the smaller value $\sqrt{T/T_S}$. This can be understood as a waste of energy by transmitting a part of the symbol (i.e. the guard interval) that is not used for detection.

Now let $g'_{kl,\tau}(t) = g'_{kl}(t - \tau)$ with $0 < \tau < \Delta$ denote a base pulse delayed by τ. By writing down the corresponding integral, we easily see that

$$\langle g_{kl}, g'_{k'l',\tau} \rangle = \sqrt{\frac{T}{T_S}}\, \mathrm{e}^{-j2\pi f_k \tau}\, \delta_{kk'}\delta_{ll'}. \tag{4.9}$$

This means that – as long $\tau < \Delta$ holds – the orthogonality between the transmit and detect pulses for different indices is still preserved, and the detector output for the same index is only affected by a frequency-dependent phase factor. Now, let $s(t)$ be an OFDM signal given by Equation (4.7) and let $r(t) = s(t - \tau)$ with $0 < \tau < \Delta$ be the receive signal, which is just a delayed version of that signal. From the above equation and Equation (4.7), we obtain

$$\langle g_{kl}, r \rangle = \sqrt{\frac{T}{T_S}}\, \mathrm{e}^{-j2\pi f_k \tau}\, s_{kl},$$

that is, the transmit symbol s_{kl} is recovered without ISI, but only rotated by the phase factor. This phase factor cancels out for differential demodulation. For coherent demodulation, it must be determined by the channel estimation.

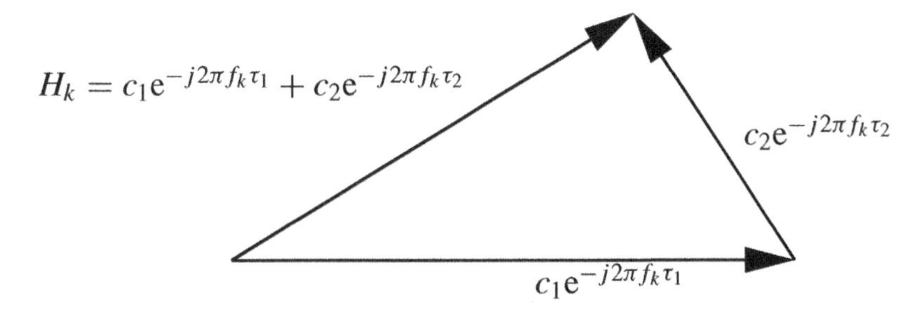

Figure 4.10 Signal plus echo.

If the received signal is the superposition of two delayed versions of the transmitted signal, that is, given by $r(t) = c_1 s(t - \tau_1) + c_2 s(t - \tau_2)$ with some complex constants c_1 and c_2, then s_{kl} will again be recovered without ISI if their delays do not exceed the guard interval. But they will be affected by a complex multiplicative factor $H_k = c_1 e^{-j2\pi f_k \tau_1} + c_2 e^{-j2\pi f_k \tau_2}$, which is the superposition of the phasors corresponding to the two echo paths (see Figure 4.10). The detector output is then given by

$$\mathcal{D}_{kl}[r] = \langle g_{kl}, r \rangle = \sqrt{\frac{T}{T_S}}\, H_k\, s_{kl}.$$

For a superposition of N such echo paths, we obtain the same expression with H_k given by

$$H_k = \sum_{n=1}^{N} c_n e^{-j2\pi f_k \tau_n}.$$

We now assume a time-variant channel given by a time-variant impulse response $h(\tau, t)$. We assume that $h(\tau, t) = 0$ for $\tau < 0$ and for $\tau > \Delta$. The corresponding time-variant transfer function $H(f, t)$ is then given by

$$H(f, t) = \int_0^{\Delta} e^{-j2\pi f \tau} h(\tau, t)\, d\tau.$$

The receive signal without noise is given by

$$r(t) = \int_0^{\Delta} h(\tau, t) s(t - \tau)\, d\tau.$$

We further assume that the channel is slowly time-variant so that it can be approximated to be time independent during the time slot number l, that is, $H(f, t) \approx H_l(f)$ and $h(\tau, t) \approx h_l(\tau)$ with

$$H_l(f) = \int_0^{\Delta} e^{-j2\pi f \tau} h_l(\tau)\, d\tau$$

during the OFDM symbol number l of length T_S. We now calculate the detector output for a base pulse transmitted over that channel, which is formally given by

$$\mathcal{D}_{kl}[h_{l'} * g'_{k'l'}] = \left\langle g_{kl}, h_{l'} * g'_{k'l'} \right\rangle.$$

This is the twofold integral

$$\mathcal{D}_{kl}[h_l * g'_{k'l'}] = \int_0^T \mathrm{d}t\, g^*_{kl}(t) \int_0^\Delta \mathrm{d}\tau\, h_{l'}(\tau) g'_{k'l'}(t-\tau)$$

with the first integral corresponding to the scalar product and the second one corresponding to the convolution. We change the order of integration and obtain

$$\mathcal{D}_{kl}[h_l * g'_{k'l'}] = \int_0^\Delta \mathrm{d}\tau\, h_{l'}(\tau) \int_0^T \mathrm{d}t\, g^*_{kl}(t) g'_{k'l'}(t-\tau).$$

The second integral is just $\langle g_{kl}, g'_{k'l',\tau}\rangle$. From Equation (4.9) we get

$$\mathcal{D}_{kl}[h_l * g'_{k'l'}] = \int_0^\Delta \mathrm{d}\tau\, h_{l'}(\tau) \sqrt{\frac{T}{T_S}}\, \mathrm{e}^{-j2\pi f_k \tau}\, \delta_{kk'} \delta_{ll'}$$

that is,

$$\mathcal{D}_{kl}[h_l * g'_{k'l'}] = \sqrt{\frac{T}{T_S}}\, H_l(f_k)\, \delta_{kk'} \delta_{ll'}.$$

The detector output at time l and frequency k for the noise-free receive signal $r(t)$ is then given by

$$\langle g_{kl}, r\rangle = \sqrt{\frac{T}{T_S}}\, H_l(f_k)\, s_{kl}. \tag{4.10}$$

We define $r_{kl} = \langle g_{kl}, r\rangle$ and $c_{kl} = H_l(f_k)$. Then, the OFDM transmission with guard interval in a noisy slowly fading channel can be described by the discrete channel model

$$r_{kl} = \sqrt{\frac{T}{T_S}}\, c_{kl} s_{kl} + n_{kl}, \tag{4.11}$$

where n_{kl} is discrete complex AWGN with variance $\sigma^2 = \mathrm{E}\left\{|n_{kl}|^2\right\} = N_0$. This is just the same as the discrete-time multiplicative fading channel that has been analyzed in Section 2.4, but with an additional second index for the frequency. The fading amplitude c_{kl} is typically modeled as Rayleigh or Ricean fading. We assume a channel transfer power normalized to one, that is, $\mathrm{E}\left\{|c_{kl}|^2\right\} = 1$. Note the factor $\sqrt{T/T_S}$, which means that there is an energy loss in performance because a part of the signal available at the receiver is not evaluated. All Euclidean distance in the expressions for error probabilities will be lowered by that factor, that is,

$$\frac{1}{2}\operatorname{erfc}\left(\sqrt{\frac{1}{N_0}\sum_{kl} |c_{kl}|^2\, |s_{kl} - \hat{s}_{kl}|^2}\right)$$

must be replaced by

$$\frac{1}{2}\operatorname{erfc}\left(\sqrt{\frac{1}{N_0}\frac{T}{T_S}\sum_{kl} |c_{kl}|^2\, |s_{kl} - \hat{s}_{kl}|^2}\right)$$

and thus all performance curves as functions of E_b/N_0 or E_S/N_0 have to be shifted by $10\log_{10}(T/T_S)$ decibels to the right. For the typical value $T/T_S = 0.8$, this loss is approximately 1 dB. Because the base pulses $g'_{kl}(t)$ are normalized according to $\left\|g'_{kl}\right\|^2 = 1$, we still have $E_S = \mathrm{E}\left\{|s_{kl}|^2\right\}$, and the SNR,

$$SNR = \frac{\mathrm{E}\left\{\left|\sqrt{\frac{T}{T_S}} s_{kl} c_{kl}\right|^2\right\}}{\mathrm{E}\left\{|n_{kl}|^2\right\}}$$

at the receiver is given by

$$SNR = \frac{T}{T_S} \cdot \frac{E_S}{N_0} = \frac{T}{T_S} \cdot R_c \log_2(M) \frac{E_b}{N_0} \tag{4.12}$$

for a modulation with $\log_2(M)$ bits per complex symbol and a code rate of R_c. This means that the performance curves *as a function of the SNR* will be left unchanged by the guard interval.

We summarize and add the following remarks:

- The time period of length T in Figure 4.9 that is used for the Fourier analysis at the detector will be called the *Fourier analysis window*. The spacing between two adjacent subcarriers is given by $\Delta f = f_k - f_{k-1} = T^{-1}$.

- $T_S = T + \Delta$ is the symbol period for each subcarrier. Thus, for each fixed index k, the symbol rate of the transmit symbols s_{kl} is given by T_S^{-1}.

- In contrast to OFDM without guard interval, the two concepts of Figures 4.2 and 4.3 are not equivalent because $f_k = 1/T \neq 1/T_S$. The FFT implementation of OFDM corresponds to that of Figure 4.3. To switch to the concept of Figure 4.2, we must multiply each modulation symbol s_{kl} by $\mathrm{e}^{-j2\pi klT/T_S}$. As there is no advantage in doing that, it is not implemented in any real system.

- The part of the signal transmitted during each time period T_S is called an *OFDM symbol*[2]. Each OFDM symbol corresponds to a number of K transmit symbols s_{kl}. Thus, the total symbol rate is given by $R_S = K\,T_S^{-1}$. Ignoring any other overhead (e.g. for synchronization), the useful bit rate is $R_b = K \cdot R_c \log_2(M) T_S^{-1}$, where we have assumed that M-PSK or M-QAM with code rate R_c is used for modulation and channel coding.

- The FFT length N is typically (but not necessarily) given by the smallest power of two that satisfies $K < N$. For example, the so-called *2k mode* of the DVB-T system has $K + 1 = 1705$, so $N = 2048$ is the smallest possible FFT length (which explains the name of the mode). However, one could also use an (I)FFT with $N = 4096$ for the Fourier analysis or synthesis. Such an oversampling may be useful for several purposes (see Subsection 4.2.1). Note that the FFT length is an implementation parameter for transmitter and receiver and it is not relevant for the description of the OFDM signal in the air.

[2]Here we have adopted the terminology used in the specification of the European DAB system.

4.2 Implementation and Signal Processing Aspects for OFDM

4.2.1 Spectral shaping for OFDM systems

In this subsection, we will discuss the implementation aspects that are related to the spectral properties of OFDM. We consider an OFDM system with subcarriers at frequency positions in the complex baseband given by $f_k = k/T$ with frequency index $k \in \{0, \pm 1, \pm 2, \ldots, \pm K/2\}$. As already discussed in Subsection 4.1.2, the subcarrier pulses in the frequency domain are shaped like sinc functions that superpose to a seemingly rectangular spectrum located between $-K/T$ and $+K/T$. However, as depicted in Figure 4.6, there is a severe out-of-band radiation outside this main lobe of the OFDM spectrum caused by the poor decay of the sinc function. That figure shows the spectrum of an OFDM signal without guard interval. The guard interval slightly modifies the spectral shape by introducing ripples into the main lobe and reducing the ripples in the side lobe. However, the statements about the poor decay remain valid. Figure 4.11 shows such an OFDM spectrum with $K = 96$. Here and in the following discussion, the guard interval length $\Delta = T/4$ has been chosen.

The number of subcarriers has a great influence on the decrease of the sidelobes. For a given main lobe bandwidth $B = K/T$, the spectrum of each individual subcarrier – including its side lobes – becomes narrower with increasing K. As a consequence,

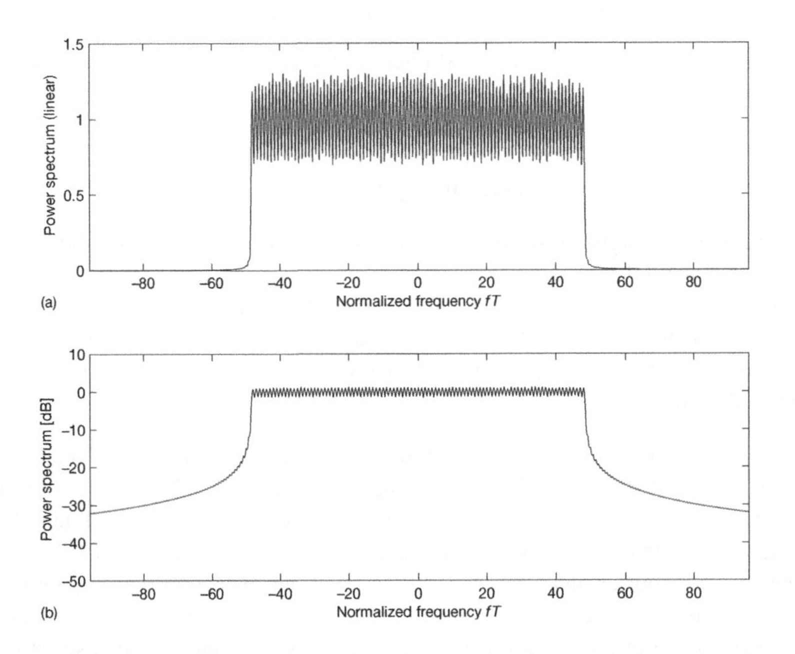

Figure 4.11 The power density spectrum of an OFDM signal with guard interval on a linear scale (a) and on a logarithmic scale (b).

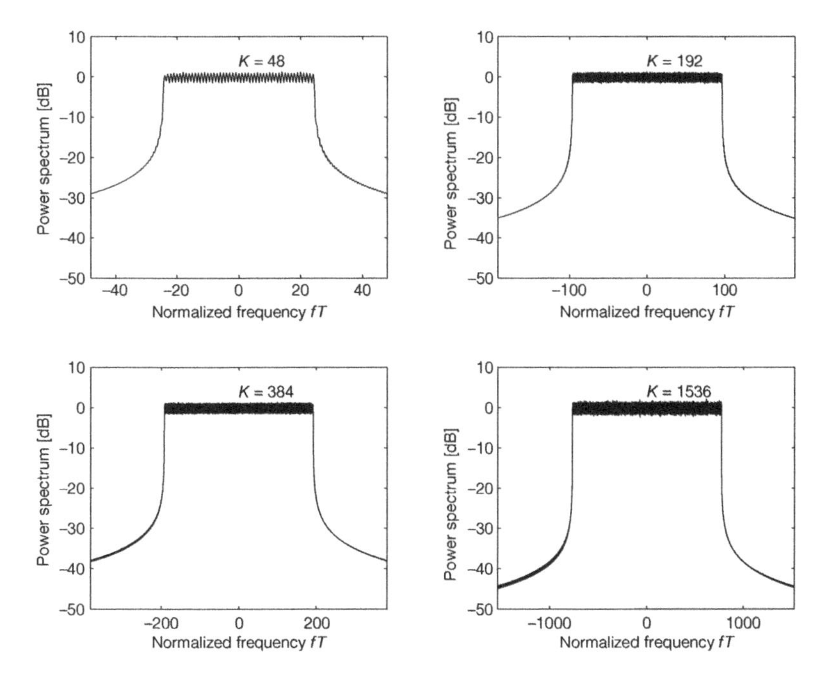

Figure 4.12 The power density spectra of an OFDM for $K = 48, 96, 384, 1536$.

the side lobes of the complete OFDM spectrum show a steeper decay and the spectrum comes closer to a rectangular shape. Figure 4.12 shows the OFDM spectra for $K = 48, 192, 384, 1536$. But, even for a high number of K, the decay may still not be sufficient to fulfill the network planning requirements. These are especially strict for broadcasting systems, where side lobe reduction in the order of -70 dB are mandatory. In that case, appropriate steps must be taken to reduce the out-of-band radiation.

We note that the spectra shown in the figures correspond to continuous OFDM signals[3].

Digital-to-analog conversion

In practice, discrete-time OFDM signals will be generated by an inverse discrete (fast) Fourier transform and then processed by a digital-to-analog converter (DAC). It is well known from signal processing theory that a discrete-time signal has a periodic spectrum from which the analog signal has to be reconstructed at the DAC by a low-pass filter (LPF) that suppresses these aliasing spectra beyond half the sampling frequency $f_s/2$. Figure 4.13(a) shows the periodic spectrum of a discrete OFDM signal with $K = 96$ and an FFT length $N = 128$, which is the lowest possible value for that number of subcarriers. The LPF must be flat inside the main lobe (i.e. for $|f| \leq 48/T$) and the side lobe must decay steeply enough so that the alias spectra at $|f| \geq 80/T$ will be suppressed. This analog filter is always a complexity item. It is a common practice to use oversampling

[3]The spectra shown above are computer simulations and not measurements of a continuous OFDM signal. However, the signal becomes quasi-continuous if the sampling rate is chosen to be high enough.

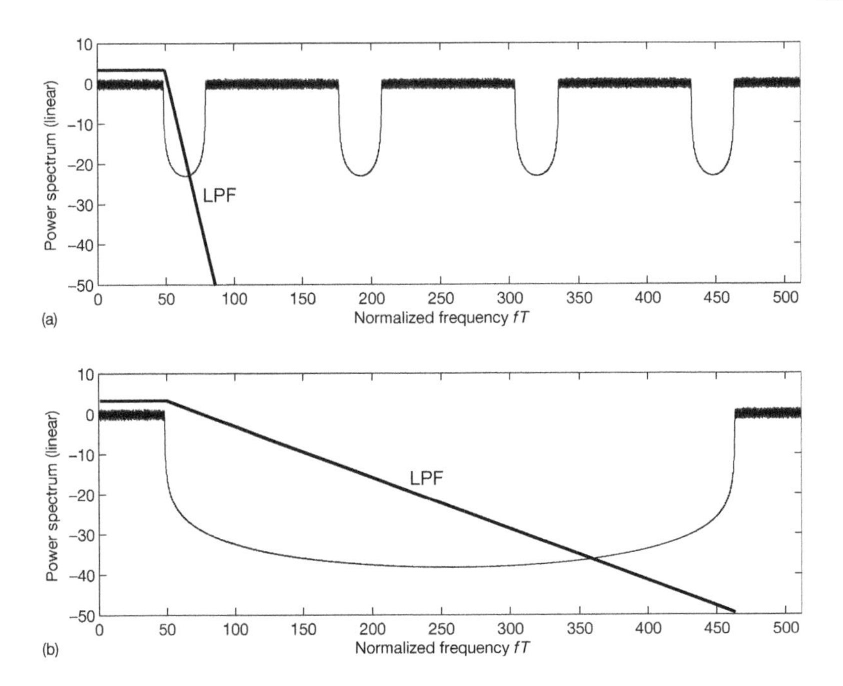

Figure 4.13 The periodic power density spectra of a discrete OFDM signal for $K = 96$ and FFT length $N = 128$ (a) and FFT length $N = 512$ (b).

to move complexity from the analog to the digital part of the system. Oversampling can be implemented by using a higher FFT length and padding zeros at the unused carrier positions[4]. Figure 4.13(b) shows the discrete spectrum for the same OFDM parameters with fourfold oversampling, that is, $N = 4 \cdot 128 = 512$ and $f_s/2 = 256/T$. Now the main lobe of the next alias spectrum starts at $|f| = 464/T$ and the requirements to the steepness of the LPF can be significantly reduced.

We finally note that since the signal is not strictly band limited, any filtering will always hurt the useful signal in some way because the sidelobes are a part of the signal, even though not the most significant.

Reduction of the out-of-band radiation

For a practical system, network planning aspects require a certain spectral mask that must not be exceeded by the implementation. Typically, this spectrum mask defined by the specification tells the maximal allowed out-of-band radiation at a given frequency. Figure 4.14 shows an example of such a spectrum mask similar to the one that is used for a wireless LAN system. The frequency is normalized with respect to the main lobe bandwidth $B = K/T$, that is, the main lobe is located between the normalized frequencies -0.5 and $+0.5$. We note that such a spectrum mask for a wireless LAN system is relatively loose compared, for example, to those for terrestrial broadcast systems like DAB and DVB-T.

[4]Alternatively, one may use the smallest possible FFT together with a commercially available oversampling circuit. This will be the typical implementation in a real system.

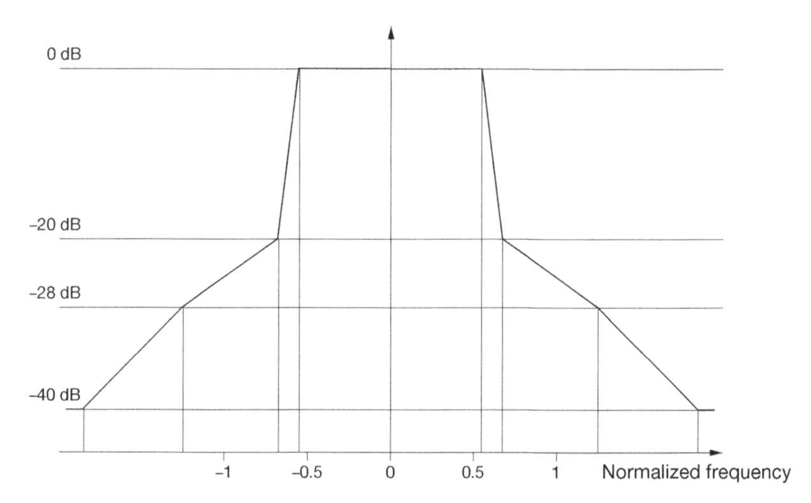

Figure 4.14 Example for the spectrum mask of an OFDM system as a function of the normalized bandwidth f/B.

To fulfill the requirements of a spectrum mask, it is often necessary to reduce the sidelobes. This can be implemented by – preferably digital – filtering.

As an example, we use a digital Butterworth filter to reduce the sidelobes of an OFDM signal with $K = 96$ and $N = 512$ (fourfold oversampling). To avoid significant attenuation or group delay distortion inside the main lobe, we choose a 3 dB filter bandwidth $f_{3\,\mathrm{dB}} = 64/T$. For this filter bandwidth, the amplitude is approximately flat and the phase is nearly linear within the main lobe. Figure 4.15 shows the OFDM spectrum filtered by a digital Butterworth filter of 5th and 10th order. As an example, let us assume that the spacing between two such OFDM signals inside a frequency band is $128/T$, that is, the lowest possible sampling frequency. Then, the main lobe of the next OFDM signal would begin at $(\pm)\,80/T$. The out-of-band radiation at this frequency is reduced from -30 to -41 dB for the 5th order filter and to -52 dB for the 10th order filter.

One must keep in mind that any filtering will influence the signal. The rectangular pulse shape of each OFDM subcarrier will be smoothened and broadened by the convolution with the filter impulse response. The guard interval usually absorbs the resulting ISI, but this reduces the capability of the system to cope with physical echoes. Thus, the effective length of the guard interval will be reduced. Figure 4.16 shows the respective impulse responses of both filters that we have used. We recall that for $N = 512$, the guard interval is $N/4 = 128$ samples long. The filter impulse responses reduce the effective guard interval length by 10–20%.

Instead of low-pass filtering, one may also form the spectral shape by smoothing the shape of the rectangular subcarrier pulse. This can be done as described in the following text. We first cyclically extend the OFDM symbol *at the end* by δ to obtain a harmonic wave of symbol length $T_S + \delta$. We then choose a smoothing window that is equal to one for $-\Delta + \delta < t < T$ and decreases smoothly to zero outside that interval (see Figure 4.17). The (cyclically extended) OFDM signal will then be multiplied by this window. The signal remains unchanged within $-\Delta + \delta < t < T$, that is, the effective guard interval will be

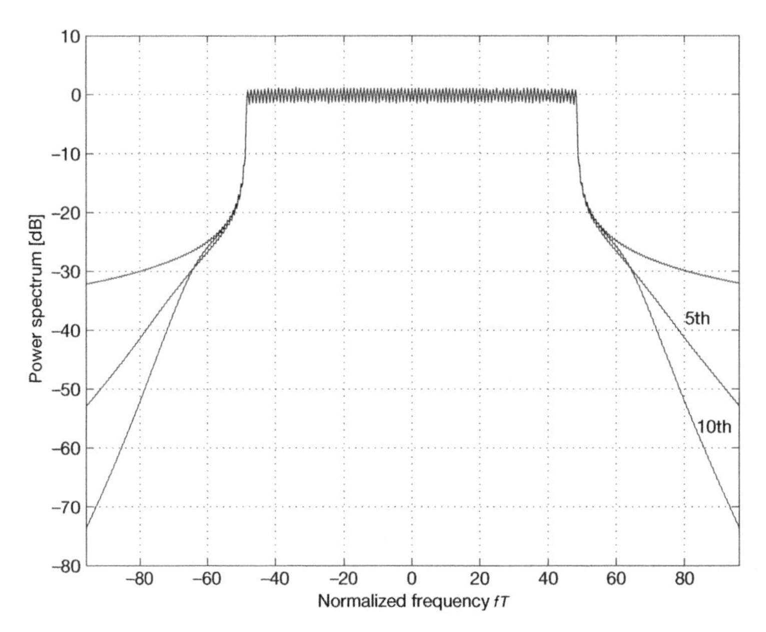

Figure 4.15 OFDM spectrum filtered by a digital Butterworth filter of 5th and 10th order.

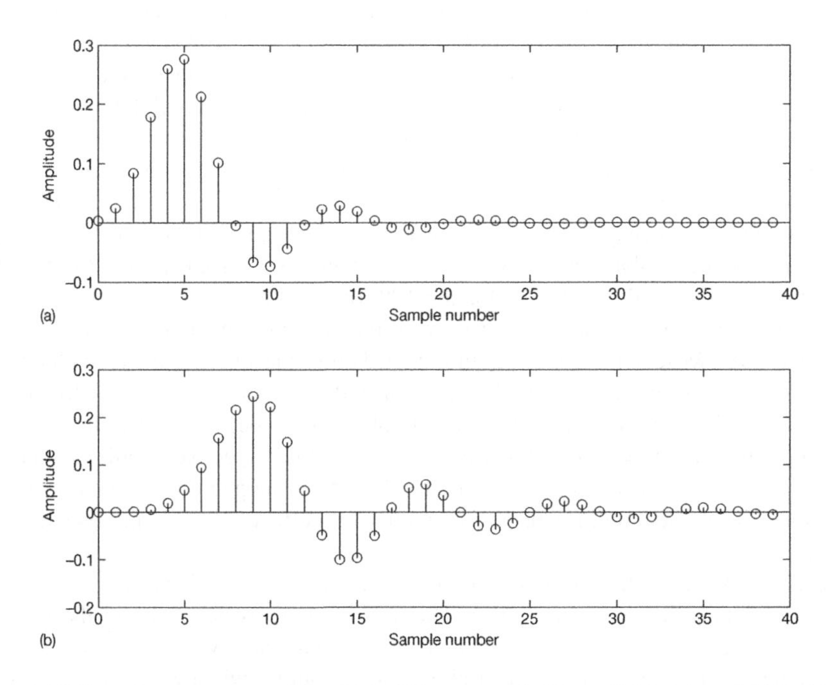

Figure 4.16 Impulse response of the digital Butterworth filter of 5th (a) and 10th (b) order.

Figure 4.17 Smoothing window for the OFDM symbol.

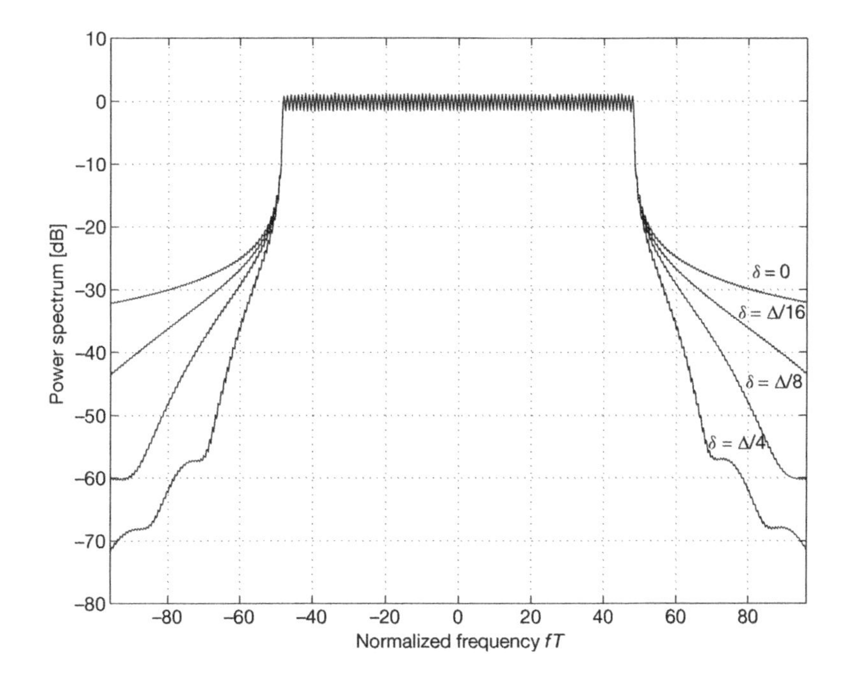

Figure 4.18 OFDM spectrum for a smoothened subcarrier pulse shape.

reduced by δ. We choose a raised-cosine pulse shape (Schmidt 2001). For the digital implementation, the flanks are just the increasing and decreasing flanks of a discrete Hanning window. Figure 4.18 shows the OFDM spectra for $\delta = 0$, $\Delta/16$, $\Delta/8$, $\Delta/4$. The out-of-band power reduction is similar to that of digital filtering.

We finally show the efficiency of the windowing method for an OFDM signal with a high number of carriers. Figure 4.19 shows the OFDM spectra for $K = 1536$ and $\delta = 0$, $\Delta/16$, $\Delta/8$, $\Delta/4$. We note a very steep decay for the out-of-band radiation. Even a small reduction of the guard interval is enough to fulfill the requirements of a broadcasting system[5]. Similar results can be achieved by digital filtering. However, this would require higher-order filters with more computational complexity and a smaller 3 dB bandwidth. Thus, the method of pulse shape smoothing seems to be the better choice.

[5]The DAB system with $K = 1536$ requires a -71 dB attenuation at $fT = 970$ for the most critical cases.

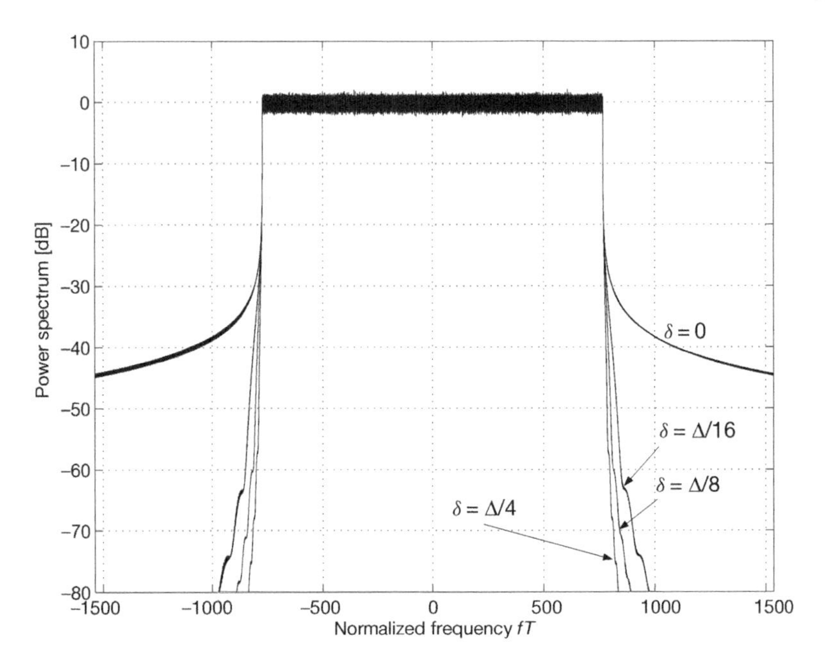

Figure 4.19 OFDM spectrum for a smoothened subcarrier pulse shape ($K = 1536$).

4.2.2 Sensitivity of OFDM signals against nonlinearities

As we have already seen, OFDM signals in the frequency domain look very similar to band-limited white noise. The same is true in the time domain. Figure 4.20 shows the inphase component $I(t) = \Re\{s(t)\}$, the quadrature component $Q(t) = \Im\{s(t)\}$ and the amplitude $|s(t)|$ of an OFDM signal with subcarriers at frequency positions in the complex baseband given by $f_k = k/T$ with $k \in \{0, \pm 1, \pm 2, \ldots, \pm K/2\}$ and $K = 96$ and the guard interval length $\Delta = T/4$. We will further use these OFDM parameters in the following discussion.

Because the inphase and the quadrature component the OFDM are superpositions of many sinoids with random phases, one can argue from the central limit theorem that both are Gaussian random processes. A *normplot* is an appropriate method to test whether the samples of a signal follow Gaussian statistics. To do this, one has to plot the (measured) probability that a sample is smaller than a certain value as a function of that value. The probability values are then scaled in such a way that a Gaussian normal distribution corresponds to a straight line. Figure 4.21 shows such a normplot for the OFDM signal under consideration. We note that the measurements fit quite well to the straight line that corresponds to the Gaussian normal distribution. However, there are deviations for high amplitudes. This is due to the fact that the number of subcarriers is not very high ($K = 96$) and the maximum amplitude of their superposition cannot exceed a certain value. For an increasing number of subcarriers, the measurements follow closely the straight line. For a lower value of K, the agreement becomes poorer. The crest factor $C_s = P_{s,\max}/P_{s,\mathrm{av}}$ is defined as the ratio (usually given in decibels) between the maximum signal power $P_{s,\max}$ and the average signal power $P_{s,\mathrm{av}}$. With $K \to \infty$, the amplitude of an OFDM signal is

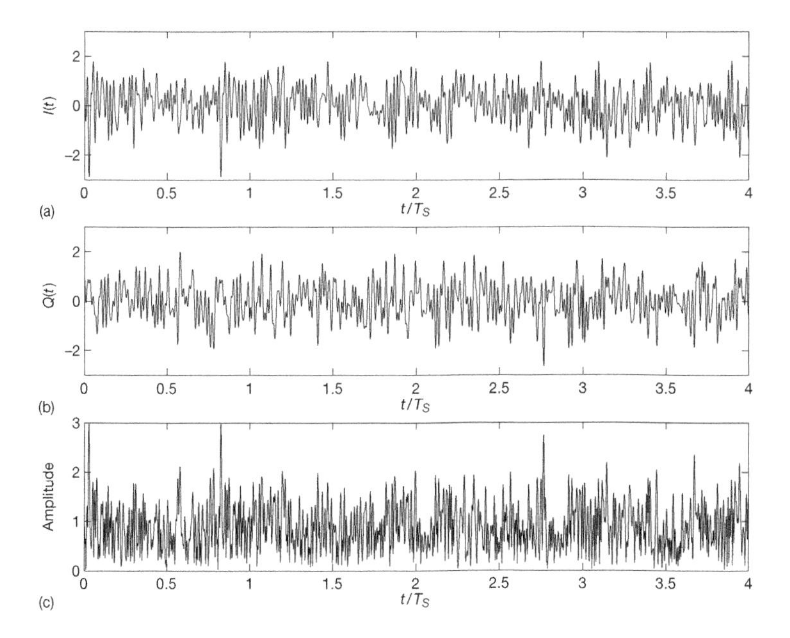

Figure 4.20 The inphase component $I(t)$ (a), the quadrature component $Q(t)$ (b) and the amplitude (c) of an OFDM signal of average power one.

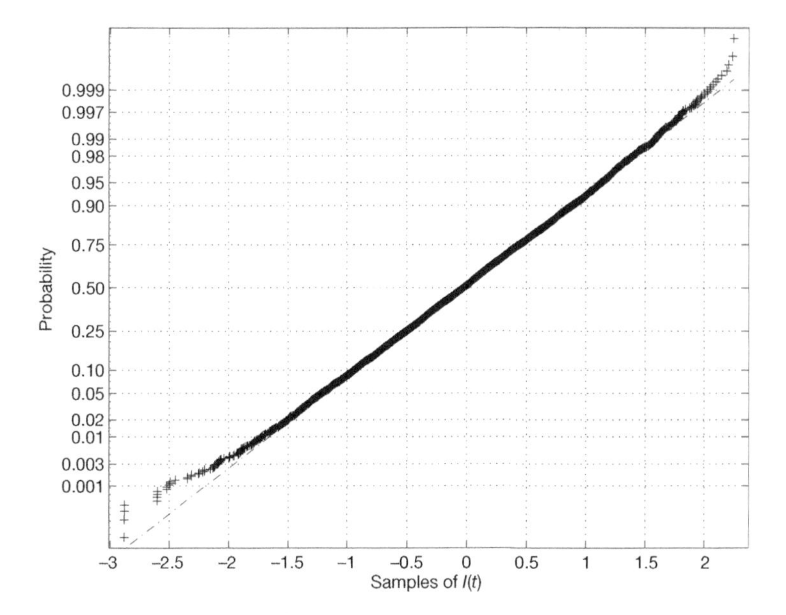

Figure 4.21 Normal probability plot for the inphase component $I(t)$ of an OFDM signal.

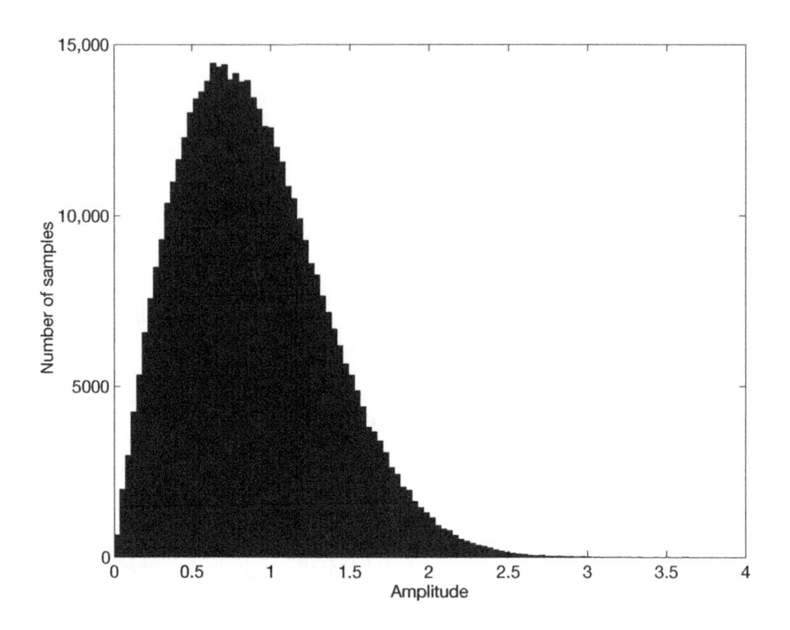

Figure 4.22 Histogram for the amplitude of an OFDM signal.

a Gaussian random variable and the crest factor becomes infinity. Even for a finite (high) number of subcarriers, the crest factor is so high that it does not make sense to use it to characterize the signal. This is because the probability of extremely high-power values decreases exponentially with increasing power.

As discussed in detail in Chapter 3, a normal distribution for the I and Q component of a signal leads to a Rayleigh distribution for the signal amplitude. Figure 4.22 shows the histogram for the amplitude of the OFDM signal under consideration.

We now consider an OFDM complex baseband signal

$$s(t) = a(t)\mathrm{e}^{j\varphi(t)}$$

with amplitude $a(t)$ and phase $\varphi(t)$ that passes a nonlinear amplifier with power saturation as depicted in Figure 4.23. For low values of the input power, the output power grows approximately linear. For intermediate values, the output power falls below that linear growth and it runs into a saturation as the input power grows higher. In addition to that *smooth* nonlinear amplifier, we consider a *clipping* amplifier. This amplifier is linear as long as the input power is smaller than a certain value $P_{\mathrm{in,max}}$ corresponding to the maximum output power $P_{\mathrm{out,max}}$. If the input exceeds $P_{\mathrm{in,max}}$, the output will be *clipped* to $P_{\mathrm{out,max}}$. As depicted in Figure 4.23, for any nonlinear amplifier with power saturation, there is a uniquely defined clipping amplifier with the same linear growth for small input amplitudes and the same saturation (maximum output). For an input signal with average power $P_{s,\mathrm{av}}$, the *input backoff* $IBO = P_{\mathrm{in,max}}/P_{s,\mathrm{av}}$ is defined as the ratio (usually given in decibels) between the power $P_{\mathrm{in,max}}$ and the average signal power $P_{s,\mathrm{av}}$.

The nonlinear amplifier output in the complex baseband model is given by

$$r(t) = F\left(a(t)\right)\mathrm{e}^{j(\varphi(t)+\Phi(a(t)))}$$

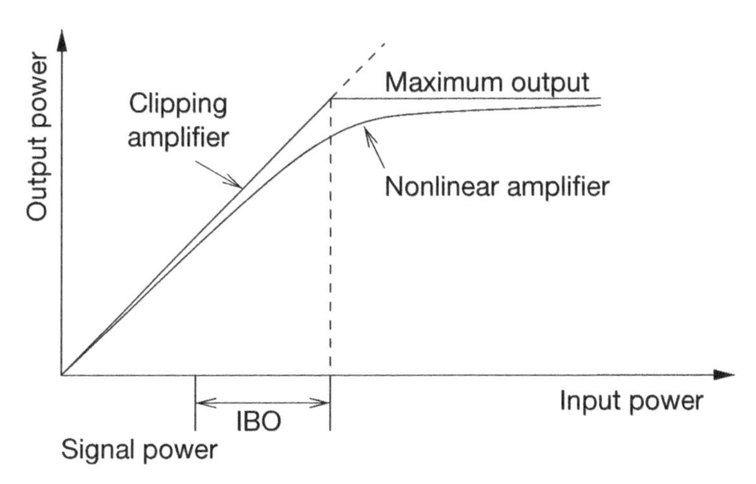

Figure 4.23 Characteristic curves for nonlinear amplifiers with power saturation.

(see (Benedetto and Biglieri 1999)). The real-valued function $F(x)$ is the characteristic curve for the amplitude distortion, and the real-valued function $\Phi(x)$ describes the phase distortion caused by the nonlinear amplifier.

To see how nonlinearities influence an OFDM signal, we consider a very simple characteristic curve $F(x)$ that is approximately linear for small values of x and runs into a saturation for $x \to \infty$. Such a behavior can be modeled by the characteristic curve (normalized to $P_{\mathrm{in,max}} = P_{\mathrm{out,max}} = 1$) given by the function

$$F_{\exp}(x) = 1 - \mathrm{e}^{-x}.$$

For $x \to \infty$, the curve runs exponentially into the saturation $F_{\exp}(x) \to 1$. For small values of x, we can expand into the Taylor series

$$F_{\exp}(x) = x - \frac{1}{2!}x^2 + \frac{1}{3!}x^3 - \frac{1}{4!}x^4 \pm \cdots$$

and observe a linear growth for small values of x. The clipping amplifier is given by the characteristic curve

$$F_{\mathrm{clip}}(x) = \min(x, 1),$$

which is linear for $x < 1$ and equal to 1 for higher values of x. For simplicity, we do not consider phase distortions.

In Figure 4.24, we see an OFDM time signal and the corresponding amplifier output for the smooth nonlinear amplifier corresponding to $F_{\exp}(x)$ and for the clipping amplifier corresponding to $F_{\mathrm{clip}}(x)$ for an IBO of 6 dB. The average OFDM signal power is normalized to one. Thus, an IBO of 6 dB means that all amplitudes with $a(t) > 2$ are clipped in part (c) of that figure.

The nonlinearity severely influences the spectral characteristics of an OFDM signal. As can be seen from the Taylor series for $F_{\exp}(x)$, mixing products of second, third and higher order occur for every subcarrier and for every pair of subcarriers. These mixing products

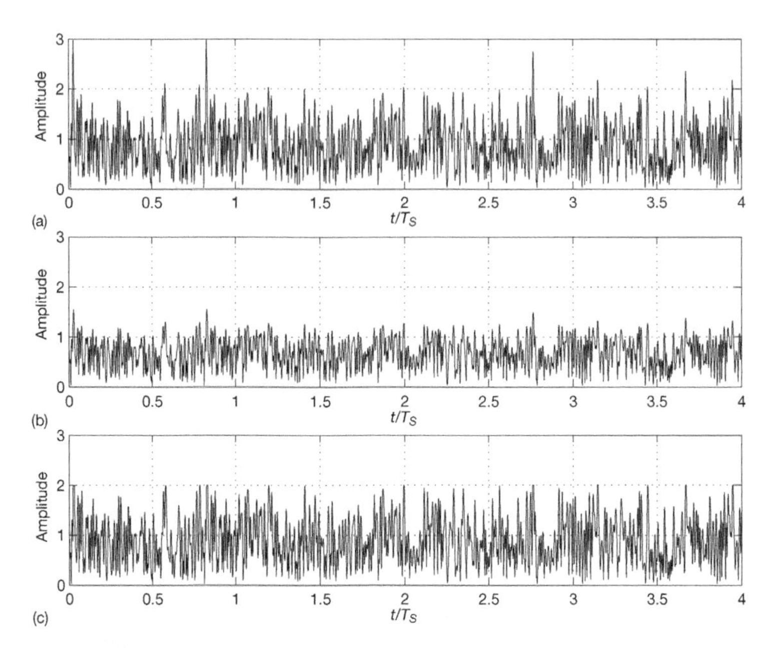

Figure 4.24 Amplitude of an OFDM signal (a) after a smooth nonlinear (b) and a clipping (c) amplifier with IBO = 6 dB.

corrupt the signal inside the main lobe and they will cause out-of-band radiation. This must not be confused with the out-of-band radiation discussed in the preceding subsection, which is caused by the rectangular pulse shape. To distinguish between these two things, we use the spectral smoothing by a raised-cosine window as described in the preceding subsection. We choose $\delta = \Delta/4$ to achieve a very fast decay of the side lobes. Figure 4.25 shows the OFDM signal corrupted by the amplifier corresponding to $F_{\mathrm{exp}}(x)$ for an IBO of 3 dB, 9 dB and 15 dB. As expected, there is a severe out-of-band radiation, and a very high IBO is necessary to reduce this radiation. Note that we have renormalized all the amplifier output signals to the same average power in order to draw all the curves in the same picture. Figure 4.26 shows the OFDM signal corrupted by the amplifier corresponding to $F_{\mathrm{clip}}(x)$ for an IBO of 3 dB, 6 dB and 9 dB. We observe that, compared to the other amplifier, we need much less IBO to reduce the out-of-band radiation.

Inside the main lobe, the useful signal is corrupted by the mixing products between all subcarriers. Simulations of the bit error rate would be necessary to evaluate the performance degradations for a given OFDM system and a given amplifier for the concrete modulation and coding scheme. For a given modulation scheme, the disturbances caused by the nonlinearities can be visualized by the constellation diagram in the signal space. Figure 4.27 shows the constellation of a 16-QAM signal for both amplifiers and the IBO values as given above. For the IBO of 3 dB, the QAM signal is severely distorted for both amplifiers. For the clipping amplifier, the distortion soon becomes smaller as the IBO increases. For the other amplifier, much more IBO is necessary to reduce the disturbance. This is what we may expect by looking at the spectra.

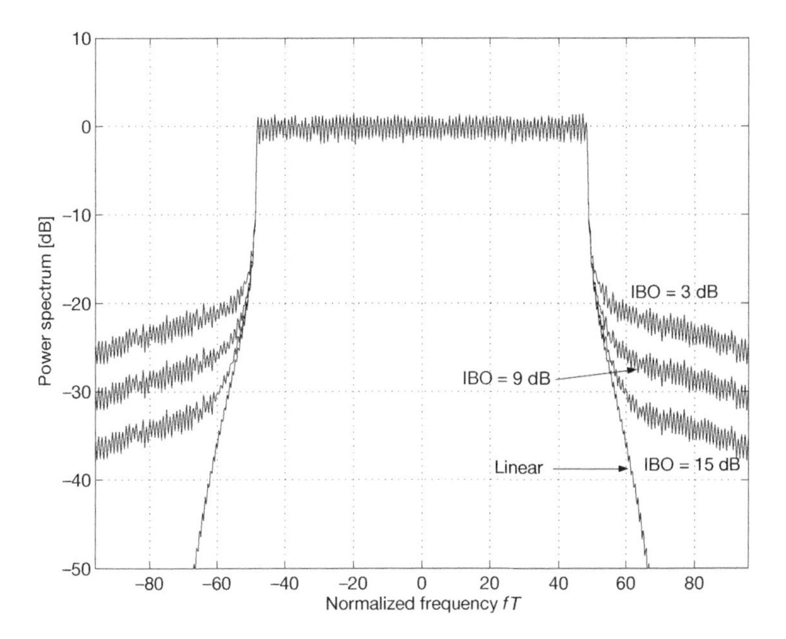

Figure 4.25 Spectrum of an OFDM signal with a (smooth exponential) nonlinear amplifier.

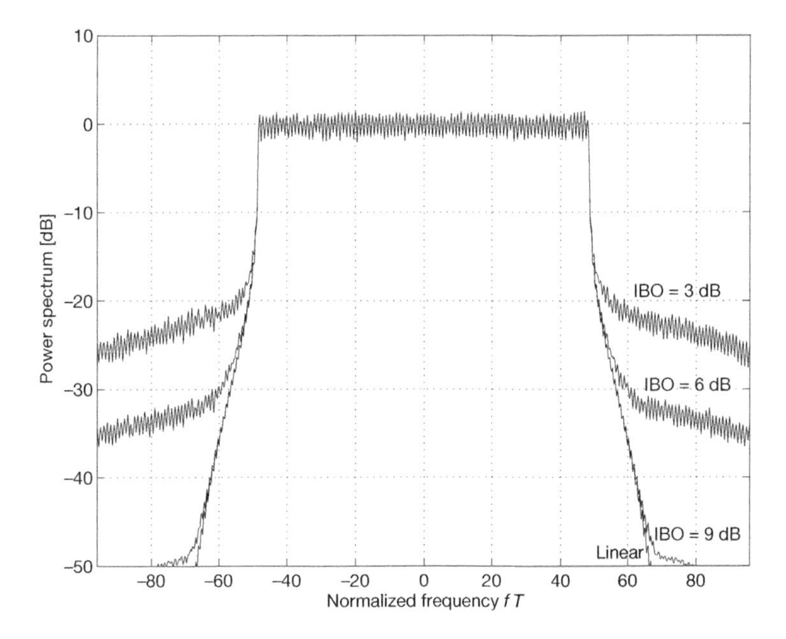

Figure 4.26 Spectrum of an OFDM signal with a (clipping) nonlinear amplifier.

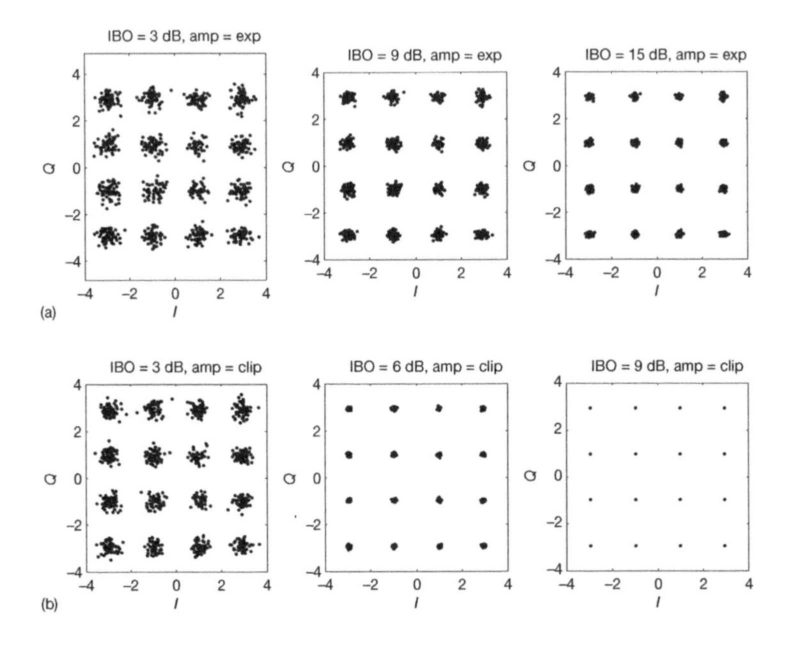

Figure 4.27 The 16-QAM constellation diagram of an OFDM signal with a smooth exponential (a) and a clipping (b) nonlinear amplifier.

Figure 4.27 gives the impression that the QAM symbols are corrupted by an additive noise-like signal. The spectra of Figures 4.25 and 4.26 agree with the picture of an additive *noise floor* that corrupts the signal. At least for the smooth amplifier with a characteristic curve $F_{\exp}(x)$ given by a Taylor series, one can heuristically argue as follows. The quadratic, cubic and higher-order terms cause mixing products of the subcarriers that interfere additively with the useful signal. Each subcarrier is affected by many mixing produces of other subcarriers. Thus, there is an additive disturbance that is the sum of many random variables. By using the central limit theorem, we can argue that this additive disturbance is a Gaussian random variable for the inphase and the quadrature component of the 16-QAM constellation diagram. Figure 4.28 shows the normplots of the error signal (samples of the real and imaginary parts) for the smooth exponential amplifier for the three IBO values.

The samples fit well to the straight line, which confirms the heuristic argument given above. We have also investigated the spectral properties of this interfering signal and found that it shows a white spectrum. Thus, the interference can be modeled as AWGN and can be analyzed by known methods (see Problem 2).

Figure 4.29 shows the normplots of the error signal for the clipping amplifier. Only for 3 dB, the statistics error signal seems to follow a normal distribution. For higher values of the IBO, there are severe deviations.

One can argue that the performance degradations caused by the interference can be neglected if the signal-to-interference ratio (SIR) is significantly higher than the signal-to-noise ratio (SNR). We have calculated the SIR for several values of the IBO. The results are depicted in Figure 4.30 . We find a rapid growth of the SIR as a function of the IBO for the clipping amplifier. For an IBO above approximately 6 dB, the SIR can be

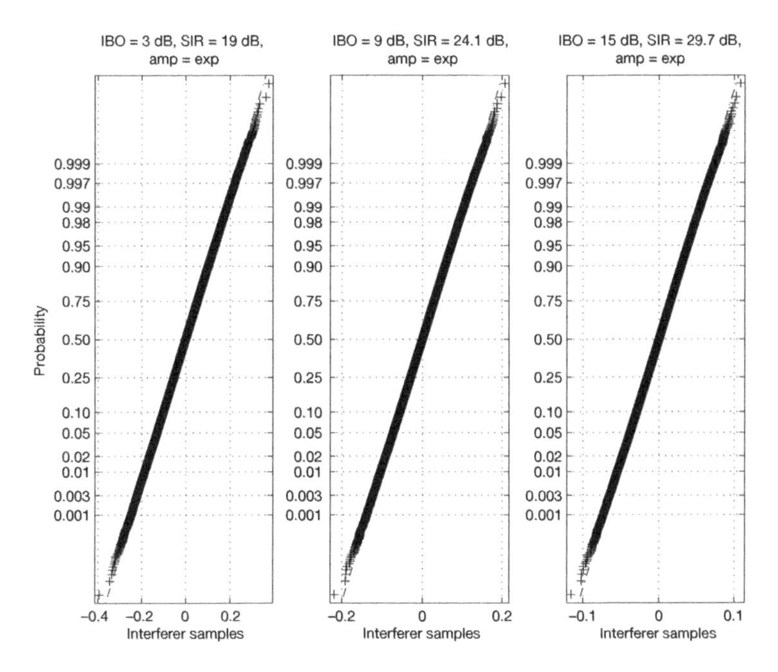

Figure 4.28 Normal probability plot of the 16-QAM error signal for a smooth exponential nonlinear amplifier.

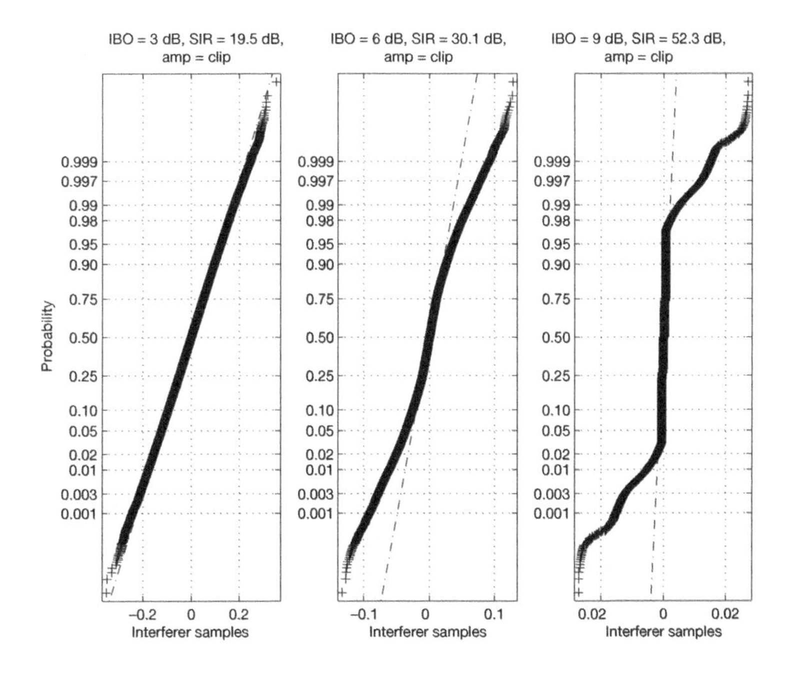

Figure 4.29 Normal probability plot of the 16-QAM error signal for a clipping amplifier.

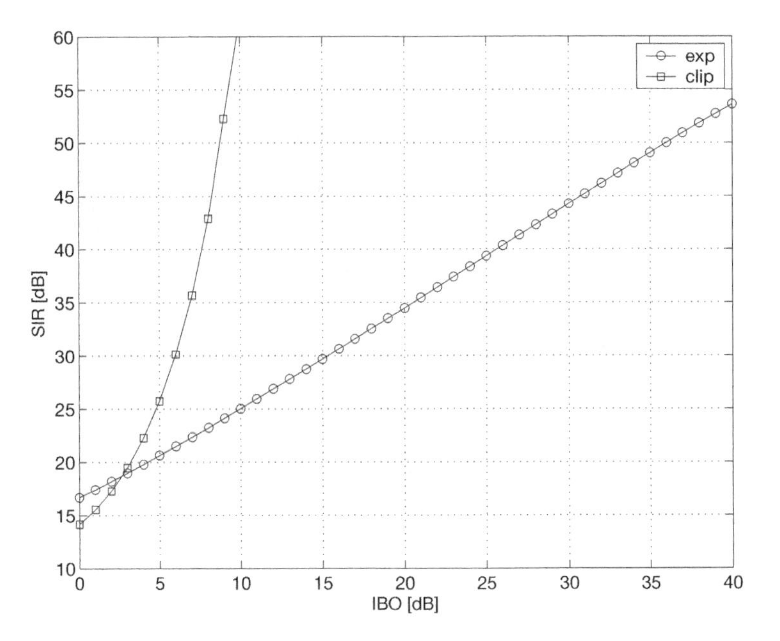

Figure 4.30 The SIR for the 16-QAM symbol for OFDM signal with a smooth exponential and a clipping nonlinear amplifier.

practically neglected. For the smooth exponential amplifier, the SIR increases very slowly as an approximately linear function at IBO values above 10 dB. One must increase the IBO by a factor of 10 for an SIR increase approximately by a factor of 10.

We summarize and add the following remarks:

- OFDM systems are much more sensitive against nonlinearities than single carrier systems. The nonlinearities of a power amplifier degrade the BER (bit error rate) performance and inflate the out-of-band radiation. The performance degradations will typically be the less severe problem because most communication systems work at an SNR well below 20 dB, while the SIR will typically be beyond that value. However, a high IBO value may be necessary to fulfill the requirements of a given spectral mask. As a consequence, the power amplifier will then work with a poor efficiency. In certain cases, it may be necessary to reduce the out-band radiation by using a filter after the power amplifier.

- There are several methods to reduce the crest factor of an OFDM signal by modifying the signal in a certain way (see (Schmidt 2001) and references therein). However, these methods are typically incompatible to existing standards and cannot be applied in those OFDM systems.

- Preferably one should separate the problem of nonlinearities and the OFDM signal processing. This can be done by a predistortion of the signal before amplification.

Analog and digital implementations are possible. After OFDM was chosen as the transmission scheme for several communication standards, there has been a considerable progress in this field (see (Banelli and Baruffa 2001; D'Andrea *et al.* 1996) and references therein).

4.3 Synchronization and Channel Estimation Aspects for OFDM Systems

4.3.1 Time and frequency synchronization for OFDM systems

There are some special aspects that make synchronization for OFDM systems very different from that for single carrier systems. OFDM splits up the data stream into a high number of subcarriers. Each of them has a low data rate and a long symbol duration T_S. This is the original intention for using multicarrier modulation, as it makes the system more robust against echoes. Consequently, the system also becomes more robust against time synchronization errors that can also be absorbed by the guard interval of length $\Delta = T_S - T$. A typical choice is $\Delta = T_S/5 = T/4$ which allows a big symbol timing uncertainty of 20% in case of no physical echoes. In practice, there will appear a superposition of timing uncertainty and physical echoes.

On the other hand, because the subcarrier spacing T^{-1} is typically much smaller than the total bandwidth, frequency synchronization becomes more difficult. Consider, for example, an OFDM system working at the center frequency $f_c = 1500$ MHz with $T = 500$ ms. The ratio between carrier spacing and center frequency is then given by $(f_c T)^{-1} = 1.33 \cdot 10^{-9}$, which is a very high demand for the accuracy of the downconversion to the complex baseband.

Once the correct Fourier analysis window is found by an appropriate time synchronization mechanism and the downconversion is carried out with sufficient accuracy, the OFDM demodulator (implemented by the FFT) produces the noisy receive symbols given by the discrete channel

$$r_{kl} = \sqrt{\frac{T}{T_S}} c_{kl} s_{kl} + n_{kl}$$

(see Subsection 4.1.4). The amplitudes and phases of the channel coefficients c_{kl} are still unknown. The knowledge of the channel is not required for systems with differential demodulation. For coherent demodulation, the channel estimation is a different task that has to be done after time and frequency synchronization. In this subsection, we focus our attention on time and frequency synchronization and follow (in parts) the discussion presented by Schmidt (2001). Channel estimation will be discussed in the subsequent subsections.

When speaking of frequency synchronization items for OFDM, there often appears some misunderstanding because for single carrier PSK systems there is a joint frequency and phase synchronization that can be realized, for example, by a squaring loop or a Costas loop (see e.g. (Proakis 2001)). As mentioned above, frequency synchronization and phase estimation are quite different tasks for OFDM systems.

Time synchronization

An obvious way to obtain time synchronization is to introduce a kind of *time stamp* into the seemingly irregular and noise-like OFDM time signal. The EU147 DAB system – which can be regarded as the pioneer OFDM system – uses quite a simple method that even allows for traditional analog techniques to be used for a coarse time synchronization. At the beginning of each transmission frame, the signal will be set to zero for the duration of (approximately) one OFDM symbol. This *null symbol* can be detected by a classical analog envelope detector (which may also be digitally realized) and tells the receiver where the frame and where the first OFDM symbol begin.

A more sophisticated time stamp can be introduced by periodically repeating a certain known OFDM reference symbol of known content. The subcarriers should be modulated with known complex symbols of equal amplitude to have a white frequency spectrum and a δ-type cyclic time autocorrelation function. Thus, as long as the echoes do not exceed the length of the guard interval, the channel impulse response can be measured by cross correlating the received and the transmitted reference symbol.

In the DAB system, the first OFDM symbol after the null symbol is such a reference symbol. It has the normal OFDM symbol duration T_S and is called the *TFPR* (*time-frequency-phase reference*) symbol. It is also used for frequency synchronization (see the following text) and it provides the phase references for the beginning of the differential demodulation. We note that the channel estimate provided by the TFPR symbol is only needed for the positioning of the Fourier analysis window, not for coherent demodulation.

In the wireless LAN systems IEEE 802.11a and HIPERLAN/2, a reference OFDM symbol of length $2T_S$ is used for time synchronization and for the estimation of the channel coefficients c_{kl} that are needed for coherent demodulation. The OFDM subcarriers are modulated with known data. The signal of length T resulting from the Fourier synthesis will then be cyclically extended to twice the length of the other OFDM symbols.

Another smart method to find the time synchronization without any time stamp is based on the guard interval. We note that an OFDM signal with guard interval has a regular structure because the cyclically extended part of the signal occurs twice in every OFDM symbol of duration T_S – this means that the OFDM signal $s(t)$ given has the property

$$s(t) = s(t + T)$$

for $lT_S - \Delta < t < lT_S$ (l integer), that is, the beginning and the end of each OFDM symbol are identical (see Figure 4.31). We may thus correlate $s(t)$ with $s(t + T)$ by using a sliding

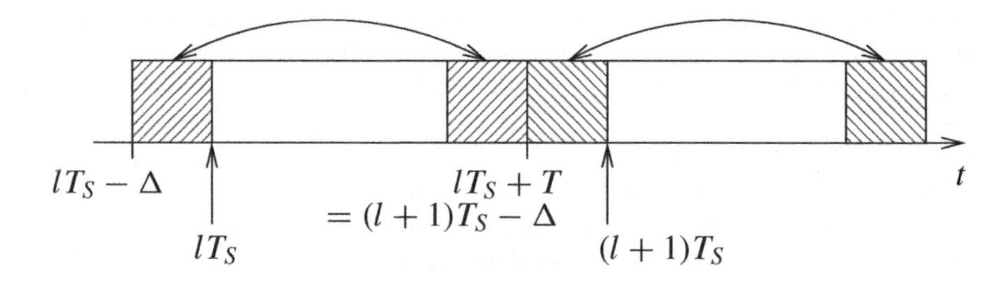

Figure 4.31 Identical parts of the OFDM symbol.

correlation analysis window of length Δ, that is, we calculate the correlator output signal

$$y(t) = \Delta^{-1} \int_{t-\Delta}^{t} \Re\left\{s(\tau)s^*(\tau + T)\right\} d\tau.$$

This correlator output can be considered as a sliding average given by the convolution

$$y(t) = h(t) * x(t).$$

Here,

$$h(t) = \Delta^{-1}\Pi\left(\frac{t}{\Delta} - \frac{1}{2}\right)$$

is the (normalized) rectangle between $t = 0$ and $t = \Delta$, and

$$x(t) = \Re\left\{s(\tau)s^*(\tau + T)\right\}$$

is the function to be averaged. The signal $y(t)$ has peaks at $t = lT_S$, that is, at the beginning of the analysis window for each symbol, (see Figure 4.32(a)). Because of the statistical nature of the OFDM signal, the correlator output is not strictly periodic, but it shows some fluctuations. But it is not necessary to place the analysis window for every OFDM symbol. Only the relative position is relevant and it must be updated from time to time. Thus, we may average over several OFDM symbols to obtain a more regular symbol synchronization signal (see Figure 4.32(b)). This averaging also reduces the impairments due to noise. In a mobile radio environment, the signal in Figure 4.32 is smeared out because of the impulse response of the channel. It is a nontrivial task to find the optimal position of the Fourier analysis window. This may be aided by using the results of the channel estimation.

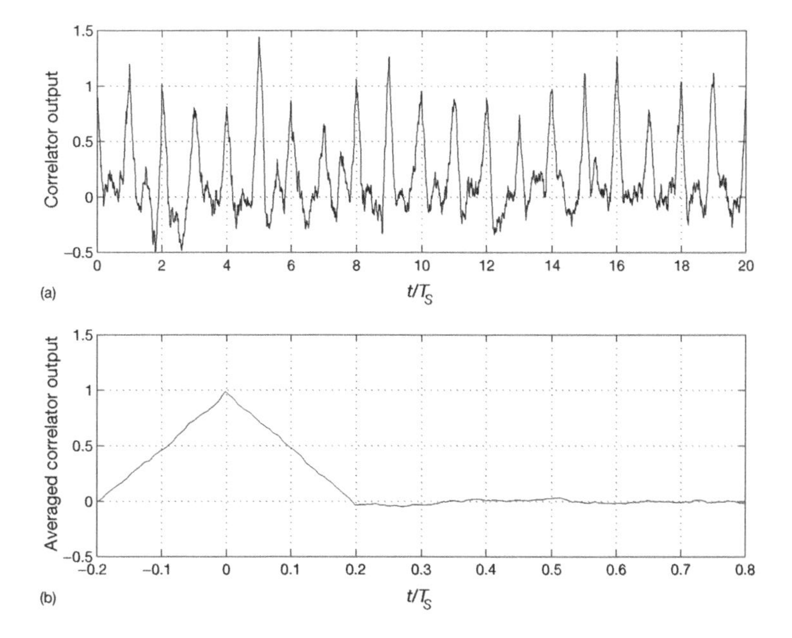

Figure 4.32 The correlator output $y(t)$ (a) and the average of it over 20 OFDM symbols (b).

Frequency synchronization

Because the spacing T^{-1} between adjacent subcarriers is typically very small, accurate frequency synchronization is an important item for OFDM systems. Such a high accuracy can usually not be provided by the local oscillator itself. Standard frequency-tracking mechanisms can be applied if measurements of the frequency deviation δf are available.

First, we want to discuss what happens to an OFDM system if there is a residual frequency offset δf that has not been corrected. There are two effects:

1. The orthogonality between transmit and receive pulses will be corrupted.
2. There is a time-variant phase rotation of the receive symbols.

The latter effect occurs for any digital transmission system, but the first is a special OFDM item that can be understood as follows. Using the notation introduced in Subsection 4.1.4, we write

$$s(t) = \sum_{kl} s_{kl} g'_{kl}(t)$$

for the transmitted OFDM signal that is modulated, for example, with complex QAM symbols s_{kl}. Here, k and l are the time and frequency indices, respectively. We assume a noise-free channel with a time variance that describes the frequency shift. The receive signal is then given by

$$r(t) = \mathrm{e}^{j2\pi\delta f\, t} s(t).$$

To study the first effect, we consider only the first OFDM symbol and drop the corresponding time index $l = 0$. The detector for the subcarrier at frequency $f_k = k/T$ is given by the Fourier analysis operation

$$\mathcal{D}_k[r] = \langle g_k, r \rangle = \int_{-\infty}^{\infty} g_k^*(t) r(t)\, \mathrm{d}t = \sqrt{\frac{1}{T}} \int_0^T \mathrm{e}^{-j2\pi f_k t}\, r(t)\, \mathrm{d}t.$$

Because of the orthogonality

$$\left\langle g_k, g'_{k'} \right\rangle = \sqrt{\frac{T}{T_S}}\, \delta_{kk'}$$

between the transmit and receive base pulses, the Fourier analysis detector recovers the undisturbed QAM symbols from the original transmit symbol, that is,

$$\mathcal{D}_k[s] = s_k.$$

The frequency offset, however, corrupts the orthogonality, leading to the detector output

$$\mathcal{D}_k[r] = \sqrt{\frac{T}{T_S}} \sum_m \gamma_{km}(\delta f)\, s_m$$

with

$$\gamma_{km}(\delta f) = \int_{-\infty}^{\infty} g_k^*(t) g_m(t) \mathrm{e}^{j2\pi\delta f\, t}\, \mathrm{d}t.$$

Typically, for small frequency offsets with $\delta = \delta f \cdot T \ll 1$, the term with $k = m$ dominates the sum, but all the other terms contribute and cause intersymbol interference that must be regarded as an additive disturbance to the QAM symbol.

We now consider an OFDM signal with running time index $l = 0, 1, 2, \ldots$. The frequency shift that is given by the multiplication with $\exp(j2\pi\delta f\, t)$ means that the QAM symbols s_{kl} are rotated by a phase angle $2\pi\delta f \cdot T_S$ between the OFDM symbols with time indices l and $l+1$. Figure 4.33 shows a 16-QAM constellation affected by that rotation and the additive disturbance. The OFDM parameters are the same as above, and a small frequency offset given by $\delta = \delta f \cdot T = 0.01$ is chosen.

In the discrete channel model, the phase rotation can be regarded as the time variance of the channel, that is, the channel coefficient shows the proportionality

$$c_{kl} \propto e^{j2\pi\delta f\, T_S l}.$$

In a coherent system with channel estimation, this time variance can be measured and the QAM constellation will be back rotated. Part (a) of Figure 4.34 shows the back-rotated 16-QAM constellation for $\delta = 0.01$, $\delta = 0.02$ and $\delta = 0.05$. The additive disturbances look similar to Gaussian noise. Indeed, a statistical analysis with a normplot fits well to a Gaussian normal distribution (see part (b) of Figure 4.34. One can therefore argue that the frequency is accurate enough if the SIR of the residual additive disturbance (after frequency tracking) is significantly below the SNR where the system is supposed to work. The latter can be obtained from the BER performance curves of the channel coding and modulation scheme.

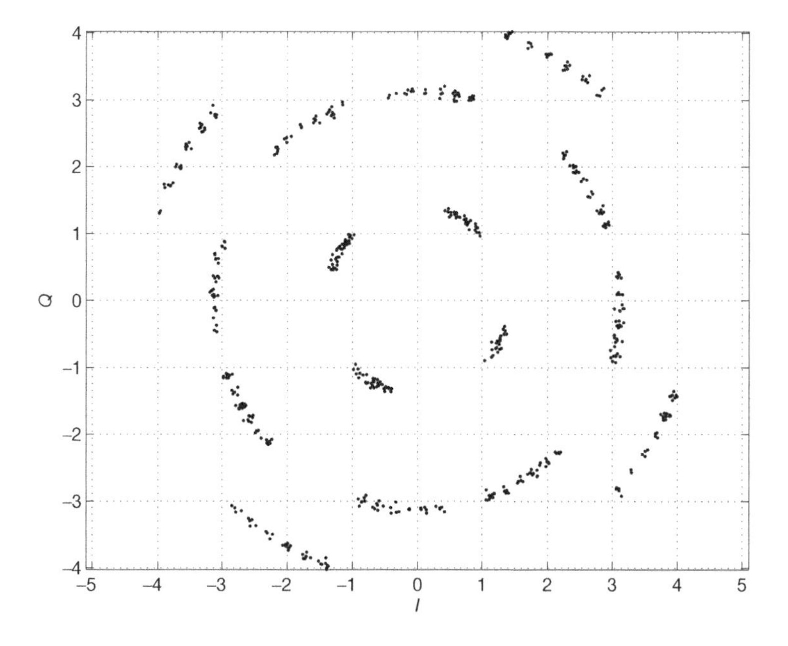

Figure 4.33 16-QAM for OFDM with frequency offset given by $\delta = \delta f \cdot T = 0.01$.

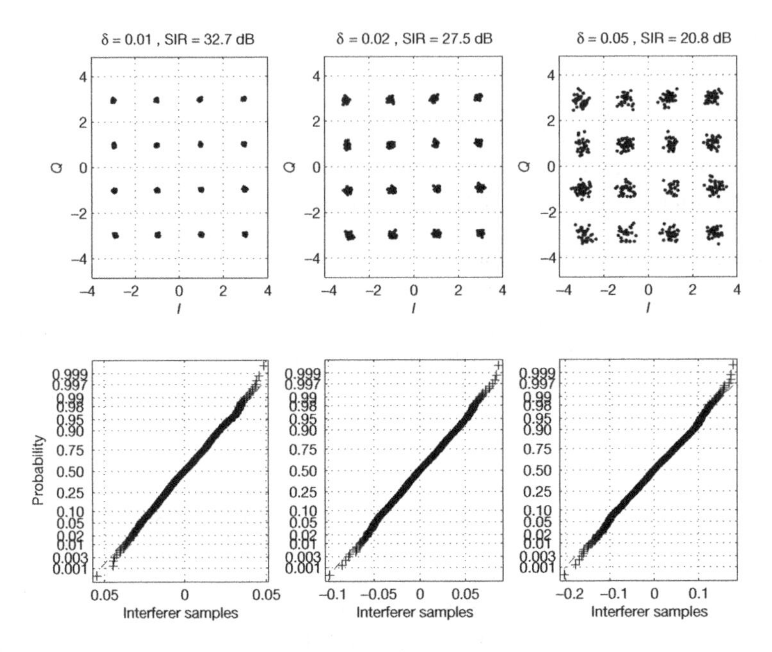

Figure 4.34 16-QAM for OFDM with frequency offset given by $\delta = \delta f \cdot T = 0.01$.

As pointed out above, an estimate for δf can be obtained from the estimated channel coefficients $\hat{c}_{kl}$. This can be done by frequency demodulation and averaging. The frequency demodulation can be implemented as follows. We note that for any complex time signal

$$z(t) = a(t)\mathrm{e}^{j\varphi(t)}$$

with amplitude $a(t)$ and phase $\varphi(t)$, the time derivative of the phase can be calculated as (see Problem 3)

$$\dot{\varphi}(t) = \Im\left\{\frac{\dot{z}(t)}{z(t)}\right\},$$

where the dot denotes the time derivative. The instantaneous frequency modulation (FM) is then given by

$$f_M(t) = \frac{1}{2\pi}\dot{\varphi}(t) = \frac{1}{2\pi}\Im\left\{\frac{\dot{z}(t)}{z(t)}\right\}.$$

For a discrete-time signal $z[n] = z(nt_s)$ that has been obtained by sampling $z(t)$ with the sampling frequency $f_s = t_s^{-1}$, the time discrete FM is

$$f_M[n] = \frac{1}{2\pi t_s}\Im\left\{\frac{z[n] - z[n-1]}{z[n]}\right\}.$$

For an OFDM system with channel estimation as discussed in the next subsection, a noisy estimate $\hat{c}_{kl}$ of the channel coefficient c_{kl} is obtained for every OFDM symbol of duration T_S at a frequency position k. The estimated instantaneous frequency deviation for time index l is then

$$\widehat{\delta f}_{kl} = \frac{1}{2\pi T_S}\Im\left\{\frac{\hat{c}_{kl} - \hat{c}_{k\,l-1}}{\hat{c}_{kl}}\right\}.$$

This noisy instantaneous estimate of the frequency deviation has to be averaged over a sufficiently large number of OFDM symbols and over the frequency index k. This average $\widehat{\delta f}$ may then be used to obtain a frequency-shift-corrected receive signal

$$\hat{r}(t) = \mathrm{e}^{-j2\pi\widehat{\delta f}\,t} r(t).$$

In a typical mobile radio channel with Doppler spread, the time variance of the channel will introduce an additional frequency modulation. The averaging of the FM will place the Doppler spectrum in such a way that its first moment vanishes. In the WSSUS model, the Doppler spectrum is the same for every subcarrier frequency. Thus, an accurate estimate for the frequency offset and for the Doppler spectrum can be obtained from the measurements at a certain number of subcarrier positions (at least one). It is a common method applied in the DVB-T system and the wireless LAN systems IEEE 802.11a and HIPERLAN/2 to use certain subcarriers as *continuous pilots*. These subcarriers that are boosted by a certain factor and carry known data will be used for frequency synchronization and estimation of the Doppler bandwidth $\nu_{\max}$. The latter will be needed for the channel estimation by Wiener filtering, as discussed in the next subsection. The Doppler spectrum can be estimated from the continuous pilots (after frequency-shift correction) by standard power spectral density estimation methods.

Wireless LAN systems require a very fast frequency synchronization at the beginning of every burst. For this purpose, a special OFDM symbol at the beginning of the burst has been defined. In this OFDM symbol, only 12 subcarriers are modulated to serve as a frequency reference.

An accurate frequency synchronization is also necessary for OFDM systems with differential demodulation. The EU147 DAB system uses DQPSK. The first symbol in every frame (after the null symbol) serves as the phase reference for the differential modulation and as a reference for time and frequency synchronization. The complex symbols are built from CAZAC (constant amplitude zero autocorrelation) sequences. They allow a frequency offset estimation by correlating in frequency direction.

4.3.2 OFDM with pilot symbols for channel estimation

Coherent demodulation requires the knowledge of the channel, that is, of the coefficients c_{kl} in the discrete-time model for OFDM transmission in a fading channel. The two-dimensional structure of the OFDM signal makes a two-dimensional pilot grid especially attractive for channel measurement and estimation. An example of such a grid is depicted in Figure 4.35. These pilots are usually called *scattered pilots* to distinguish them from the continuous pilots discussed in the preceding subsection.

At certain positions in time and frequency, the modulation symbols s_{kl} will be replaced by known pilot symbols. At these positions, the channel can be measured. Figure 4.35 shows a rectangular grid with pilot symbols at every third frequency and every fourth time slot. The pilot density is thus 1/12, that is, 1/12 of the whole capacity is used for channel estimation. This lowers not only the data rate, but also the available energy E_b per bit. Both must be taken into account in the evaluation of the spectral and the power efficiency of such a system.

The density of the grid has to be matched to the incoherency of the channel, that is, to the time-frequency fluctuations described by the scattering function. To illustrate this by a

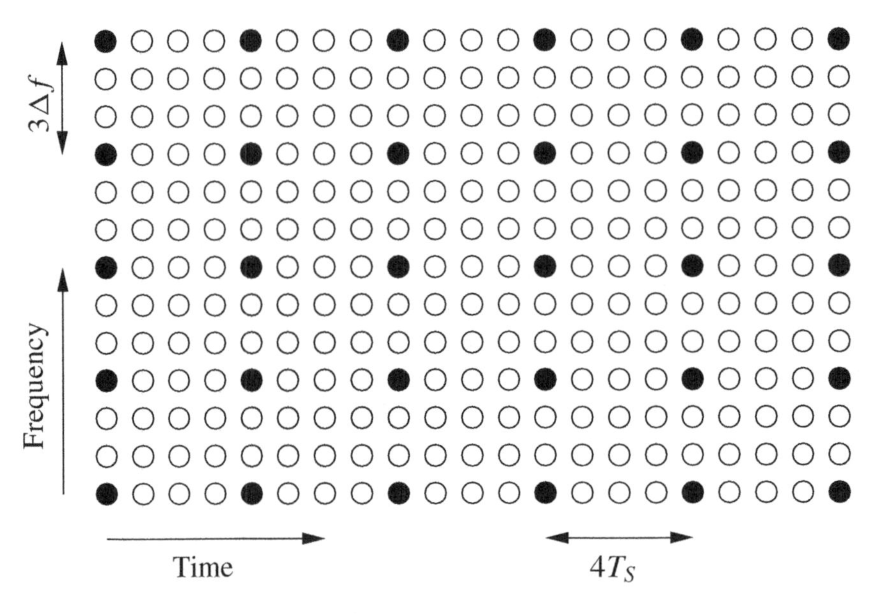

Figure 4.35 Example of a rectangular pilot grid.

numerical example, we consider the grid of Figure 4.35 for an OFDM system with carrier spacing $\Delta f = 1/T = 1$ kHz and symbol duration $T_S = 1250$ µs. At every third frequency, the channel will be measured once in the time $4T_S = 5$ ms, that is, the unknown signal (the time-variant channel) is sampled at the sampling frequency of 200 Hz. For a noise-free channel, we can conclude from the sampling theorem that the signal can be recovered from the samples if the maximum Doppler frequency $\nu_{\max}$ fulfills the condition

$$\nu_{\max} < 100\,\text{Hz}.$$

More generally, for a pilot spacing of $4T_S$, the condition

$$\nu_{\max} T_S < 1/8$$

must be fulfilled.

In frequency direction, the sample spacing is 3 kHz. From the (frequency domain) sampling theorem, we conclude that the delay power spectrum must be inside an interval of the length of 333 µs. Since the guard interval already has the length 250 µs, this condition is automatically fulfilled if we can assume that all the echoes lie within the guard interval. We can now start the interpolation (according to the sampling theorem) either in time or in frequency direction and then calculate the interpolated values for the other direction. Simpler interpolations are possible and may be used in practice for a very coherent channel, for example, linear interpolation or piecewise constant approximation. However, for a really time-variant and frequency-selective channel, these methods are not adequate. For a noisy channel, even the interpolation given by the sampling theorem is not the best choice because the noise is not taken into account. The optimum linear estimator will be derived in the next subsection.

In some systems, the pilot symbols are *boosted*, that is, they are transmitted with a higher energy than the modulation symbols. In that case, a rectangular grid as shown in

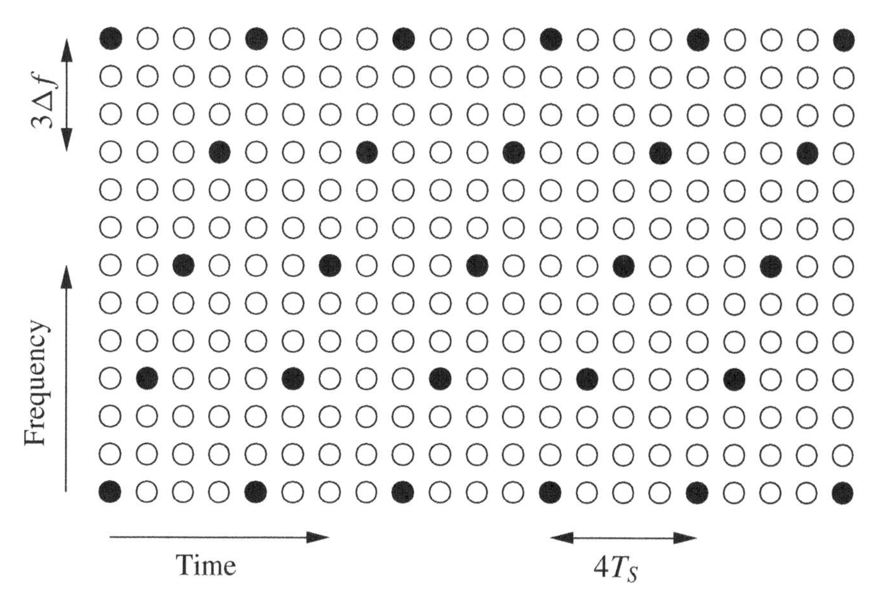

Figure 4.36 Example of a diagonal pilot grid.

Figure 4.35 would cause a higher average power for every fourth OFDM symbol, which is not desirable for reasons of transmitter implementation. In that case, a diagonal grid will be chosen. Figure 4.36 shows such a diagonal grid as it is used for the DVB-T system.

4.3.3 The Wiener estimator

Consider a complex discrete stochastic process with samples y_l that have to be estimated. For our application, we think of the complex fading amplitudes of a discrete channel model. Recall that in Section 4.1.4 we derived a discrete channel model for OFDM that could be written as

$$r_{kl} = \sqrt{\frac{T}{T_S}}\, c_{kl} s_{kl} + n_{kl}.$$

Here, c_{kl} is the complex fading amplitude of the time-frequency discrete channel model with frequency index k and time index l. This is a stochastic process in two dimensions. We may keep either the time index or the frequency index fixed and consider only one dimension. For the treatment of the two-dimensional stochastic process, we may rearrange the numbering similar to a parallel–serial conversion so that one can work with only one index. This makes the formalism more clear. The samples y_l of the process under consideration must be estimated from measurements x_m that are samples of another stochastic process. For our application, these processes are closely related: the x_m are the noisy channel measurements at the pilot positions. We look for a linear estimator, that is, we assume that the estimates $\hat{y}_l$ of the process y_l can be written as

$$\hat{y}_l = \sum_m b_{lm} x_m \tag{4.13}$$

with properly chosen estimator coefficients b_{lm}. The sum can be finite or infinite. To simplify the formalism, we assume that only a finite number of L samples y_l must be estimated from a finite number M of measurements x_m. We may then write the linear estimator as

$$\hat{\mathbf{y}} = \mathbf{B}\mathbf{x} \tag{4.14}$$

with the vectors $\hat{\mathbf{y}} = (\hat{y}_1, \ldots, \hat{y}_L)^T$ and $\mathbf{x} = (x_1, \ldots, x_M)^T$ and the estimator matrix

$$\mathbf{B} = \begin{pmatrix} b_{11} & b_{12} & \cdots & b_{1M} \\ b_{21} & b_{22} & \cdots & b_{2M} \\ \vdots & \vdots & \ddots & \vdots \\ b_{L1} & b_{L2} & \cdots & b_{LM} \end{pmatrix}.$$

Let $e_l = y_l - \hat{y}_l$ be the error of the estimate for the sample number l. The *ansatz* of the Wiener estimator is to minimize the mean square error (MMSE) for each sample, that is,

$$\mathrm{E}\left\{|e_l|^2\right\} = \min.$$

The *orthogonality principle* (or *projection theorem*) of probability theory (Papoulis 1991; Therrien 1992) says that this is equivalent to the orthogonality condition

$$\mathrm{E}\left\{e_l x_m^*\right\} = 0. \tag{4.15}$$

This orthogonality principle becomes intuitively clear and it can easily be visualized by means of the vector space structure of random variables. Then $\mathrm{E}\left\{e_l x_m^*\right\}$ is the scalar product of the random variables (vectors) e_l and x_m, and $\mathrm{E}\{|e_l|^2\} = \mathrm{E}\{|y_l - \hat{y}_l|^2\}$ is the squared distance between the vectors y_l and $\hat{y}_l$. Equation (4.13) says that $\hat{y}_l$ lies in the plane that is spanned by the random variables (vectors) $x_1, \ldots, x_l$. Then, as depicted in Figure 4.37, this distance (length of the error vector) becomes minimal if $\hat{y}_l$ is just the orthogonal projection of y_l on that plane. In that case, $e_l = y_l - \hat{y}_l$ is orthogonal to every vector x_m, that is, Equation (4.15) holds.

It is convenient to write Equation (4.15) in vector notation as

$$\mathrm{E}\left\{\mathbf{e} \cdot \mathbf{x}^\dagger\right\} = 0,$$

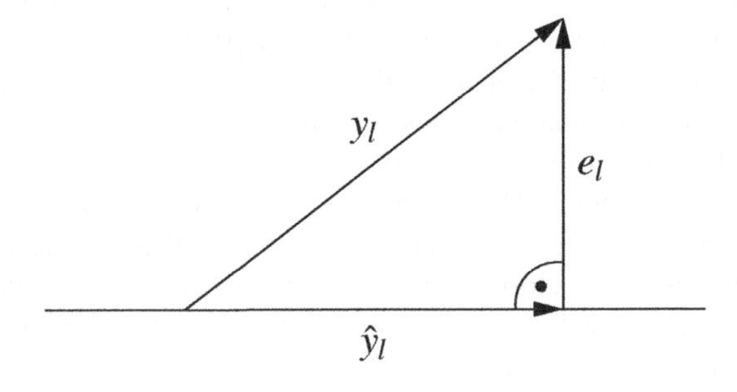

Figure 4.37 Illustration of the orthogonality principle.

that is, the $L \times M$ cross-correlation matrix between the error vector $\mathbf{e} = (e_1, \ldots, e_M)^T$ and the vector of measurements $\mathbf{x} = (x_1, \ldots, x_M)^T$ vanishes. Writing $\mathbf{e} = \mathbf{y} - \hat{\mathbf{y}}$, we obtain

$$\mathrm{E}\left\{(\mathbf{y} - \hat{\mathbf{y}}) \cdot \mathbf{x}^{\dagger}\right\} = 0,$$

and, employing Equation (4.14)

$$\mathrm{E}\left\{\mathbf{y} \cdot \mathbf{x}^{\dagger}\right\} = \mathrm{E}\left\{\mathbf{B}\mathbf{x} \cdot \mathbf{x}^{\dagger}\right\}.$$

This *Wiener–Hopf equation* can be written as

$$\mathbf{R}_{\mathbf{yx}} = \mathbf{B}\mathbf{R}_{\mathbf{xx}}, \tag{4.16}$$

where

$$\mathbf{R}_{\mathbf{xx}} = \mathrm{E}\left\{\mathbf{x} \cdot \mathbf{x}^{\dagger}\right\} \tag{4.17}$$

denotes the autocorrelation matrix of $\mathbf{x}$ and

$$\mathbf{R}_{\mathbf{yx}} = \mathrm{E}\left\{\mathbf{y} \cdot \mathbf{x}^{\dagger}\right\} \tag{4.18}$$

the cross-correlation matrix between $\mathbf{y}$ and $\mathbf{x}$. The Wiener–Hopf equation can be solved by matrix inversion, that is,

$$\mathbf{B} = \mathbf{R}_{\mathbf{yx}}\mathbf{R}_{\mathbf{xx}}^{-1}.$$

Estimation error

The estimation error of a linear predictor can be derived as follows. We define a mean square error (MSE) matrix $\mathbf{E}$ by

$$\mathbf{E} = \mathrm{E}\left\{\mathbf{e} \cdot \mathbf{e}^{\dagger}\right\} = \mathrm{E}\left\{(\mathbf{y} - \hat{\mathbf{y}}) \cdot (\mathbf{y} - \hat{\mathbf{y}})^{\dagger}\right\}.$$

The diagonal elements $\mathrm{E}\left\{|e_l|^2\right\}$ of that matrix are the MSE for the estimates. For the linear estimator of Equation (4.14), we get

$$\mathbf{E} = \mathrm{E}\left\{(\mathbf{y} - \mathbf{B}\mathbf{x}) \cdot (\mathbf{y} - \mathbf{B}\mathbf{x})^{\dagger}\right\}$$

and

$$\mathbf{E} = \mathrm{E}\left\{\mathbf{y} \cdot \mathbf{y}^{\dagger} - \mathbf{B}\mathbf{x} \cdot \mathbf{y}^{\dagger} - \mathbf{y} \cdot (\mathbf{B}\mathbf{x})^{\dagger} + \mathbf{B}\mathbf{x} \cdot (\mathbf{B}\mathbf{x})^{\dagger}\right\}.$$

With Equations (4.17) and (4.18) we get

$$\mathbf{E} = \mathbf{R}_{\mathbf{yy}} - \mathbf{B}\mathbf{R}_{\mathbf{yx}}^{\dagger} - \left(\mathbf{R}_{\mathbf{yx}} - \mathbf{B}\mathbf{R}_{\mathbf{xx}}\right)\mathbf{B}^{\dagger}.$$

This is the general expression for any linear estimator $\mathbf{B}$. If $\mathbf{B}$ is the solution of the Wiener–Hopf equation, the expression in parentheses vanishes and we get the MMSE error matrix

$$\mathbf{E} = \mathbf{R}_{\mathbf{yy}} - \mathbf{B}\mathbf{R}_{\mathbf{yx}}^{\dagger}.$$

4.3.4 Wiener filtering for OFDM

For our application, the stochastic process corresponds to the complex fading amplitudes at certain times and frequencies, that is,

$$y_i = H(f_i, t_i),$$

where $H(f, t)$ is the time-variant channel transfer function, see Section 4.1.4. If only channel estimation in time direction is required (at a given frequency), all the values of f_i are identical. If only channel estimation in frequency direction is required (at a given time), all the values of t_i are identical. But, in general, we have to deal with an arbitrary set of points in the time-frequency plane. The measurements will be taken at some pilot positions given by $\left\{\left(f_{i_m}, t_{i_m}\right)\right\}_{m=1}^{M}$. The measurements x_m are noisy channel samples, that is,

$$x_m = H\left(f_{i_m}, t_{i_m}\right) + n_i,$$

where n_i is complex AWGN with variance $\sigma^2 = E_S/N_0$ and E_S is the energy of the pilot symbols. As discussed in detail in Chapter 2, we assume the WSSUS model for $H(f, t)$ with two-dimensional autocorrelation function

$$\mathrm{E}\left\{H(f, t)H^*(f', t')\right\} = \mathcal{R}(f - f', t - t').$$

We assume that the noise and the fading are statistically independent. Then the matrix elements of $\mathbf{R_{xx}}$ are given by

$$(\mathbf{R_{xx}})_{km} = \mathcal{R}\left(f_{i_k} - f_{i_m}, t_{i_k} - t_{i_m}\right) + \delta_{km}\sigma^2, \tag{4.19}$$

and the matrix elements of $\mathbf{R_{yx}}$ are

$$\left(\mathbf{R_{yx}}\right)_{lm} = \mathcal{R}\left(f_l - f_{i_m}, t_l - t_{i_m}\right). \tag{4.20}$$

Channel estimation in time direction

Consider a fixed carrier of an OFDM signal. In that case, all the frequency samples are equal to the subcarrier frequency f_k and we only have to deal with samples of the multiplicative fading process $c(t) = H(f_k, t)$ for that frequency. This wide-sense stationary process has the autocorrelation function

$$\mathcal{R}(0, t) = \mathcal{R}_c(t),$$

which is given as the inverse Fourier transform of the Doppler spectrum. For the Jakes spectrum, it is given by

$$\mathcal{R}_c(t) = \mathrm{J}_0\left(2\pi\nu_{\max}t\right).$$

For a rectangular Doppler spectrum between $-\nu_{\max}$ and $\nu_{\max}$, it is given by

$$\mathcal{R}_c(t) = \mathrm{sinc}\left(2\nu_{\max}t\right).$$

For some applications, the Gaussian Doppler spectrum

$$\mathcal{S}_c(\nu) = \frac{1}{\sqrt{2\pi\sigma_D^2}} \exp\left(-\frac{1}{2\sigma_D^2}\nu^2\right)$$

of width σ_D is an appropriate model. It has the autocorrelation function

$$\mathcal{R}_c(t) = \exp\left(-\frac{1}{2}(2\pi\sigma_D t)^2\right).$$

The autocorrelation matrix $\mathbf{R_{xx}}$ is given by the elements

$$(\mathbf{R_{xx}})_{km} = \mathcal{R}_c\left(t_{i_k} - t_{i_m}\right) + \delta_{km}\sigma^2.$$

The matrix elements of the cross correlation $\mathbf{R_{yx}}$ are given by

$$\left(\mathbf{R_{yx}}\right)_{lm} = \mathcal{R}\left(t_l - t_{i_m}\right).$$

To give a concrete example, we assume that every fourth symbol in time direction is a pilot symbol, as it is the case for DVB-T for certain subcarrier frequencies. The channel at the positions in between and at the pilot positions itself must be estimated from the channel measurements taken at these pilot positions. In practice, only a finite number of measurements can be used. We illustrate the channel estimation for the case that channel measurements are taken at the five positions $t = -8T_S, -4T_S, 0, +4T_S, +8T_S$ for the estimation at the four positions $t = 0, T_S, 2T_S, 3T_S$. This corresponds to the grid as depicted in Figure 4.35 or Figure 4.36 at a frequency where pilots are located. Note that measurements are taken at five adjacent pilot positions to estimate the channel at four adjacent time slots. One of them is the pilot position in the middle and the other three are those between this pilot and the next one. Obviously, this is a noncausal estimator, that is, a delay has to be taken into account. The measurements are given by the random vector

$$\mathbf{x} = \begin{pmatrix} c(-8T_S) & c(-4T_S) & c(0) & c(4T_S) & c(8T_S) \end{pmatrix}^T + \mathbf{n},$$

where $\mathbf{n}$ is a vector of five AWGN samples, each with variance σ^2. The random vector to be estimated is given by

$$\mathbf{y} = \begin{pmatrix} c(0) & c(T_S) & c(2T_S) & c(3T_S) \end{pmatrix}^T,$$

and the autocorrelation matrix is given by

$$\mathbf{R_{xx}} = \begin{pmatrix} \mathcal{R}_c(0)+\sigma^2 & \mathcal{R}_c(4T_S) & \mathcal{R}_c(8T_S) & \mathcal{R}_c(12T_S) & \mathcal{R}_c(16T_S) \\ \mathcal{R}_c(-4T_S) & \mathcal{R}_c(0)+\sigma^2 & \mathcal{R}_c(4T_S) & \mathcal{R}_c(8T_S) & \mathcal{R}_c(12T_S) \\ \mathcal{R}_c(-8T_S) & \mathcal{R}_c(-4T_S) & \mathcal{R}_c(0)+\sigma^2 & \mathcal{R}_c(4T_S) & \mathcal{R}_c(8T_S) \\ \mathcal{R}_c(-12T_S) & \mathcal{R}_c(-8T_S) & \mathcal{R}_c(-4T_S) & \mathcal{R}_c(0)+\sigma^2 & \mathcal{R}_c(4T_S) \\ \mathcal{R}_c(-16T_S) & \mathcal{R}_c(-12T_S) & \mathcal{R}_c(-8T_S) & \mathcal{R}_c(-4T_S) & \mathcal{R}_c(0)+\sigma^2 \end{pmatrix}.$$

We note that the additive SNR term σ^2 on the diagonal ensures that the matrix is nonsingular. We will see in Subsection 4.4.3 that the eigenvalues of this matrix for $\sigma^2 = 0$ represent the diversity branches of the equivalent independent fading channel. It may happen (and is often the case in practice) that the channel does not have the full diversity degree. This corresponds to a singular matrix. To ensure that $\mathbf{R_{xx}}$ can be inverted, one should always set $\sigma^2 > 0$. In practice, a very rough estimate of the noise is sufficient. For example, one can iteratively improve the noise estimate by comparing the measured channel values at the pilot position with their estimates. One starts that procedure with some reasonable SNR value, where the system will typically work.

The cross-correlation matrix is given by

$$\mathbf{R_{yx}} = \begin{pmatrix} \mathcal{R}_c(8T_S) & \mathcal{R}_c(4T_S) & \mathcal{R}_c(0) & \mathcal{R}_c(-4T_S) & \mathcal{R}_c(-8T_S) \\ \mathcal{R}_c(9T_S) & \mathcal{R}_c(5T_S) & \mathcal{R}_c(T_S) & \mathcal{R}_c(-3T_S) & \mathcal{R}_c(-7T_S) \\ \mathcal{R}_c(10T_S) & \mathcal{R}_c(6T_S) & \mathcal{R}_c(2T_S) & \mathcal{R}_c(-2T_S) & \mathcal{R}_c(-6T_S) \\ \mathcal{R}_c(11T_S) & \mathcal{R}_c(7T_S) & \mathcal{R}_c(3T_S) & \mathcal{R}_c(-T_S) & \mathcal{R}_c(-5T_S) \end{pmatrix}.$$

The estimator matrix

$$\mathbf{B} = \mathbf{R_{yx}}\mathbf{R_{xx}^{-1}}$$

has the shape

$$\mathbf{B} = \begin{pmatrix} b_{11} & b_{12} & b_{13} & b_{14} & b_{15} \\ b_{21} & b_{22} & b_{23} & b_{24} & b_{25} \\ b_{31} & b_{32} & b_{33} & b_{34} & b_{35} \\ b_{41} & b_{42} & b_{43} & b_{44} & b_{45} \end{pmatrix}.$$

Now, denote the noisy channels measurements by the vector

$$\mathbf{x} = \begin{pmatrix} \tilde{c}(-8T_S) & \tilde{c}(-4T_S) & \tilde{c}(0) & \tilde{c}(4T_S) & \tilde{c}(8T_S) \end{pmatrix}^T$$

and the estimates by the vector

$$\hat{\mathbf{y}} = \begin{pmatrix} \hat{c}(0) & \hat{c}(T_S) & \hat{c}(2T_S) & \hat{c}(3T_S) \end{pmatrix}^T.$$

Then, the estimator is given by

$$\begin{pmatrix} \hat{c}(0) \\ \hat{c}(T_S) \\ \hat{c}(2T_S) \\ \hat{c}(3T_S) \end{pmatrix} = \begin{pmatrix} b_{11} & b_{12} & b_{13} & b_{14} & b_{15} \\ b_{21} & b_{22} & b_{23} & b_{24} & b_{25} \\ b_{31} & b_{32} & b_{33} & b_{34} & b_{35} \\ b_{41} & b_{42} & b_{43} & b_{44} & b_{45} \end{pmatrix} \begin{pmatrix} \tilde{c}(-8T_S) \\ \tilde{c}(-4T_S) \\ \tilde{c}(0) \\ \tilde{c}(4T_S) \\ \tilde{c}(8T_S) \end{pmatrix}.$$

Since the random process $c(t)$ is wide-sense stationary and the pilot positions are periodic with period $4T_S$, any time shift of the whole setup by $4iT_S$, $i = 1, 2, 3, \ldots$ result in the same estimator, that is,

$$\begin{pmatrix} \hat{c}(4iT_S) \\ \hat{c}((4i+1)T_S) \\ \hat{c}((4i+2)T_S) \\ \hat{c}((4i+3)T_S) \end{pmatrix} = \begin{pmatrix} b_{11} & b_{12} & b_{13} & b_{14} & b_{15} \\ b_{21} & b_{22} & b_{23} & b_{24} & b_{25} \\ b_{31} & b_{32} & b_{33} & b_{34} & b_{35} \\ b_{41} & b_{42} & b_{43} & b_{44} & b_{45} \end{pmatrix} \begin{pmatrix} \tilde{c}(4(i-2)T_S) \\ \tilde{c}(4(i-1)T_S) \\ \tilde{c}(4iT_S) \\ \tilde{c}(4(i+1)T_S) \\ \tilde{c}(4(i+2)T_S) \end{pmatrix}.$$

The estimate can now be interpreted as the convolution of the measurements with the left–right flipped columns of the estimator matrix. To see this, we define the four discrete-time signals

$$\hat{c}_l[i] = \hat{c}((4i+l)T_S), \quad l = 0, 1, 2, 3$$

for the estimates and the discrete-time signal

$$\tilde{c}[m] = \tilde{c}(4mT_S)$$

for the measurements. We define four impulse responses $b_l[m]$, $l = 0, 1, 2, 3$ of the estimator by rewriting the estimator matrix as

$$\mathbf{B} = \begin{pmatrix} b_0[2] & b_0[1] & b_0[0] & b_0[-1] & b_0[-2] \\ b_1[2] & b_1[1] & b_1[0] & b_1[-1] & b_1[-2] \\ b_2[2] & b_2[1] & b_2[0] & b_2[-1] & b_2[-2] \\ b_3[2] & b_3[1] & b_3[0] & b_3[-1] & b_3[-2] \end{pmatrix}.$$

The estimator can now be written as the noncausal filtering

$$\hat{c}_l[i] = \sum_{m=-2}^{2} b_l[m]\tilde{c}[i-m]$$

of the measurements.

To keep the treatment more easy, we have chosen a small filter with a fixed number of five taps. The generalization to more taps is straightforward. In practice, a figure in the order of 20 taps is a reasonable choice.

Channel estimation in frequency direction

We now consider the time slot of a fixed OFDM symbol. In that case, all the time samples are equal and we only have to deal with frequency samples of the transfer function. This uncorrelated scattering (i.e. frequency-shift invariant) process $H(f)$ has the frequency autocorrelation function

$$\mathcal{R}(f, 0) = \mathcal{R}_H(f),$$

which is given as the Fourier transform of the delay power spectrum. For the exponential delay power spectrum, we have

$$\mathcal{R}_H(f) = \frac{1}{1 + j2\pi f \tau_m}.$$

For a rectangular delay power spectrum between 0 and $\tau_{\max}$, it is given by

$$\mathcal{R}_H(f) = e^{-j\pi f \tau_{\max}} \cdot \operatorname{sinc}(f\tau_{\max}).$$

The autocorrelation matrix $\mathbf{R_{xx}}$ has the elements

$$(\mathbf{R_{xx}})_{km} = \mathcal{R}_H\left(f_{i_k} - f_{i_m}\right) + \delta_{km}\sigma^2,$$

and the cross-correlation matrix $\mathbf{R_{yx}}$ has the elements

$$\left(\mathbf{R_{yx}}\right)_{lm} = \mathcal{R}_H\left(f_l - f_{i_m}\right).$$

Again, we consider a concrete example that is inspired by the pilot structure of DVB-T. We assume that every third symbol in frequency direction is a pilot symbol. The channel at the positions in between and at the pilot positions itself must be estimated from the channel measurements taken at these pilot positions. We assume that channel measurements are taken at the five positions $f = -6\Delta f, -3\Delta f, 0, +3\Delta f, +6\Delta f$ for the estimation at the four positions $f = 0, \Delta f, 2\Delta f, 3\Delta f$. Here $\Delta f = 1/T$ is the OFDM carrier spacing, and

$f = 0$ corresponds to the center frequency. The measurements are given by the random vector

$$\mathbf{x} = \left(H(-6\Delta f) \quad H(-2\Delta f) \quad H(0) \quad H(2\Delta f) \quad H(6\Delta f)\right)^T + \mathbf{n},$$

where $\mathbf{n}$ is a vector of five AWGN samples with variance σ^2. The random vector to be estimated is

$$\mathbf{y} = \left(H(0) \quad H(\Delta f) \quad H(2\Delta f)\right)^T.$$

For the autocorrelation matrix, we have

$$\mathbf{R_{xx}} = \begin{pmatrix} \mathcal{R}_H(0)+\sigma^2 & \mathcal{R}_H(3\Delta f) & \mathcal{R}_H(6\Delta f) & \mathcal{R}_H(9\Delta f) & \mathcal{R}_H(12\Delta f) \\ \mathcal{R}_H(-3\Delta f) & \mathcal{R}_H(0)+\sigma^2 & \mathcal{R}_H(3\Delta f) & \mathcal{R}_H(6\Delta f) & \mathcal{R}_H(9\Delta f) \\ \mathcal{R}_H(-6\Delta f) & \mathcal{R}_H(-3\Delta f) & \mathcal{R}_H(0)+\sigma^2 & \mathcal{R}_H(3\Delta f) & \mathcal{R}_H(6\Delta f) \\ \mathcal{R}_H(-9\Delta f) & \mathcal{R}_H(-6\Delta f) & \mathcal{R}_H(-3\Delta f) & \mathcal{R}_H(0)+\sigma^2 & \mathcal{R}_H(3\Delta f) \\ \mathcal{R}_H(-12\Delta f) & \mathcal{R}_H(-9\Delta f) & \mathcal{R}_H(-6\Delta f) & \mathcal{R}_H(-3\Delta f) & \mathcal{R}_H(0)+\sigma^2 \end{pmatrix},$$

and the cross-correlation matrix is given by

$$\mathbf{R_{yx}} = \begin{pmatrix} \mathcal{R}_H(6\Delta f) & \mathcal{R}_H(3\Delta f) & \mathcal{R}_H(0) & \mathcal{R}_H(-3\Delta f) & \mathcal{R}_H(-6\Delta f) \\ \mathcal{R}_H(7\Delta f) & \mathcal{R}_H(4\Delta f) & \mathcal{R}_H(\Delta f) & \mathcal{R}_H(-2\Delta f) & \mathcal{R}_H(-5\Delta f) \\ \mathcal{R}_H(8\Delta f) & \mathcal{R}_H(5\Delta f) & \mathcal{R}_H(2\Delta f) & \mathcal{R}_H(-\Delta f) & \mathcal{R}_H(-4\Delta f) \end{pmatrix}.$$

The estimator matrix

$$\mathbf{B} = \mathbf{R_{yx}}\mathbf{R_{xx}}^{-1}$$

has the shape

$$\mathbf{B} = \begin{pmatrix} b_{11} & b_{12} & b_{13} & b_{14} & b_{15} \\ b_{21} & b_{22} & b_{23} & b_{24} & b_{25} \\ b_{31} & b_{32} & b_{33} & b_{34} & b_{35} \end{pmatrix}.$$

Now, denote the noisy channel measurements by the vector

$$\mathbf{x} = \left(\tilde{H}(-6\Delta f) \quad \tilde{H}(-3\Delta f) \quad \tilde{H}(0) \quad \tilde{H}(3\Delta f) \quad \tilde{H}(6\Delta f)\right)^T$$

and the estimates by the vector

$$\hat{\mathbf{y}} = \left(\hat{H}(0) \quad \hat{H}(\Delta f) \quad \hat{H}(2\Delta f)\right)^T.$$

Then, the estimator is given by

$$\begin{pmatrix} \hat{H}(0) \\ \hat{H}(\Delta f) \\ \hat{H}(2\Delta f) \end{pmatrix} = \begin{pmatrix} b_{11} & b_{12} & b_{13} & b_{14} & b_{15} \\ b_{21} & b_{22} & b_{23} & b_{24} & b_{25} \\ b_{31} & b_{32} & b_{33} & b_{34} & b_{35} \end{pmatrix} \begin{pmatrix} \tilde{H}(-6\Delta f) \\ \tilde{H}(-3\Delta f) \\ \tilde{H}(0) \\ \tilde{H}(3\Delta f) \\ \tilde{H}(6\Delta f) \end{pmatrix}.$$

Since the random process $H(f)$ is *uncorrelated scattering* (i.e. wide-sense stationary in frequency direction) and the pilot positions are periodic with period $3\Delta f$, any frequency shift by $3i/T$, $i = 0, \pm 1, \pm 2, \ldots$ of the whole setup will result into the same estimator.

Then, we can use the same arguments as for the estimation in time direction to show that the estimate can be interpreted as the convolution of the measurements with the left–right flipped columns of the estimator matrix. We define the three (frequency-) discrete signals

$$\hat{H}_l[i] = \hat{H}\left((3i + l)\,\Delta f\right),\ l = 0, 1, 2$$

for the estimates and the discrete signal

$$\tilde{H}[m] = \tilde{H}\left(3mT_S\right)$$

for the measurements. We define three impulse responses $b_l[m]$, $(l = 0, 1, 2)$ of the estimator by rewriting the estimator matrix as

$$\mathbf{B} = \begin{pmatrix} b_0[2] & b_0[1] & b_0[0] & b_0[-1] & b_0[-2] \\ b_1[2] & b_1[1] & b_1[0] & b_1[-1] & b_1[-2] \\ b_2[2] & b_2[1] & b_2[0] & b_2[-1] & b_2[-2] \end{pmatrix}.$$

The estimator can now be written as the noncausal filtering

$$\hat{H}_l[i] = \sum_{m=-2}^{2} b_l[m]\tilde{H}[i - m]$$

of the measurements.

Even though this filtering in the frequency domain is formally the same as filtering in the time domain, there is an essential difference in practice. One can easily think of an infinite time duration of the transmission, but the frequency domain is always hard limited by the signal bandwidth, corresponding to a finite number of OFDM subcarriers. This number may be quite small (approximately 50 for WLAN systems or approximately 200 for DRM) or large (more than 6000 for the 8K mode for DVB-T). In any case, there will be edge effects. At both edges, there are some measurements not available because there are no pilots outside the band. If one ignores these terms in the sum, it will severely degrade the performance. This is especially severe for a low number of carriers. For WLAN systems and a filter of approximately length 10, the channel estimation would be correct for only approximately 40 subcarriers. Especially for higher-level QAM modulation, this results in an unacceptable error, even for a high number of subcarriers. To cope with this, the estimator must be modified at the edges. It is not a convolution there, but a matrix estimator that makes use of, for example, the 20 closest available pilots.

In contrast to time domain estimation, the estimation in frequency domain is always a problem with a finite number of measurements and a finite number of estimates. It is therefore worth considering the use of the full matrix estimator to solve this problem. This needs more processor power, but it provides us with the optimal MMSE estimate for the given number of measurements that are available.

In that case, the matrix elements of $\mathbf{R_{xx}}$ are calculated for all pairs of pilot positions f_{i_k}, f_{i_m} as

$$(\mathbf{R_{xx}})_{km} = \mathcal{R}_H\left(f_{i_k} - f_{i_m}\right) + \delta_{km}\sigma^2$$

and the matrix elements of $\mathbf{R_{yx}}$ as

$$\left(\mathbf{R_{yx}}\right)_{lm} = \mathcal{R}_H\left(f_l - f_{i_m}\right),$$

where the f_l are the positions of the estimates. The estimator matrix

$$\mathbf{B} = \mathbf{R}_{\mathbf{yx}}\mathbf{R}_{\mathbf{xx}}^{-1}$$

then has as much columns as there are pilot positions and as many rows as there are estimate positions.

Channel estimation in time and frequency direction

In most situations, a channel estimation has to be performed both in time and frequency direction. In that case, the full autocorrelation matrix according to Equation (4.19) and the cross correlation according to Equation (4.20) must be calculated. The pilot positions (f_{i_k}, t_{i_k}) are arbitrary. Typically, they are positioned on some grid. Figure 4.35 shows such a rectangular grid corresponding to the above examples for time and frequency.

The pilot positions are numbered in some order. In most typical channel models, the scattering function and thus the two-dimensional autocorrelation function factorizes to

$$\mathcal{R}(f, t) = \mathcal{R}_H(f)\mathcal{R}_c(t).$$

From this, the estimator matrix **B** can be calculated. The estimation is time- and frequency-shift invariant because of the WSSUS property. As a consequence, the estimation can be written as a two-dimensional convolution. Of course, the edges of the frequency have to be taken into account as discussed above.

Instead of that rather involved 2-D filtering one can perform a suboptimal 1-D $\times$ 1-D filtering without losing significantly in performance (Hoeher 1991; Hoeher *et al.* 1997). One may first perform a 1-D channel estimation, for example, in frequency direction (with the method described above) at those time slots where the pilots are located. After that, at these time slots, there is a channel estimate available for every frequency. Now a 1-D channel estimation in time direction can be performed and an estimate for all time-frequency positions is available. One can also first do a 1-D channel estimation in time direction and then in frequency direction. The order can be shown to be arbitrary due to linearity of the estimation and due to the fact that the rectangular constellation is the Cartesian product of two one-dimensional ones.

We finally note that in practical systems as DVB-T and DRM a diagonal grid has been chosen rather than a rectangular one. The procedure is similar, but the order of channel estimation is no longer arbitrary.

4.4 Interleaving and Channel Diversity for OFDM Systems

4.4.1 Requirements of the mobile radio channel

As already discussed in Chapter 3, channel coding is one important method to introduce diversity into the mobile radio transmission. To achieve the full diversity gain of the code, the transmitted bits must be affected by independent fading. Independent or uncorrelated fading amplitudes[6] can only be realized by a physical separation of the parts of the signal corresponding to the different bits. Bits that are closely related by the code should not be

[6]Here we do not distinguish between uncorrelated and independent, keeping in mind that we will later focus on the case where both are the same.

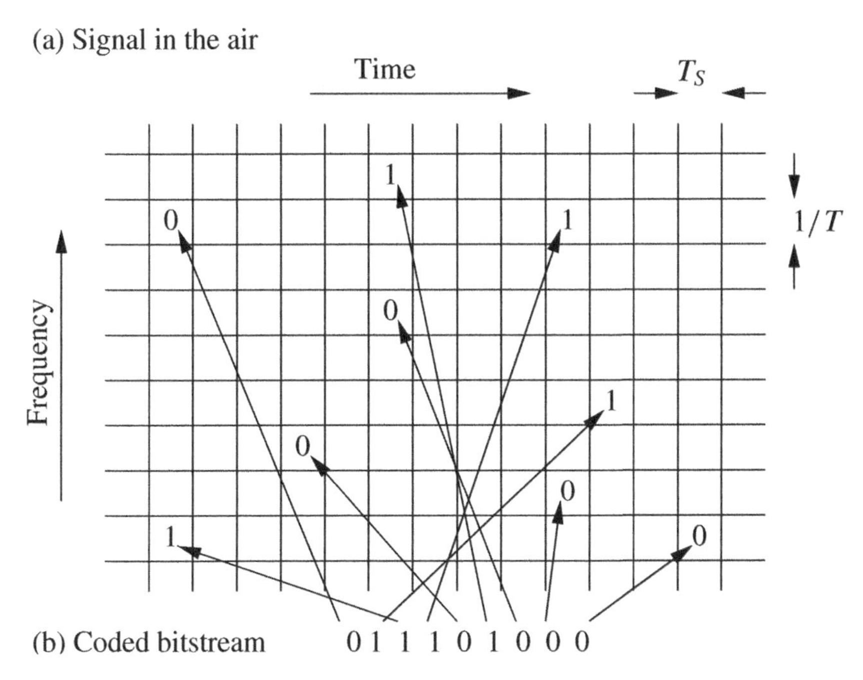

Figure 4.38 Time and frequency interleaving for OFDM with symbol duration T_S and Fourier analysis window T.

transmitted at closely related locations of the channel. For a block code, bits are closely related by the code if they are part of the same code word. For a convolutional code, they are closely related unless there are many constraint lengths between them. For a moving receiver (or transmitter), the separation of closely related bits can be realized in the physical dimension of time. This separation is called *time interleaving*. Obviously, one must wait some time until all closely related bits are received and can be decoded. Thus, time interleaving introduces a *decoding delay*.

For multicarrier transmission, there is an obvious additional way to realize a physical separation in frequency direction. This is called *frequency interleaving*. Thus, as depicted in Figure 4.38, multicarrier transmission provides us with two degrees of freedom to separate the information in the physical transmission channel: time and frequency. The physical reasons that cause decorrelation in time and in frequency direction are independent, so that multicarrier transmission with time *and* frequency interleaving is a very powerful technique for mobile radio transmission systems.

The decorrelation in time has its origin in the time variance of the channel due to the Doppler spread that is caused by multipath reception for a moving vehicle. It can equivalently be interpreted as caused by the motion of a vehicle through a spatial interference pattern. The spatial correlation length is $x_{\text{corr}} = \lambda$, where $\lambda = c/f_0$ is the wavelength, c is the velocity of light and f_0 is the radio frequency. The corresponding correlation time is $t_{\text{corr}} = x_{\text{corr}}/v = \nu_{\max}^{-1}$, where v is the vehicle speed and

$$\nu_{\max} = \frac{v}{c} f_0$$

is the maximum Doppler frequency. The correlation time is then given by

$$t_{\text{corr}} = \frac{c}{f_0 v} \approx \frac{1080 \text{ MHz}}{f_0} \frac{\text{km/h}}{v} \text{ seconds.}$$

The time separation of closely related bits should be significantly larger that this correlation time. We give a numerical example.

Example 7 (Time interleaving for DAB) *The European DAB system has been introduced in many regions in Europe and in Canada. It uses Band III (174–240 MHz) and L-band (1452–1492 MHz) frequencies. In many European regions, the TV channel 12 (223–230 MHz) has been allocated for DAB. For* $f_0 = 225$ *MHz and* $v = 100$ *km/h, the above formula yields a correlation time* $t_{\text{corr}} = 48$ *ms. We will later see that the DAB time interleaver separates two adjacent encoded bits by exactly 24 ms, that is, twice the correlation time. Thus, for Band III transmission and typical vehicle speeds, time interleaving alone does, by far, not guarantee a sufficient statistical independence of the fading amplitude of closely related bits. In the DAB system – as in many other communication systems – a trade-off has to be made between the requirements of channel coding (leading to a huge interleaver) and the restrictions of decoding delay given by the application. For multicarrier systems as DAB, the frequency interleaving is an appropriate method to come out of this dilemma.*

The decorrelation in frequency direction has its origin in the frequency selectivity of the channel due to the different travel times caused by the multipath reception. The frequency correlation length f_{corr} – or coherence bandwidth – as introduced in Chapter 2 is given by $f_{\text{corr}} = \tau_m^{-1}$, where τ_m is the delay spread of the channel. An electromagnetic wave travels 300 m in 1 μs. Thus, in an environment with path differences of a few hundred meters, the system bandwidth should be significantly greater than 1 MHz to achieve an efficient frequency interleaving.

4.4.2 Time and frequency interleavers

Interleaving can be implemented by different techniques. A *block interleaver* always takes a *block* of K coded symbols[7] of the data stream and changes the order of symbols within this block. After this permutation, the symbols are sent to the channel. At the receiver, the permutation is inverted. A *convolutional interleaver* – similar to a convolutional code – has no block-oriented structure and acts on the flowing symbol stream. Frequency interleavers are typically block interleavers as there is a block structure involved by the number of subcarriers. Convolutional interleavers are often regarded as the best choice for the time interleaver as they have desirable properties concerning decoding delay.

Block interleavers

The simplest way to implement a block interleaver is to apply a pseudorandom permutation to each block of K encoded symbols. Owing to the random nature of the interleaver, this, by construction, does not guarantee any minimum separation of symbols on the channel.

[7]This can be encoded bits or bytes (for RS codes) or even complex modulation symbols.

However, any randomly chosen permutation can easily be analyzed numerically and its performance for a given transmission system can be evaluated by simulations.

A matrix block interleaver offers a more constructive approach. We take a block of $K = N \cdot B$ symbols and write them row-wise into an $N \times B$ matrix. Then we read them out column-wise and send them to the channel. Take, as an example, $N = 12$, $B = 4$, and $K = 48$. We write the 48 symbols $a_0, a_1, a_2, \ldots, a_{48}$ row-wise into a 12×4 matrix as

$$\mathrm{r} \downarrow \begin{pmatrix} a_0 & a_1 & a_2 & a_3 \\ a_4 & a_5 & a_6 & a_7 \\ a_8 & a_9 & a_{10} & a_{11} \\ a_{12} & a_{13} & a_{14} & a_{15} \\ \vdots & \vdots & \vdots & \vdots \\ a_{44} & a_{45} & a_{46} & a_{47} \\ & \mathrm{w} & \rightarrow & \end{pmatrix}$$

and read them out column-wise. The bit stream on the channel is then given by

$$(a_0, a_4, a_8, a_{12}, a_{16}, a_{20}, a_{24}, \ldots, a_{44}\, a_1, a_5, \ldots, a_{39}, a_{43}, a_{47}).$$

At the receiver, the deinterleaver writes column-wise into the same matrix and reads out row-wise.

Let us call each matrix row a *subblock*. Assume that the symbols of each subblock are closely related, but symbols of different subblocks are unrelated because each subblock is a code word. We may think, for example, of a WH(4, 2, 2) code and write each code word of length $B = 4$ into a row. Then, on the channel, any two bits of the same code word are separated on the channel by at least $N = 12$ symbol positions. To extend this example further, we think of OFDM transmission with $K = 48$ subcarriers with carrier spacing T^{-1}. We choose BPSK modulation and map the symbols in the above order on these carriers. Then each two bits of the same code word are separated in the frequency dimension by $12/T$[8]. Thus, for two symbols of the same subblock, this matrix block interleaver guarantees a separation in frequency by $N \cdot T^{-1}$ when used as frequency interleaver for OFDM, or in time by $N \cdot T_S$ when used as a time interleaver. The deinterleaver at the receiver simply writes column-wise into the matrix and reads out row-wise. The matrix at the (de) interleaver can be implemented by a RAM.

Let us summarize. We call each matrix column a *frame*. A frame is a sequence of N adjacent symbols on the channel. The $N \times B$ matrix block interleaver with frame length N and subblock length B has the following properties:

1. All symbols of one subblock of length B are transmitted in different frames.

2. If one frame of length N is completely corrupted by an error burst, this will only affect one symbol in each subblock of length B.

3. The overall decoding delay (interleaver plus deinterleaver together) is $2BN$ symbol clocks.

[8]For QPSK transmission, half of the code words are mapped on the inphase and the other half on the quadrature component. We may thus simply double the matrix in horizontal direction. The left part will then be mapped on the inphase, the right on the quadrature component.

We note that, by construction, such a matrix block interleaver guarantees a certain physical separation. However, the very regular structure of this interleaver may sometimes lead to unwanted properties.

Convolutional interleavers

It is possible to construct a convolutional interleaver with the same two first properties as above but only half the decoding delay of the corresponding block interleaver. Because decoding delay is a critical item for most communication systems, convolutional interleavers are often preferred in practical systems.

We consider a serial symbol stream $a_0, a_1, a_2, \ldots$. With this symbol stream we do a serial-to-parallel (S/P) conversion that results in a stream of subblocks of B parallel symbols. Let $i = 0, 1, 2, \ldots, B-1$ denote the position of each symbol of a subblock. Then the symbol with position i in a subblock is delayed by $i \cdot M$ parallel symbol clocks, where M is a certain integer number that must be chosen to adjust the interleaver properties. After that delay, the parallel symbol stream will be P/S converted and then transmitted (see Figure 4.39).

For a better understanding of the properties of the convolutional interleaver, we may think that M subblocks of length B are grouped together to a *data frame* of length $N = B \cdot M$. The interleaver output will be grouped into *transmission frames* of the same structure, but, by the action of the interleaver, the symbols are transmitted in different frames. Since the serial symbol clock and the parallel symbol clock are related by the factor B, the symbol with position i in each subblock will be delayed by $i \cdot N$ serial symbol clocks, that is, by i frames, thereby retaining the relative position within a frame.

The deinterleaver at the receiver has the same structure as the interleaver, but the symbol with position $i = 0, 1, 2, \ldots, B-1$ in a subblock will be delayed by $(B - i - 1) \cdot M$ parallel symbol clocks (see Figure 4.40).

It is evident from both figures that the deinterleaver inverts the operation of the interleaver, thereby introducing an overall decoding delay of $(B-1) \cdot N$ symbol clocks, that is, $B-1$ frames.

The action of an (N, B) convolutional interleaver can be visualized as shown in Figure 4.41 for $N = 12$ and $B = 4$. As depicted in part (a) of the figure, we arrange

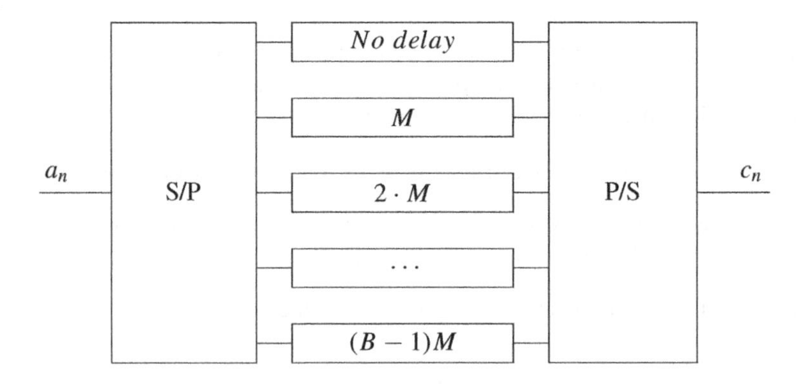

Figure 4.39 Block diagram for the convolutional interleaver.

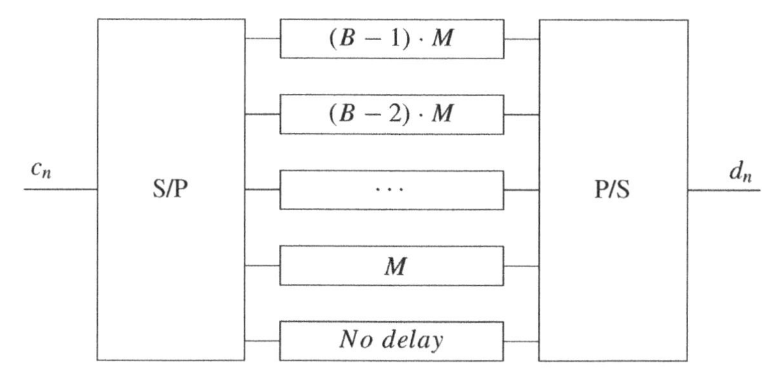

Figure 4.40 Block diagram for the convolutional deinterleaver.

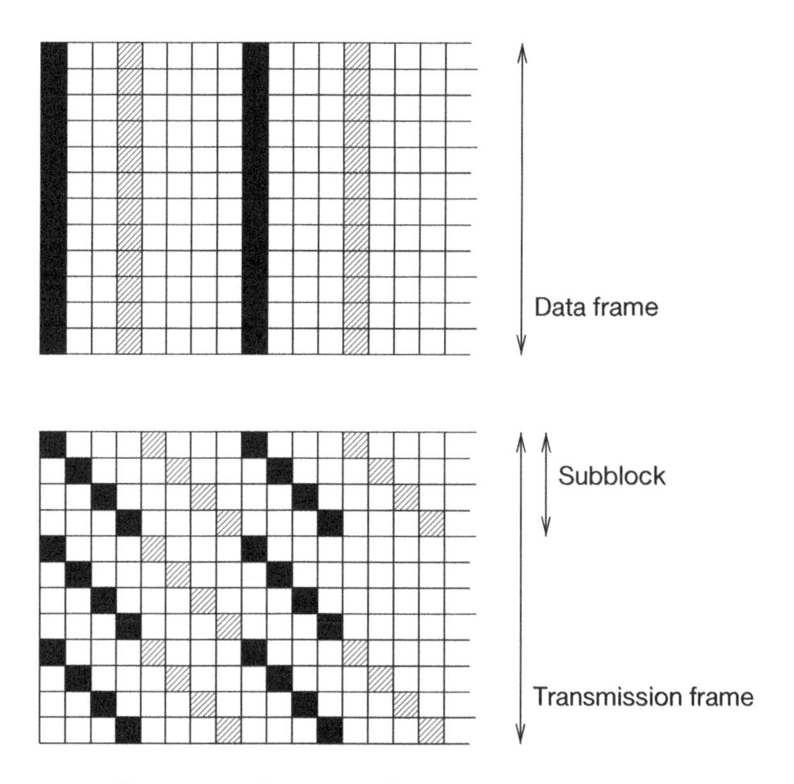

Figure 4.41 The action of a convolutional interleaver.

the sequence of frames as the columns of a matrix. If we read out this matrix column-wise, adjacent positions in the same row will be read out with a time difference of one frame duration, that is, N symbol clocks. We may thus realize the actions of the interleaver by shifting the row of position number i in the subblock to the right by i horizontal positions (see part (b) of the figure). Obviously, the B bits within the same subblock will be shifted to B different columns that correspond to different transmission frames when we read out the matrix. We summarize the properties of a (N, B) convolutional interleaver.

1. All symbols of one subblock of length B are transmitted in different frames.

2. If one transmission frame of length N is completely corrupted by an error burst, this affects only one symbol in each subblock of length B in the data frame.

3. If one subblock of length B inside a transmission frame of length N is completely corrupted by an error burst, this will only affect one symbol in a data frame of length N.

4. The overall decoding delay (interleaver plus deinterleaver together) is $(B - 1) \cdot N$ symbol clocks.

The first two properties are clear from the above discussion. They are identical to the first two properties of the $N \times B$ matrix block interleaver. The third property has no such correspondence. It is clear from the figure and it can be understood from the fact that one can interchange the interleaver and the deinterleaver without altering the properties. Using this argument, the third property immediately follows from the first one.

Figures 4.39 and 4.40 correspond to a shift register implementation of the delays. However, Figure 4.41 suggests a matrix representation realized by a RAM. This will be favorable for implementation.

To do so, we need a matrix with N rows. For a moment, we allow the matrix to have an infinite number of columns. The interleaver can then be implemented by diagonal writing for each subblock. For $N = 12$, $B = 4$, $M = 3$, we write the matrix

$$\left(\begin{array}{cccc|cccc}
a_0 & a_{12} & a_{24} & a_{36} & a_{48} & a_{60} & a_{72} & \cdots \\
\cdot & a_1 & a_{13} & a_{25} & a_{37} & a_{49} & a_{61} & \cdots \\
\cdot & \cdot & a_2 & a_{14} & a_{26} & a_{38} & a_{50} & \cdots \\
\cdot & \cdot & \cdot & a_3 & a_{15} & a_{27} & a_{39} & \cdots \\
a_4 & a_{16} & a_{28} & a_{40} & a_{52} & a_{64} & a_{76} & \cdots \\
\cdot & a_5 & a_{17} & a_{29} & a_{41} & a_{53} & a_{65} & \cdots \\
\cdot & \cdot & a_6 & a_{18} & a_{30} & a_{42} & a_{54} & \cdots \\
\cdot & \cdot & \cdot & a_7 & a_{19} & a_{31} & a_{43} & \cdots \\
a_8 & a_{20} & a_{32} & a_{44} & a_{56} & a_{68} & a_{80} & \cdots \\
\cdot & a_9 & a_{21} & a_{33} & a_{45} & a_{57} & a_{69} & \cdots \\
\cdot & \cdot & a_{10} & a_{22} & a_{34} & a_{46} & a_{58} & \cdots \\
\cdot & \cdot & \cdot & a_{11} & a_{23} & a_{35} & a_{47} & \cdots
\end{array}\right)$$

and then read it out column-wise. At the receiver, the deinterleaver has the same structure, but the write and read operations are interchanged. The matrix will be initialized with dummy symbols. Because of the diagonal writing, there will be some dummy symbols left at the beginning of the matrix. A real RAM is not a matrix with an infinite number of rows, but this is not necessary. At positions where the data are read out, new data can be written. It is easy to see that a matrix with B columns is sufficient. The RAM can then be addressed cyclically. All data in the right part of the matrix above can be written into the left part as the symbols there have already been read out.

4.4.3 The diversity spectrum of a wideband multicarrier channel

In this subsection, we address the question how much interleaving is necessary to have enough statistical independence for the channel code to work. We further present a method to analyze the correlations of a wideband channel with interleaving. In particular, we discuss the question how much bandwidth is needed to allow the channel code to exploit its diversity degree that is given by the Hamming distance d_H (or free distance d_{free}).

To start with a simple example, we first consider a transmission channel given by four frequencies f_1, f_2, f_3, f_4 that are sufficiently separated so that their Rayleigh fading amplitudes can be regarded as independent. We encode a bit stream by a repetition code of rate $R_c = 1/4$ with Hamming distance $d_H = 4$ and transmit each of the four bits in a code word on another frequency, where we may use, for example, BPSK modulation. Of course, this is nothing else but simple frequency diversity, but we regard it as RP(4, 1, 4) coding with frequency interleaving and multicarrier transmission. The pairwise error probability (PEP) that the code word (0000) is transmitted but the receiver decides for (1111), decreases asymptotically as

$$P_{\text{err}} \sim \left(\frac{E_b}{N_0}\right)^{-L} \tag{4.21}$$

with $L = d_H = 4$. We now consider the same transmission setup with the Walsh–Hadamard code of length 4, rate $R_c = 1/2$ and Hamming distance $d_H = 2$. The decay of the PEP for each error event is given by the power law of Equation (4.21) with $L = d_H = 2$. We may say that the channel has a diversity degree of four – because of the four independently fading subcarriers – and the two codes have diversity degrees (i.e. Hamming distance d_H) four and two, respectively. We are interested in the question whether a fading channel provides enough diversity so that the code can exploit its full diversity, that is, Equation (4.21) holds with $L = d_H$. It is obvious that the channel diversity must not be smaller than d_H. However, equality of both diversity degrees typically does not guarantee that the diversity of the code can be fully exploited, as it can easily be seen by the example of the WH(4, 2, 2) code whose code words are given by the rows of the matrix

$$\begin{pmatrix} 0 & 0 & 0 & 0 \\ 0 & 1 & 0 & 1 \\ 0 & 0 & 1 & 1 \\ 0 & 1 & 1 & 0 \end{pmatrix}.$$

If we use only the frequencies f_1, f_2 and transmit the bits numbers 1 and 3 on f_1 and the bits numbers 2 and 4 on f_2, we have different power laws for different error events. Let (0000) be the transmitted code word. The power law for probability that the receiver decides for (0101) is given by Equation (4.21) with $L = 1$, because the two bits in which the code words differ are transmitted on the same frequency. This is not the case for the other two error events corresponding to the code words (0011) and (0110) for which the power law with $L = 2$ holds.

An obviously sufficient condition for a (block) code to exploit its full diversity is to transmit each bit of a code word at another frequency, for example, if the diversity degree of the channel is at least the length of the code. The condition is not necessary. One can easily see that three frequencies would be sufficient for the WH(4, 2, 2) code. For a convolutional code, a detailed analysis is more difficult because the number of possible error events is infinite and their length is growing to infinity. However, one should intuitively expect that the diversity degree of the code will be exploited if the diversity degree of the channel significantly exceeds the free distance d_{free}.

We add the following remarks:

- The two example codes we have chosen are quite weak so they will not be used in practice. Indeed, they have no coding gain in an AWGN channel because $d_H R_c = 1$ in either case.

- A transmission channel that splits up into a given number of K independently fading channels ($K = 4$ in the above example) is called a *block fading* channel. It is of practical relevance, for example, for frequency hopping systems, where K different frequencies are used subsequently during different time slots. The question, what number of K must be chosen to allow the code to exploit its diversity is of great practical relevance. For the simple convolutional code with memory 2, the most probable error event corresponds to the code word (111011000...). We conclude that we should hop at least between six different frequencies for that code.

Up to now, we have considered independently fading (sub) carrier frequencies. In practice, the fading of adjacent subcarriers in a multicarrier system is highly correlated. Take as an example an OFDM system with the (typical) ratio $T/\Delta = 4$ between the Fourier analysis window and the guard interval. Since Δ must be chosen to be larger than the maximum path delay, the delay spread τ_m should be significantly smaller than Δ. We take as an example $\tau_m = \Delta/5$, which is already quite a frequency-selective channel. We then have $T = 20\,\tau_m$. For this figure, the frequency correlation length (or coherency bandwidth) $f_{\text{corr}} = \tau_m^{-1}$ exceeds the frequency separation T^{-1} of the subcarriers by a factor of 20, which means that up to 20 neighboring subcarriers are highly correlated. The number of subcarriers in an OFDM system must therefore significantly exceed this number to guarantee some decorrelation that is necessary to exploit the frequency diversity in a channel coded and frequency-interleaved OFDM system. Only in that case we may legitimately call this a wideband system from the physical system point of view. A proper code design needs some information on the diversity that can be provided by the channel. In a real multicarrier system, there are always significant correlations. It is therefore desirable to find a quantity to characterize the diversity of a correlated fading channel.

In the following discussion, we will present a method to characterize the diversity of a wideband channel with correlated fading. We consider a set of channel samples

$$c_i = H(f_i, t_i),\ i = 1, \ldots, K$$

in the time-frequency plane. We assume a WSSUS Rayleigh process with average power $\mathrm{E}\left\{|c_i|^2\right\} = 1$. Thus, the channel samples c_i are zero mean complex Gaussian random variables that can be completely characterized by their autocorrelation properties. Writing the channel samples as a channel vector $\mathbf{c} = (c_1, \ldots, c_K)^T$, the autocorrelation matrix is

$$\mathbf{R} = \mathrm{E}\left\{\mathbf{c}\mathbf{c}^\dagger\right\}$$

with elements $R_{ik} = \mathrm{E}\left\{c_i c_k^*\right\}$ given by

$$R_{ik} = \mathcal{R}(f_i - f_k, t_i - t_k),$$

where $\mathcal{R}(f, t)$ is the two-dimensional autocorrelation function of the GWSSUS process $H(f, t)$. Since $\mathcal{R}(f, t) = \mathcal{R}^*(-f, -t)$, the matrix $\mathbf{R}$ is Hermitian. From matrix theory, we know that every Hermitian matrix can be transformed to a diagonal matrix: there exists a unitary matrix $\mathbf{U}$ (i.e. $\mathbf{U}^{-1} = \mathbf{U}^{\dagger}$) such that

$$\mathbf{URU}^{\dagger} = \mathbf{D},$$

where $\mathbf{D} = \text{diag}(\lambda_1, \ldots, \lambda_K)$ is the diagonal matrix of the eigenvalues of $\mathbf{R}$. We may write this as

$$\mathrm{E}\left\{\mathbf{Uc}(\mathbf{Uc})^{\dagger}\right\} = \mathbf{D}$$

and set

$$\mathbf{b} = \mathbf{Uc}.$$

This is a vector of mean zero Gaussian random variables with the diagonal autocorrelation matrix $\mathbf{D}$, that is, the coefficients of $\mathbf{b} = (b_1, \ldots, b_K)^T$ are uncorrelated

$$\mathrm{E}\left\{b_i b_k^*\right\} = \lambda_i \delta_{ik}$$

and, because they are Gaussian, even independent. We may regard this unitarily transformed channel vector as the *equivalent independently fading channel*. To explain this name, we consider as a simple example the multicarrier BPSK modulation with a K-fold repetition code, that is, the same BPSK symbol $s \in \left\{\pm\sqrt{E_S}\right\}$ will be transmitted at K different positions in the time-frequency plane. This is again simple frequency diversity combined with time diversity, but with correlated fading amplitudes. We recall from subsection 2.4.6 that the conditional PEP is given by

$$P(s \mapsto \tilde{s}|\mathbf{c}) = \frac{1}{2}\text{erfc}\left(\sqrt{\frac{E_S}{N_0}\sum_{i=1}^{K}|c_i|^2}\right).$$

Because the matrix $\mathbf{U}$ is unitary, it leaves the vector norm invariant, that is,

$$\|\mathbf{b}\|^2 = \|\mathbf{Uc}\|^2 = \|\mathbf{c}\|^2$$

or

$$\sum_{i=1}^{K}|b_i|^2 = \sum_{i=1}^{K}|c_i|^2.$$

This means that the transfer power of the equivalent channel is the same. $P(s \mapsto \tilde{s}|\mathbf{c})$ can then be expressed as

$$P(s \mapsto \tilde{s}|\mathbf{c}) = P(s \mapsto \tilde{s}|\mathbf{U}^{-1}\mathbf{b}) = \frac{1}{2}\text{erfc}\left(\sqrt{\frac{E_S}{N_0}\sum_{i=1}^{K}|b_i|^2}\right).$$

Since the transformed fading amplitudes b_i are independent, we can apply the same method as in Subsection 2.4.6 to perform the average for

$$P(s \mapsto \tilde{s}) = \mathrm{E}\left\{P(s \mapsto \tilde{s}|\mathbf{U}^{-1}\mathbf{b})\right\}$$

and eventually obtain the expression

$$P(s \mapsto \tilde{s}) = \frac{1}{\pi} \int_0^{\pi/2} \prod_{i=1}^{L} \frac{1}{1 + \frac{\lambda_i}{\sin^2 \theta} \frac{E_S}{N_0}} \, d\theta.$$

Using $E_b = K E_S$, we obtain the tight Chernoff-like bound

$$P(s \mapsto \tilde{s}) \leq \frac{1}{2} \prod_{i=1}^{K} \frac{1}{1 + \frac{\lambda_i}{K} \frac{E_b}{N_0}}. \tag{4.22}$$

For independent fading, we have $\lambda_i = 1$ for all values of i and we obtain the power law of Equation (4.21) with $L = K$. For correlated fading they are different, but because of

$$\mathrm{E}\left\{\sum_{i=1}^{K} |b_i|^2\right\} = \mathrm{E}\left\{\sum_{i=1}^{K} |c_i|^2\right\},$$

their sum

$$\sum_{i=1}^{K} \lambda_i = K$$

is always the same. Even though – in case that $\lambda_i \neq 0$ for all i – Equation (4.22) will asymptotically approach the power law of Equation (4.21) with $L = K$ for large E_b/N_0, many of the eigenvalues may be very small so that they will not contribute significantly to the product for relevant values of E_b/N_0. Only those eigenvalues λ_i of significant size contribute, but there is no natural threshold. The diversity that can be achieved by the channel is thus characterized by the whole eigenvalue spectrum $\{\lambda_i\}_{i=1}^{K}$ of the autocorrelation matrix of the fading. We thus call it the *diversity branch spectrum* of the channel.

For the following numerical example, we restrict ourselves to the frequency direction and assume an exponential delay power spectrum

$$S_D(\tau) = \frac{1}{\tau_m} \mathrm{e}^{-\tau/\tau_m} \epsilon(\tau),$$

where τ_m is the mean delay and $\epsilon(\tau)$ is the Heaviside function. The corresponding frequency autocorrelation function is given by

$$\mathcal{R}_f(f) = \frac{1}{1 + j2\pi f \tau_m}.$$

Figure 4.42 shows the first 16 eigenvalues for $K = 64$ and different values of the bandwidth B. We define a *normalized* bandwidth $X = B\tau_m$. We have assumed that the BPSK symbols are equally frequency spaced over the bandwidth. We see that for a small bandwidth (e.g. $X = 1$ corresponding to 1 MHz for $\tau = 1$ μs), the equivalent independent fading channel has only a low number of diversity branches of significant power. We found that the *diversity branch spectrum* as shown in Figure 4.42 is nearly independent of K if K is significantly greater than X. It is therefore a very useful quantity to characterize the diversity that can be provided by the channel. A look at the eigenvalues gives a first glimpse at how many diversity branches of the equivalent independent fading channel contribute

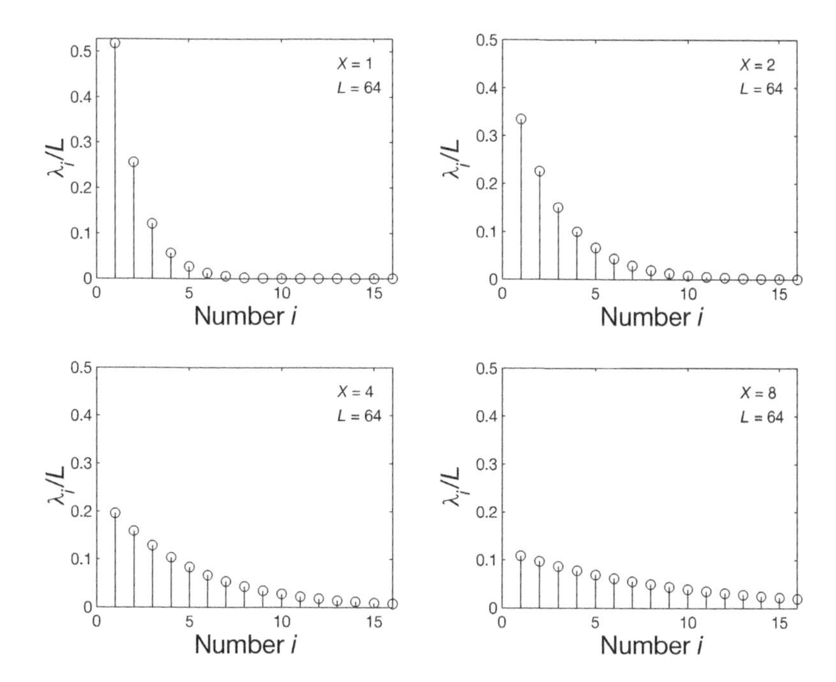

Figure 4.42 Diversity branches of the equivalent independent fading channel and normalized bandwidth $X = B\tau_m = 1, 2, 4, 8$.

significantly to the transmission. It finds its reflection in the performance curves. Figure 4.43 shows the pairwise (=bit) error probability for $K = 32$ and $X = B\tau_m = 0.5, 1, 2, 4, 8, 16$. The high diversity degree of the repetition code ($K = 32$) can show a high diversity gain if the equivalent channel has enough independent diversity branches of significant power. This is the case for $X = 16$, but not for $X = 1$ or $X = 2$. For low X, a lower repetition rate K would have been sufficient.

Figure 4.44 shows the bit error probability for $K = 10$ and the same values of X. For low X, the curves of Figure 4.43 and 4.44 are nearly identical. For higher X, the curves of Figure 4.44 run into a saturation that is given by the performance curve of the independent Rayleigh fading. For $X = 8$, this limit is practically achieved. There is still a gap of nearly 2 dB in the AWGN limit at the bit error rate of 10^{-4}.

For BPSK and any linear code, the probability for an error event corresponding to a Hamming distance d is given by

$$P_d = \frac{1}{\pi}\int_0^{\pi/2} \prod_{i=1}^{d} \frac{1}{1 + \frac{\lambda_i}{\sin^2\theta}\frac{E_S}{N_0}}\, \mathrm{d}\theta, \tag{4.23}$$

which can be upper bounded by

$$P_d \leq \frac{1}{2} \prod_{i=1}^{d} \frac{1}{1 + \lambda_i \frac{E_S}{N_0}}.$$

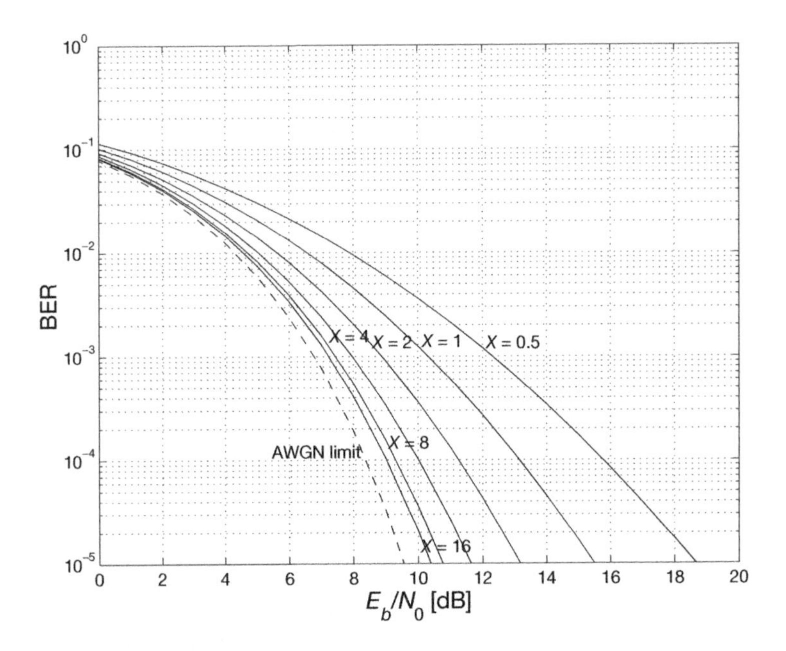

Figure 4.43 Bit error probabilities for 32-fold repetition diversity with $X = 0.5$, 1, 2, 4, 8, 16.

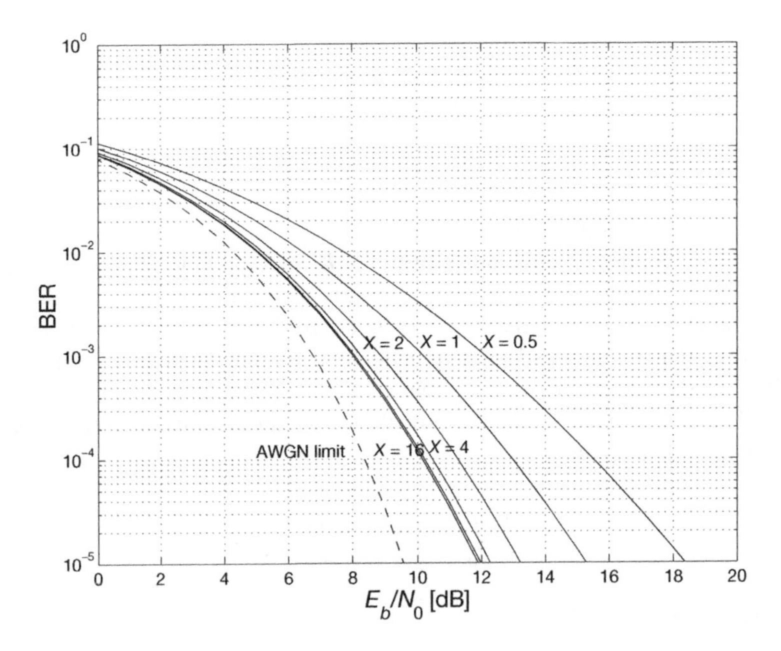

Figure 4.44 Bit error probabilities for 10-fold repetition diversity with $X = 0.5$, 1, 2, 4, 8, 16.

For the region of reasonable E_S/N_0, those factors with $\lambda_i \ll 1$ do not contribute significantly to the product. Thus, it is not possible to obtain tight union bounds like

$$P_d \leq \sum_{d=d_{\text{free}}}^{\infty} c_d P_d$$

because P_d does not decrease as $(E_S/N_0)^{-d}$ if d is greater than the diversity degree of the channel, that is, the number of significant eigenvalues λ_i. The c_d values grow with d and thus the union bound will typically diverge.

However, the diversity branch spectrum may serve as a good indicator of whether the time-frequency interleaving for a coded OFDM system is sufficient. Consider for example a system with a convolutional code[9] with free distance $d_{\text{free}} = 10$ like the popular NASA code $(133, 171)_{\text{oct}}$. The probability for the most likely error event is given by Equation (4.23) with $d = d_{\text{free}} = 10$. This probability will decrease as $(E_S/N_0)^{-10}$ only if the 10 eigenvalues λ_i, $i = 1, \ldots, 10$ are of significant size. Let us consider an OFDM system with a pseudorandom time-frequency interleaver over the time T_{frame} of one frame and over a bandwidth B. We consider a GWSSUS model scattering function given by

$$S(\tau, \nu) = S_{\text{Delay}}(\tau) S_{\text{Doppler}}(\nu)$$

as a product of a delay power spectrum $S_{\text{Delay}}(\tau)$ and a Doppler spectrum $S_{\text{Doppler}}(\nu)$. As a consequence, the time-frequency autocorrelation function also factorizes into

$$\mathcal{R}(f, t) = \mathcal{R}_f(f)\mathcal{R}_t(t).$$

We assume an exponential power delay spectrum with delay time constant τ_m that has a frequency autocorrelation function

$$\mathcal{R}_f(f) = \frac{1}{1 + j2\pi f \tau_m}$$

and an isotropic Doppler spectrum (Jakes spectrum) with a maximum Doppler frequency $\nu_{\max}$ that has a time autocorrelation function given by

$$\mathcal{R}_t(t) = \mathrm{J}_0\left(2\pi \nu_{\max} t\right).$$

The correlation lengths in frequency and time are given by $f_{\text{corr}} = \tau_m^{-1}$ and $t_{\text{corr}} = \nu_{\max}^{-1}$, respectively.

The $d_{\text{free}} = 10$ time-frequency positions (t_i, f_i) of the BPSK symbols corresponding to the most likely error event are spread randomly over the time T_{frame} and the bandwidth B. Thus, the diversity branch spectrum is a random vector. To eliminate this randomness, we average over an ensemble of 100 such vectors, which turns out to be enough for a stable result. To justify this procedure, we recall that error probabilities are averaged quantities.

Figure 4.45 shows the diversity branch spectrum $\{\lambda_i\}_{i=1}^{10}$ for frequency interleaving only (i.e. $T_{\text{frame}}/t_{\text{corr}} = 0$) and values $B/f_{\text{corr}} = 1, 2, 4, 8, 16, 32$ for the normalized bandwidth. It can be seen that even for $B/f_{\text{corr}} = 32$, the full diversity is not reached because the size

[9] Similar considerations apply for linear block codes.

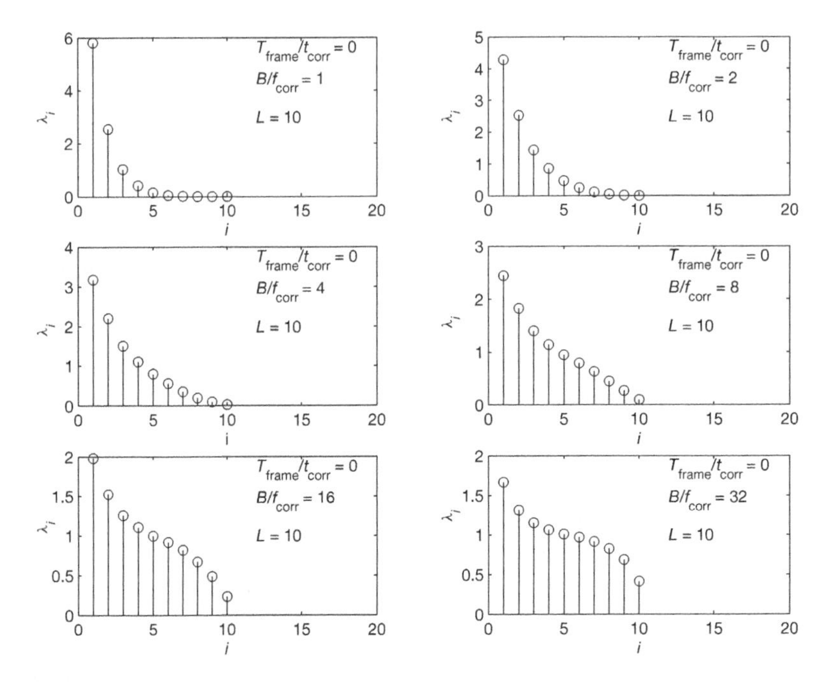

Figure 4.45 Diversity branch spectrum for $d = 10$ and frequency interleaving only.

of normalized eigenvalues is very different and the greatest values dominate the product. As shown in Figure 4.46, the same is true if only time interleaving is applied. The figure shows the spectra for time interleaving over a normalized length of $T_{\text{frame}}/t_{\text{corr}} = 1, 2, 4, 8, 16, 32$. Note that, due to the different autocorrelation in time and frequency domain, both diversity branch spectra show a different shape. Figure 4.47 shows the diversity branch spectrum for combined frequency-time interleaving. It can be seen that both mechanisms help each other, and for a wideband system with long time interleaving, all eigenvalues contribute to the product. However, the interleaving can be considered to be ideal only if all eigenvalues are of nearly the same size. As shown in Figure 4.48, a huge time-frequency interleaver is necessary to achieve this.

We may say that an OFDM system is a *wideband system* if the system bandwidth B is large enough compared to f_{corr} so that the frequency interleaver works properly. For a well-designed OFDM system, the guard interval length Δ must be matched to the maximum echo length. Assume, for example, a channel with $\tau_m = \Delta/5$ and a guard interval of length $\Delta = T/4$. Using $B = K/T$, where K is the number of carriers and T is the Fourier analysis window length, we obtain the relation

$$K = 20B\tau_m.$$

With a look at the figures we may speak of a wideband system, for example, for $B/f_{\text{corr}} = B\tau_m = 32$, which leads to $K = 640$. There may of course occur flat fading channels with $\tau_m \ll \Delta$, where the frequency interleaving fails to work. But we may conclude that an OFDM system may be called a *wideband* system relative to the channel parameters only

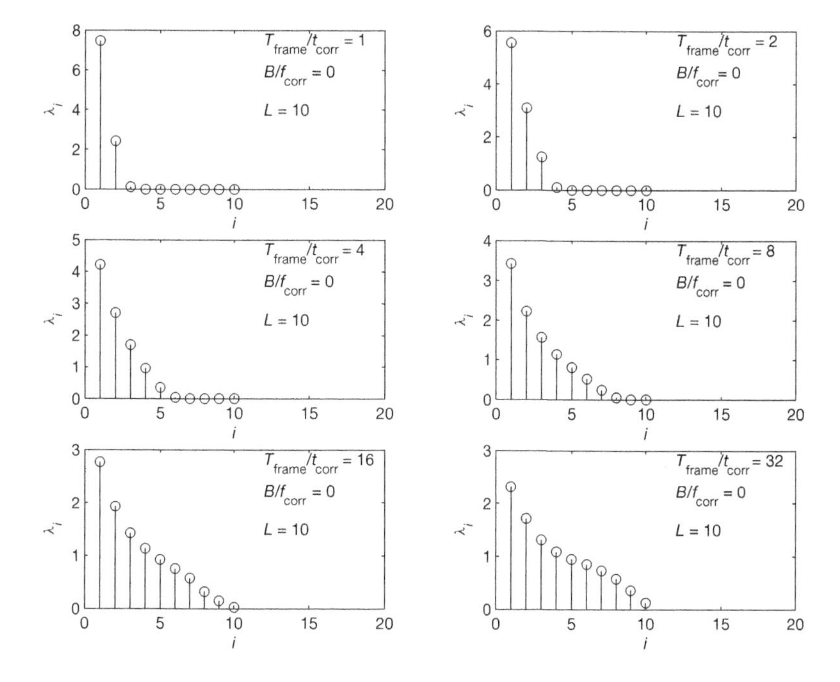

Figure 4.46 Diversity branch spectrum for $d = 10$ and time interleaving only.

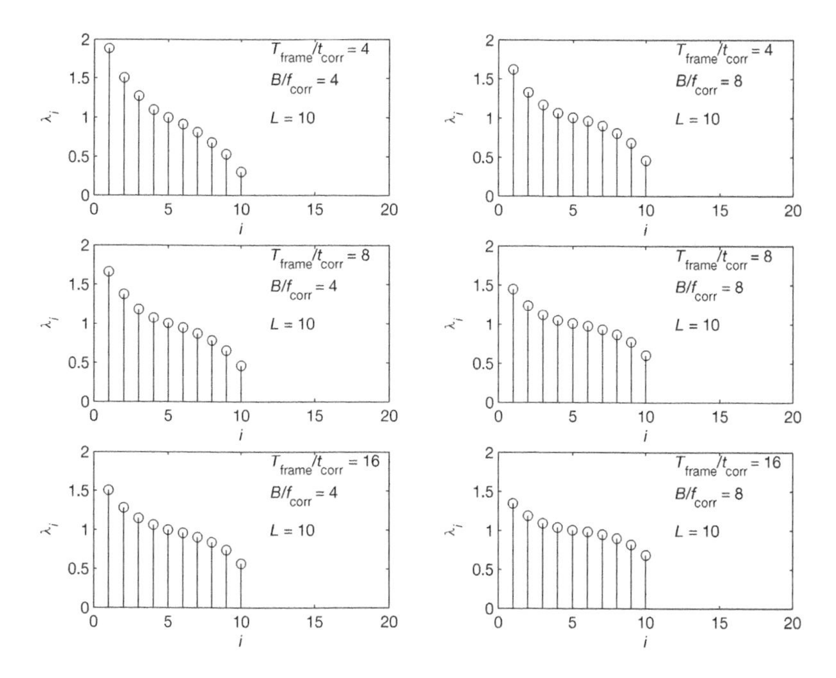

Figure 4.47 Diversity branch spectrum for $d = 10$ for moderate time-frequency interleaving.

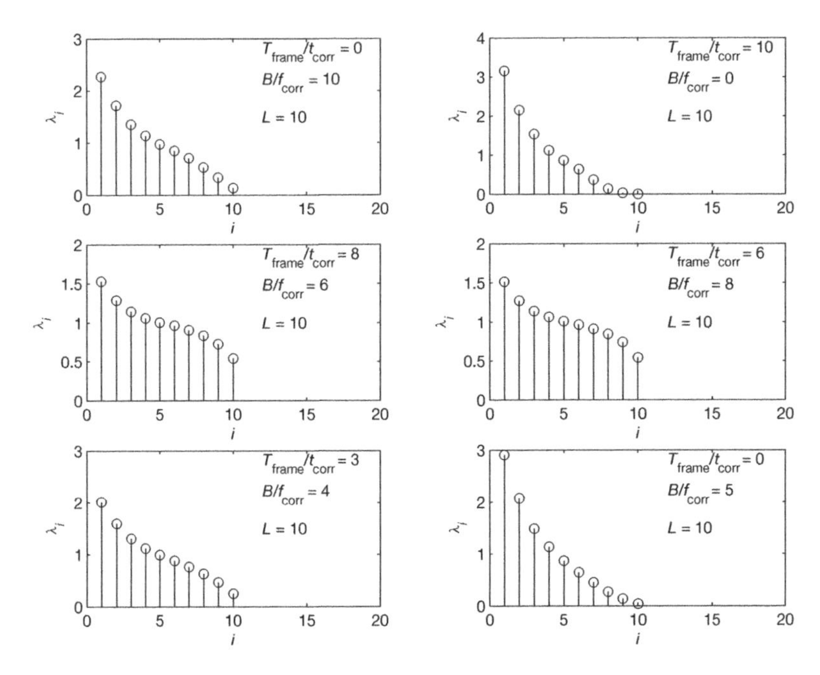

Figure 4.48 Diversity branch spectrum for $d = 10$ and small and huge time-frequency interleavers.

if at least several hundred subcarriers are used. This is the case for the digital audio and video broadcasting systems DAB and DVB-T. It is not the case for the WLAN systems IEEE 802.11a and HIPERLAN/2 with only 48 carriers.

Time interleaving alone is often not able to provide the system with sufficient diversity. A certain vehicle speed can, typically, not be guaranteed in practice. For the DAB system working at 225 MHz, a vehicle speed of 48 km/h leads to a Doppler frequency that is as low as 10 Hz. For such a Doppler frequency, sufficient time interleaving alone would lead to a delay of several seconds, which is not tolerable in practice. It is an attractive feature of OFDM that the time and frequency mechanisms together may often lead to a good interleaving. However, there will always be situations where the correlations of the channel must be taken into account.

4.5 Modulation and Channel Coding for OFDM Systems

4.5.1 OFDM systems with convolutional coding and QPSK

In this subsection, we present theoretical performance curves for OFDM systems with QPSK modulation, both with differential and coherent demodulation. These curves are of great relevance for the performance analysis of existing practical systems. Fortunately, most practical OFDM systems use essentially the same convolutional code, at least for the inner code. And most of these systems use QPSK modulation, at least as one of several possible options. DAB always uses differential QPSK, and DVB-T as well as the WLAN systems (IEEE 802.11a and HIPERLAN/2) use QAM, where QPSK is a special case. These WLAN systems also have the option to use BPSK. The performance curves for coherent BPSK are the same as those for QPSK when plotted as a function of E_b/N_0. When plotted as a function of *SNR*, there is a gap of 3.01 dB between the BPSK and the QPSK curves. The performance of higher-level QAM will be discussed in a subsequent subsection.

The channel coding of all the above-mentioned systems is based on the so-called NASA planetary standard, the rate 1/2, memory 6 convolutional code with generator polynomials $(133,\ 171)_{\mathrm{oct}}$, that is,

$$\mathbf{g}(D) = \begin{pmatrix} 1 + D^2 + D^3 + D^5 + D^6 \\ 1 + D + D^2 + D^3 + D^6 \end{pmatrix}.$$

This code can be punctured to get higher code rates. For the DAB system, lower code rates are needed, for example, to protect the most sensitive bits in the audio frame, and two additional generator polynomials are introduced. The generator polynomials of this code $R_c = 1/4$ are given by $(133,\ 171,\ 145,\ 133)_{\mathrm{oct}}$, that is,

$$\mathbf{g}(D) = \begin{pmatrix} 1 + D^2 + D^3 + D^5 + D^6 \\ 1 + D + D^2 + D^3 + D^6 \\ 1 + D + D^4 + D^6 \\ 1 + D^2 + D^3 + D^5 + D^6 \end{pmatrix}.$$

This encoder is depicted in Figure 4.49. The shift register is drawn twice to make it easier to survey the picture. For DVB-T and the wireless LAN systems, only the part of the code corresponding to the upper shift register is used.

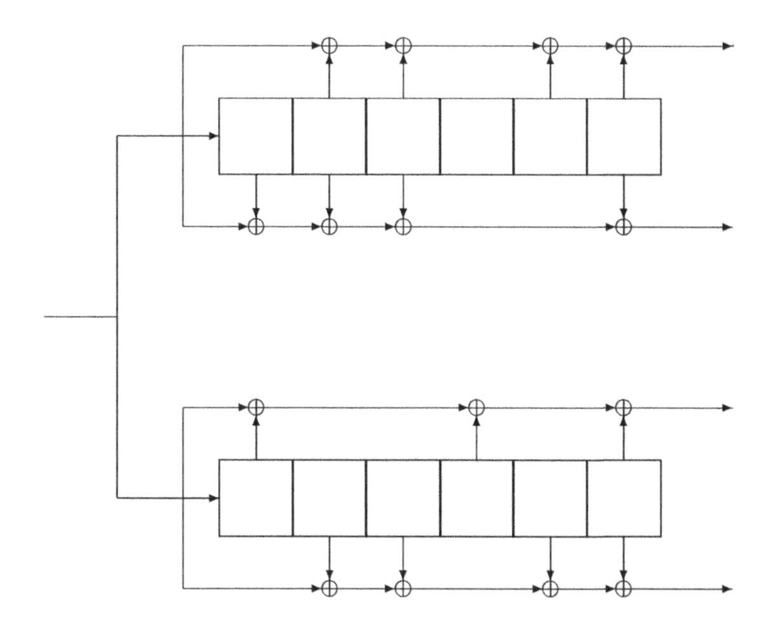

Figure 4.49 The DAB convolutional encoder.

The bit error rates for a convolutional code can be upper bounded by the union bound

$$P_d \leq \sum_{d=d_{\text{free}}}^{\infty} c_d P_d. \tag{4.24}$$

Here, P_d is the PEP for d-fold diversity as given by the expressions in Subsection 2.4.6. The coefficient c_d is the error coefficient corresponding to all the error events with Hamming distance d. We note that c_d depends only on the code, while P_d depends only on the modulation scheme and the channel. The union bound given in Equation (4.24) is valid for any channel. For an AWGN channel, the error event probability is simply given by

$$P_d = \frac{1}{2}\text{erfc}\left(\sqrt{d\frac{E_S}{N_0}}\right),$$

where $E_S = |s|^2$ is the energy of the PSK symbol s. For the independently fading Rayleigh channel, the expressions for the error event probabilities P_d were discussed in Subsection 2.4.6. All the curves asymptotically decay as

$$P_d \sim \left(\frac{E_S}{N_0}\right)^{-d}.$$

The union bound is also valid for the correlated fading channel, but it does not tightly bound the bit error rate. It may even diverge. This is because the degree of the channel diversity is limited and the pairwise error probabilities for diversity run into a saturation for $d \to \infty$, while the coefficients c_d grow monotonically.

The c_d values can be obtained by the analysis of the state diagram of the code. In Hagenauer's paper about RCPC (rate compatible punctured convolutional) codes (Hagenauer 1988), these values have been tabulated for punctured codes of rate $R_c = 8/N$ with $N \in \{9, 10, 11, \ldots, 24\}$. These punctured codes have been implemented in the DAB system. In the other systems, some different code rates are used. However, their performance can be estimated from the closest code rates of that paper. We now discuss the performance of these codes for (D)QPSK in a Rayleigh fading channel.

First we consider DQPSK and an ideally interleaved Rayleigh fading channel with the isotropic Doppler spectrum of maximum Doppler frequency $\nu_{\max}$. The P_d values depend on the product $\nu_{\max} T_S$. High values of this product cause a loss of coherency between adjacent symbols, which degrades the performance of differential modulation. We first consider the ideal case $\nu_{\max} T_S = 0$. In practice, this is of course a contradiction to the assumption of ideal interleaving. But we may think of a very huge (time and frequency) interleaver and the limit of very low vehicle speed. Figure 4.50 shows the union bounds of the performance curves in that case for several code rates. We have plotted the bit error probabilities as a function of the *SNR*, not as a function of E_b/N_0. The latter is better suited to compare the power efficiencies, but for practical planning aspects the *SNR* is the relevant physical quantity. Both are related by

$$SNR = \frac{T}{T_S} R_c \log_2(M) \frac{E_b}{N_0}$$

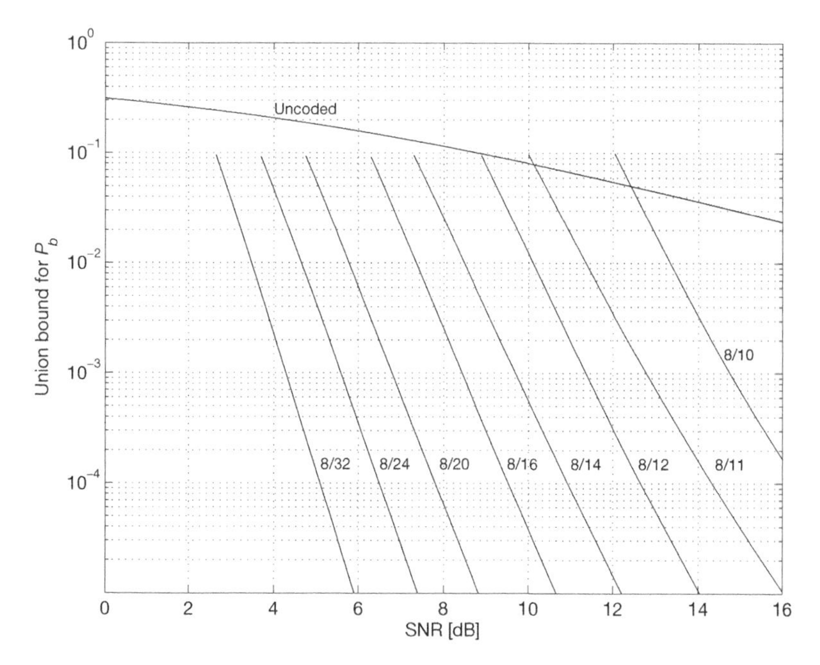

Figure 4.50 Union Bounds for the bit error probability for DQPSK and $\nu_{\max} T_S = 0$ for $R_c = 8/10,\ 8/11,\ 8/12,\ 8/14,\ 8/16,\ 8/20,\ 8/24,\ 8/32$.

with $M = 4$ for (D)QPSK. Another reason to plot the different performance curves together as a function of the SNR is that different parts of the data stream may be protected by different code rates as it is the case for the DAB system discussed in Subsection 4.6.1. Here, all parts of the signal are affected by the same SNR. For example, the curves of Figure 4.50 are the basis for the design of the unequal error protection (UEP) scheme of the DAB audio frame, where the most important header bits are better protected than the audio scale factors that are better protected than the audio samples. For more details, see (Hoeg and Lauterbach 2003; Hoeher *et al.* 1991). The curves show that there is a high degree of flexibility to choose the appropriate error protection level for different applications. Note that there are still intermediate code rates in between that have been omitted in order not to overload the picture. Figure 4.51 shows the union bounds for the performance curves for the same codes, but with a higher Doppler frequency corresponding to $\nu_{\max} T_S = 0.02$. For the DAB system (Transmission Mode I) with $T_S \approx 1250$ µs working at 225 MHz, this corresponds to a moderate vehicle speed of approximately 80 km/h. One can see that the curves become less steep, and *flatten out*. This effect is greater for the weak codes, and it is nearly neglectible for the strong codes. In any case, this degradation is still small. Figure 4.52 shows the union bounds for the performance curves for the same codes, but with a higher Doppler frequency corresponding to $\nu_{\max} T_S = 0.05$. For the DAB system (Transmission Mode I) with $T_S \approx 1250$ µs working at 225 MHz, this corresponds to a high vehicle speed of approximately 190 km/h. The curves flatten out significantly; the loss is approximately 1.5 dB at $P_b = 10^{-4}$ for $R_c = 8/16$, and it is more than 3 dB for $R_c = 8/12$.

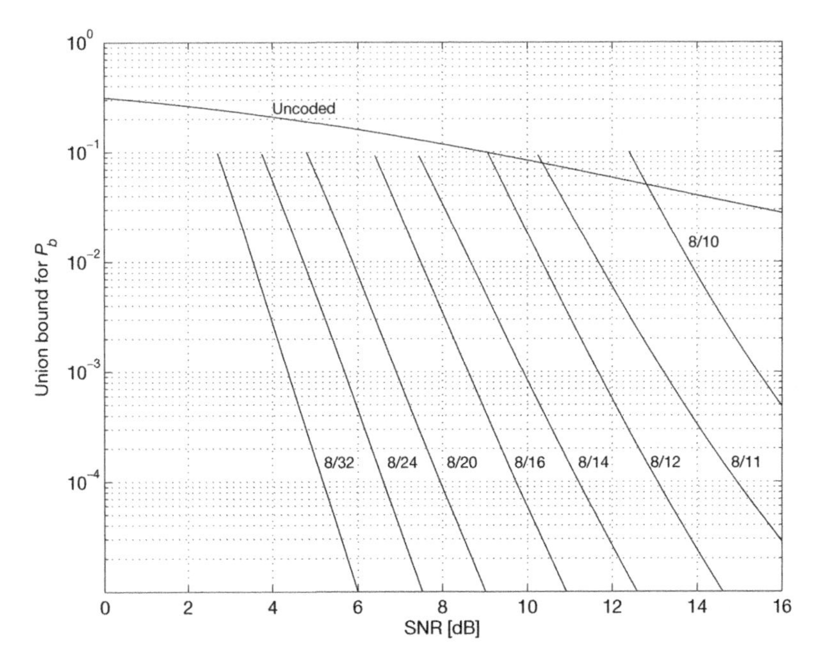

Figure 4.51 Union Bounds for the bit error probability for DQPSK and $\nu_{\max} T_S = 0.02$ for $R_c = 8/10,\ 8/11,\ 8/12,\ 8/14,\ 8/16,\ 8/20,\ 8/24,\ 8/32$.

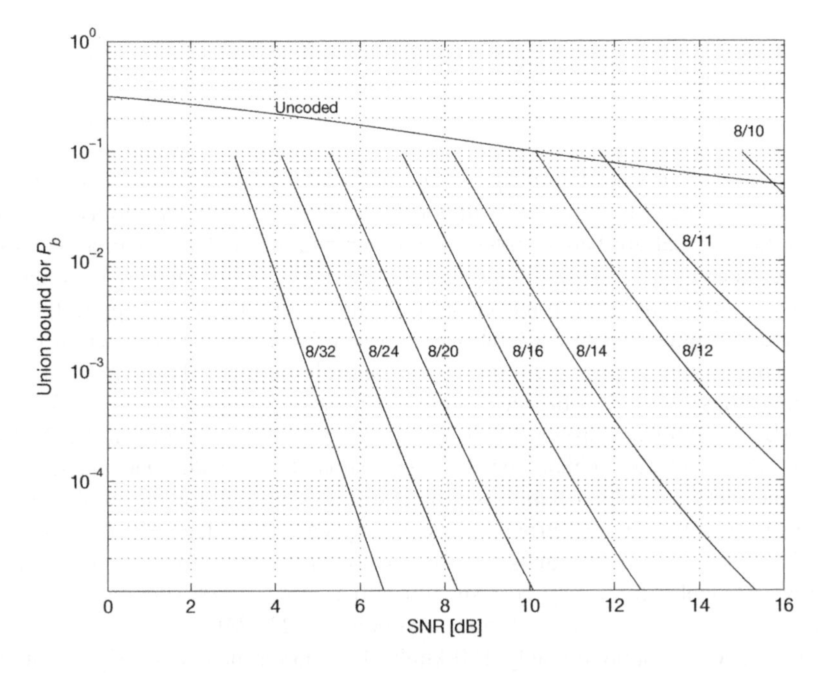

Figure 4.52 Union Bounds for the bit error probability for DQPSK and $\nu_{\max} T_S = 0.05$ for $R_c = 8/10,\ 8/11,\ 8/12,\ 8/14,\ 8/16,\ 8/20,\ 8/24,\ 8/32$.

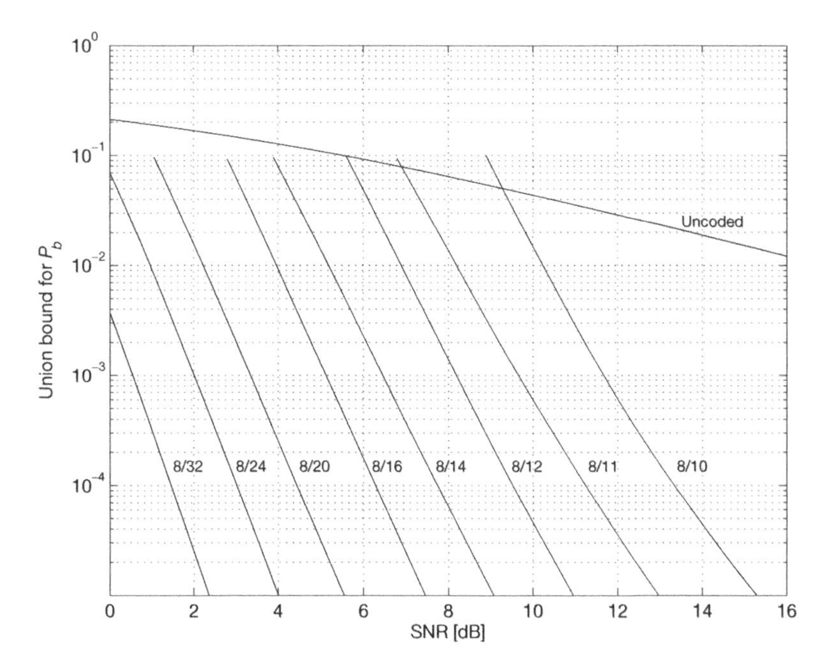

Figure 4.53 Union Bounds for the bit error probability for QPSK and $R_c = 8/10$, 8/11, 8/12, 8/14, 8/16, 8/20, 8/24, 8/32.

As long as the interleaving is sufficient, all these curves fit quite well to computer simulations. We will show some DQPSK performance curves for simulations of the DAB system in a subsequent section. One must keep in mind that high Doppler frequencies also effect the orthogonality of the subcarriers, which will cause additional degradations. However, this effect turns out to be significantly smaller than the DQPSK coherency loss for each single subcarrier.

Figure 4.53 shows the union bounds of the performance curves for QPSK and the same code rates. QPSK is not affected directly by the Doppler spread. However, the loss of orthogonality will also degrade QPSK. In practice, the most significant loss due to high Doppler frequencies turns out to be due to degradations in the channel estimation. In fact, it was generally believed for many years that for this reason, in practice, coherent QPSK is not really superior to differential QPSK, because this channel estimation loss approximately compensates the gain. In a subsequent section, we will discuss this item and we will show that this is not true.

4.5.2 OFDM systems with convolutional coding and M^2-QAM

In this subsection, we analyze the performance of OFDM systems with M^2-QAM modulation, as it is used for DVB-T as well as the WLAN systems IEEE 802.11a and HIPERLAN/2. The channel coding of these systems is based on the same coding scheme as discussed in the preceding subsection.

Applications with higher data rate than audio broadcasting motivated system designers to consider higher-level QAM modulation schemes for OFDM systems. Examples are the terrestrial digital video broadcasting system DVB-T and the wireless LAN standards IEEE 802.11a and HIPERLAN/2. Coding and modulation are closely connected in such systems and they must be carefully fitted together. For the above-mentioned systems, an approach has been chosen, which uses standard convolutional coding and QAM modulation with conventional Gray mapping and a bit interleaver in between. Such an approach is called bit *interleaved coded modulation* (BICM) in the literature, and these systems are probably the first applications of BICM. There are two arguments for using BICM rather than trellis-coded modulation with Ungerböck codes and set partitioning:

1. The implementation aspect: it is possible to use standard components for the Viterbi decoder.

2. The performance aspect: for the performance in a fading channel, the Hamming distance is more important than the Euclidean distance. A high Hamming distance can be achieved by choosing sufficiently strong codes.

The system model

Because only those are applied in the above-mentioned systems, we restrict ourselves to square M^2-QAM constellations that are Cartesian products of two M-ASK constellations for the I (inphase) and the Q (quadrature) component. We regard it as convenient to interpret a QAM symbol as a two-dimensional real symbol instead of a one-dimensional complex symbol. Each ASK symbol is labeled by $m = \log_2 M$ bits. The block diagram for the transmitter is shown in Figure 4.54.

The useful data bit stream a_i will be encoded by a convolutional encoder with code rate R_c to produce an encoded bit stream b_i. Between the encoder and the symbol mapper, bit interleaving will be applied to avoid closely neighboring bits in the code word to be mapped onto the same QAM symbol. For the theoretical analysis, this bit interleaver will be modeled as a pseudorandom permutation π of the time index together with a pseudorandom serial–parallel (S/P) conversion for the symbol mapping of m parallel bits of one ASK symbol. Both are assumed to be statistically independent. This block is given by a random index map $\pi : i \to (k, l)$ that chooses for each time index i of the encoded bit b_i a new time position with index l and a labeling position $k \in \{0, \ldots, m-1\}$ for the Gray labeling ($k = 0$ means LSB, $k = m - 1$ means MSB). For each time index l, the m bits

$$c_l^{(k)} = b_i = b_{\pi^{-1}(k,l)},\ k = 0, 1, \ldots, m-1$$

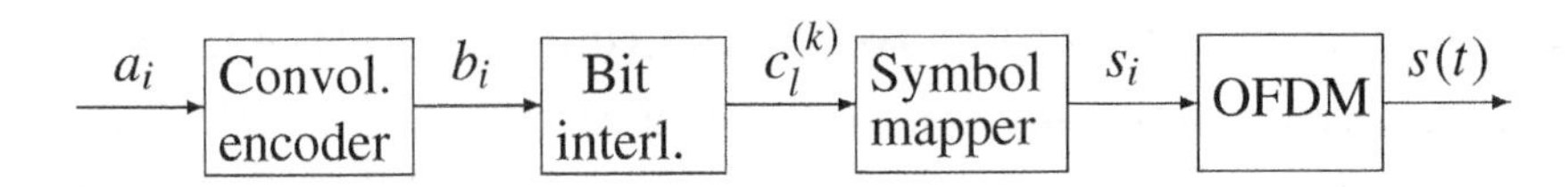

Figure 4.54 Transmitter block diagram for an OFDM system with convolutional coding and QAM.

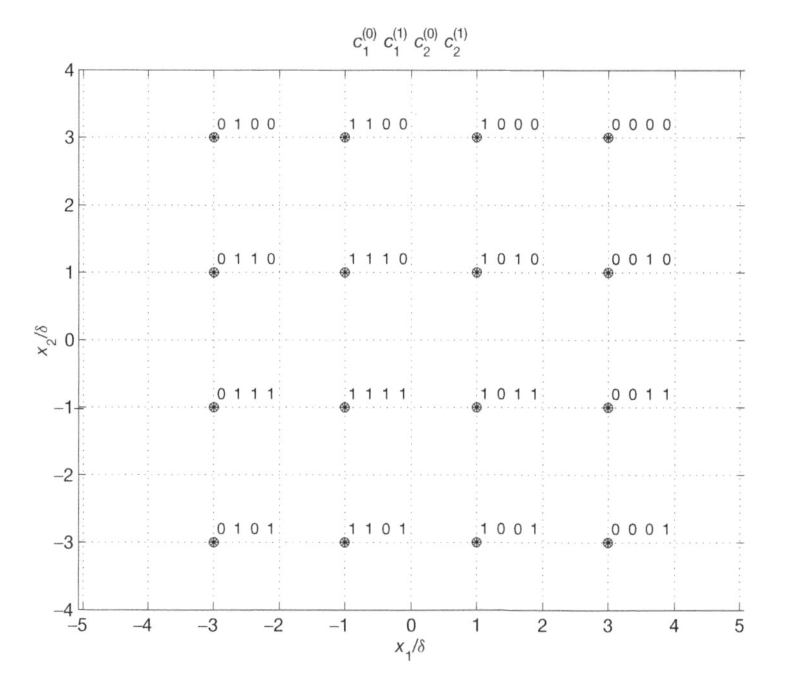

Figure 4.55 16-QAM with Gray mapping.

determine which ASK symbol will be transmitted. Let x_l denote the sequence of ASK symbols. Then $x_1, x_3, x_5, \ldots$ is the sequence of inphase symbols and $x_2, x_4, x_6, \ldots$ is the sequence of the quadrature component symbols. Each ASK symbol can take the values

$$x_l \in \mathcal{C} := \{\pm\delta, \pm 3\delta, \ldots, \pm(M-1)\delta\}$$

of the signal constellation $\mathcal{C}$. Here, we introduce a distance unit δ that is related to the symbol energy. The symbol mapper X maps $m = \log_2 M$ bits $c_l^{(0)}, c_l^{(1)}, \ldots, c_l^{(m-1)}$ on a real symbol x_l. Two subsequent M-ASK symbols x_{2i-1} and x_{2i} are composed to a M^2-QAM symbol $s_i = x_{2i-1} + jx_{2i}$. Figure 4.55 shows a 16–QAM configuration with this mapping. Note that we write $c_l^{(0)} c_l^{(1)}$ and thus the LSB is in the leftmost position. The QAM symbols will be processed by the OFDM unit, which will typically include a symbol interleaving in frequency direction, as it is the case for the above-mentioned systems. Finally, the OFDM signal $s(t)$ will be transmitted over the channel.

The receiver

At the receiver, the signal will first be processed by the OFDM unit to produce output symbols $r_i = y_{2i-1} + jy_{2i}$. We thereby assume perfect back rotation of the phase so that we can work with the discrete real channel model as given by

$$y_l = a_l x_l + n_l$$

or, in vector notation, by

$$\mathbf{y} = \mathbf{A}\mathbf{x} + \mathbf{n}.$$

Here, the sequence of ASK symbols x_l is written as a vector $\mathbf{x} = (x_1, x_2, x_3, \ldots)^T$. $\mathbf{y}$ is the vector of received symbols and $\mathbf{n}$ is the real AWGN vector with variance $\sigma^2 = N_0/2$ in each component. The fading is described by the diagonal matrix $\mathbf{A} = \mathrm{diag}(a_1, a_2, a_3, \ldots)^T$ of (real) fading amplitudes. The fading amplitudes are normalized to average power one. We assume independent Ricean fading amplitudes. We denote the energy per (two-dimensional) QAM symbol by E_S and the energy per data bit by E_b. The relation between both is given by

$$E_S = R_c \log_2(M^2) E_b.$$

One can easily show that $E_S = 2\delta^2, 10\delta^2, 42\delta^2, \ldots$ for 4-QAM, 16-QAM, 64-QAM, ..., and so on. For OFDM with guard interval $\Delta = T_S - T$, the relation to the RF SNR is given by

$$SNR = \frac{T}{T_S} \frac{E_S}{N_0}.$$

From the real receive symbols y_l, soft metric values must be calculated by the metric computation unit (MCU) to obtain soft metric values μ_i as the input for the Viterbi decoder. For each received symbol y_l, the MCU generates the metric values $\nu_l^{(k)}$, that is, the soft decision values of the bits $c_l^{(k)}$ corresponding to labeling positions $k \in \{0, 1, \ldots, m-1\}$. The soft metric values $\nu_l^{(k)}$ are deinterleaved by the inverse permutation $\pi^{-1} : (k, l) \mapsto i = \pi^{-1}(k, l)$. The deinterleaved metric values are then given by

$$\mu_i = \mu_{\pi^{-1}(k,l)} = \nu_l^{(k)}.$$

The decoder DEC is a Viterbi decoder that decides for the bit sequence $\mathbf{b} = (b_1, b_2, \ldots)$ that maximizes

$$\mu(\mathbf{b}) = \sum_i \mu_i (-1)^{b_i}.$$

Metric calculation

There are several possibilities to obtain $m = \log_2 M$ metric values $\nu_l^{(k)}$ (i.e. soft decision variables) from the receive symbol y_l. Such a metric value must be positive if the bit value $c_l^{(k)} = 0$ is more likely, and negative if otherwise. Its absolute value should be a measure for the reliability of the bit value. Obviously, the LLR value $L(c_l^{(k)} = 0 | y_l)$ would be the best choice, and it can easily be obtained from the LLR formalism. However, we will first construct a suboptimal simple *threshold metric* (TR) in a geometrical illustrative way without using this formalism.

We note that for Gray mapping, the MSB $c_l^{(m-1)}$ just indicates the sign of the symbol. Thus, even though the amplitudes are different and therefore it is not exactly the same case as bipolar signaling (2-ASK or BPSK), it seems to be an appropriate choice to obtain a soft decision variable in the same way and take y_l directly as the metric value for the MSB in the AWGN channel. For the fading channel, we set $\nu_l^{(m-1)} = a_l y_l$, that is, the bipolar receive value has to be weighted by the channel amplitude to form a decision variable. For the next less significant bit $c_l^{(m-2)}$, there is a threshold at $x_l = +\frac{M}{2}\delta$ if $c_l^{(m-1)} = 0$ and at $x_l = -\frac{M}{2}\delta$ if $c_l^{(m-1)} = 1$ (see Figure 4.56) for 4-ASK[10] with $a_l = 1$. Thus, if we discard

[10]Note again that we write $c_l^{(0)} c_l^{(1)}$ and thus the LSB is in the leftmost position.

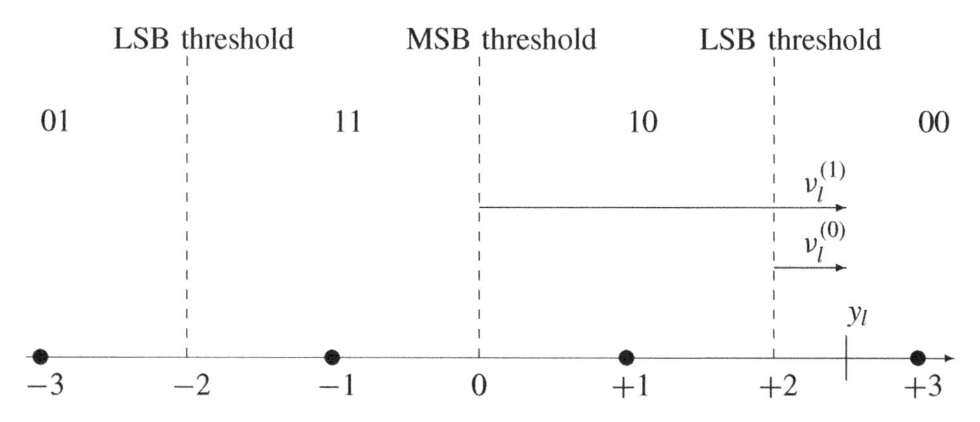

Figure 4.56 Threshold metric calculation for 4-ASK.

the information of the MSB by taking the absolute value, we again have the situation of bipolar signaling with a threshold shifted by $\frac{M}{2}\delta$ for the AWGN channel and shifted by $a_l\frac{M}{2}\delta$ for the fading channel. Thus, for the bit $k = m - 2$ we take $|y_l| - a_l\frac{M}{2}\delta$ instead of y_l for the MSB. Thus, we use the metric $v_l^{(m-2)} = a_l|y_l| - a_l^2\frac{M}{2}\delta$. We proceed in the same way for the next bits until the LSB and obtain the metric values by the recursion

$$v_l^{(m-1-\kappa)} = |v_l^{(m-\kappa)}| - a_l^2\frac{M}{2^\kappa}\delta \tag{4.25}$$

with $v_l^{(m-1)} = a_l y_l$.

In this construction, we have used decision thresholds that depend on the fading amplitude a_l since the size of signal constellation is attenuated by this factor. From the engineering point of view, it seems to be natural to compensate this by means of an *equalizer* that divides by the amplitude and then computes the decision variables as in the case of a channel without fading. However, this equalizer will inflate the noise because the noise samples of very unreliable receive symbols with small a_l will be amplified more than others and will corrupt the decision through an inappropriately high magnitude. One easily sees that a_l^2 is the appropriate weight factor for the decision variables: one must multiply by a_l to rescind the noise inflation done by the equalizer and then multiply again by a_l, which is the appropriate weight factor for antipodal decisions in a fading channel. The setup is depicted in Figure 4.57.

First, the received symbols are equalized. The equalized receive symbols $\eta_l = a_l^{-1}y_l$ are then used to calculate the decision variables $\theta_l^{(k)}$ given by $\theta_l^{(m-1)}$ and

$$\theta_l^{(m-1-\kappa)} = |\theta_l^{(m-\kappa)}| - \frac{M}{2^\kappa}\delta \tag{4.26}$$

for an AWGN channel. These AWGN metric values are then weighted to build the appropriate fading channel metric values

$$v_l^{(k)} = a_l^2\theta_l^{(k)}.$$

One can easily see that these values for $v_l^{(k)}$ are the same as those calculated by Equation (4.25).

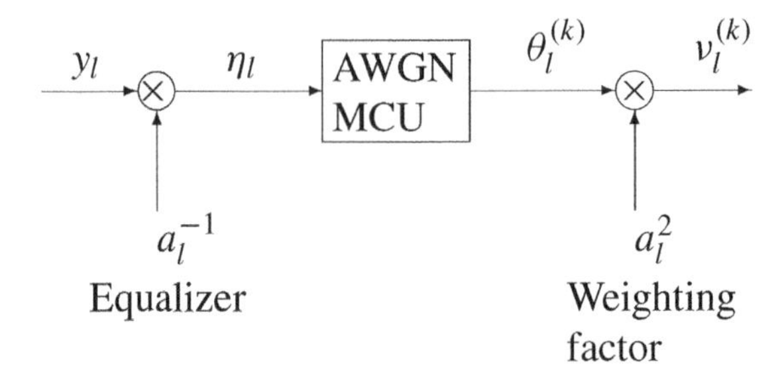

Figure 4.57 Metric computation for QAM by using an equalizer.

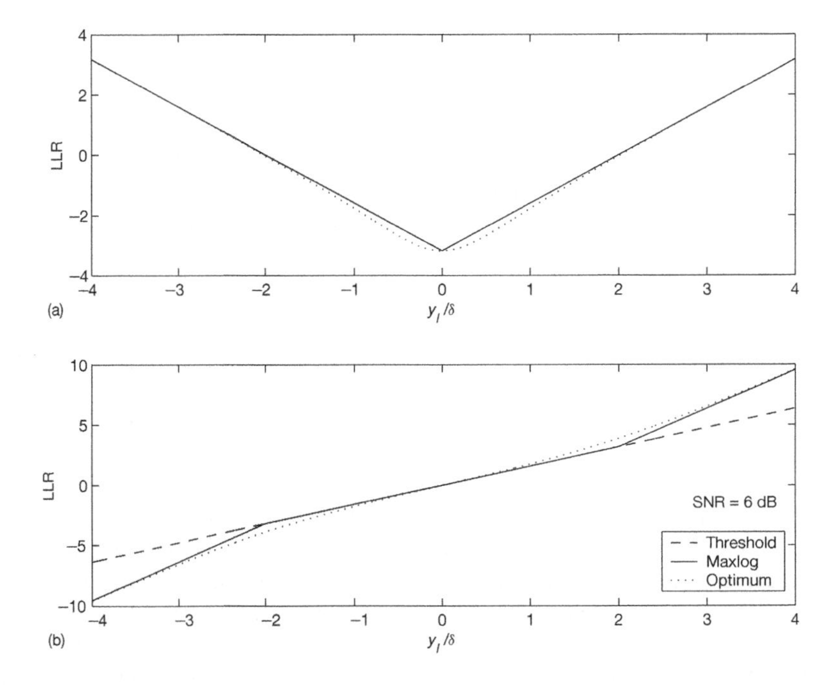

Figure 4.58 Comparison of metric expression for 4-ASK at SNR = 6 dB: LSB (a) and MSB (b).

This intuitively convincing threshold receiver has not been derived from general principles. It is suboptimum, but, as we shall see, the performance loss compared to the optimum is not severe. Given a receive symbol y_l, the optimum receiver is the one that calculates the LLR of a certain bit and uses it as the optimum metric for $\mu_i = \nu_l^{(k)}$ to be fed into the Viterbi decoder. This LLR is given by (see Subsection 3.1.4)

$$L\left(c_l^{(k)} = 0\,|y_l\right) = \log\left(\frac{\Pr(c_l^{(k)} = 0|y_l)}{\Pr(c_l^{(k)} = 1|y_l)}\right).$$

The probability $\Pr(c_l^{(k)} = c|y_l)$ that the transmitted bit at label k has the value c under the condition that y_l has been received may depend on hard or soft decision values from other bits $c_l^{(k')}$, $k' \neq k$ that are known from the preceding decoding steps. This means that the constellation points x_l may have different *a priori* probabilities $\Pr(x_l)$. Let $\mathcal{C}_0^{(k)}$ and $\mathcal{C}_1^{(k)}$ be the subset of the constellation corresponding to $c_l^{(k)} = 0$ and $c_l^{(k)} = 1$, respectively, and

$$p(y_l|a_l x_l) = \frac{1}{\sqrt{\pi N_0}} \exp\left(-\frac{1}{N_0}|y_l - a_l x_l|^2\right) \tag{4.27}$$

the probability density for y_l under the condition that x_l was transmitted over a channel with (ideally known) fading amplitude a_l. The LLR is then given by

$$L\left(c_l^{(k)} = 0\,|\,y_l\right) = \log\left(\frac{\sum_{x_l \in \mathcal{C}_0^{(k)}} p(y_l|a_l x_l) P(x_l)}{\sum_{x_l \in \mathcal{C}_1^{(k)}} p(y_l|a_l x_l) P(x_l)}\right). \tag{4.28}$$

In practice, the *maxlog* approximation

$$L\left(c_l^{(k)} = 0\,|\,y_l\right) \approx \max_{x_l \in \mathcal{C}_0^{(k)}} \log(p(y_l|a_l x_l) P(x_l)) - \max_{x_l \in \mathcal{C}_1^{(k)}} \log(p(y_l|a_l x_l) P(x_l)). \tag{4.29}$$

can be used. If no *a priori* information is available, $\Pr(x_l) = 1/M$ for all x_l. For this case – which corresponds to the first decoding step – the metric values obtained from the three different methods (TR, LLR, maxlog) are depicted in Figure 4.58 for $M = 4$ and SNR = 6 dB. The maxlog approximation is very close to the optimum and becomes practically the same for higher SNR values. For the LSB, the maxlog and TR curves are identical. For the MSB, the TR metrics underestimates the reliability for the most reliable values of y_l.

If – in the next decoding step – hard decision values for the other bits are fed back, we must set $P(x_l) = 1/2$ for exactly one point in the subset of the constellation corresponding to $c_l^{(k)} = 0$ and for exactly one point in the subset of the constellation corresponding to $c_l^{(k)} = 1$. For all other points, we have $\Pr(x_l) = 0$. The LLR is then given by

$$L\left(c_l^{(k)} = 0\,|\,y_l\right) = \frac{2}{N_0} a_l (x_l^0 - x_l^1)\left(y_l - a_l \frac{x_l^0 + x_l^1}{2}\right). \tag{4.30}$$

The distance $|x_l^0 - x_l^1|$, however, is time varying because the values of the other bits change for different l. This time-varying distance behaves just like another fading amplitude. The knowledge of this quantity (if available) occurs in the metric as a multiplicative weighting factor and improves the performance, just like the knowledge of the channel state information for fading channels. Soft outputs of the decoder can be incorporated into the LLR in a straightforward manner. We will not investigate this further since hard feedback already gives very good results.

Error probabilities

There is hardly a chance to obtain analytical expressions for error probabilities by using the rather complicated metric expressions given by Equations (4.28) and (4.29). For the threshold metric, the task is easier and we will subsequently see (for the case of 16-QAM) how error event probabilities can be calculated at least for the AWGN channel.

However, for iterative decoding, error event probabilities for the optimum metric can be derived by available methods if we assume that all the metric calculations in the second iteration step are based on correctly fed back bits from the first iteration. In that case, we have to analyze the metric expression given by Equation (4.30), which is just a decision variable between two constellation points, but with a multiplicative random variable $a_l \Delta_l$ with

$$\Delta_l = \frac{1}{2}\left|x_l^0 - x_l^1\right|.$$

If no iterative decoding is applied, one can use union-bound techniques to upper bound the error event probability by the sum of all pairwise error probabilities that correspond to the same bit error sequence. This overestimates the error probability, but the overestimate can be reduced by an *expurgated* union bound. One can show (Caire *et al.* 1998) that only the nearest constellation corresponding to the erroneous bit needs to be taken into account. For the MSB of the 4-ASK constellation, for example, and the correct transmit symbol $x_l = 3\delta$ (MSB = 0), only the nearest neighbor $x_l = -\delta$ corresponding to the wrong MSB (MSB = 1) needs to be taken into account, while the event corresponding to $x_l = -3\delta$ can be omitted in the sum.

For both the ideal iterative decoding (ID) and the expurgated union-bound (EX) approach, the pairwise error probabilities $P_d^{\mathrm{ID/EX}}$ corresponding to an error event of Hamming distance d can be obtained from Equation (2.44). To apply that equation, we must average over the random variables Δ_l. We assume that the Δ_l are independent, identically distributed random variables with respective expectation values $\mathrm{E}_\Delta^{\mathrm{ID/EX}}\{\cdot\}$ for the ID and the EX case. For ID and EX, the random variable Δ has its specific statistics. We obtain the expression

$$P_d^{\mathrm{ID/EX}} = \frac{1}{\pi}\int_0^{\pi/2} \mathrm{E}_\Delta^{\mathrm{ID/EX}}\left\{R_K\left(\frac{1}{N_0}\cdot\frac{\Delta^2}{\sin^2\theta}\right)\right\}^d d\theta. \tag{4.31}$$

We first illustrate our result for the example $M = 4$ (16-QAM) where $x_l \in \{\pm\delta, \pm 3\delta\}$. The bit under consideration is either the MSB $b_l^{(1)}$ or the LSB $b_l^{(0)}$, both with the same probability 1/2. Consider an MSB of value 0. Then $x_l = +3\delta$ for $b_l^{(0)} = 0$ and $x_l = +\delta$ for $b_l^{(0)} = 1$. For the ID case ($b_l^{(0)}$ known), the erroneous symbols are $\hat{x}_l = -3\delta$ for $b_l^{(0)} = 0$ and $\hat{x}_l = -\delta$ for $b_l^{(0)} = 1$, respectively, corresponding to $\Delta = 3\delta$ and $\Delta = \delta$, both with equal probability. For the EX case, we need to consider only the nearest erroneous symbol, which is $\hat{x}_l = -\delta$ in any case, leading to $\Delta = 2\delta$ and $\Delta = \delta$, both with equal probability. If the bit under consideration is the LSB, $\Delta = \delta$ for ID and EX and any value of the MSB. It follows that $\Delta = \delta$ with probability 3/4 (ID and EX) and $\Delta = 3\delta$ (ID) or $\Delta = 2\delta$ (EX) with probability 1/4. We thus have

$$\mathrm{E}_\Delta^{\mathrm{ID}}\left\{R_K\left(\frac{\Delta^2}{\alpha^2}\right)\right\} = \frac{3}{4}R_K\left(\frac{\delta^2}{\alpha^2}\right) + \frac{1}{4}R_K\left(\frac{9\delta^2}{\alpha^2}\right) \tag{4.32}$$

for the ID case and

$$\mathrm{E}_\Delta^{\mathrm{EX}}\left\{R_K\left(\frac{\Delta^2}{\alpha^2}\right)\right\} = \frac{3}{4}R_K\left(\frac{\delta^2}{\alpha^2}\right) + \frac{1}{4}R_K\left(\frac{4\delta^2}{\alpha^2}\right) \tag{4.33}$$

for the EX case. Here we use the abbreviation

$$\alpha^2 := N_0 \sin^2 \theta.$$

Utilizing these expressions, Equation (4.31) can now be easily evaluated numerically.

We now show that, for the AWGN channel, a closed-form expression can be found. For $K \to \infty$, we insert

$$R_\infty = \exp(-\gamma)$$

into Equations (4.32) and (4.33) and expand the dth powers of these expressions using binomial coefficients, and insert into Equation (4.31). Using again the polar form of the Gaussian probability integral, we finally get the formulas

$$P_d^{\text{ID,AWGN}} = \left(\frac{1}{4}\right)^d \sum_{e=0}^{d} \binom{d}{e} 3^{d-e} \cdot \frac{1}{2} \operatorname{erfc}\left(\sqrt{(d+8e) \cdot \frac{\delta^2}{N_0}}\right) \tag{4.34}$$

and

$$P_d^{\text{EX,AWGN}} = \left(\frac{1}{4}\right)^d \sum_{e=0}^{d} \binom{d}{e} 3^{d-e} \cdot \frac{1}{2} \operatorname{erfc}\left(\sqrt{(d+3e) \cdot \frac{\delta^2}{N_0}}\right). \tag{4.35}$$

We note that these expressions can also be obtained without the polar form of the Gaussian probability integral. We only need some probabilistic analysis. To do so, we consider an error event where the two code words differ in d positions with time indices $l = 1, \ldots, d$. The squared Euclidean distance is then given by

$$4 \sum_{l=1}^{d} \Delta_l^2.$$

For the ID case, Δ_l is a random variable that takes the value $\Delta_i = \delta$ with probability 3/4 and $\Delta_i = 3\delta$ with probability 1/4. We calculate

$$P_d = \mathrm{E}_{\Delta_l}\left\{\frac{1}{2} \operatorname{erfc}\left(\sqrt{\frac{1}{N_0} \sum_{l=1}^{d} \Delta_l^2}\right)\right\}, \tag{4.36}$$

where E_{Δ_l} means averaging over all random variables Δ_l. To perform the average, we count all the possible events. Assume the event that a fixed sequence of e symbols with $\Delta_l = 3\delta$ and $d - e$ symbols with $\Delta_l = \delta$ has been transmitted so that the squared Euclidean distance equals

$$\sum_{l=1}^{d} \Delta_l^2 = 4(d + 8e)\delta^2. \tag{4.37}$$

There are $\binom{d}{e}$ such sequences, and each occurs with probability $\left(\frac{1}{4}\right)^e \left(\frac{3}{4}\right)^{d-e}$. Averaging over all possible sequences, we get Equation (4.34). For the EX case, the same method leads to Equation (4.35).

Using arguments of the same type, we are now able to derive expression for P_d for the soft threshold receiver in the AWGN channel for decoding without additional information about the other bit(s). Consider an error event where the two code words differ in d positions. For $l = 1, \ldots, d$, let ξ_l be the difference of transmit symbol x_l to the decision threshold. For simplicity and without loss of generality, we assume $\xi_l \geq 0$ for all l. ξ_l is a random variable that takes the value $\xi_l = \delta$ with probability 3/4 and the value $\xi_l = 3\delta$ with probability 1/4. Let η_l be the difference of the receive symbol y_l to the decision threshold, that is, $\eta_l = \xi_l + n_l$, where n_l is the real AWGN with variance $N_0/2$. Given a fixed transmit vector, an error occurs if the random variable

$$Y = \sum_{l=1}^{d} \eta_l \tag{4.38}$$

becomes negative. Assume the event that a fixed sequence of e symbols with $\xi_l = 3\delta$ and $d - e$ symbols with $\xi_l = \delta$ was transmitted. For such a fixed sequence, Y is a Gaussian random variable with mean value

$$\mu_Y = e3\delta + (d - e)\delta = (d + 2e)\delta \tag{4.39}$$

and variance

$$\sigma_Y^2 = d\frac{N_0}{2}. \tag{4.40}$$

The probability that this random variable becomes negative is given by

$$P(Y < 0) = \frac{1}{2}\operatorname{erfc}\left(\sqrt{\frac{(d+2e)^2}{d} \cdot \frac{\delta^2}{N_0}}\right). \tag{4.41}$$

Averaging over all the $\binom{d}{e}$ sequences with their respective probabilities $\left(\frac{1}{4}\right)^e \left(\frac{3}{4}\right)^{d-e}$ leads to

$$P_d^{\mathrm{TR,AWGN}} = \left(\frac{1}{4}\right)^d \sum_{e=0}^{d} \binom{d}{e} 3^{d-e} \cdot \frac{1}{2}\operatorname{erfc}\left(\sqrt{\frac{(d+2e)^2}{d} \cdot \frac{\delta^2}{N_0}}\right). \tag{4.42}$$

Comparing Equations (4.34), (4.35), and (4.42), it follows from

$$d + 8e \geq (d + 2e)^2/d \geq d + 3e \tag{4.43}$$

that

$$P_d^{\mathrm{ID,AWGN}} \leq P_d^{\mathrm{TR,AWGN}} \leq P_d^{\mathrm{EX,AWGN}}. \tag{4.44}$$

We note that the reason the ID receiver is superior compared to TR receiver is because it makes use of the known value of Δ_l as a weighting factor.

We have used these three formulas for P_d in the AWGN channel to obtain union bounds of the type

$$P_b \leq \sum_{d=d_{\mathrm{free}}}^{\infty} c_d P_d \tag{4.45}$$

for the bit error rate P_b of RCPC coded transmission with the rate 1/3 memory and 6 mother code $(133, 171, 145)_{\mathrm{oct}}$ and the error coefficients tabulated by Hagenauer (1988). Figure 4.59 shows the BER curves for the three bounds and code rates $R_c = 8/24$, 8/16,

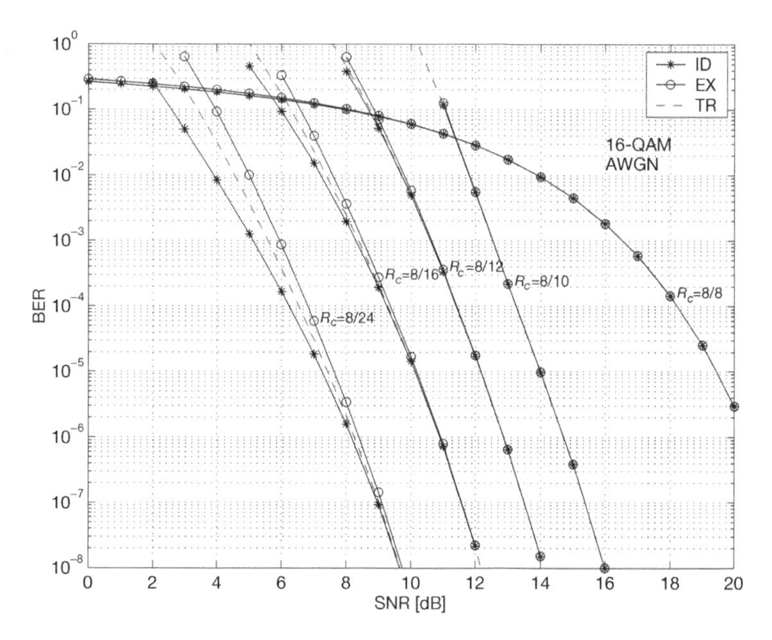

Figure 4.59 Bit error rates (union bounds) for 16-QAM for different code rates in the AWGN channel.

8/12, 8/10, and for uncoded transmission. We observe that the three bounds lie very close together at relevant BER values for code rate 8/16 and higher.

For fading channels, the integral can be evaluated numerically. Figure 4.60 shows the BER curves in a Rayleigh fading channel for the ID and EX bounds and the same code rates as above. We note that for this channel, the gap between both curves becomes larger, indicating that iterative decoding may give some noticeable gain in performance. Figure 4.61 shows the same curves for a Ricean channel with Rice factor $K = 6$ dB. Figure 4.62 shows the ID and EX curves for 64 QAM and the Rayleigh fading channel for code rates $R_c = 8/24,\ 8/16,\ 8/12,\ 8/10$. The gap between the ID and EX curves becomes larger than for 16-QAM. This can be understood from the fact that in a larger constellation, more useful information can be gained from the successful decoding of the other bits.

The question arises if the ID curves for iteration with *ideal* knowledge of the other bits reflect a real situation where there can be bit errors that may influence further iteration steps. One must also ask how many iterations are necessary. We have carried out numerical simulations for several code rates and several values of M^2. Figure 4.63 shows as an example a simulation for 64-QAM and code rate $R_c = 1/2$ in comparison with the theoretical ID and EX curves. We have simulated the first decoding step without iteration (stars), and then one additional iterative decoding step using only the decoded hard decision values for the information from the other bits (circles). A third curve shows the iterative decoding with ideal knowledge of the other bits (squares). The first curve is tightly bounded by the theoretical EX curve, but there is still an observable gap of about 0.3 dB at $P_b = 10^{-4}$. The third curve is extremely tightly bounded by the theoretical ID curve at relevant BERs (much less than 0.1 dB below $P_b = 10^{-3}$). The effective gain due to iterative decoding is

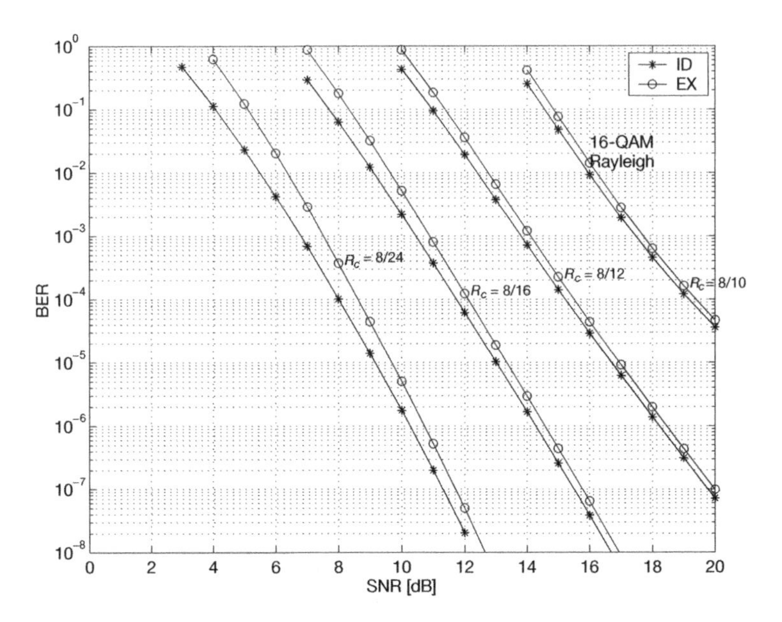

Figure 4.60 Bit error rates (union bounds) for 16-QAM for different code rates in the Rayleigh fading channel.

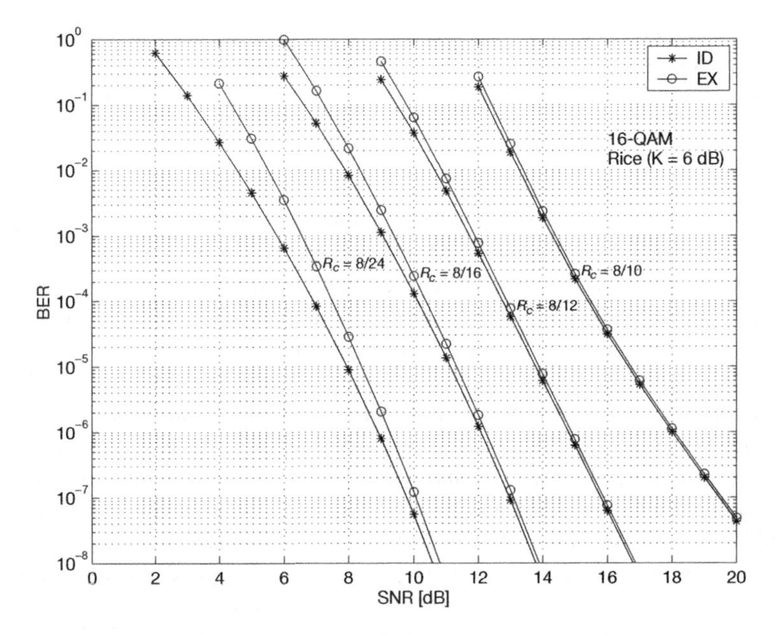

Figure 4.61 Bit error rates (union bounds) for 16-QAM for different code rates in the Ricean fading channel (Rice factor $K = 6$ dB).

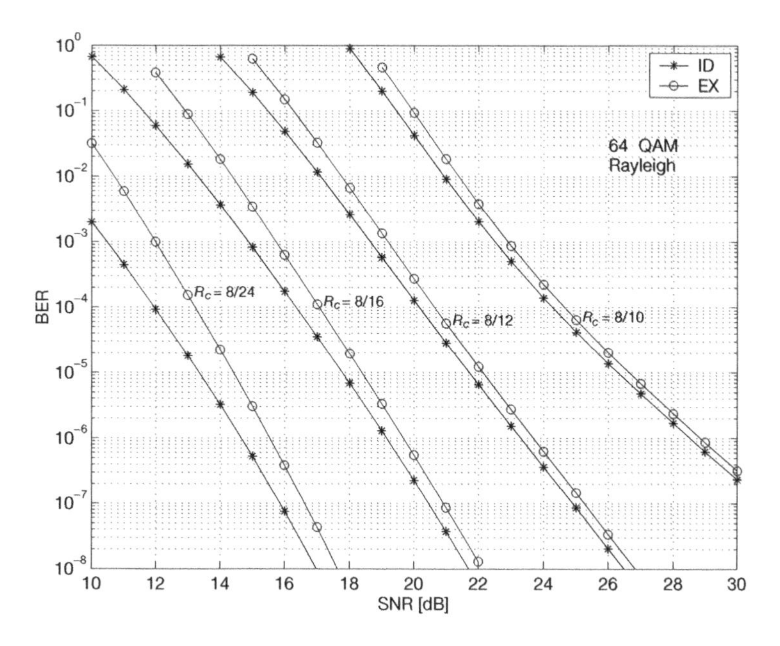

Figure 4.62 Bit error rates (union bounds) for 64-QAM for different code rates in the Rayleigh fading channel.

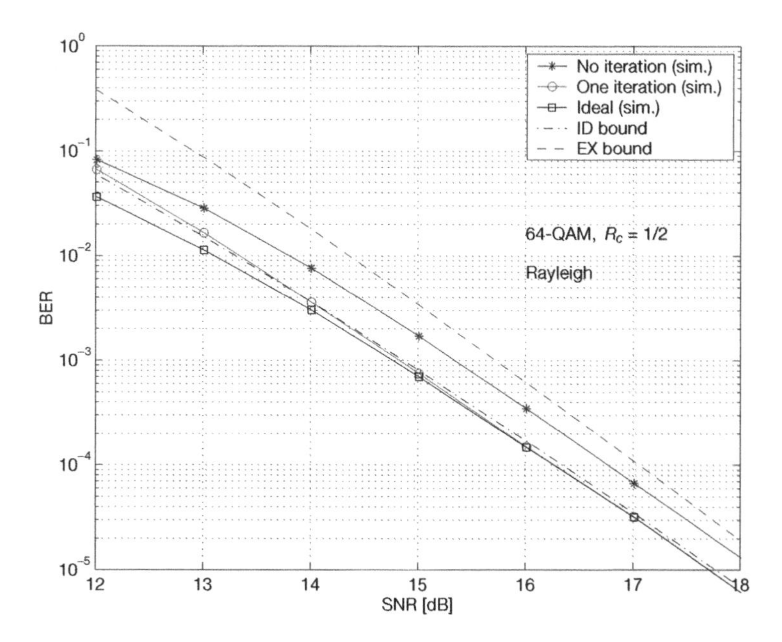

Figure 4.63 Comparison of the theoretical bounds with simulation results for 64-QAM and $R_c = 1/2$ and the Rayleigh fading channel.

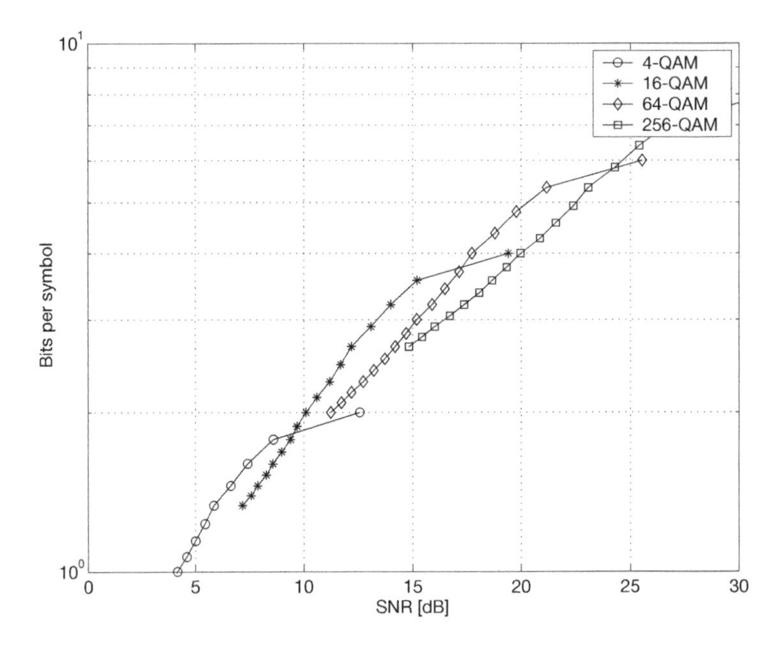

Figure 4.64 SNR needed for BER = 10^{-5} for different spectral efficiencies (bits per symbol) in the AWGN channel.

about 0.5 dB at $P_b = 10^{-4}$, which seems to be worth enough to carry out this one iteration. Our most important simulation result is that the second curve with one hard decision iteration lies between these curves at relevant BERs. We conclude that indeed the theoretical ID curves may serve as a guidance for practical system design for the lowest complexity iterative decoding with only one hard decision step. For our simulations, we have used the *maxlog* approximation of the optimum metric given by Equation (4.29). We have also simulated the soft threshold receiver (TR) metric that leads to a very slight degradation in the performance.

We have evaluated the ID bounds for code rates $8/24, 8/23, \ldots, 8/9$, and for uncoded transmission to get a diagram of the spectral efficiency (in bits per QAM symbol) as a function of the SNR that is needed for a certain BER. Figure 4.64 shows this diagram for the AWGN channel and a required $P_b = 10^{-5}$ and $M^2 = 4, 16, 64, 256$. We note that the lower-level modulation scheme performs better than the higher-level scheme at almost all spectral efficiencies. At 2 bits per symbol, there is a gain of approximately 2.5 dB for the rate 1/2 coded 16-QAM compared to the uncoded 4-QAM. But this gain is quite poor compared to even the simplest TCM schemes. This is not surprising, since BICM does not maximize the squared Euclidean distance that is needed to get an optimized transmission scheme for the AWGN channel.

Things become different for the Rayleigh fading channel. Figure 4.65 shows this diagram for the Rayleigh channel and a required $P_b = 10^{-4}$ and $M^2 = 4, 16, 64, 256$. We observe that 16-QAM performs always better than 4-QAM at the spectral efficiencies under consideration. For 1.33 bits per symbol, the lowest spectral efficiency for 16-QAM that can be achieved with our code family ($R_c = 8/24$), there is still a gain of approximately 1.6 dB

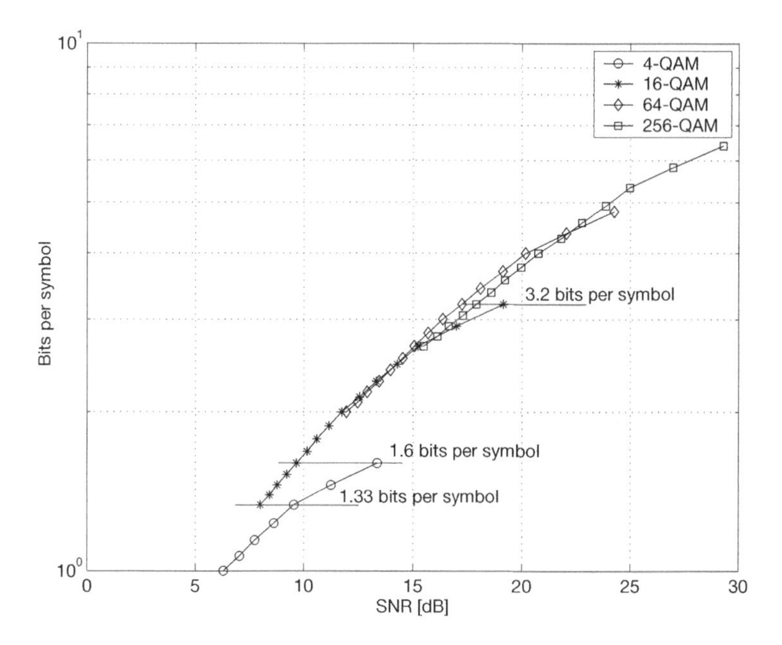

Figure 4.65 SNR needed for BER $= 10^{-4}$ for different spectral efficiencies (bits per symbol) in the Rayleigh fading channel.

compared to 4-QAM with $R_c = 8/12$. Even without iterative decoding, using the EX bound, we see from Figure 4.42 that 16-QAM is still at least 1 dB better than 4-QAM. At 1.6 bits per symbol, that is, 16-QAM with $R_c = 8/20$ and 4-QAM with $R_c = 8/10$, the gain of 3.7 dB is even more significant. We conclude that it is always favorable to use 16-QAM instead of 4-QAM for spectral efficiencies above 1 bit per symbol. As a rule of thumb we can state that one should avoid weak codes like $R_c = 8/10$. It is better to use a higher-level modulation scheme instead. For example, at 3.2 bits per symbol, 64-QAM with $R_c = 8/15$ performs 2 dB better than 16-QAM with $R_c = 8/10$. Even 64-QAM with $R_c = 8/16$ performs more than 0.5 dB better than 16-QAM with $R_c = 8/11$ at a slightly better spectral efficiency. We note that 256-QAM has no advantage for less than 5 bits per symbol.

4.5.3 Convolutionally coded QAM with real channel estimation and imperfect interleaving

Up to now, we have only considered the theoretical performance for the case of ideal interleaving. For coherent demodulation, we have further assumed perfect knowledge of the channel, that is, ideal channel estimation. Powerful interleaving requires a sufficient decorrelation of symbols, that is, an incoherent channel. On the other hand, for coherent demodulation, the channel must be coherent enough so that the channel estimation with pilot symbols can work. For differential demodulation, it must be slow enough to allow the comparison of two subsequent phases. In this section, we concentrate on coherent demodulation. Differential demodulation will be discussed in a subsequent subsection in the light of the DAB system.

OFDM, coding and modulation parameters

We consider an OFDM system with coded QAM and *real* channel estimation with a Wiener filter to see how much is lost due to the *frequency and time incoherence* of the channel. Coding and interleaving are included to see the degradations due to the *coherency* of the channel. Instead of using the fixed parameters of a specified system, we consider a *scalable model system* defined by the ratios between symbol length, echo length and correlation time of the channel so that the physical parameter T_S of the simulated system is matched to different applications.

We investigate a standard OFDM transmission system with guard interval Δ. Let T be the useful symbol duration and $T_S = T + \Delta$ the total length of the OFDM symbol. We concentrate on the popular choice $\Delta = T/4$, which is used in most existing OFDM system. For coherent demodulation, we need to insert pilot symbols. The number of carriers K is chosen to be an integer multiple of 3. For channel estimation, we use a rectangular grid of pilots as depicted in Figure 4.35. For certain OFDM symbols, pilots are inserted at every third carrier, that is, at number $k = 0, 3, 6, \ldots, K - 1$. A pilot at $k = K$ is added so that the band is bounded by pilots on both sides. The time distance between two pilots on the same frequency is given by $4T_S$, so the pilot density is 1/12 like in DVB-T. All pilots are boosted to twice the power of the information symbols for which 64-QAM modulation with conventional gray mapping has been applied. We use the convolutional code with generators $(133, \ 171)_{\text{oct}}$ (optionally punctured to have higher rates) as described in Subsection 3.2.1. We use a symbol interleaver in frequency *and* time direction that permutes pseudorandomly the QAM symbols of K carriers and L OFDM symbols. Thus, the interleaving is over a bandwidth $B = K/T$ and a time frame of length $T_F = LT_S$. We use only *one* big pseudorandom interleaver for both time and frequency together. This is very natural because OFDM is a 2-D transmission scheme. As discussed in the preceding subsection, a bit interleaver between the different bit streams of the QAM symbols is also necessary. This can be rather small compared to the symbol interleaver.

For simplicity, we do not consider iterative decoding. The additional gains due to iterative decoding have been discussed in Subsection 4.5.2.

Channel parameters

We assume a Rayleigh fading channel with an isotropic (Jakes) Doppler spectrum with maximum Doppler frequency $\nu_{\max}$ and exponential delay power spectrum with time constant τ_m as discussed in Subsection 4.4.3. The correlation length in time and frequency is defined by $t_{\text{corr}} = \nu_{\max}^{-1}$ and $f_{\text{corr}} = \tau_m^{-1}$, respectively. The guard interval Δ must be chosen large enough to guarantee $\tau_{\max} < \Delta$ for all relevant conditions. For channel estimation for a given time slot in frequency direction, the Sampling Theorem requires a minimum pilot distance of $4/T$ because $\Delta = T/4$ is the maximum possible delay. Since we have chosen a distance of $3/T$, we have an oversampling factor 4/3. The pilot distance in frequency direction is $4T_S$. The maximum Doppler frequency is then limited by the Sampling Theorem to $\nu_{\max} T_S < 1/8$.

For coded transmission in a fading channel, interleaving is indispensable. How much interleaving is necessary? In Subsection 4.4.3, we discussed the *diversity spectrum* of the channel as an indicator to find a rule of thumb to answer this question. We have seen that with interleaving in one (time or frequency) direction only, the interleaving length must significantly exceed the correlation length, that is, $T_F \gg t_{\mathrm{corr}}$ or $B \gg f_{\mathrm{corr}}$ respectively. As a rule of thumb, we found that a factor of 32 was a reasonable figure. We have further shown that the condition $B \gg f_{\mathrm{corr}}$ of a *wideband system* (wideband relative to the coherency bandwidth) can only be fulfilled for a great number of subcarriers (several hundreds).

Channel estimation

The theoretical background and practical details of channel estimation for OFDM systems were discussed in Section 4.2. Here we apply the 1-D × 1-D channel estimator, that is, first an 1-D channel estimation will be done in frequency direction and then in time direction (or vice versa). For 64-QAM – in contrast to QPSK – the edge effects discussed in Section 4.2 are a critical issue. This is due to the fact that the system works at a much higher SNR level, and residual noise in the channel estimation can severely corrupt the performance. We propose to use the optimal *matrix estimator* in frequency direction, that is, we use the information of *all pilots* available for the channel estimation of *every carrier*. We do not assume that the delay power spectrum is known and design the Wiener estimator for the case that the echoes are uniformly distributed inside the guard interval. In time direction, however, there are no edges and the estimator is given by a convolution. Since the Doppler spectrum can be measured by continual pilots, we assume that $\nu_{\max}$ is known to the receiver. The (noncausal) filter in time direction has chosen to have 16 taps.

Simulation results

Figure 4.66 shows the simulated BER for a wideband OFDM system with $K = 1536$ and $L = 80$ with 16-QAM and $R_c = 1/2$ in a fast and frequency-selective Rayleigh fading channel with time variance given by $\nu_{\max} T_S = 0.05$ and frequency selectivity given by $\tau_m = \Delta/5$. Comparing the curves with ideal and real channel estimation (CE) we observe a loss that is less than 1 dB. We note that this channel is already so fast that it causes a much more severe performance loss for DQPSK (see Figure 4.52). From this fact, we may conclude that coherent demodulation is more robust against the time incoherency of the channel, even if spectrally more efficient higher-level modulation schemes are applied. We note that the time interleaver with $T_F = 4t_{\mathrm{corr}}$ is quite small, but the frequency interleaver with $B = 76.8 f_{\mathrm{corr}}$ guarantees a sufficient interleaving.

We now consider a channel with half the time variance and a small delay spread corresponding to $B = 2 f_{\mathrm{corr}}$.. Such a situation may occur in a typical urban situation. Figure 4.67 shows the simulation for that channel with all other parameters being left unchanged. Even though the code cannot fully exploit its diversity for that channel (see Subsection 4.4.3), the degradations for the ideal CE curve are quite small. With real CE the performance becomes even better because the channel estimator takes profit from the higher coherency of the channel. However, the insufficient interleaving severely influences the error structure. To show this, we have plotted the error rates of the individual frames. Comparing Figures 4.68 and 4.69, we see that the error rates (real CE) in one frame for the first channel vary at least by a factor of 2, but for the second one there is a variation by a factor of 100. Depending on the application, this may lead to severe performance losses. This will be the case for a concatenated coding system with an outer block code, as it is applied in the DVB-T system.

We have seen that a channel with $\nu_{\max} T_S = 0.05$ does not cause severe problems to the channel estimation. However, this is more than a factor of 2 below the threshold 1.25

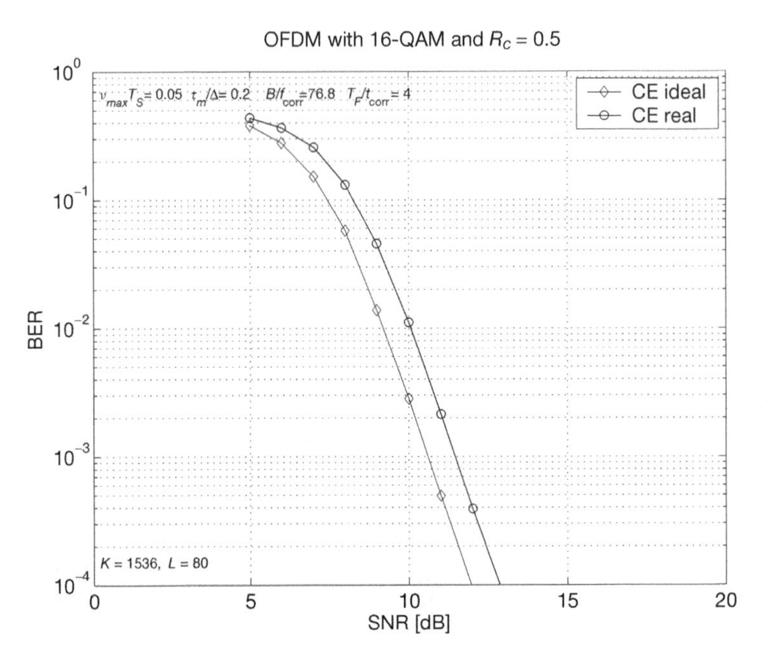

Figure 4.66 Simulation of a wideband OFDM system with 16-QAM for a fast and frequency-selective channel.

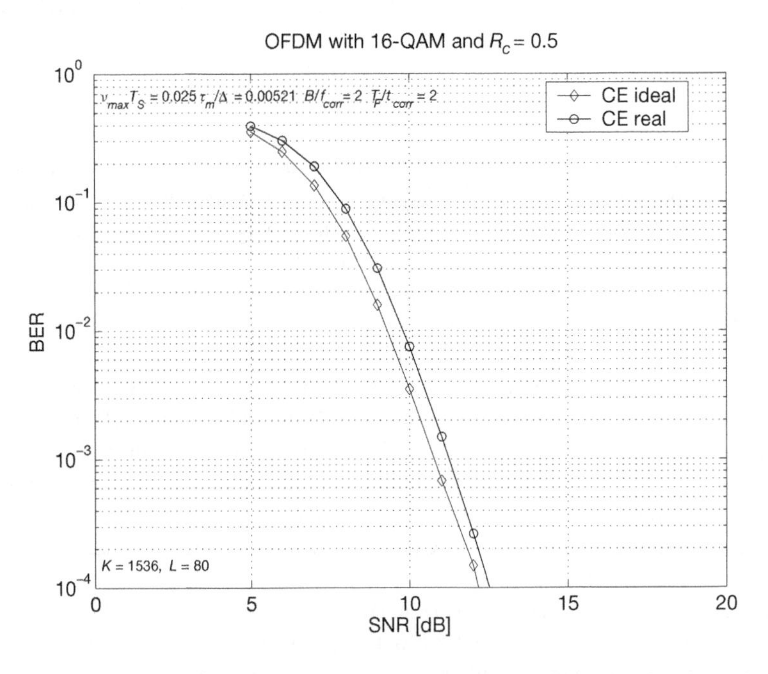

Figure 4.67 Simulation of a wideband OFDM system with 16-QAM for a moderately fast and frequency nonselective channel.

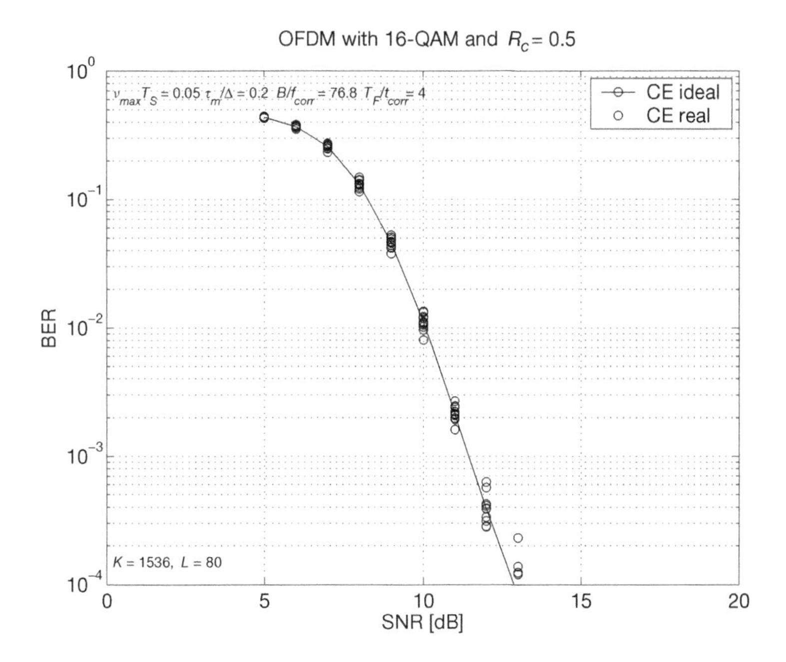

Figure 4.68 The BERs for Figure 4.66 for the individual frames.

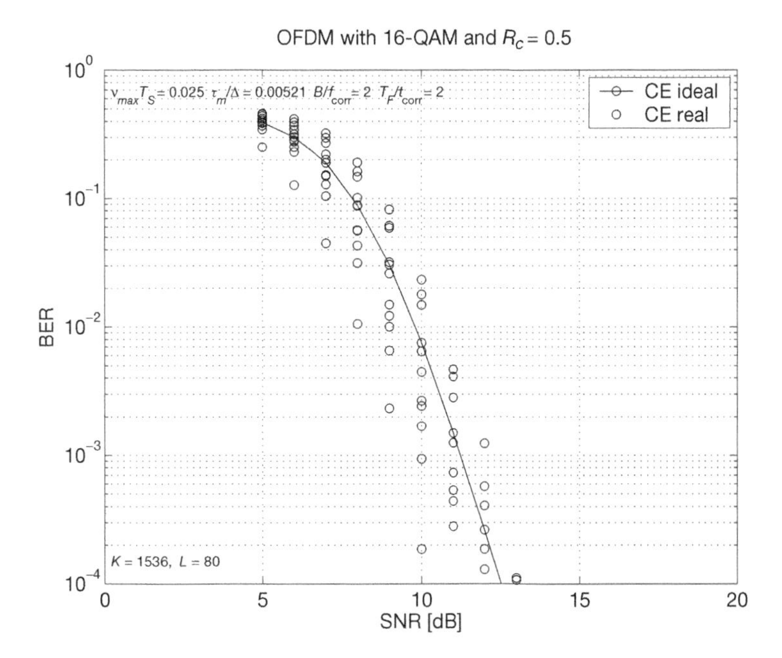

Figure 4.69 The BERs for Figure 4.67 for the individual frames.

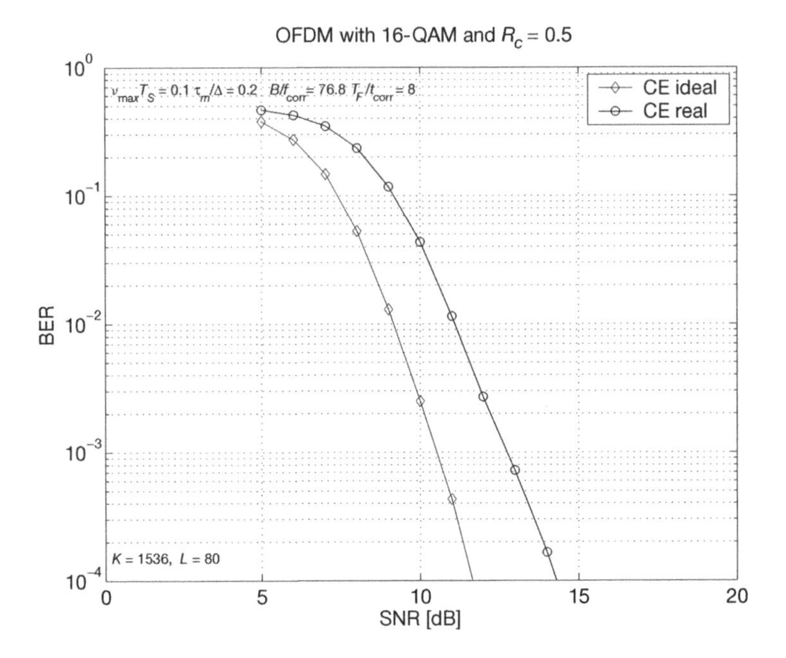

Figure 4.70 Simulation of a wideband OFDM system with 16-QAM for a very fast and frequency-selective channel.

given by the Sampling Theorem. We may now ask how fast the channel can be before the channel estimation fails to work. Figure 4.70 shows a simulation for $\nu_{\max} T_S = 0.1$ with all other parameters being the same as in Figure 4.66. Now the degradation relative to the ideal CE that is due to time incoherency increases from one to approximately 2 dB which seems to be tolerable.

We now consider 64-QAM. Figure 4.71 shows a 64-QAM simulation with all other parameters being the same as in Figure 4.66. We observe that the difference between ideal and real CE is not much more than 1 dB. We conclude that 64-QAM can be used in this quite fast channel, where DQPSK already suffers severely from time incoherency.

However, it is not surprising that 64-QAM is less robust than 16-QAM. Figure 4.72 shows a 64-QAM simulation with $\nu_{\max} T_S = 0.07$ with all other parameters being the same as in Figure 4.66. The degradations due to time incoherency are now between 3 and 4 dB. The ideal CE curves of these two 64-QAM simulation of an OFDM system with fast fading are the same as the 64-QAM curve of Figure 4.63 for an ideal Rayleigh channel. This means that the loss of orthogonality due to fast fading is practically not relevant.

We finally look at an OFDM system with a relatively low number of carriers, which is therefore not a wideband system compared to the coherency bandwidth of the channel. We choose $K = 48$ (the same number as for the WLAN systems) $L = 80$ and $\nu_{\max} T_S = 0.01$. With $B = 0.5 f_{\mathrm{corr}}$ and $T_F = 0.8 t_{\mathrm{corr}}$, the interleaving is rather poor. Figures 4.73 and 4.74 show the performance curves. Figure 4.73 shows that there is a loss of about 3 dB due to poor interleaving. More important is the extreme variation of the BER in the individual frames, as depicted in the second figure.

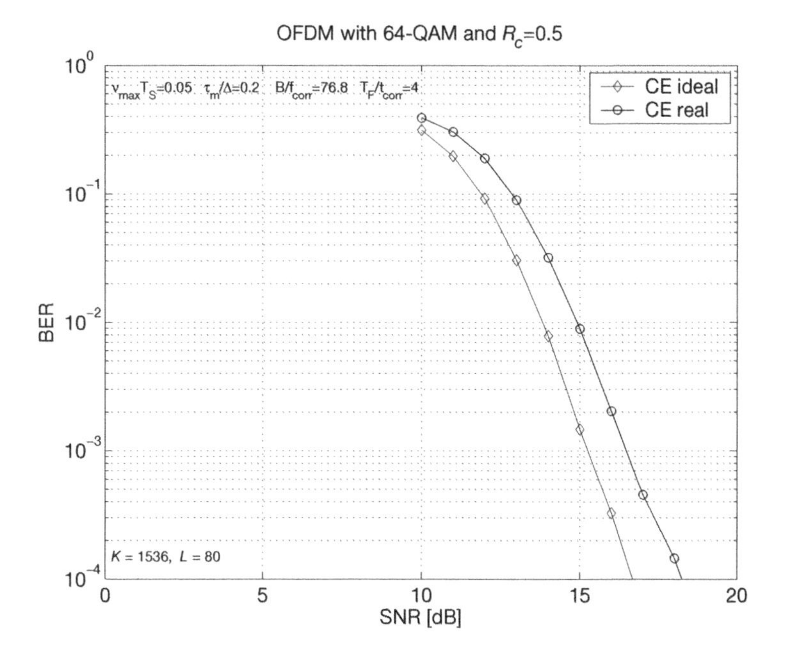

Figure 4.71 Simulation of a wideband OFDM system with 64-QAM for a fast and frequency-selective channel.

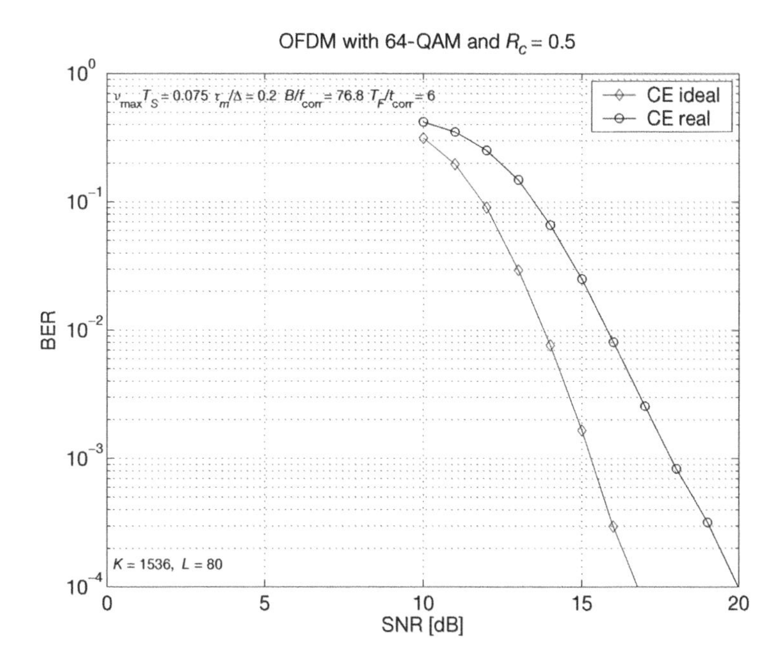

Figure 4.72 Simulation of a wideband OFDM system with 64-QAM for a very fast and frequency-selective channel.

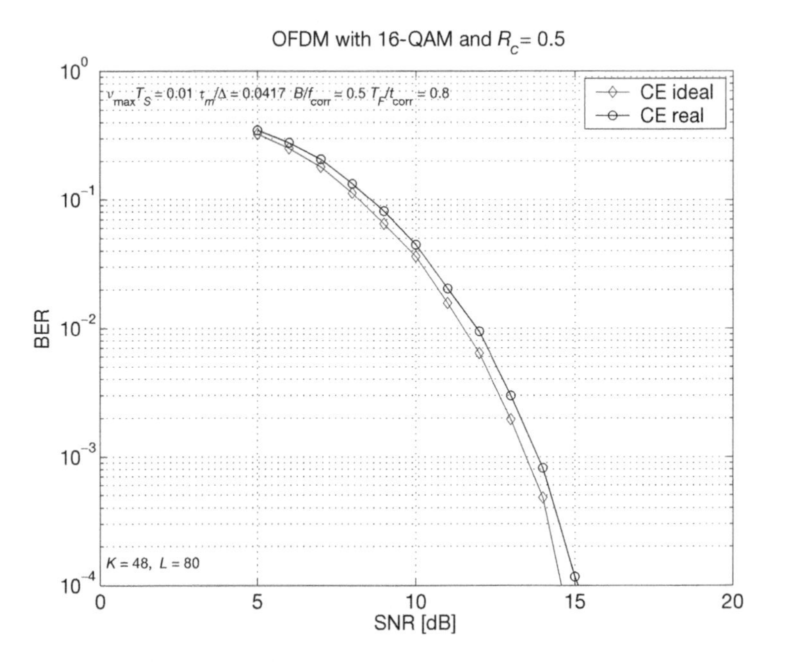

Figure 4.73 Simulation of narrowband OFDM for a slow and flat channel.

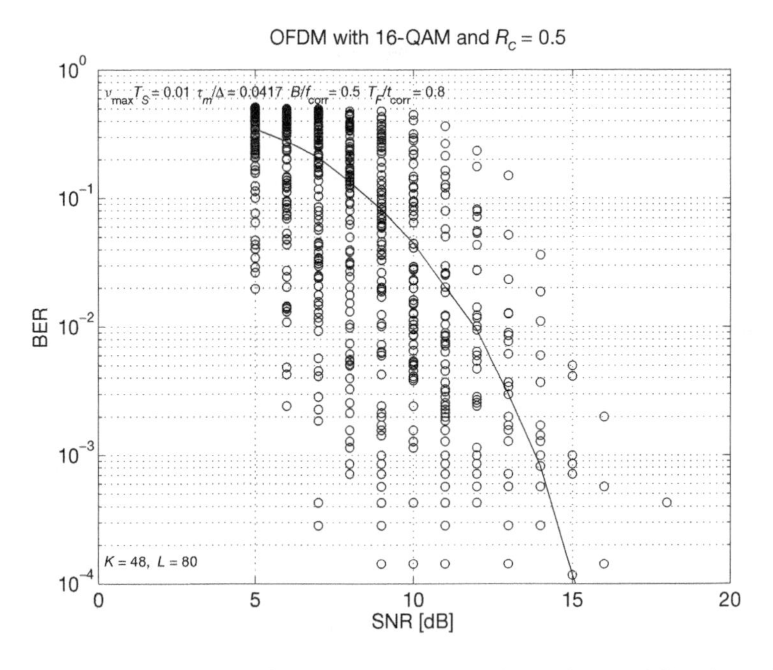

Figure 4.74 Simulation of narrowband OFDM for a slow and flat channel.

Power efficiencies

To compare power efficiencies, we look at the BERs at a given E_b/N_0 instead of the SNR. As discussed in Subsection 4.1.4, for an OFDM system with analysis window T and total symbol length T_S , code rate R_c and M^2-QAM, both are related by

$$SNR = \frac{T}{T_S}\frac{E_S}{N_0} = \frac{T}{T_S}R_c \log_2(M^2)\frac{E_b}{N_0}.$$

In this formula, the loss of useful energy due to the pilot symbols has not yet been taken into account. For pilots with the same energy E_S (without boosting), only 11 of 12 transmitted symbols carry useful information, that is, $E_S = \frac{11}{12}R_c \log_2(M^2)E_b$. If the pilots are boosted by a factor of two, we transmit the energy $13\,E_S$ of 13 symbols, but only 11 of them are useful, that is, $E_S = \frac{11}{13}R_c \log_2(M^2)E_b$, which leads to

$$SNR = \frac{T}{T_S}\frac{E_S}{N_0} = \frac{11}{13}\frac{T}{T_S}R_c \log_2(M^2)\frac{E_b}{N_0}.$$

From this equation, we conclude that the SNR curves[11] for $R_c = 1/2$ have to be shifted by

- 1.69 dB for 4-QAM
- −1.31 dB for 16-QAM
- −3.08 dB for 64-QAM

From Figure 4.66, we conclude that a BER of 10^{-4} can be reached at $E_b/N_0 \approx 10.5$ dB for ideal CE and at $E_b/N_0 \approx 11.5$ dB for real CE. We want to compare this with the DQPSK curves and $R_c = 1/2$ curves that are presented in Subsection 4.4.1. Because of the guard interval, the SNR curves depicted there must be shifted by 1 dB to the right to obtain the curves for E_b/N_0. From Figure 4.50, we conclude that $E_b/N_0 \approx 11.5$ dB is needed for a BER of 10^{-4} for ideal channel coherency between two adjacent symbols. This is 1 dB worse than 16-QAM with ideal CE. From Figure 4.52, we see that for a fast fading channel with $\nu_{\max}T_S = 0.05$, we need $E_b/N_0 \approx 13.5$ dB to reach that error rate. This is 2 dB worse than 16-QAM with real CE under the same condition. Thus, 16-QAM performs significantly better than DQPSK at nearly[12] twice the spectral efficiency.

4.5.4 Antenna diversity for convolutionally coded QAM multicarrier systems

As we have seen in the preceding subsection, the performance of a coded multicarrier system can suffer severely from nonsufficient interleaving, which means that the code cannot exploit its diversity as the diversity degree of the channel is too small. A remedy is the use of antenna diversity in addition to the existing channel coding. This can be done at the receiver or, for example, by using the Alamouti scheme, at the transmitter. We will investigate the question of how diversity can improve the performance of a convolutionally coded multicarrier QAM system as described in the preceding subsections.

[11] We note that this SNR in the abscissae of the figures must be understood as the SNR for the QAM symbols (after the OFDM demodulator), which is the relevant quantity for the BER performance. As a consequence of the boosting, the SNR on the analog physical channel is higher by a factor of 13/12. E_S is the energy of one useful transmit symbol including the guard interval.

[12] The spectral efficiency is slightly below two bits per complex symbol because of the pilots.

Receive antenna diversity

We first consider receive antenna diversity. As discussed in Subsection 2.4.5, the vector $\mathbf{r} = (r_1, \ldots, r_L)^T$ of receive symbols r_l at the L different antennas can be written as

$$\mathbf{r} = s\mathbf{c} + \mathbf{n},$$

where s is the QAM symbol and the channel vector $\mathbf{c}$ with Ricean fading coefficients c_l given by

$$\mathbf{c} = (c_1, \ldots, c_L)^T$$

and $\mathbf{n}$ is L-dimensional complex AWGN with variance $\sigma^2 = N_0$ in each (complex) dimension. As discussed in that subsection, the maximum ratio combiner (MRC) acts on the receive vector by calculating the complex scalar product

$$\mathbf{c}^\dagger \mathbf{r} = |\mathbf{c}|^2 s + \mathbf{c}^\dagger \mathbf{n}.$$

By using a sufficient statistics argument, we have seen that this MRC output provides us with the full information that is necessary for an optimal symbol decision on s. It can be shown (Problem 4) that $\mathbf{c}^\dagger \mathbf{n}$ is the one-dimensional complex AWGN with variance $\sigma^2 = |\mathbf{c}|^2 N_0$. It is therefore convenient to renormalize that equation and divide it by

$$a = \sqrt{a_1^2 + \cdots + a_L^2}, \tag{4.46}$$

where $a_l = |c_l|$ is the absolute value of the lth fading coefficient. This leads to the equation

$$a^{-1}\mathbf{c}^\dagger \mathbf{r} = as + a^{-1}\mathbf{c}^\dagger \mathbf{n}.$$

We define

$$u = a^{-1}\mathbf{c}^\dagger \mathbf{r}$$

as the normalized MRC output and write

$$u = as + n. \tag{4.47}$$

Now, n is the one-dimensional complex AWGN with variance $\sigma^2 = N_0$. This receiver is depicted in part (a) of Figure 4.75. This equivalent fading channel model as depicted in part (b) of that figure is convenient because now we have only one composed real-fading coefficient amplitude a given by Equation (4.46) corresponding to the power sum of the individual diversity branches. For square QAM constellations, it is convenient to split up the real and imaginary parts of Equation (4.47) and work with a real-valued channel model. Now u is the input of a metric computation unit as described in Subsection 4.5.2. For convolutionally coded QAM and an error event of Hamming distance d, we may now apply the formalism introduced in that subsection to a channel described by Equation (4.49). We then note that a is already a composed fading amplitude of L diversity branches. As a consequence, we must replace d by dL in the formulas for the error probabilities derived in that subsection.

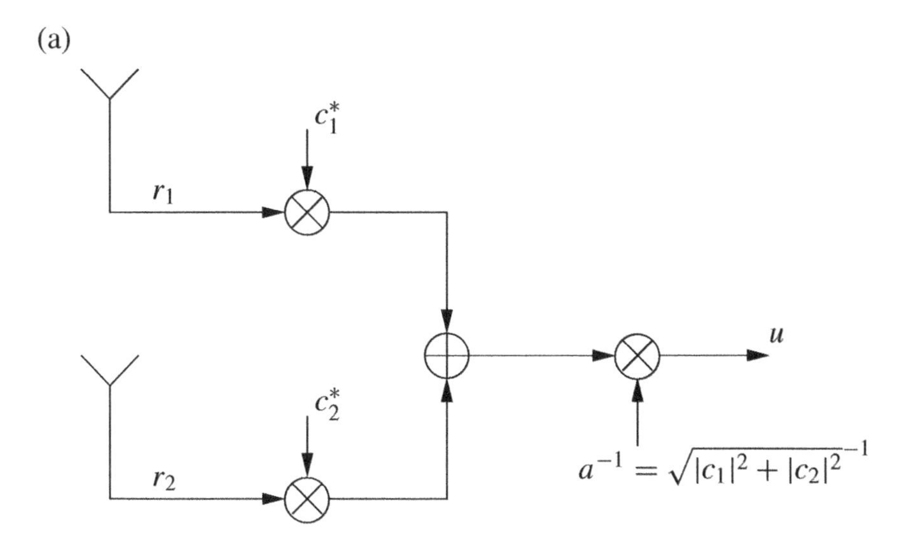

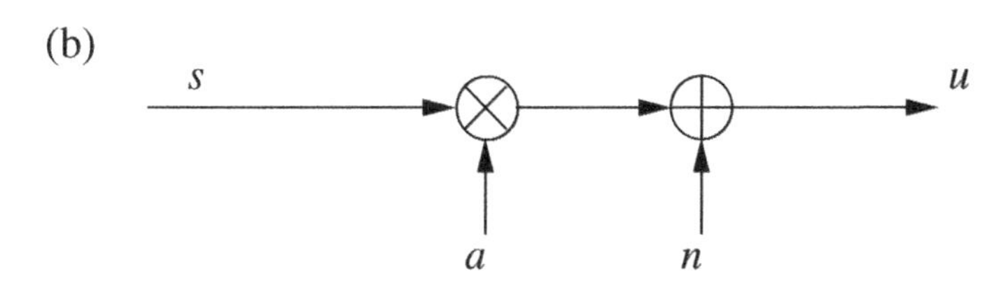

Figure 4.75 Equivalent diversity channel model.

Transmit antenna diversity

We now consider Alamouti's twofold transmit antenna diversity. As shown in Subsection 2.4.7, the receive symbol vector $\mathbf{r} = (r_1, r_2)^T$ for the symbols r_1 and r_2 received at the two time slots can be written as

$$\mathbf{r} = \mathbf{Cs} + \mathbf{n}.$$

$\mathbf{s} = (s_1, s_2)^T$ is a pair of complex (QAM) symbols, $\mathbf{n}$ is complex two-dimensional AWGN with $\sigma^2 = N_0$ in each (complex) dimension. The fading channel is described by the matrix

$$\mathbf{C} = \begin{pmatrix} c_1 & c_2 \\ -c_2^* & c_1^* \end{pmatrix}.$$

The fading coefficients c_1 and c_2 correspond to the two transmit antennas. As discussed in that subsection, the Alamouti diversity combiner acts on the receive vector by calculating

$$\mathbf{C}^\dagger \mathbf{r} = \left(|c_1|^2 + |c_2|^2\right) \mathbf{s} + \mathbf{C}^\dagger \mathbf{n}.$$

This is due to the property

$$\mathbf{C}^{\dagger}\mathbf{C} = \left(|c_1|^2 + |c_2|^2\right)\mathbf{I}_2.$$

Again, we argue that the combiner output provides us with the full information that is necessary for an optimal symbol decision on $\mathbf{s}$. This is evident because $\mathbf{C}$ just acts as a rotation on the signal vector, together with a real multiplicative factor. We may also find a sufficient statistics argument: the two rows of $\mathbf{C}$ are an orthogonal transmit base (not yet normalized) and thus the components of $\mathbf{C}^{\dagger}\mathbf{r}$ are (up to a factor) the components of $\mathbf{r}$ in the direction of those base vectors. It can be shown (Problem 5) that $\mathbf{C}^{\dagger}\mathbf{n}$ is two-dimensional complex AWGN with variance $\sigma^2 = \left(|c_1|^2 + |c_2|^2\right) N_0$ in each dimension. It is therefore convenient to renormalize that equation and divide it by

$$a = \sqrt{a_1^2 + a_2^2}, \tag{4.48}$$

where $a_l = |c_l|$ is the absolute value of the lth fading coefficient. This leads to the equation

$$a^{-1}\mathbf{C}^{\dagger}\mathbf{r} = a\mathbf{s} + a^{-1}\mathbf{C}^{\dagger}\mathbf{n}.$$

We define

$$\mathbf{u} = a^{-1}\mathbf{C}^{\dagger}\mathbf{r}$$

as the normalized combiner output and write

$$\mathbf{u} = a\mathbf{s} + \mathbf{n}. \tag{4.49}$$

Now $\mathbf{n}$ is the two-dimensional complex AWGN with variance $\sigma^2 = N_0$ in each dimension. We may split up this vector equation into two scalar equations

$$u_1 = as_1 + n_1$$

and

$$u_2 = as_2 + n_2.$$

Obviously, each of the two QAM symbols s_1 and s_2 is affected by the same transmission channel given by Equation (4.47), where we have to set $L = 2$ in Equation (4.46). Thus, both setups are equivalent.

In the two preceding subsections, we have discussed convolutionally coded QAM transmission. The transmission channel for each QAM symbol was also of the form given by Equation (4.47), but with a fading amplitude a with Rayleigh or Ricean statistics. Here, a is a composed fading amplitude given by Equation (4.48), where each of both components has such a statistics. Let us assume that the fading amplitudes a_1 and a_2 are identical independent random variables. Then, since the performance depends on the sum of all squared fading amplitudes, we conclude that twofold antenna diversity just doubles the diversity degree for each error event. Thus, we then have to replace P_d by P_{2d} in the union bound and write

$$P_b \leq \sum_{d=d_{\text{free}}}^{\infty} c_d P_{2d}.$$

We note that at the Viterbi decoder input, the combiner output has to be weighted with the composed fading amplitude.

The transmission scheme

Our goal is to combine Alamouti's transmit diversity scheme with OFDM and convolutionally coded QAM. The transmit diversity will be implemented in time direction on each subcarrier for a pair of subsequent OFDM symbols. Let $s_l[k] = s_{kl}$ denote the complex QAM symbols with time index $l = 1, 2$. For each frequency index k, the symbols will be multiplexed to the two transmit antennas TX1 and TX2 according to Table 4.1. In principle, the OFDM signal generation by IFFT will then have to be done for both antennas and for the two time slots, that is, four IFFTs of length N have to be performed. However, because of well-known Fourier transform relations, only two transforms are necessary. Let us write $x_l[n]$ for the time domain signal (i.e. the IFFT) corresponding to the frequency domain signal values $s_l[k]$. Then the transmit diversity multiplexer in the time domain can be written as shown in Table 4.2. Thus, only two IFFTs must be performed during two time slots and the Alamouti transmit diversity scheme combined with OFDM does not require additional IFFT computation complexity. After multiplexing, the guard interval has to be inserted to the four signals of Table 4.2. After digital-to-analog conversion, the signals of the two branches will be sent to the respective antennas. The complete system setup is shown in Figure 4.76. The data bits will be encoded by a standard convolutional encoder of constraint length 7 with generators $(133,\ 171)_{\text{oct}}$. Optionally, higher code rates can be obtained by puncturing. The coded bit stream is interleaved by a block interleaver that performs a random interleaving over $\log_2(M)I$ bits, corresponding to I complex QAM symbols and I/K OFDM symbols, where K is the number of OFDM subcarriers. The block of interleaved bits are mapped on the $M-$QAM symbols with conventional Gray mapping.

Table 4.1 Transmit diversity multiplexer in the frequency domain

Antenna	TX1	TX2
Time slot $l = 1$	$s_1[k]$	$s_2[k]$
Time slot $l = 2$	$s_2^*[k]$	$-s_1^*[k]$

Table 4.2 Transmit diversity multiplexer in the time domain

Antenna	TX1	TX2
Time slot $l = 1$	$x_1[n]$	$x_2[n]$
Time slot $l = 2$	$x_2^*[N-n]$	$-x_1^*[N-n]$

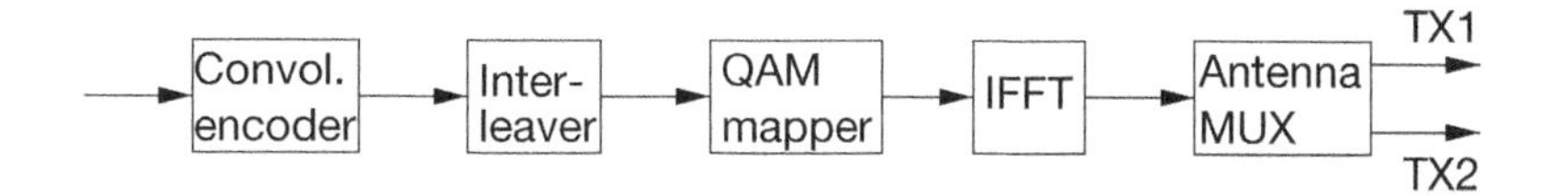

Figure 4.76 System setup block diagram.

The complex QAM symbols will be processed blockwise by an IFFT device of FFT length N $(N \geq K)$, resulting into a stream of complex time domain samples that are grouped into vectors $x_i[n], n = 0, 1, \ldots, N-1$ of length N, each corresponding to one OFDM symbol numbered by i. These vectors are grouped into pairs, and each pair is processed by the transmit diversity multiplexer according to Table 4.2. After this antenna multiplexing, the guard interval will be inserted separately to the resulting OFDM symbols for the respective antennas.

We note that if we switch off TX2, except for the complex conjugate in the second symbol, our setup is just a standard transmission setup like it is used in the standards of DVB-T, HIPERLAN/2 and DRM (with some modifications concerning the QAM modulation).

Channel simulations

We denote the OFDM symbol duration by $T_S = T + \Delta$, where T is the useful OFDM symbol duration, and $\Delta = T/4$ is the guard interval. We simulate a time and frequency-selective Rayleigh fading channel with an isotropic Doppler spectrum with maximum shift $\pm\nu_{\max}$ and an exponential delay power spectrum characterized by the delay spread τ_m.

The time variance of the channel is characterized by $\nu_{\max} T_S$, the frequency variance by $\tau_m B$, where $B = K/T$ is the transmission bandwidth. For a fixed ratio τ_m/Δ, and because $T = 4\Delta$, the number of carriers K can directly be regarded as a measure for the frequency selectivity.

Fast fading corresponding to high values of $\nu_{\max} T_S$ may cause degradations because the complex channel amplitudes vary too much over two time slots. On the other hand, slow (small $\nu_{\max} T_S$) and flat (small K) fading may cause insufficient interleaving.

In our simulations, we always assume ideal channel estimation. We consider a channel with moderate (but not extremely small) delay spread compared to the guard interval with $\tau_m = 0.2\Delta = T_S/20$.

First, we assume a moderately fast fading channel with $\nu_{\max} T_S = 0.01$ for which coded QAM systems are known to work well (see above).

Figure 4.77 shows the BER as a function of E_b/N_0 for such a channel and 16-QAM transmission with code rate $R_c = 1/2$ and a huge time and frequency interleaving over the bandwidth of $K = 768$ carriers and the time of $I/K = 360$ OFDM symbols. The curves fit well to the union bounds for the ideally interleaved Rayleigh channel. Here, we have used the expurgated union bounds as described in Subsection 4.5.2. The gain due to antenna diversity lies between 1 and 2 dB in the region of interest, which is much smaller than in the uncoded case (see Figure 2.12).

Figure 4.78 shows the performance curve for the same channel with a small relative bandwidth ($K = 48$) and no time interleaving. This system, which can be compared with a HIPERLAN/2 system, shows severe degradations of 5.5 dB at $BER = 10^{-4}$ because of insufficient interleaving if no diversity has been applied. We conclude that such a low number of carriers means that the OFDM system is a narrowband system compared to the channel. Figure 4.78 shows that only one additional transmit antenna can compensate approximately 4 dB of that narrowband loss. If we increase the bandwidth by a factor of 16 to $K = 768$, similar simulations show that the interleaving only in frequency direction means only a loss of 0.8 dB for one TX antenna and 0.4 dB for 2 TX antennas. We may call this a wideband system. However, smaller values of τ_m will cause severe degradations and antenna diversity would be very helpful in such situations.

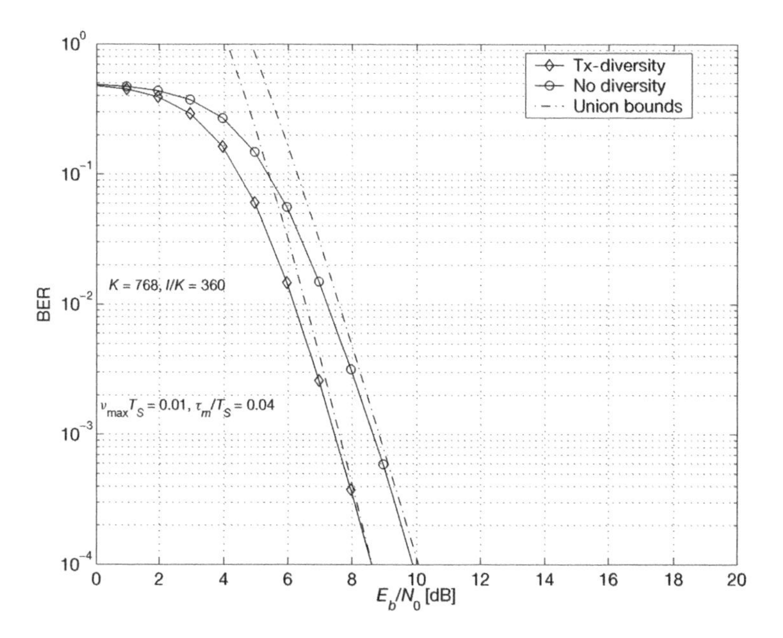

Figure 4.77 Bit error rates for the coded system with and without diversity for $\nu_{\max}T_S = 0.01$ and $\tau_m/\Delta = 0.2$ (huge interleaver).

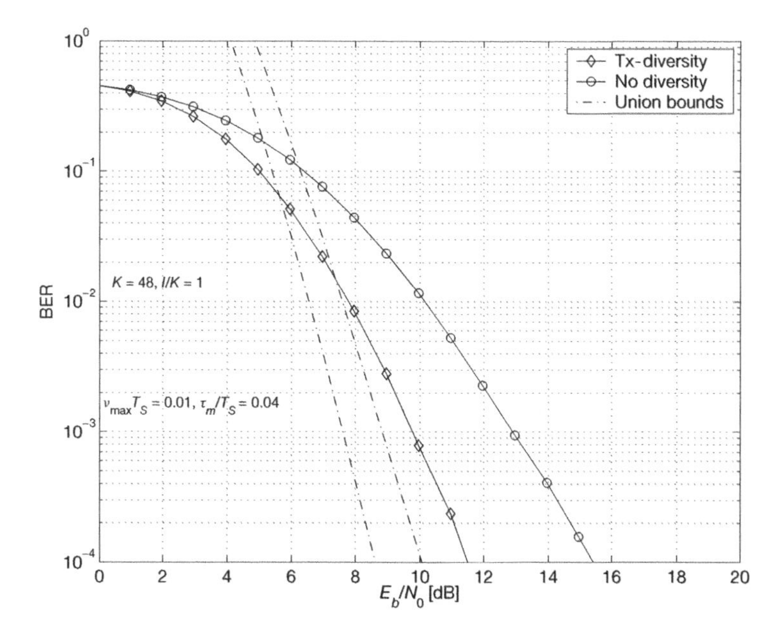

Figure 4.78 Bit error rates for the coded system with and without diversity for $\nu_{\max}T_S = 0.01$ and $\tau_m/\Delta = 0.2$ (small interleaver).

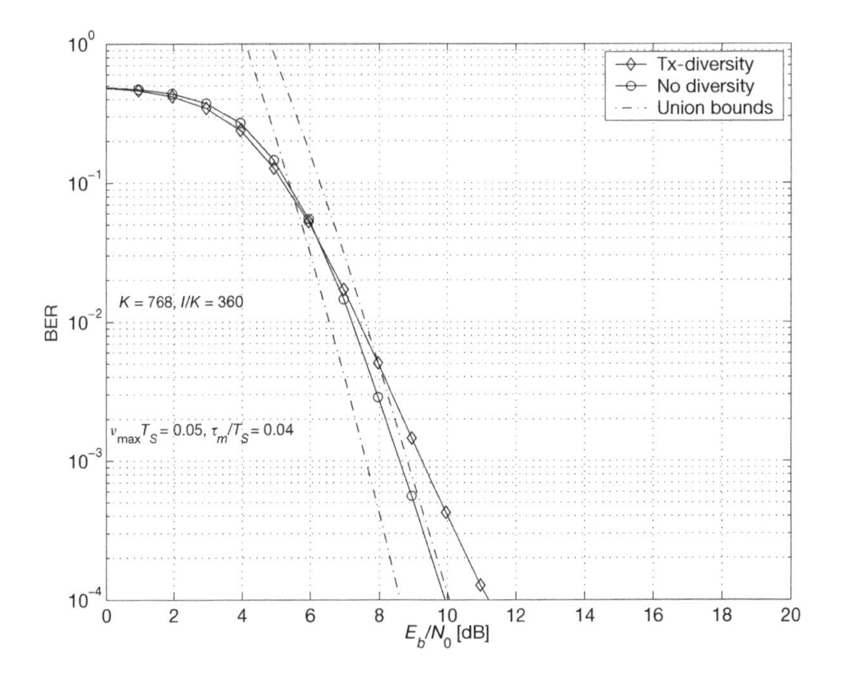

Figure 4.79 Bit error rates for the coded system with and without diversity for $\nu_{\max}T_S = 0.05$ and $\tau_m/\Delta = 0.2$ (huge interleaver).

Figure 4.79 shows a simulations with the same parameters as in Figure 4.77, but with a fast fading channel with $\nu_{\max}T_S = 0.05$. In that case, the time variance of the channel causes a coherency loss between the two symbols, which is so severe that the diversity system performs worse than the one antenna system. This sensitivity to time variance must be carefully considered by designing a transmit diversity system.

We finally recall that (except the sensitivity to time variance) a two transmit antenna diversity system is equivalent to a two receive antenna diversity system if the performance curves are drawn as functions of E_b/N_0. This means that Figures 4.77 and 4.78 are also valid for a two receive antenna system.

4.6 OFDM System Examples

4.6.1 The DAB system

The European DAB system (EN300401 2001a) uses OFDM with differential QPSK modulation together with punctured convolutional codes as described in the preceding section. As already discussed in detail, the OFDM parameters must be carefully adjusted to the requirements of the channel. A long symbol length T_S with a long guard interval Δ makes the system robust against long echoes, but sensitive against high Doppler frequencies. To achieve a certain flexibility in service planning, four different *transmission modes* have been defined (see Table 4.3). In this table, a time unit $t_s = f_s^{-1} = \frac{1}{2048}$ ms

Table 4.3 The OFDM parameters for the four DAB transmission modes

Mode	K	T^{-1}	T_S	Δ	Max. frequency
TM I	1536	1 kHz	$2552\,t_s \approx 1.246$ ms	$504\,t_s \approx 246$ μs	≈375 MHz
TM IV	768	2 kHz	$1276\,t_s \approx 623$ μs	$252\,t_s \approx 123$ μs	≈750 MHz
TM II	384	4 kHz	$638\,t_s \approx 312$ μs	$126\,t_s \approx 62$ μs	≈1500 MHz
TM III	192	8 kHz	$319\,t_s \approx 156$ μs	$63\,t_s \approx 31$ μs	≈3000 MHz

has been introduced. The frequency $f_s = 2048$ kHz is the sampling frequency if the smallest possible (power of two) FFT length will be used for each transmission mode. Thus, the integer numbers in the third and fourth column correspond to the number of samples.

Transmission mode I with the very long guard interval of nearly 250 μs has been designed for large-area coverage, where long echoes are possible. It is suited for single frequency networks (Hoeg and Lauterbach 2003) with long artificial echoes; 200 μs correspond to a distance of 60 km, which is a typical distance between transmitters. If all transmitters of the same coverage area are exactly synchronized and send exactly the same OFDM signal, no signal of relevant level and delay longer than the guard interval should be received. Since the OFDM symbol length T_S is very long, transmission mode I is sensitive against rapid phase fluctuations and should only be used in the VHF region.

Transmission mode II can cope with echoes that are typical for most topographical situations. However, in mountainous regions, problems may occur. This mode is suited for VHF/UHF transmission as well as for transmission in the L-band at 1.5 GHz.

Transmission mode III has been designed for satellite transmission. It may be suited also for terrestrial coverage, if no long echoes are to be expected.

The parameters of TM IV lie just between mode I and II. It was included later in the specification to take into account the special conditions of the broadcasting situation in Canada. It will be used there even at 1.5 GHz. This is possible for limited vehicle speed and a direct line of sight, so that a Ricean channel (instead of a Rayleigh channel) can be assumed.

The transmission modes have different numbers of subcarriers within the same bandwidth of approximately 1.5 MHz, and thus different carrier spacings and symbol lengths. The maximum transmit frequency in the last column of the table corresponds to the Doppler frequency that leads to $\nu_{\max} T_S = 0.05$ for a vehicle speed of 120 km/h. As we have seen in Subsection 4.5.1, in a Rayleigh fading channel this value already leads to a significant performance degradation for DQPSK modulation and should therefore not be exceeded.

The OFDM signal has K modulated subcarriers at frequencies $f_k = k/T$, with subcarrier index $k \in \{\pm 1, \pm 2, \pm 3, \ldots, \pm K/2\}$. The center subcarrier corresponding to the DC component in the complex baseband signal is not modulated for reasons of receiver implementation. For all transmission modes, the spacing between the highest and the lowest subcarrier is exactly $f_{K/2} - f_{-K/2} = 1536$ kHz. For transmission mode I, the Fourier analysis window is $T = 1$ ms. The carrier spacing is $T^{-1} = 1$ kHz. The minimum FFT length for transmission mode I that is a power of two is $N_{\text{FFT}} = 2048$, that is, the smallest power of two that is larger than the number of subcarriers $K = 1536$. The total OFDM symbol length is given by $T_S = 2552\,t_s \approx 1.246$ ms. The guard interval length is $\Delta = 504\,t_s \approx 246$ ms. Thus, we have the ratio $T/T_S \approx 0.8$. This means that loss in spectral power and spectral efficiency due to the guard interval is approximately 20%. The OFDM parameters for transmission modes IV, II, III (in that order[13]) can be derived from those of transmission mode I by successively halving all time periods and the FFT length and doubling all frequencies.

Transmission frames

For each transmission mode, a *transmission frame* is defined on the physical signal level as a periodically repeating set of OFDM symbols. It fulfills certain tasks for the data stream. It is an important feature of the DAB system (and in contrast to the DVB system) that the time periods on the physical level and on the logical (data) level are matched together. The period T_F of the transmission frame is either the same as the audio frame length of 24 ms or an integer multiple of it. As a consequence, the audio data stream does not need its own synchronization. This ensures a better synchronization stability especially for mobile reception.

The structure for TM II is the simplest and will thus be described first. The frame length is 24 ms. The first two OFDM symbols of the transmission frame build up the *synchronization channel* (SC). The next three OFDM symbols carry the data of the *fast information channel* (FIC) that contains information about the multiplex structure and transmitted programmes. The next 72 OFDM symbols carry the data of the *main service channel* (MSC). The MSC carries useful information, such as audio data or other services. Figure 4.80 shows the transmission frame structure. This figure is also valid for TMs I and IV, but the respective frame lengths are different.

All the OFDM symbols in a transmission frame of TM II have the same duration $T_S = 638\,t_s \approx 312$ µs, except the first one. This so-called null symbol of length $T_{\mathrm{Null}} = 664\,t_s \approx 324$ µs is to be used for rough time synchronization. The signal is set to zero (or nearly to zero) during this time to indicate the beginning of a frame on the physical layer. The second OFDM symbol of the SC is called the *TFPR* (*Time-Frequency-Phase Reference*) symbol. The complex Fourier coefficients s_k of that OFDM symbol have been chosen in a sophisticated way so that it serves as a frequency reference as well as for echo estimation for the fine tuning of the time synchronization to find the position of the Fourier analysis window. Furthermore, it is the start phase for the differential phase modulation. Each of the following OFDM symbols carries 384 DQPSK symbols corresponding to 768 bits (including

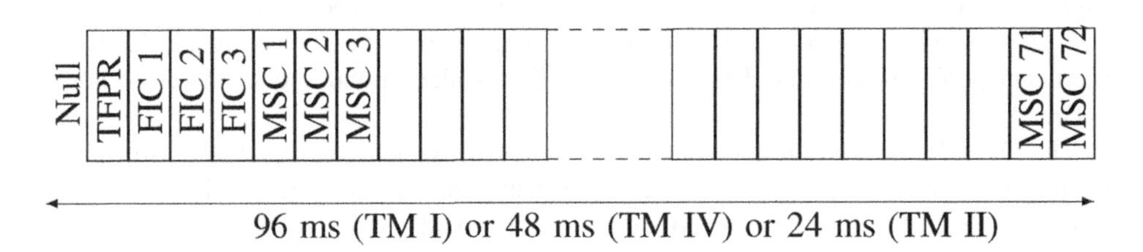

Figure 4.80 The DAB transmission frame.

[13]This seemingly strange order is due to the fact that TM IV, which lies between I and II, has been introduced subsequently.

redundancy for error protection, see the following text). The three OFDM symbols of the FIC carry 2304 bits. Since they are highly protected with a rate $R_c = 1/3$ code, only 768 data bits remain. The FIC data of each transmission frame can be decoded immediately without reference to the data of other transmission frames, because this most important information must not be delayed. The 72 OFDM symbols of the MSC carry 55,296 bits, including error protection. This corresponds to a (gross) data rate of 2.304 Mbit/s. The data capacity of 55,296 bits in each 24 ms time period is divided into 864 *capacity units* (CUs) of 64 bits. In the MSC, several audio programmes and other useful data services are multiplexed together. Since each of them has its own error protection, it is not possible to define a fixed net data rate of the DAB system.

The transmission frames of TMs IV and I, respectively, have exactly the same structure. Since the OFDM symbols are longer by a factor of two or four, respectively, the transmission frame length is 48 or 96 ms. The number of bits in the FIC and MSC increases by the same factor, but the data rate is always the same.

For TM III, the frame duration is $T_F = 24$ ms. Eight OFDM symbols carry the FIC, and 144 OFDM symbols carry the MSC. The data rate of the FIC is higher by a factor of 4/3 compared to the other modes. The MSC always has the same data rate.

For all four transmission modes, the MSC transports 864 CUs in 24 ms. There is a data frame of 864 CUs = 55,296 bits common for all transmission modes, which is called the *common interleaved frame* (CIF). For TMs II and III, there is exactly one CIF inside the transmission frame. For TM I, there are four CIFs inside one transmission frame of 96 ms. Each of them occupies 18 subsequent OFDM symbols of the MSC. The first is located in the first 18 symbols, and so on. For TM IV, there are two CIFs inside one transmission frame of 48 ms. Each of them occupies 36 subsequent OFDM symbols of the MSC.

Channel coding

The DAB system allows great flexibility in the choice of the proper error protection for different applications and for different physical transmission channels. Using RCPC codes introduced by (Hagenauer, 1988), it is possible to use codes of different redundancy without the necessity for different decoders. One has a family of RCPC codes originated by a convolutional code of low rate that is called the *mother code*. The daughter codes will be generated by omitting specific redundancy bits. This procedure is called *puncturing*. The receiver must know which bits have been punctured. Only one Viterbi decoder for the mother code is necessary.

The RCPC code family is the one that we have already discussed in detail in Subsection 4.5.1. It is based on the rate $R_c = 1/4$ code defined by the generators (133, 171, 145, 133) in octal notation. For RCPC codes, it is possible to switch the code rate even inside the same data stream. Thus, RCPC codes offer the possibility of *unequal error protection* (UEP) of a data stream: some bits in the data stream may require a very low BER, others may be less sensitive against errors. Using RCPC codes, it is possible to save capacity and add just as much redundancy as necessary. UEP is especially useful for audio data. Figure 4.81 illustrates the idea by an example of a 192 kbit/s audio data stream. The inverse code rate R_c^{-1} is drawn as a function of the bit index inside the frame. Thus, the area inside the rectangles corresponds to the number of encoded bits. It is quite obvious from the picture that it is very efficient to use UEP because the group of bits with the weakest code is

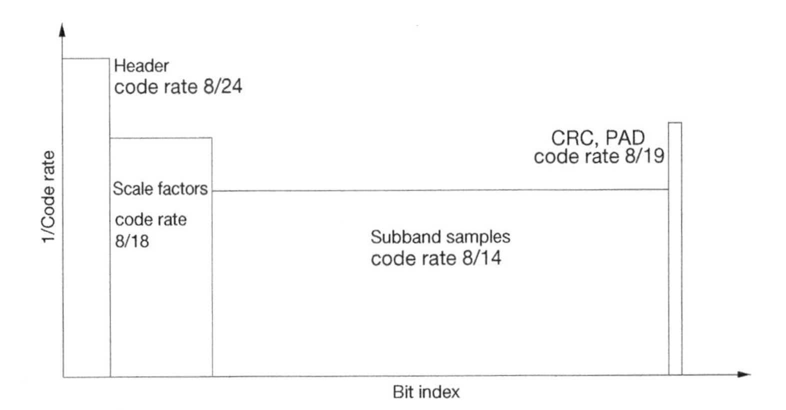

Figure 4.81 Example for an error protection profile for the audio data rate 192 kbit/s.

by far the biggest one. The first bits inside a frame are the header, the bit allocation (BAL) table, and the scale factor select information (SCFSI). An error in this group would make the whole frame useless. Thus, it is necessary to use a strong (low-rate) code here. The next group consists (mainly) of scale factors. Errors will cause annoying sounds (so-called *birdies*), but these can be concealed up to a certain point on the audio level if they are detected by a proper mechanism. The third group is the least sensitive one. It consists of subband samples. Subband sample errors cause a kind of gurgling sound. Often this will not even be noticed in a noisy car environment. A last group consists of programme-associated data (PAD) and the cyclic redundancy check (CRC) for error detection in the scale factors (of the following frame). This group requires approximately the same protection as the second one. The distribution of the redundancy over the audio frame defines such an *error protection profile*. For DAB audio transmission, 64 different protection profiles have been specified (ETS 300 401) that correspond to different audio data rates from 32 kbit/s and 384 kbit/s and allow 5 different *protection levels* from PL1 (the strongest) to PL5 (the weakest) corresponding to five average code rates. Each of them requires (approximately) the same SNR for distortion-free audio reception. Table 4.4 gives the detailed definition of the protection profile corresponding to Figure 4.81. The last column shows the number of encoded bits. Note that for each frame, the trellis will be closed by tail bits. These are

Table 4.4 Example for an error protection profile profile (PL3) for the audio data rate 192 kbit/s

	Audio data bits	Code rate	Encoded bits
Group 1	352	$R_c = 8/24$	1056
Group 2	768	$R_c = 8/18$	1758
Group 3	3392	$R_c = 8/14$	5936
Group 4	96	$R_c = 8/19$	228
Tail bits	6	$R_c = 8/16$	12

always six zero bits that are encoded by $R_c = 1/2$. In this example, the total number of encoded bits per frame is 8960. This corresponds to 140 capacity units of 64 bits (see the following table).

For data transmission, eight different protection levels with *equal error protection* (EEP) have been specified with code rates $R_c = 1/4$, $R_c = 3/8$, $R_c = 4/9$, $R_c = 1/2$, $R_c = 4/7$, $R_c = 3/4$, and $R_c = 4/5$. The code rates 3/8 and 3/4 are constructed by a composition of two adjacent RCPC code rates. The EEP protection profiles allow fixed data rates that are integer multiples of 8 kbit/s or 32 kbit/s.

The paper (Hoeher *et al.* 1991) gives some insight into how the channel coding for DAB audio has been developed. It reflects the state of the research work on this topic a few months before the parameters were fixed.

We finally note that the UEP protection profiles for audio have been designed in such a way that one has a kind of *graceful degradation*. This means that if the reception becomes worse, the listener first hears the gurgling sound from the sample errors before the reception is lost. These errors can be noticed at a BER slightly above 10^{-4} with headphones in a silent environment. In the noisy environment of a car, up to 10^{-3} may be occasionally tolerated.

Multiplexing

All the UEP and EEP channel coding profiles are based on a frame structure of 24 ms. These frames are called *logical frames*. They are synchronized with the transmission frames, and, for audio data subchannels, with the audio frames. At the beginning of one logical frame, the coding starts with the shift registers in the all-zero state. At the end, the shift register will be forced back to the all-zero state by appending six additional bits (tail bits) to the useful data for the traceback of the Viterbi decoder. After encoding, such a 24 ms logical frame builds up a punctured code word. It always contains an integer multiple of 64 bits, which is an integer number of CUs. Whenever necessary, some additional puncturing is done to achieve this. A data stream of subsequent logical frames that is coded independently of other data streams is called a *subchannel*. For example, an audio data stream of 192 kbit/s is such a possible subchannel. A PAD data stream is always only a part of a subchannel. After the channel encoder, each subchannel will be time-interleaved independently as described in the next subsection. After time interleaving, all subchannels are multiplexed together into the MSC (see Figure 4.82 for an example). There is an elementary 24 ms time period in the MSC that is called a *common interleaved frame* (CIF). For TM II and TM III, each transmission frame carries one CIF. For TM I and TM IV, each transmission frame carries four or two subsequent CIFs, respectively.

The multiplex configuration of the DAB system is extremely flexible. For each subchannel, the appropriate source data rate and the error protection can be individually chosen. The total capacity of 864 will be shared by all these subchannels. Table 4.5 shows an example (taken from reality) of how the capacity may be shared by different subchannels (which are loosely called *programmes* in that table).

Time interleaving

For DAB, time and frequency interleaving has been implemented. To spread the coded bits over a wider time span, time interleaving is applied for each subchannel. It is based on the

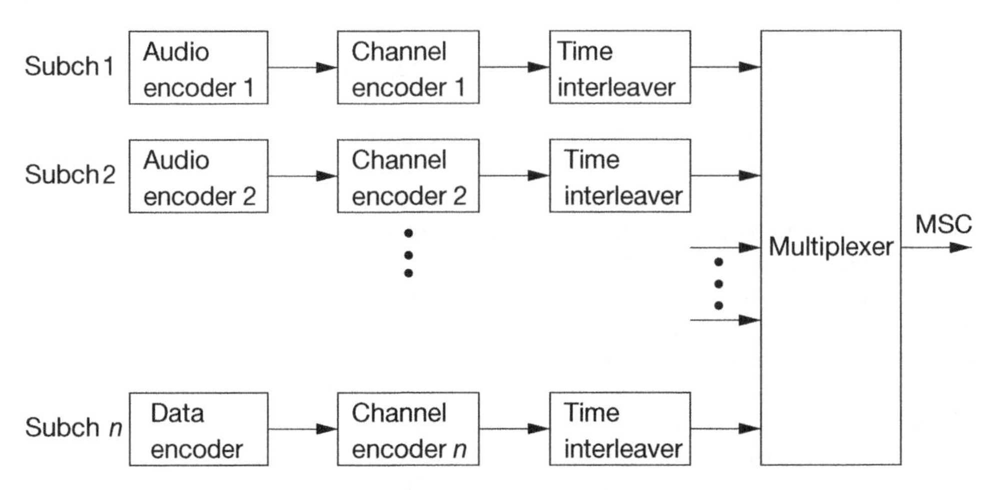

Figure 4.82 Block diagram for the DAB multiplex.

Table 4.5 Example for multiplex configuration

Programme	Content	Bit rate	Capacity	Protection
Audio 1	Pop music	160 kbit/s	116 CU	PL3
Audio 2	Classical music	192 kbit/s	140 CU	PL3
Audio 3	Classical music	224 kbit/s	168 CU	PL3
Audio 4	Traffic info	80 kbit/s	58 CU	PL3
Data 1	Visual service	72 kbit/s	54 CU	PL3
Audio 5	Information	192 kbit/s	116 CU	PL4
Audio 6	Information	128 kbit/s	96 CU	PL3
Audio 7	Pop music	160 kbit/s	116 CU	PL3
Sum			864 CU	

convolutional interleaver as explained in Subsection 4.4.2. With the notation introduced in that subsection, $B = 16$ has been chosen and N is the number of coded bits of one logical frame. First, the code word (i.e. the bits of one logical frame) will be split up into small groups of 16 bits. The bits with number 0 to 15 of each group will be permuted according to the bit reverse law (i.e. $0 \mapsto 0$, $1 \mapsto 8$, $2 \mapsto 4$, $3 \mapsto 12$, $\ldots$, $14 \mapsto 7$, $15 \mapsto 15$). Then, in each 16 bit group, bit number 0 will be transmitted without delay, bit number 1 will be transmitted with a delay of N serial bit periods T_S, that is, by the duration of one logical frame of $T_L = NT_S$ =24 ms. Bit number 2 will be transmitted with a delay of $2T_L = 2 \cdot 24$ ms, and so on, until bit number 15 will be transmitted with a delay of $15T_L = 15 \cdot 24$ ms. At the receiver side, the deinterleaver works as follows. In each group, bit number 0 will be delayed by $15T_L = 15 \cdot 24$ ms, bit number 1 will be delayed by $14T_L = 14 \cdot 24$ ms, $\ldots$, bit number 14 will be delayed by $T_L = 24$ ms and bit number 15 will not be delayed. Afterwards, the bit reverse permutation will be inverted. The deinterleaver restores the bit stream in the proper order, but the whole interleaving and deinterleaving procedure results in an overall decoding delay of $15T_L = 15 \cdot 24$ ms $= 360$ ms. This is a price that has to be paid for a better distribution of errors. A burst error on the physical channel will be broken up by the deinterleaver, because a long burst of adjacent (unreliable) bits before the deinterleaver will be broken up so that two bits of a burst have a distance of at least 16 after the deinterleaver and before the decoder.

The time interleaving is defined individually for each subchannel. This has been done because the receiver usually will decode only one subchannel and should therefore not process any data that belong to other subchannels. At the transmitter, it is more convenient to process all the subchannels together. The DAB system has been designed in such a way that both are possible. It is an important fact that the size of the capacity unit of 64 bits is an integer multiple of the period of $B = 16$ bits. As a consequence, each subchannel has a logical frame size N that is an integer multiple of $B = 16$ bits. Thus, we may interchange the order of time interleaving and multiplexing in Figure 4.82 and get the same bit stream for the MSC.

The time interleaving will only be applied to the data of the MSC. The FIC has to be decoded without delay and will therefore only be frequency interleaved.

Frequency interleaving and modulation

Because the fading amplitudes of adjacent OFDM subcarriers are highly correlated, the modulated complex symbols will be frequency interleaved. This will be done with the QPSK symbols before the differential modulation. We explain it by an example for TM II with $K = 384$ subcarriers: A block of $2K = 768$ encoded and time-interleaved bits have to be mapped onto the 384 complex modulation symbols for one OFDM symbol of duration T_S. The first 384 bits will be mapped to the real parts of the 384 QPSK symbols, the last 384 bits will be mapped to the imaginary parts. To write it down formally, the bits $p_{i,l}$ $(i = 0, 1, \ldots, 2K - 1)$ of the block corresponding to the OFDM symbol with time index l will be mapped onto the QPSK symbols $q_{i,l}$ $(i = 0, 1, \ldots, K - 1)$ according to the rule

$$q_{i,l} = \frac{1}{\sqrt{2}}\left[\left(1 - 2p_{i,l}\right) + j\left(1 - 2p_{i+K,l}\right)\right],\; i = 0, 1, \ldots, K - 1.$$

The frequency interleaver is simply a renumbering of the QPSK symbols according to a fixed pseudorandom permutation. The QPSK symbols after renumbering are denoted by $x_{k,l}$ $(k = \pm 1, \pm 2, \pm 3, \ldots, \pm K/2)$. Then the frequency-interleaved QPSK symbols will be differentially modulated according to the law

$$s_{k,l} = s_{k,l-1} \cdot x_{k,l}.$$

The complex numbers $s_{k,l}$ are the Fourier coefficients of the OFDM with time index l in the frame.

Performance considerations

Sufficient interleaving is indispensable for a coded system in a mobile radio channel. Error bursts during deep fades will cause the Viterbi decoder to fail. As already discussed in detail, OFDM is very well suited for coded transmission over fading channels because it allows time and frequency interleaving. Both interleaving mechanisms work together.

An efficient interleaving requires some incoherency of the channel to achieve uncorrelated or weakly correlated errors at the input of the Viterbi decoder. This is in contrast to the requirement of the demodulation. A fast channel makes the time interleaving more efficient, but causes degradations because of fast phase fluctuations. As discussed in the example at the end of Subsection 4.4.1, the benefit of time interleaving is very small for Doppler frequencies below 40 Hz. On the other hand, this is already the upper limit for the DQPSK demodulation for TM I. For even lower Doppler frequencies corresponding to moderate or low car speeds and VHF transmission, the time interleaving does not help very much. In this case, the performance can be saved by an efficient frequency interleaving. Long echoes ensure efficient frequency interleaving. As a consequence, SFNs (single frequency networks) support the frequency interleaving mechanism. If, on the other hand, the channel is slowly and frequency-flat fading, severe degradations may occur even for a seemingly sufficient reception power level.

To compare with the theoretical DQPSK bit error rates discussed in Subsection 4.4.1, we performed several simulations of the DAB system. For the delay power spectrum DAB HT2 that was defined during the evaluation process is based on real channel measurements. It is the superposition of three exponential delay power spectra delayed by $\tau_1 = 0$ µs, $\tau_2 = 20$ µs, $\tau_3 = 40$ µs with normalized powers $P_1 = 0.2$, $P_2 = 0.6$, $P_3 = 0.2$ and respective delay spreads $\tau_{m1} = 1$ µs, $\tau_{m2} = 5$ µs, $\tau_{m3} = 2$ µs. The overall delay spread is $\tau_m \approx 14$ µs.

Figure 4.83 shows BER simulations for the DAB transmission mode II system with a 256 kbit/s data stream with EEP compared with the DQPSK union bounds. The maximum Doppler frequency for the isotropic spectrum is 64 Hz, which leads to $\nu_{\max} T_S = 0.02$. Time

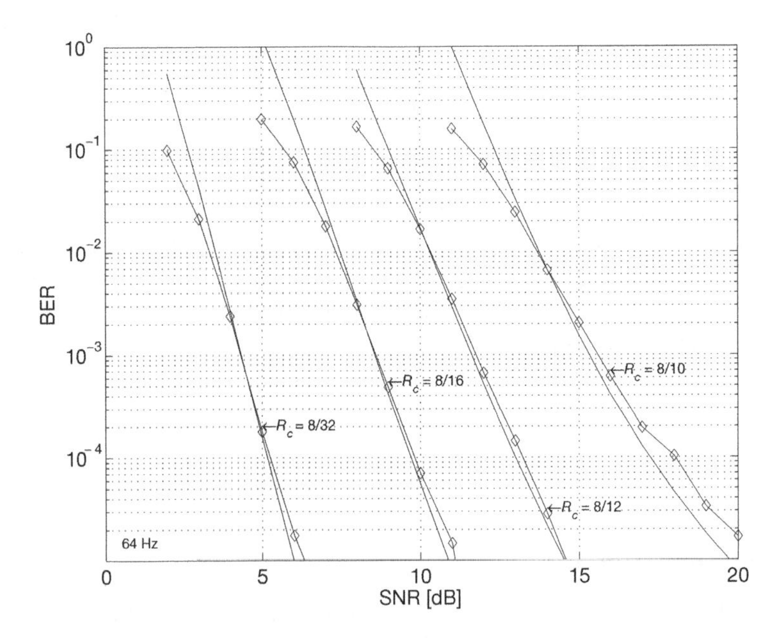

Figure 4.83 Simulated BER for the DAB system for $\nu_{\max} T_S = 0.02$ and $R_c = 8/10$, $8/12$, $8/16$, $8/32$ and a frequency-selective channel.

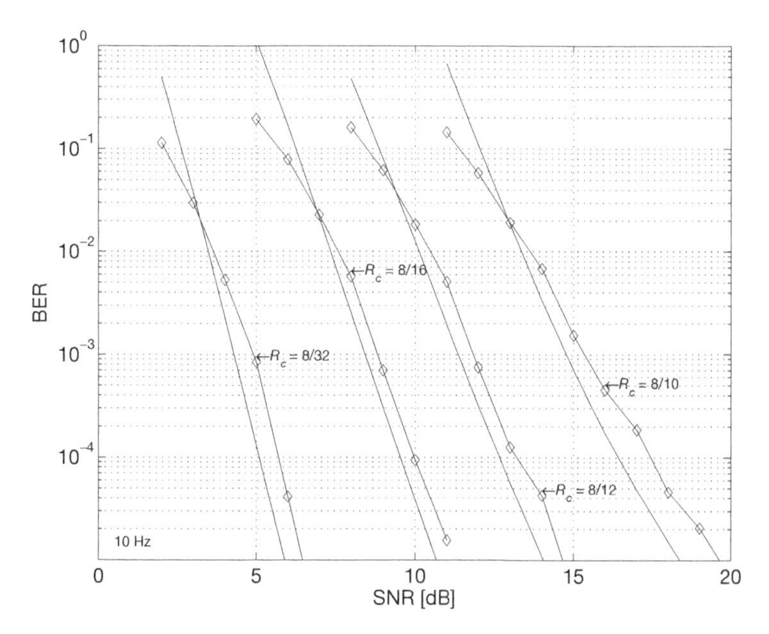

Figure 4.84 Simulated BER for the DAB system for $\nu_{\max} T_S = 0.003$ and $R_c = 8/10, 8/12, 8/16, 8/32$ and a frequency-flat channel.

interleaving alone cannot be sufficient because closely related bits are only separated by 24 ms. To separate them, the Doppler frequency would have to exceed, significantly, 40 Hz, which would lead to unacceptable high values of $\nu_{\max} T_S$. The simulated curves fit quite well with the theoretical curves, which indicates that both interleaving mechanisms together lead to a sufficient separation of the bits on the physical channel. The weakest protection profile shows some degradations. This can be understood by the fact that the DAB EEP profiles have exactly those fractional code rates *including* the coded tail bits. The tail bits are coded by $R_c = 1/2$. The corresponding 12 coded bits are saved by using the next weakest code for the last 96 bits in the data stream, which leads to a poorer performance there. It can be verified by computer simulations that this effect becomes smaller for higher data rates and more severe for lower data rates.

Figure 4.84 shows BER simulations for the DAB transmission mode II system with a 256 kbit/s data stream with EEP compared with the DQPSK union bounds. The maximum Doppler frequency for the isotropic spectrum is 10 Hz, which leads to $\nu_{\max} T_S = 0.003$. For a radio frequency of 230 MHz, this corresponds to a vehicle speed of 48 km/h. The delay power spectrum is the GSM *typical urban* spectrum, which is of exponential type with $\tau_m = 1\ \mu$s. Neither time interleaving nor frequency interleaving is sufficient for this channel. Significant degradations compared to the other channel can be observed.

4.6.2 The DVB-T system

The European Digital Video Broadcasting (DVB) system splits up into three different transmission systems[14] corresponding to three different physical channels: a cable system

[14]Further extensions are currently being defined. We do not discuss them here.

(DVB-C), a satellite system (DVB-S) and a terrestrial system (DVB-T). Because the requirements of the three channels are very different, different coding and modulation schemes have been implemented. Common to all three systems is an (outer) Reed–Solomon (RS) code to achieve the extremely low bit error rates that are required for the video data stream and that cannot be reached efficiently by convolutional coding alone. For the DVB-C standard, an AWGN channel with very high SNR can be assumed so that the Reed–Solomon code alone is sufficient. Both DVB-S and DVB-T need an inner convolutional code. This is necessary for the first one because of the severe power limitation of the satellite channel. For the second one, the terrestrial channel is typically a fading channel for which convolutional codes are usually the best choice because they can take benefit from the channel state information. All three systems use QAM modulation. For DVB-S, only 4-QAM (= QPSK) is used for reasons of power efficiency. Both other systems have higher-level QAM as possible options. DVB-C and DVB-S use conventional single carrier modulation. DVB-T uses OFDM to cope with long echoes and to allow SFN coverage. We concentrate on the discussion of the terrestrial system.

The physical channel is similar to that of the DAB system. We may have runtime differences of the signal of several ten microseconds, which are due to echoes caused by the topographical situation. For both systems, SFNs are a requirement at least as one possible option. One significant difference in the requirements is that the DAB system has been especially designed for mobile reception. For the DVB-T system, portable – but not mobile – reception was required when the system parameters were chosen.

DVB-T is intended to replace existing analog television signals in the same channels. Depending on the country and the frequency band (VHF or UHF band), there exist TV channels of 6 MHz, 7 MHz and 8 MHz nominal bandwidth. The DVB-T system can match the signal bandwidth to these three cases. Similar to the DAB system, *transmission modes* have been specified to deal with different scenarios. For each of the three different bandwidth options, there exist two such parameter sets. They are called *8k mode* and *2k mode,* corresponding to the smallest possible (power of two) FFT length 8192 and 2048, respectively. The OFDM symbol length of the 8k mode is similar to that of the DAB transmission mode I and thus intended for SFN coverage. Because of the long symbol duration, it is more sensitive against high Doppler frequencies. The OFDM symbol length of the 2k mode is similar to that of the DAB transmission mode II. It is suited for typical terrestrial broadcasting situations, but not for SFNs. It may thus preferably be used for local coverage. Let us denote again the OFDM Fourier analysis window by T, the total symbol length by T_S and the guard interval by Δ. In contrast to the DAB system, there exist several options for the length of the guard interval: $\Delta = T/4$, $\Delta = T/8$, $\Delta = T/16$ and $\Delta = T/32$. Table 4.6 shows the OFDM symbol parameters for the 8k mode and Table 4.7 for the 2k mode, both with the

Table 4.6 OFDM Parameters for the DVB-T 8k mode and $\Delta = T/4$

Channel	t_s	T	T_S	Δ	Max. frequency
		$8192\,t_s$	$10,240\,t_s$	$2024\,t_s$	
8 MHz	$7/64$ μs	896 μs	1120 μs	224 μs	≈800 MHz
7 MHz	$1/8$ μs	1024 μs	1280 μs	256 μs	≈700 MHz
6 MHz	$7/48$ μs	≈1195 μs	≈1493 μs	≈299 μs	≈600 MHz

Table 4.7 OFDM Parameters for the DVB-T 2k mode and $\Delta = T/4$

Channel	t_s	T	T_S	Δ	Max. frequency
		$2048\,t_s$	$2560\,t_s$	$512\,t_s$	
8 MHz	7/64 µs	224 µs	280 µs	56 µs	≈3200 MHz
7 MHz	1/8 µs	256 µs	320 µs	64 µs	≈2800 MHz
6 MHz	7/48 µs	≈299 µs	≈373 µs	≈75 µs	≈2400 MHz

guard interval length $\Delta = T/4$. All time periods are defined as a multiple of the sampling period $t_s = f_s^{-1}$ that is different for the three different TV channel bandwidths. For each mode, the three bandwidths can be obtained by a simple scaling of that sampling frequency.

The number of carriers is given by $K + 1 = 6817$ for the 8k mode and by $K + 1 = 1705$ for the 2k mode. The spacing $f_{K/2} - f_{-K/2}$ between the highest and the lowest subcarrier is approximately given by 7607 kHz for the 8 MHz channel, 6656 kHz for the 7 MHz channel and by 5705 kHz for the 6 MHz channel.

The frequency in the last column is the optimistic upper limit for the maximum frequency that can be used for a vehicle speed of 120 km/h if a very powerful channel estimation with Wiener filtering has been implemented and if an appropriately strong channel coding and modulation scheme has been chosen. The pilot grid for DVB-T is the diagonal one of Figure 4.36. The parameters of the 7 MHz system correspond approximately to the numerical example given in Subsection 4.3.2. For the 8k mode, according to that example, the channel will be sampled with a sampling frequency of approximately 200 Hz. Owing to the sampling theorem, the limit for the Doppler frequency is then given by 100 Hz. This corresponds to 900 MHz radio frequency for a vehicle speed of 120 km/h. In practice, one should be well below the limit given by the sampling theorem. For a good channel estimation, 700 MHz should be possible. This value corresponds to approximately 78 Hz Doppler frequency or $\nu_{\max} T_S = 0.1$. As we have seen in Subsection 4.5.3, this value can be tolerated, for example, for 16-QAM and code rate $R_c = 1/2$, but not for higher spectral efficiencies. For 64-QAM and code rate $R_c = 1/2$, the maximum frequency should be 25% lower.

Because the DAB transmission modes I and II have similar symbol length as the 8k and 2k modes of DVB-T, a direct comparison of the sensitivity against high Doppler frequencies are possible. We conclude that the DVB-T system allows approximately twice the carrier frequency (or vehicle speed) compared to the DAB system. From the discussion in Subsection 4.5.3, we further conclude that at the highest possible value for the DAB system, the DVB-T system with 16-QAM has a similar performance as the DAB system at approximately twice the spectral efficiency. In both cases, $R_c = 1/2$ has been assumed.

Baseline transmission system

The baseline DVB-T transmission system is depicted in Figure 4.85. Packets of 188 bytes length will first be encoded to code words of length 204 by the outer RS(204, 188, 17) code. This code has Hamming distance 17 and can thus correct up to eight byte errors. This shortened RS code has been obtained from a RS(255, 239, 17) code by setting the first 51 systematic bytes to zero and not transmitting them. The code words are interleaved by a convolutional byte interleaver as described in Subsection 4.4.2 with parameters $B = 12$ and $M = 17$. Thus, $N = BM = 204$ is just the block length of one code word. The interleaver

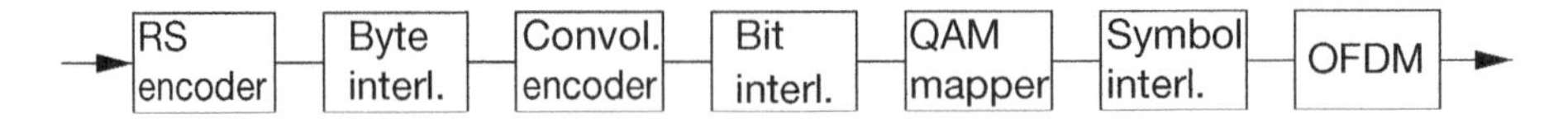

Figure 4.85 Simplified block diagram for the DVB-T signal generation.

works in such a way that the byte number zero (i.e. the first one) in a block stays at the same position and in the same block. The byte number one is delayed by the block length N, that is, it will be transmitted in the next block at the same position within the block. The byte number two is delayed by $2N$, that is, two blocks, and so forth until byte number 12, which stays inside the block at the same position and the whole procedure, will continue that way. This outer byte interleaver is necessary because at the receiver the inner decoder produces error bursts. These error bursts must be distributed over several code words because more than 8 bytes in one code word cannot be corrected. Following the discussion in Subsection 4.4.2 we observe that an error burst of 12 bytes (= 96 bits) length after the inner decoder will result in only one corresponding byte error inside one code word. The RS code can correct up to eight byte errors, that is, the error bursts may be eight times longer.

The *bit* stream of the byte interleaved code words will be encoded by an inner encoder for the standard $(133, 171)_{\text{oct}}$ convolutional code and then modulated as discussed in Subsections 4.5.1 and 4.5.2. With optional puncturing, the code rates $R_c = 1/2$, $R_c = 2/3$, $R_c = 3/4$, $R_c = 5/6$ and $R_c = 7/8$ are possible. The output bit stream of the convolutional encoder will be interleaved by a (small) pseudorandom permutation and mapped on complex QAM symbols by a symbol mapper. Thus, exactly the concept of bit-interleaved coded modulation has been implemented in the DVB-T system. The options 4-QAM, 16-QAM and 64-QAM are possible. The QAM symbols are OFDM modulated. Each OFDM symbol carries 6048 QAM symbols in the 8k mode and 1512 QAM symbols in the 2k mode, respectively. The other complex symbols serve as pilot symbols for channel estimation. In addition to the diagonal grid of *scattered pilot* of Figure 4.36, there are *continuous pilots* that serve as references for frequency synchronization. All pilots are boosted by a factor of 4/3 in the amplitude compared to the QAM symbols. Table 4.8 shows the possible coding and modulation options and the corresponding data rates for $\Delta = T/4$ and the 8 MHz system.

To exploit the channel diversity in frequency direction, for each OFDM symbol, the QAM symbols are frequency interleaved by a pseudorandom permutation of length 6048 or 1512, respectively. In contrast to the DAB system, no time interleaving is applied. This is due to the fact that originally no mobile reception was intended.

A set of 68 OFDM symbols are grouped together to a transmission frame, and four such frames build a hyperframe. There are some significant differences to the DAB system. First, there is no correspondence between certain parts of the data stream and certain OFDM symbols in the frame. DAB allows different code rates for different parts of the signal. This is not possible for DVB-T. The information that is necessary to identify the overall code rate and the guard interval length are transmitted on special TPS (transmission parameter signaling) carriers.

Channel coding aspects

The DVB-T channel coding scheme consists of an inner convolutional code and an outer Reed–Solomon code. The outer symbol interleaver is a frequency interleaver that has the

Table 4.8 Transmission options and data rates for DVB-T for guard interval length $\Delta = T/4$

Modulation	Code rate	Bits per symbol	R_b	Useful R_b
QPSK	$R_c = 1/2$	1	5.4 Mbit/s	4.98 Mbit/s
QPSK	$R_c = 2/3$	1.33	7.2 Mbit/s	6.64 Mbit/s
QPSK	$R_c = 3/4$	1.5	8.1 Mbit/s	7.46 Mbit/s
QPSK	$R_c = 5/6$	1.67	9.0 Mbit/s	8.29 Mbit/s
QPSK	$R_c = 7/8$	1.75	9.45 Mbit/s	8.71 Mbit/s
16-QAM	$R_c = 1/2$	2	10.8 Mbit/s	9.95 Mbit/s
16-QAM	$R_c = 2/3$	2.67	14.4 Mbit/s	13.27 Mbit/s
16-QAM	$R_c = 3/4$	3	16.2 Mbit/s	14.93 Mbit/s
16-QAM	$R_c = 5/6$	3.33	18.0 Mbit/s	16.59 Mbit/s
16-QAM	$R_c = 7/8$	3.5	18.9 Mbit/s	17.42 Mbit/s
64-QAM	$R_c = 1/2$	3	16.2 Mbit/s	14.93 Mbit/s
64-QAM	$R_c = 2/3$	4	21.6 Mbit/s	19.91 Mbit/s
64-QAM	$R_c = 3/4$	4.5	24.3 Mbit/s	22.39 Mbit/s
64-QAM	$R_c = 5/6$	5	27.0 Mbit/s	24.88 Mbit/s
64-QAM	$R_c = 7/8$	5.25	28.4 Mbit/s	26.13 Mbit/s

purpose to break up the correlations of the channel and provide the inner code with the diversity that can be obtained from the frequency selectivity of the channel. No similar mechanism is intended to take advantage from time variance of the channel. The bit interleaved coded modulation needs a (small) bit interleaver between the convolutional encoder and the symbol mapper. This is necessary in order to avoid closely related bits of the code word being affected by the same noise sample. Of course, it would have been possible to use a bigger bit interleaver for both purposes together. The outer byte interleaver has the purpose to break up long error bursts resulting from erroneous convolutional decoding. The combination of a convolutional inner code together with an outer RS code with an interleaver in between is a very powerful combination. The RS code is very efficient for burst error decoding as long as the bursts are not too long. It takes advantage from the fact that more than one bit error is inside one erroneous byte. Let P be the byte error probability and P_b the bit error probability after the Viterbi decoder. We note that the worst case of only one average bit error in one erroneous byte corresponds to $P = 8P_b$, two bit errors correspond to $P = 4P_b$ and four bit errors correspond to $P = 2P_b$. The assumption of ideal interleaving means that the byte errors are uniformly distributed. The block error probability analysis of Subsection 3.1.2 can be generalized to the case that we deal with bits rather than with bytes. The probability for the block code word error probability is then given by

$$P_{\text{Block}} = \sum_{i=t+1}^{N} \binom{N}{i} P^i (1-P)^{N-i}.$$

In these equations, $N = 204$ is the length of the code word, and $t = 8$ is the error correction capability. To obtain the residual bit error probability, we can argue as we did in Subsection 3.1.2. We take into account that, for a given bit inside a byte, 128 of 255 possible byte

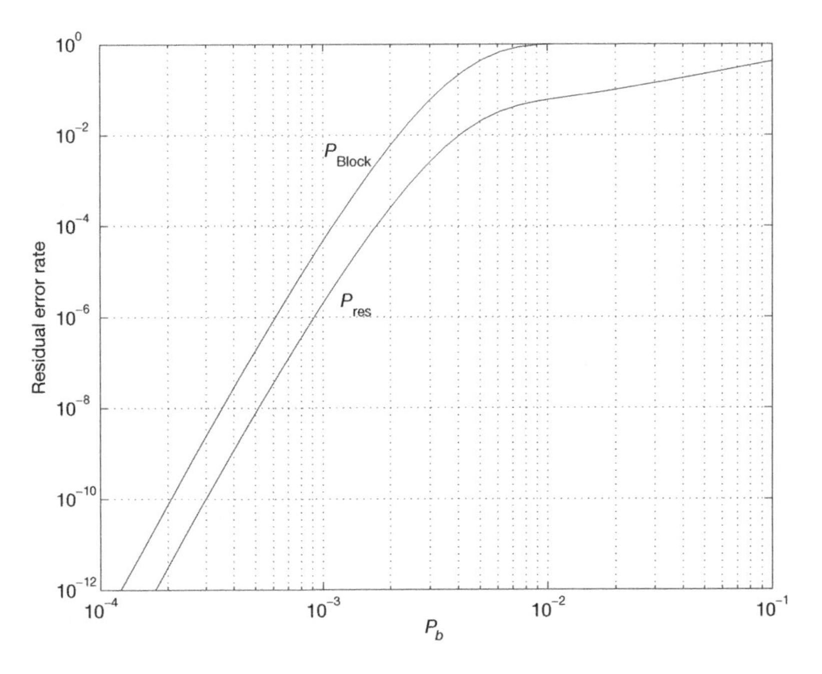

Figure 4.86 Block error rate and residual bit error rate for the RS code.

errors would lead to a bit error. The residual bit error rate is then upper bounded by

$$P_{\text{res}} \leq \frac{128}{255} \sum_{i=t+1}^{N} \frac{\min(t+i, N)}{N} \binom{N}{i} P^i (1-P)^{N-i},$$

For the worst case $P = 8P_b$, these curves are plotted in Figure 4.86. If ideal interleaving can be assumed for all interleaving mechanisms, the curve for P_{Block} in conjunction with the bit error curves for the convolutionally coded QAM can be used to conclude from the channel SNR to the error event frequency for the video signal. We discuss the line of thought on the basis of Figure 4.87. First, the QAM symbols are deinterleaved in frequency direction by the symbol interleaver. From the QAM symbols, the MCU calculated the metric expressions (i.e. *soft bits*) as described in Subsection 4.2.1. These soft bit values are deinterleaved before they are passed to the Viterbi decoder. The Viterbi decoder produces burst errors, that is, there are more or less long sequences inside the bit stream of unreliable bits. For the following RS decoder, it is favorable that the bit errors are grouped close together in the same bytes, but long sequences of byte errors must be avoided. The purpose of the byte interleaver is to break up such long sequences.

To give a concrete numerical example, we start with $P_b = 2 \cdot 10^{-4}$ for the required BER after the Viterbi decoder. This is a requirement that can be found in many papers because it is stated by the DVB-T developers that this would guarantee a *virtual error-free channel* after the RS decoder. From the theoretical analysis of convolutionally coded QAM, we know what SNR is needed to achieve this BER. As an example, for 64-QAM and $R_c = 1/2$ in an ideally interleaved Rayleigh fading channel, we infer from Figure 4.63

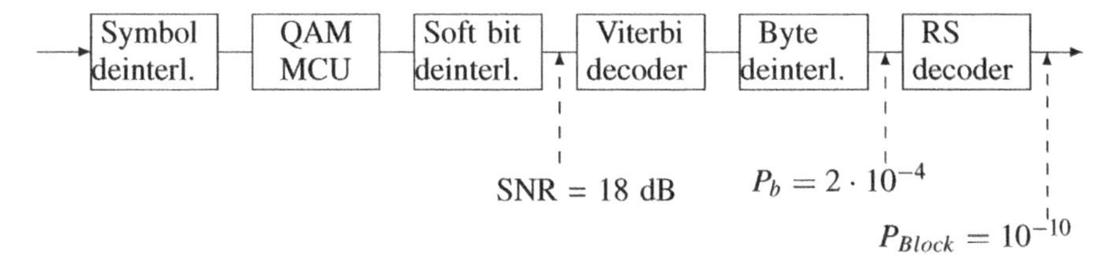

Figure 4.87 The DVB-Decoder.

an SNR between 16 dB and 17 dB. If we take into account some loss that is due to channel estimation, we may regard 18 dB as a reasonable figure. From Figure 4.86, we infer a block error rate $P_{\mathrm{Block}} = 10^{-10}$ after RS decoding. To interpret this, we assume as an example a low video data rate of approximately 3 Mbit/s. Recall that one block has 188 useful bytes corresponding to 1504 useful bits. This means that approximately 2000 blocks are transmitted per second. For $P_{\mathrm{Block}} = 10^{-10}$, the average time between two error events is $5 \cdot 10^6$ seconds, which corresponds to 58 days. For a high video data rate of 30 Mbit/s, this reduces to six days, which can still be regarded as virtually error-free reception. We note that not every error event will lead to perceptible errors in the picture. Furthermore, a powerful RS decoder is able to detect a large amount of uncorrectable code words and will send a flag to an error concealment mechanism. Thus, the time between perceptible picture errors may be much larger.

For the DVB-T system, the concept of a *virtually error-free channel* has been introduced as a reception with the residual bit error rate of $P_{\mathrm{res}} = 10^{-11}$. For uniformly distributed bit errors, this corresponds to approximately one bit error per hour for 30 Mbit/s. However, this figure is misleading because the RS decoder does not produce uniformly distributed bit errors, but block errors with many bit errors inside. The most probable error event corresponds to code words at the Hamming distance, that is, typically there are 17 wrong bytes or 68 wrong bits in average. This means that a burst of typically 68 bit errors occur every 68 hours ($\approx$3 days) and not one single bit error per hour[15]. However, because the BER curves for the concatenated coding system are very steep, a weakening of these requirements for the virtual error-free channel would only result in a small SNR gain. Much more important is the fact that the curves are based on the assumption of ideal interleaving.

Mobile reception

Even though mobile reception was originally not required, this item has become more and more important for the practical application. Is the DVB-T system suited for mobile reception? Taking into account the results of the preceding sections, we can make the following statements:

1. The modulation scheme of DVB-T with coherent modulation together with the channel estimation concept is very well suited for fading channels if the interleaving can

[15]The factor of 2 between these three days and the six days of the preceding analysis has its origin that the DVB-T figures are bases on the error rates for the unshortened RS(255, 239, 17) code, for which $P_b = 2 \cdot 10^{-4}$ leads to $P_{\mathrm{res}} = 10^{-11}$. In Figure 4.86, we find $P_{\mathrm{res}} \approx 5 \cdot 10^{-12}$ for the same P_b.

be assumed to be sufficient. The coherent QAM is by far superior in robustness and spectral efficiency compared to the differential demodulation as applied by DAB. However, only a very restricted number of the combinations of Table 4.8 are suited for mobile reception. Only the lowest possible code rates can be recommended. For low data rates, $R_c = 1/3$ also should have been included.

2. Since the number of subcarriers is very large, the DVB-T system can be considered as a wideband system if the channel is not too flat. Unfortunately, time interleaving has not been included. For frequency-flat channels with insufficient interleaving, burst errors will corrupt the whole concatenated coding scheme. However, receive antenna diversity may help in such situations.

3. In a mobile radio channel, the concept of a virtually error-free channel does not make sense because the conditions may change severely during a short period of time that is much less than one hour. In mobile reception practice, there will always be situations where the system approaches its limits. The system design must take this into account. In contrast to the DAB system, nothing has been done for this case. There is no unequal error protection or graceful degradation or error detection in the scale factors. This may result in annoying perturbations of the audio quality.

4.6.3 WLAN systems

OFDM with a guard interval is applied within two systems for wireless communications between computers in a local area network. The corresponding standards for these Wireless Local Area Networks (WLAN) are called:

- the HIPERLAN/2 standard released by the European Telecommunications Standards Institute (ETSI) in 2000;
- the IEEE 802.11a and IEEE 802.11g standard released by the Institute of Electrical and Electronics Engineers (IEEE) in 1999 and in 2003, respectively.

While HIPERLAN/2 and IEEE 802.11a operate in the 5 GHz band, IEEE 802.11g uses a frequency band at about 2.4 GHz, which is also occupied by other systems like Bluetooth and another variant of the IEEE 802.11 standard, namely, the IEEE 802.11b variant using the spread spectrum and code keying techniques as the basic transmission scheme (see Subsection 5.5.1). The OFDM parameters as well as the main modulation and channel coding parameters of IEEE 802.11a and IEEE 802.11g are absolutely identical. There are only some differences with respect to the header and the preamble of the physical data bursts since the coexistence of IEEE 802.11b and 802.11g mode within one frequency band requires special means. In the following text, we focus on the IEEE 802.11a variant, nevertheless the considerations may be transferred directly to the IEEE 802.11g variant. Also, the parameters of the physical layer of IEEE 802.11a and HIPERLAN/2 have been harmonized to a high degree by the corresponding standardization groups. However, there are some fundamental differences concerning the format of a physical burst and especially concerning the multiple access technique. While HIPERLAN/2 uses a time division multiple access (TDMA) scheme with a fixed TDMA frame length of 2 ms and a centralized resource allocation, the multiple access within all IEEE 802.11 modes is based on carrier sense

Table 4.9 The OFDM Parameters for HIPERLAN/2 and IEEE 802.11a

K	T	T_S	Δ
52	3.2 μs	4 μs	0.8 μs

multiple access (CSMA). CSMA is a decentralized multiple access scheme known from wired LANs (IEEE 802.3: Ethernet) which does not use a fixed time slot structure, but data packets of a variable length.

Modulation and coding parameters

Let us again denote the OFDM Fourier analysis window length by T, the total symbol length by T_S, the guard interval length by Δ and the number of carriers by K. Table 4.9 shows the values of these parameters. The $K = 52$ subcarrier frequency positions are given by $f_k = k/T$ with $k \in \{\pm 1, \pm 2, \ldots, \pm K/2\}$, that is, similar to the DAB system, the center subcarrier position is left empty. The four subcarriers with index $k \in \{\pm 7, \pm 21\}$ are used as continuous pilots for frequency synchronization. The spacing between the highest and the lowest subcarrier is given by $f_{K/2} - f_{-K/2}$ =16.25 MHz. The guard interval length $\Delta = 0.8$ μs is able to absorb path length differences up to 240 m. For an environment with shorter echoes, $\Delta = 0.4$ μs is a possible option. In that case, all possible data rates can be increased by 11%.

For both systems, BPSK, QPSK, 16-QAM and 64-QAM are possible modulation schemes. For channel coding, the same $(133, 171)_{\text{oct}}$ convolutional code is used as in the systems described above. To achieve higher code rates, puncturing will be applied. For HIPERLAN/2, the possible code rates are $R_c = 1/2$, $R_c = 9/16$, and $R_c = 3/4$. For IEEE 802.11a, the possible code rates are $R_c = 1/2$, $R_c = 2/3$, and $R_c = 3/4$. Table 4.10 shows the possible coding and modulation options for both systems. Note that the only difference between both systems is that 24 Mbit/s and 48 Mbit/s are only used in the IEEE 802.11a system, while 27 Mbit/s is used only in the HIPERLAN/2 system.

Performance considerations

Since the systems have not been designed for mobile reception, only frequency interleaving has been applied, together with a small bit interleaver. The system can be considered as a BICM system as discussed in Subsection 4.5.2. However, in contrast to the DVB-T system, we do not have a real wideband system relative to the coherence bandwidth of the channel. As a consequence, the performance curves derived there cannot be applied directly because frequency interleaving alone cannot allow for sufficient decorrelation for such a low number of subcarriers. However, the results of ideal interleaving may serve as a hint for the system evaluation and may allow a comparison of the combinations of code rate and modulation scheme.

First we note that – as discussed in detail before – BPSK always has (for the AWGN and a multiplicative fading channel) the same power efficiency as QPSK. This means that, for both schemes, we need the same energy E_b per bit which is just the power per bit rate. BPSK transmission allows only half the bit rate compared to QPSK, and thus the power can

Table 4.10 Transmission options for HIPERLAN/2 and IEEE 802.11a

R_b	Modulation	Code rate	Bits per symbol	
6 Mbit/s	BPSK	$R_c = 1/2$	0.5	
9 Mbit/s	BPSK	$R_c = 3/4$	0.75	
12 Mbit/s	QPSK	$R_c = 1/2$	1	
18 Mbit/s	QPSK	$R_c = 3/4$	1.5	
24 Mbit/s	16-QAM	$R_c = 1/2$	2	IEEE only
27 Mbit/s	16-QAM	$R_c = 9/16$	2.25	HIPERLAN only
36 Mbit/s	16-QAM	$R_c = 3/4$	3	
48 Mbit/s	64-QAM	$R_c = 2/3$	4	IEEE only
54 Mbit/s	64-QAM	$R_c = 3/4$	4.5	

be reduced by a factor of 2 corresponding to a 3 dB lower SNR. Looking at Figure 4.65, we observe that this corresponds to an SNR reduction from 6 dB to 3 dB at a bit error rate of 10^{-4} for $R_c = 1/2$ and the ideally interleaved Rayleigh channel. From that figure, we also conclude that the increase of the code rate from $R_c = 1/2$ to $R_c = 3/4$ will require at least 5 dB more SNR. Thus, the 12 Mbit/s (QPSK, $R_c = 1/2$) mode will require less SNR than the 9 Mbit/s mode (BPSK, $R_c = 3/4$). Thus, in a Rayleigh fading channel, the 9 Mbit/s mode is obsolete. We further conclude from that figure and the corresponding discussion in Subsection 4.5.2 that at approximately 1.5 bits per symbol, 16-QAM with a low code rate would be a much better choice than 4-QAM (QPSK). Thus, in a Rayleigh fading channel, 16-QAM would be a better candidate for the 18 Mbit/s mode. For 36 Mbit/s, 64-QAM with $R_c = 1/2$ performs better than the parameter combination (16-QAM, $R_c = 3/4$) that has been chosen for the wireless LAN systems. We note that these statements apply for a Rayleigh fading channel. But this is of course the worst case.

We note that the BER is not really the adequate measure for the performance of a data communication system. Since errors can be tolerated in such a system (in contrast to an audio broadcasting system), an error detection scheme is necessary. In the systems under consideration, a CRC (cyclic redundancy check) has been implemented. If an error occurs in a packet of 432 bits, the packet will be retransmitted. Therefore, the *packet error rate* (PER) rate is more adequate than the BER. Since the available data rate will be lowered by the PER, the resulting effective data rate as a function of the SNR is the adequate performance measure for which the modulation and coding schemes have to be compared.

For each burst of N_{sym} OFDM symbols, the shift register of the convolutional code will be reset to the zero state by adding tail bits (see Subsection 3.2.1). To retain the exact ratio of the code rate, a technique similar to that in the DAB system has been introduced (see Subsection 4.6.1).

Physical burst (frame) structure

As mentioned above, HIPERLAN/2 and IEEE 802.11a use different burst formats and multiple access schemes. Hence, with respect to these topics the systems have to be discussed separately. We start with HIPERLAN/2.

HIPERLAN/2 is a TDMA system. Uplink and downlink share different time slots at the same frequency. A physical TDMA burst has the length of exactly 2 ms, which corresponds to the duration of 500 OFDM symbols. A physical burst starts with a *preamble* that is used for synchronization. After that, a variable number N_{sym} of OFDM symbol form the so-called *payload*. There are five different bursts with different preamble length:

1. The Broadcast burst: Preamble of length 16 μs. The payload consists of $N_{sym} = 496$ OFDM symbols.
2. The Downlink burst: Preamble of length 8 μs. The payload consists of $N_{sym} = 498$ OFDM symbols.
3. Uplink burst with short preamble: Preamble of length 12 μs. The payload consists of $N_{sym} = 497$ OFDM symbols.
4. Uplink burst with long preamble: Preamble of length 16 μs. The payload consists of $N_{sym} = 496$ OFDM symbols.
5. Direct link burst: Preamble of length 16 μs. The payload consists of $N_{sym} = 496$ OFDM symbols.

The last 8 μs of the preamble is common to all bursts and serves as a reference for the channel estimation that is necessary for the coherent demodulation. It consists of an OFDM reference symbol of length $2T_S = 8$ μs, which is BPSK modulated with a known pseudorandom sequence of length 52 that is modulated on the subcarriers with index $k \in \{\pm 1, \pm 2, \ldots, \pm K/2\}$. The resulting OFDM symbol (without guard interval) of length T is cyclically extended to the length $2T_S$ by a guard interval of length $T + 2\Delta$. Equivalently, one can say that the OFDM symbol of length T will be repeated and the resulting symbol of length $2T$ is cyclically extended (into the past) by a guard interval of length 2Δ to absorb the echoes. In the first part of the preamble, only 12 carriers are modulated, leading to shorter OFDM symbols. This part is used for coarse synchronization and as a reference for the automatic gain control (AGC).

A physical frame of the IEEE 802.11a system has a variable length and may carry some thousands of bytes. The header provides information on the length of the frame and on the modulation and channel coding scheme applied to the payload part. The header consisting of 24 bits is transmitted using the 6 Mbit/s mode, that is, it is transmitted as one OFDM symbol. The preamble in front of the physical frame has a length of 16 μs, where two different types of training sequences are transmitted as within the HIPERLAN/2 system. Error detection at the physical layer is only applied for the header using one parity bit; error detection of the payload is performed by higher layers using a CRC of 4 bytes.

4.7 Bibliographical Notes

The idea of multicarrier transmission goes back to the 1960s (Chang 1966; Chang and Gibby 1968; Saltzberg 1967). The original idea was indeed a physical realization of the concept of Figure 4.2 by using a large number of oscillators. The idea to simplify the implementation by using Fourier transform techniques goes back to (Weinstein and Ebert 1971) and was further developed by Hirosaki (1981). For a long time, however, the implementation of multicarrier transmission by digital circuits for high-speed data communication was still out of question. Thus, these fundamental ideas were widely unknown not only for practical engineers but even for the scientific community. It was pointed out by Cimini (1985) that OFDM with guard interval is especially suited for the mobile radio channel. This paper seems to be an inspiration for people at the French telecommunication and broadcasting research institute, CCETT, to propose OFDM as a digital broadcasting transmission system for mobile receivers (Alard and Lassalle 1987). It was the merit of these engineers to recognize that the time of OFDM had come and its realization by digital circuits had become a distinct possibility. In the European Digital Audio Broadcasting project, this system proposal became a very serious candidate and, at the end of the project, an OFDM system was standardized in 1993 (see (EN300401 2001a) for a recent update of the standard). An exhaustive treatment of the DAB system that is also very helpful for the practical engineer can be found in (Hoeg and Lauterbach 2003). A comprehensive overview about multicarrier modulation and its history can be found in (Bingham 1990) and in (Gitlin *et al.* 1993).

The DAB system can be regarded as the OFDM pioneer system. One of the authors (Henrik Schulze) became involved in the DAB project in 1987 (at Bosch Company in Hildesheim) and came in touch with OFDM through an internal project paper that was a draft version of (Alard and Lassalle 1987). At that time, very few people understood that concept and thus OFDM was regarded as a wonder cure against everything by its supporters, and it was regarded as pure fantasy by its antagonists. Even though it is mathematically evident that OFDM should work in principle, it soon became obvious that indeed some practical implementation problems are more severe than for traditional systems. These topics are discussed in Sections 4.2 and 4.3. That treatment was partly inspired by the Ph.D. thesis of (Schmidt 2001), which provides an interesting overview of several OFDM aspects. Another problem for the DAB system design was the proper choice of the guard interval because, as pointed out by Schulze (1988), echoes longer than the guard interval lead to severe degradations. Extensive measurements of the mobile radio broadcasting channel were done by the German PTT in cooperation with the Bosch Company and lead to the choice of the OFDM parameters for the four DAB transmission modes.

Differential QPSK modulation together with convolutional coding was chosen for DAB. At that time, no appropriate channel estimation technique for OFDM was available, and DQPSK was the favorite choice because it was widely believed to be the most robust modulation scheme in a mobile radio channel. About one year after the decisions were made about the system parameters, it was shown by Hoeher (1991) that a coherent modulation scheme with a suitable channel estimation using Wiener filtering outperformes DQPSK. For an introduction to Wiener filtering, we refer to (Haykin 1996). These ideas became part of the DVB-T system concept (EN300744 2001b). One should keep in mind that the preparatory work for that system had already been done inside the DAB project. For example, the proper choice of the OFDM symbol length could be taken over from DAB. The 8k Mode of DVB-T corresponds to DAB Transmission Mode I, and the 2k Mode of DVB-T corresponds to DAB Transmission Mode II. The outer channel coding is also very similar.

DVB-T was originally not intended for mobile reception. There is no unequal error protection adjusted to the audio data stream, and there is no time interleaving. The channel estimation is very robust, and DVB-T can cope with higher Doppler bandwidths than DAB. Higher car velocities become a problem because DVB-T will typically be located at higher frequencies than DAB.

In 1997, two working groups were established separately by the IEEE and ETSI to develop standards for Wireless LANs exceeding the data rate of former versions significantly. To achieve this goal, OFDM has been introduced as the basis for the transmission techniques. Intensive discussion between these two groups led to widely harmonized parameters for OFDM, modulation and channel coding. The corresponding IEEE 802.11a standard (IEEE 802.11a 1999) and HIPERLAN/2 standard (EN101475 2001) were released in 1999 and 2000, respectively.

The channel coding schemes of all the OFDM systems discussed in Section 4.6 are very closely related. They are essentially based on the same convolutional code of constraint length 7. Section 4.5 is devoted to the channel coding and modulation for OFDM systems. It partly follows the discussion presented in (Schulze 2003b,c). The concept of the diversity degree of a multicarrier system presented in Section 4.4 follows the discussion in (Schulze 2001).

4.8 Problems

1. Let $g(t)$ be a pulse that is time limited to the symbol duration $T_S = (1+\alpha)T$ with the property

$$T|g(t)|^2 = \begin{cases} 1 & : \quad 2|t|/T \leq 1-\alpha \\ \frac{1}{2}\left(1 - \sin\left(\frac{\pi}{\alpha}\left(|t|/T - \frac{1}{2}\right)\right)\right) & : \quad 1-\alpha \leq 2|t|T \leq 1+\alpha \\ 0 & : \quad 2|t|/T \geq 1+\alpha \end{cases}.$$

   In this equation, T is a time constant and the rolloff factor α has the property $0 \leq \alpha \leq 1$. We define

$$g_k(t) = \exp\left(j2\pi\frac{k}{T}t\right)g(t).$$

   Show that

$$\langle g_k, g_l \rangle = \delta_{kl}$$

   holds.

2. Consider OFDM without guard interval and a smooth nonlinear amplifier as discussed in Subsection 4.2.2. Assume that the self interference caused by the nonlinearity may be modeled by a Gaussian random variable. We require that the maximal allowed performance degradation measured in E_b/N_0 is 1 dB. How much SIR is necessary (relative to E_b/N_0) for BPSK, QPSK, 16-QAM and 64-QAM? How much is necessary if a guard interval of length $\Delta = T/4$ is introduced? How much is necessary if a convolutional code of rate $R_c = 1/2$ is introduced?

3. Consider a complex signal

$$s(t) = a(t)e^{j\varphi(t)}$$

with amplitude $a(t)$ and phase $\varphi(t)$. Show that the time derivative of the phase is given by

$$\dot{\varphi}(t) = \Im\left\{\frac{\dot{s}(t)}{s(t)}\right\}.$$

4. Let $\mathbf{n} = (n_1, \ldots, n_L)^T$ be L-dimensional complex AWGN with variance $\sigma^2 = N_0$ in each dimension and $\mathbf{u} = (u_1, \ldots, u_L)^T$ be a vector of length $|\mathbf{u}| = 1$ in the L-dimensional complex space. Show that $n = \mathbf{u}^\dagger \mathbf{n}$ is a complex Gaussian random variable with mean zero and variance $\sigma^2 = N_0$.

5. Let $\mathbf{n} = (n_1, \ldots, n_L)^T$ be L-dimensional complex AWGN with variance $\sigma^2 = N_0$ in each dimension and $\mathbf{U}$ be a unitary $L \times L$ matrix. Show that $n = \mathbf{U}^\dagger \mathbf{n}$ is also L-dimensional complex AWGN with variance $\sigma^2 = N_0$ in each dimension.

5

CDMA

5.1 General Principles of CDMA

Code division multiple access (CDMA) is a multiple access technique where different users share the same physical medium, that is, the same frequency band, at the same time. The main ingredient of CDMA is the *spread spectrum* technique, which uses high rate signature pulses to enhance the signal bandwidth far beyond what is necessary for a given data rate. The concept of spreading is explained in more detail in Subsection 5.1.1.

In a CDMA system, the different users can be identified and, hopefully, separated at the receiver by means of their characteristic individual *signature pulses* (sometimes called the *signature waveforms*), that is, by their individual *codes*. Subsection 5.1.3 briefly discusses the main types of codes and some of their essential properties.

Nowadays, the most prominent applications of CDMA are mobile communication systems like cdmaOne (IS-95), UMTS or cdma2000, which are explained in detail in Section 5.5. To apply CDMA in a mobile radio environment, specific additional methods are required to be implemented in all these systems. Methods such as power control and soft handover have to be applied to control the interference by other users and to be able to separate the users by their respective codes. Basics of mobile radio networks are presented in Subsection 5.1.2, and methods of controlling the interference are discussed in Subsection 5.1.4.

5.1.1 The concept of spreading

Spread spectrum means enhancing the signal bandwidth far beyond what is necessary for a given data rate and thereby reducing the power spectral density (PSD) of the useful signal so that it may even sink below the noise level. One can imagine that this is a desirable property for military communications because it helps to hide the signal and it makes the signal more robust against intended interference (*jamming*). Spreading is achieved – loosely speaking – by a multiplication of the data symbols by a *spreading sequence* of pseudorandom signs. These sequences are called *pseudonoise* (PN) sequences or code signals. We

Theory and Applications of OFDM and CDMA Henrik Schulze and Christian Lüders

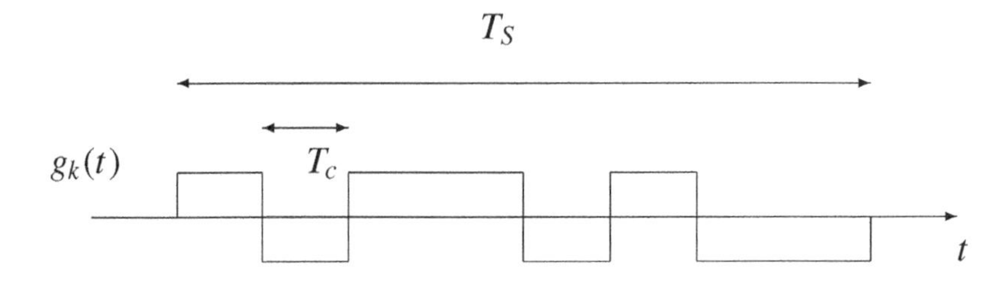

Figure 5.1 Signature pulse with $N = 8$ rectangular chips.

illustrate the method by an example; more details on codes for spreading can be found in Subsection 5.1.3.

Consider a rectangular transmit pulse

$$g(t) = \frac{1}{\sqrt{T_S}} \Pi\left(\frac{t}{T_S} - \frac{1}{2}\right)$$

of length T_S. We divide the pulse into N subrectangles, referred to as *chips*, of length $T_c = T_S/N$ and change the sign of the subrectangles according to the sign of the pseudorandom spreading sequence. Figure 5.1 shows the resulting transmit pulse $g_k(t)$ of user number k for $N = 8$. Here, the spreading sequence for user k is given by $(+, -, +, +, -, +, -, -)$. When it is convenient (e.g. for the performance analysis) the sign factors shall be appropriately normalized. We note that in practice smooth pulse shapes (e.g. raised cosine pulses) will be used rather than rectangular ones.

The increase of the signaling clock by a factor N from T_S^{-1} to T_c^{-1} leads to an increase of bandwidth by a factor of T_S/T_c (see Figure 5.2). For this reason, $N = T_S/T_c$ is called the *spreading factor* or, more precisely, the *spreading factor of the signature pulse*. This spreading is due to multiplication by the code sequence. While within the specification documents for CDMA mobile communication systems the spreading factor is often denoted by *SF*, formulas are kept simpler by using the symbol N. Hence, we use both notations.

Later we may have different spreading mechanisms that work together, especially in the context of channel coding. Therefore, we reserve the notion of the *effective* spreading factor. As discussed in detail in Chapter 3, it is often not uniquely defined where channel coding ends and where modulation starts and thus it may be ambiguous to speak of a bit rate after channel coding. We regard it as convenient to define the *effective spreading factor* by

$$SF_{\text{eff}} = \frac{R_{\text{chip}}}{R_b}, \tag{5.1}$$

where R_b is the useful bit rate and $R_{\text{chip}} = 1/T_c$ the chip rate. Obviously, this spreading factor is approximately the inverse of the spectral efficiency for a single user.

The objective of spreading is – loosely speaking – a waste of bandwidth for the single user to achieve more robustness against multiple access interference (MAI). It would thus be a contradiction to this objective to use bandwidth-efficient higher-level modulation schemes. Any modulation scheme that is more efficient than BPSK would reduce the spreading factor. Therefore, BPSK and QPSK are used as the basic modulation schemes in most

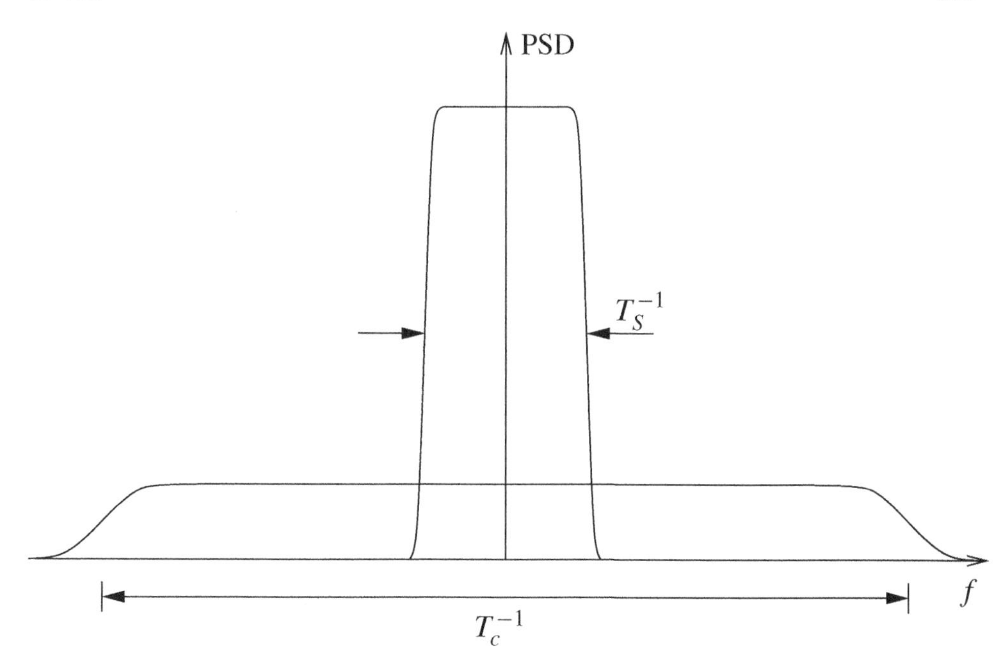

Figure 5.2 Power spectral density (PSD) for DS-CDMA.

practical communication systems. Nevertheless, higher-order modulation techniques like 8-PSK and 16-QAM also are applied as additional transmission options to offer a high-speed packet transfer at good propagation conditions. Furthermore, we point out the special role of channel coding. Channel coding usually means that a higher power efficiency has to be paid by a lower spectral efficiency. Thus, channel coding can be interpreted as an additional spreading mechanism. In the extreme case, all the spreading can be done by channel coding, and the PN sequences serve only for user separation (see e.g. (Frenger *et al.* 2000)). Keeping this in mind, we can interpret the conventional spreading by a PN-sequence as a repetition code combined with the repeated transmit symbols multiplied by a pseudorandom sign. The symbol will be repeated N times at a clock rate increased by the factor $N = T_S/T_c$ and scrambled by a random sign. Equivalently, this is time delay diversity. Because of the time dispersion of the channel, we may get a *multipath diversity* gain. The appropriate diversity combiner is the RAKE receiver, which is sketched here and will be discussed in more detail in Subsection 5.4.1.

The name RAKE receiver originates from the fact that there are some similarities to a garden rake. As illustrated in Figure 5.3, the receiver consists of a certain number of correlators (called *RAKE fingers*) correlating the received signal to the used code signal. One of the correlators (the so-called *search finger*) has the task to determine the propagation delay values τ_i $(i = 1, 2, \ldots)$ of the most relevant propagation paths. These values are used within the other correlators (fingers) to adjust the exact timing for the respective multipath components. By this method, the multipath components can be detected separately (if the codes have a good autocorrelation property); subsequently, they can be combined by a maximum ratio combiner. It should be noted that multipath components can only be resolved

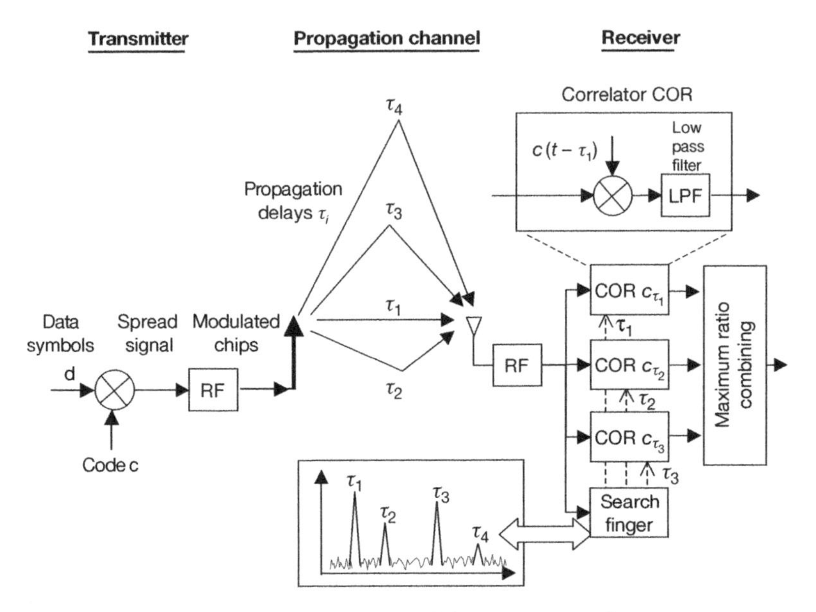

Figure 5.3 Illustration of the RAKE receiver.

if their delay difference is higher than about a quarter of the chip duration T_c. Furthermore, the number of RAKE fingers is usually restricted to 4–6 (including the search finger). The cdma2000 specification requires, for example, that there are at least four processing elements (including the search finger).

It must be emphasized that spreading by itself does not provide any performance gain in the AWGN channel[1]. As shown in Figure 5.2, for a single user, spreading means nothing but choosing a spectrally rather inefficient waveform that smears the spectral power over an *SF* times higher bandwidth. Thus, for the same signal power, the *SNR* decreases by a factor of *SF*. The necessary power per bit rate, which equals the energy per bit, E_b, does not depend on the pulse shape. We therefore avoid the popular but misleading word *processing gain* for the factor *SF*. Originally, for a single user, it is nothing but a waste of bandwidth.

Example 8 (Processing Gain) *We compare BPSK transmission employing a given pulse shape with a spread spectrum transmission that uses this pulse as a chip pulse and the spectrum will be spread by this factor of SF. Consider, for example, BPSK transmission with a Nyquist base and roll-off factor 1, which occupies a bandwidth of 200 kHz to transmit a bit rate of* $R_b = 100$ *kbit/s. For BPSK in an AWGN channel, we need* $E_b/N_0 = 9.6$ *dB to achieve a bit error rate of* 10^{-5}*. For BPSK and the Nyquist base, we have* $SNR = E_b/N_0$ *(see Subsection 1.5.1). We now compare this system with a spread spectrum system of the same data rate, a spreading factor of* $SF = 100$*, and the same roll-off factor for the chip pulse. As we have seen in Chapter 1, the performance of a linear modulation scheme does not depend on the pulse shape. Thus, we still need* $E_b/N_0 = 9.6$ *dB to achieve a bit error rate of* 10^{-5}*, that is, the power that is needed to transmit a given bit rate of* $R_b = 100$ *kbit/s is the same. However, the bandwidth has now been increased by a factor of SF and the signal occupies*

[1]In a fading channel, a diversity gain can be achieved.

20 MHz. Because the same signal power is spread over this higher bandwidth, the signal is now completely below the noise and we have SNR $= -10.4$ *dB. This fictitious mystery that a signal below the noise level can be completely recovered has its simple explanation in the fact that we have just wasted bandwidth by using a spectrally inefficient pulse shape, which does not influence the power efficiency. Thus, the processing gain is just a virtual gain.*

The reason for using spreading is not this virtual processing gain. A real gain of spreading concerning the range of data transmission can be achieved in a frequency-selective fading environment. The increased bandwidth of a spread signal provides us with an increased frequency diversity as compared to a narrowband FDMA system (see the discussion in Subsection 4.4.3). Such a frequency diversity can only be exploited if the signaling bandwidth significantly exceeds the correlation frequency (i.e. the coherency bandwidth) of the channel. In that case, we speak of a *wideband* CDMA system (Milstein 2000)[2]. However, as discussed in Chapter 4, such diversity can also be achieved by increasing the carrier bandwidth by multiplexing different users to a frequency carrier using a time division multiplex scheme. Comparing the spread spectrum and the TDMA technique at the same signal bandwidth, at the same data rate and mean transmission power (energy per transmitted bit), roughly the same performance will result since the receive E_b/N_0 is the same. Nevertheless, having in mind the discussion on electromagnetic compatibility of mobile phones, spreading may have an advantage since it uses a continuous transmission while transmission in time multiplex systems is pulsed. For this reason, sometimes the peak transmission power of systems is limited by regulatory bodies. Obviously, at equal peak transmit power the performance of spread spectrum systems is higher than that of TDMA systems.

5.1.2 Cellular mobile radio networks

Network architecture

The frequency spectrum assigned to a mobile radio network usually is separated into several frequency carriers which themselves may further be divided by a time or code multiplex scheme into a set of radio channels. Since in mobile radio networks there are many millions of subscribers but only some hundreds of radio channels, the coverage area is divided into cells and the same frequency carriers are reused in many cells. This is the principle of cellular radio networks.

As shown in Figure 5.4, radio coverage within a cell is accomplished by a *base station* (BS). Each BS may serve many *mobile stations* (MS). The transmission direction from the BS to the MS is called the *downlink* (DL) or *forward link*, the direction from the MS to the BS is called the *uplink* (UL) or *reverse link*. A group of base stations is connected via leased lines or microwave equipment to a network element, which is called *base station controller* (BSC, e.g. in GSM) or radio network controller (RNC, e.g. in UMTS). The connection between two subscribers is established by the mobile switching center (MSC).

[2]We note that the definition of *wideband* is the same as that introduced in Chapter 4 for OFDM. Furthermore, it should not be confused with one transmission mode of UMTS, which is also often called *Wideband CDMA*. Its transmission bandwidth of about 5 MHz may be viewed as wide in some urban environments, but not in any indoor environment.

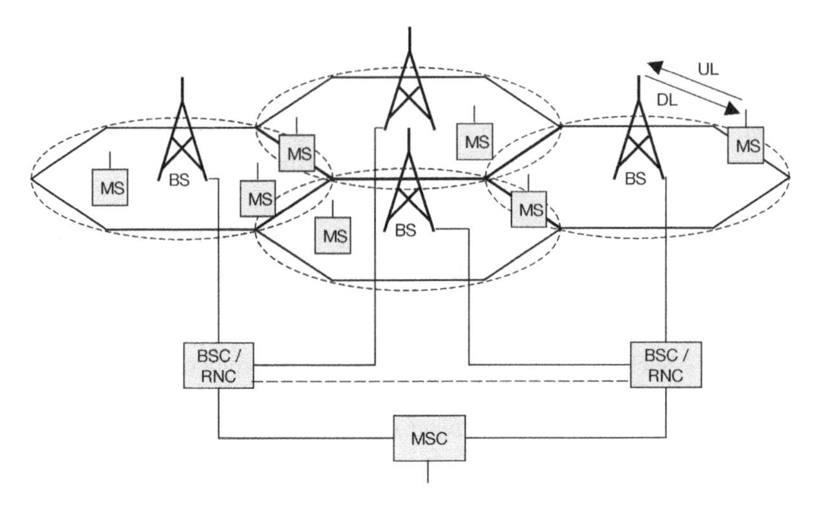

Figure 5.4 Architecture of a cellular mobile radio network.

Handover

When an MS moves from one cell to another, a *handover* occurs. One distinguishes between *hard* and *soft handover*. For a hard handover, as it is performed, for example, in GSM networks, the MS releases the old channel before connecting to the new BS via the new channel; hence, there is a short interruption of the connection. For a soft handover, which usually is performed in CDMA systems, an MS at the cell border may have several connections to the corresponding base stations at the same time so that there is a smooth transition between the cells without any interruption. To manage the soft handover between cells belonging to different RNCs, additional interconnections between the RNCs are required (in contrast to GSM).

In many cases, the handover decision is based upon the received signal level. A handover where at every moment the MS is served by the BS from which the maximum signal level is received is called an *ideal power budget handover*. Owing to fading effects, such an ideal power budget criterion would cause very frequent forward and backward handovers between different cells. For an architecture managing soft handover, there is no problem for switching the connection between the different base stations immediately (on a millisecond timescale); the signals to and from different base stations may even be combined. Because of the short interruption phases and signaling effort, frequent hard handovers should be avoided. This is usually achieved by introducing an averaging of the signal level and a hysteresis margin, that is, a hard handover is only performed when the averaged signal level of a neighboring cell exceeds one of the current serving cells by this hysteresis margin of a few decibels.

Antennas and radio propagation

Concerning the cell layout one may distinguish between *omni cells* and *sector cells*. An omni cell is served by one BS in the middle of the cell using an omni directional antenna,

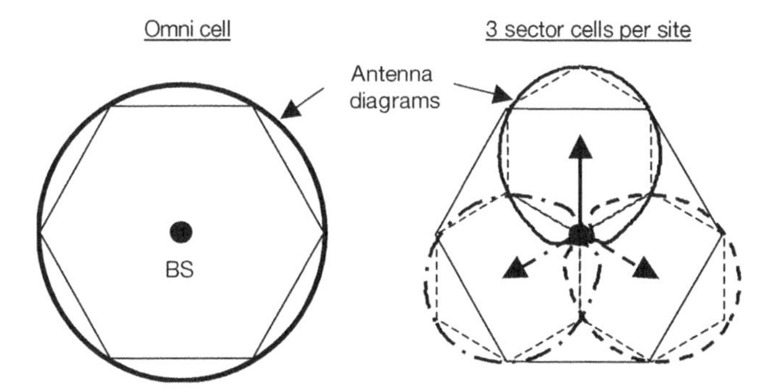

Figure 5.5 Sectorization of an omni cell into three sector cells using directional antennas with a half power beam width of 65°.

that is, an antenna with an isotropic characteristic in horizontal direction. Sectorization as illustrated in Figure 5.5 is usually accomplished by dividing an omni cell into three cells by directional antennas with a horizontal half power beam width of about 60°–120°.

The model for the radio propagation loss – which is usually written in a decibel notation – $L = L_D + L_L + L_S$ consists of three parts:

- the short-term fading L_S caused by multipath propagation;
- the long-term fading L_L caused by shadowing due to buildings or other obstacles;
- the deterministic path loss L_D mainly depending on the distance r between MS and BS.

Short-term fading was explained in detail in Chapter 2. Hence, this subsection focuses on the discussion of the other parts of the propagation model.

Most of the well-accepted models for the deterministic propagation loss L_D (e.g. free-space propagation, Okumura–Hata model, Walfish–Ikegami model, Xia–Bertoni model) are described in decibel notation by an equation of the form (ETSI TR 101 112 1998; Parsons 2000):

$$L_D = A + B \cdot \log_{10} r\ [\text{km}]. \tag{5.2}$$

The propagation parameters A and B mainly depend on the frequency f, the MS and BS antenna installation heights and on the heights, distances and orientations of the surrounding obstacles (buildings). For free-space propagation and, for example, $f = 2$ GHz, one has $A = 98.4$ dB and $B = 20$; these parameter values can be derived from the free-space formula of Problem 1.9 by converting it into a decibel notation and by using the relation $\lambda = c/f$ between wavelength and frequency as well as the correct units. For line-of-sight conditions, propagation in pedestrian areas and street canyons may be modeled by free-space propagation up to a certain distance (of some hundreds of meters) which is called the *break point* distance. Afterwards, the propagation loss increases more strongly according to a parameter of $B = 40$. In non-line-of-sight conditions, an additional diffraction loss at street corners is taken into account. The considerations below focus on scenarios where the

BS antennas are installed near or above the mean rooftop level of the buildings. For such scenarios, typical values for A and B in urban environments are:

- $A = 110 \ldots 130$ dB, $B = 30 \ldots 40$ for $f = 1$ GHz,
- $A = 120 \ldots 140$ dB, $B = 30 \ldots 40$ for $f = 2$ GHz.

Obviously, these simple but nevertheless very useful deterministic models can only describe the mean path loss as a function of the distance without reflecting all details of the buildings. Because of shadowing by buildings the calculated loss will deviate from the mean value measured in a specific area of some square meters. This deviation is usually modeled by a Gaussian distributed random variable with a standard deviation σ that typically lies in the region of $\sigma = 4 \ldots 10$ dB. The correlation of the long-term fading values at positions with a distance d decreases approximately exponentially with d and a decay constant of some tens of meters (the typical scale of obstacles).

Figure 5.6 shows a typical diagram of the received level as a function of the distance r taking into account all three parts of the propagation model. It can be seen that the received level varies in a region of about 70 dB.

Quality of service, coverage and handover gain

The objective of network planning is to guarantee a certain quality of service. The quality of service is usually quantified by setting an upper limit to the bit error rate BER, for example,

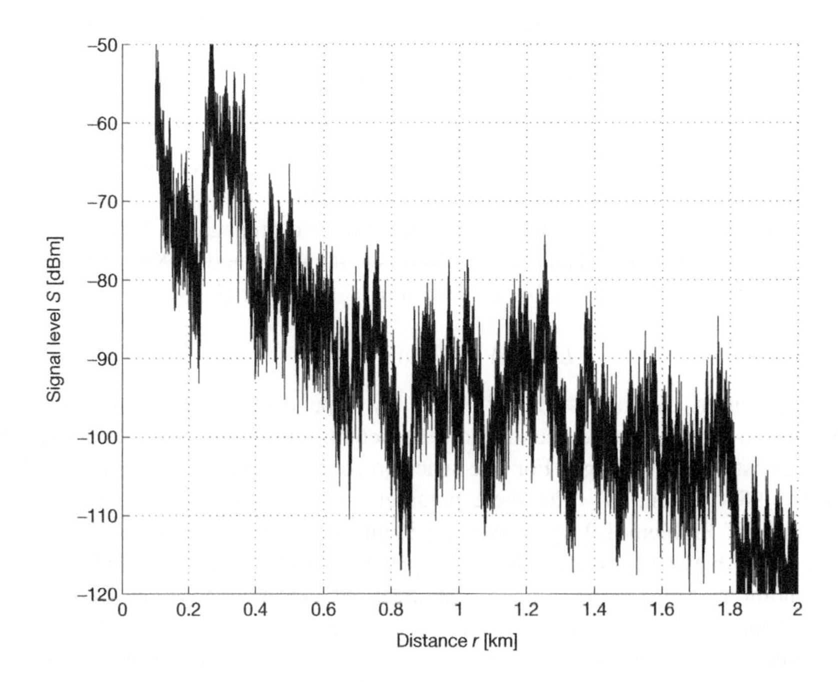

Figure 5.6 Example for the received signal level as a function of the distance between MS and BS.

$BER < 10^{-3}$ for speech services. Owing to uncertainties of radio propagation this *BER* threshold may be exceeded in some parts of the network. Therefore, a coverage probability c_p of typically 90–98% is introduced, that is, the *BER* condition has to be fulfilled in, for example, $c_p = 95\%$ of the network area. Since the *BER* depends on the E_b/N_0 or the *SNR*, the network planner has to guarantee that the *SNR* is above the corresponding threshold for $c_p = 95\%$ of the network area. The *SNR* threshold is usually derived taking into account a kind of margin against short-term fading. Considering, for example, a BPSK in a Rayleigh fading channel with twofold diversity ($L = 2$), one obtains from Figure 2.11 and Equation 2.33 a threshold of about $SNR = 11$ dB.

Hence, while planning the network, one usually considers only the deterministic part of the propagation model and the random long-term fading explicitly whereas the short-term fading is included in the *SNR* threshold. The received level value resulting only from the deterministic and the long-term fading part is called the *local mean* value.

The procedure discussed above is illustrated by the following example that by the way also shows the effect of the long-term fading and the effect of the handover mechanism on the network performance. Assuming a noise level of $N = -111$ dBm and an *SNR* threshold of 11 dB, a signal level of $S = -100$ dBm is required to guarantee a sufficient quality of service. Figure 5.7 shows the probability functions of the signal level for three different scenarios. The diagram was obtained by Monte–Carlo simulations using $A = 140$, $B = 40$ and $\sigma = 8$ dB as propagation parameters. The dotted line shows the probability function of the signal level in a cell without any fading. The probability $p(S \leq -100 \text{ dBm})$ – the so-called *outage probability* o_p – is 0, that is, the coverage probability is 100%. Including long-term fading in an isolated cell (i.e. neglecting neighbor cells), the received level is better in some regions and is worse in other region than without long-term fading. This is reflected by the dashed curve crossing the dotted one. Because of long-term fading the

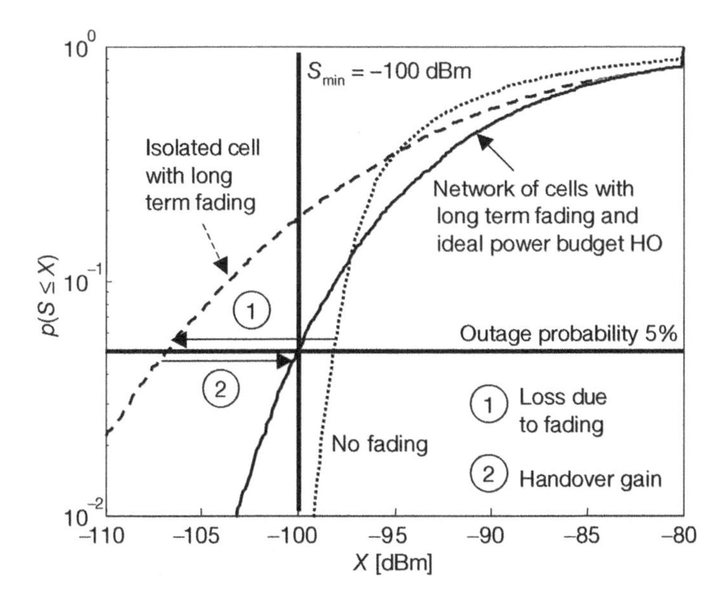

Figure 5.7 Probability functions of the received signal level S for different scenarios.

outage probability increases to an unacceptable value of about 0.2, that is, one has $c_p = 0.8$. To achieve the required coverage of $c_p = 0.95$, one has to decrease the cell radius or the value for the needed signal level by about 8.5 dB. However, coverage also gets better due to *macrodiversity* when considering a network of several cells: if the signal level with respect to one BS breaks down due to long-term fading, a handover to a BS with a better signal level may be possible. While antenna diversity only gives microdiversity against short-term fading, handover between base stations at different sites offers macrodiversity against long-term fading (and even against short-term fading if the handover is executed fast enough as in the case of a soft handover). For producing the solid line probability function of Figure 5.7, an ideal power budget handover has been used selecting the BS with the best local mean signal level. Looking at $c_p = 0.95$, that is, $o_p = 0.05$, a gain of about 7 dB is obtained by the handover method. This gain is called *handover gain* or macrodiversity gain. When using a kind of maximum ratio combining of the macrodiversity paths, the gain is higher than for the pure selection combining used in Figure 5.7. A lower handover gain is achieved by a hard handover where a hysteresis margin is necessary since in this case the MS is not connected to the best BS at each time.

Homogeneous hexagonal networks and frequency reuse

In a widely used model for simulating and comparing mobile networks, base stations are placed on a regular hexagonal grid. Assuming omnidirectional BS antennas and a radio propagation loss depending only on the distance between BS and MS, the cell area served by a BS has the form of a regular hexagon.

For these hexagonal networks, various very symmetrical frequency reuse patterns exist; some examples are illustrated in Figure 5.8: The total number of carriers N is divided into a number of disjoint sets, and this number is called the *cluster size* K. One of these frequency sets is allocated to each cell and each set is reused at each Kth cell. Cells using the same frequency set are called *cochannel* cells. In Figure 5.8, cells using the frequency set 1 are highlighted by a gray shading for $K = 3$ and $K = 7$. In a cluster of size $K = 1$, neighboring cells use the same frequencies. Owing to the hexagonal symmetry there are six cochannel cells at the reuse distance D causing the highest interference to each cell. The interference caused by cochannel cells is called *intercell interference* I_r in contrast to *intracell interference* I_a caused by connections within the same cell. Intercell interference occurs in every mobile radio network with a frequency reuse, whereas intracell interference is avoided in pure FDMA and TDMA systems by using a sufficient carrier separation and guard periods. Even for systems using frequency hopping, intracell interference is avoided by selecting noncolliding hopping sequences within one cell. However, in CDMA systems where many connections share the same frequency carrier at the same time, intracell interference has to be considered since the different connections cannot be separated perfectly by their codes. In this subsection, the discussion focuses on intercell interference whereas the effects of intracell interference are explained in Subsection 5.1.4.

Obviously, with decreasing cluster size K the number of frequencies per cell and therefore the cell capacity increases. On the other hand, the mutual interference between cochannel cells also increases. Hence, the art of network planning is to find the smallest cluster size K so that the connection quality is above a required level. The main indicator for the connection quality in the case of interference is the signal-to-interference ratio *SIR*, that is, the ratio between the (useful) signal power S and the interference power I (which

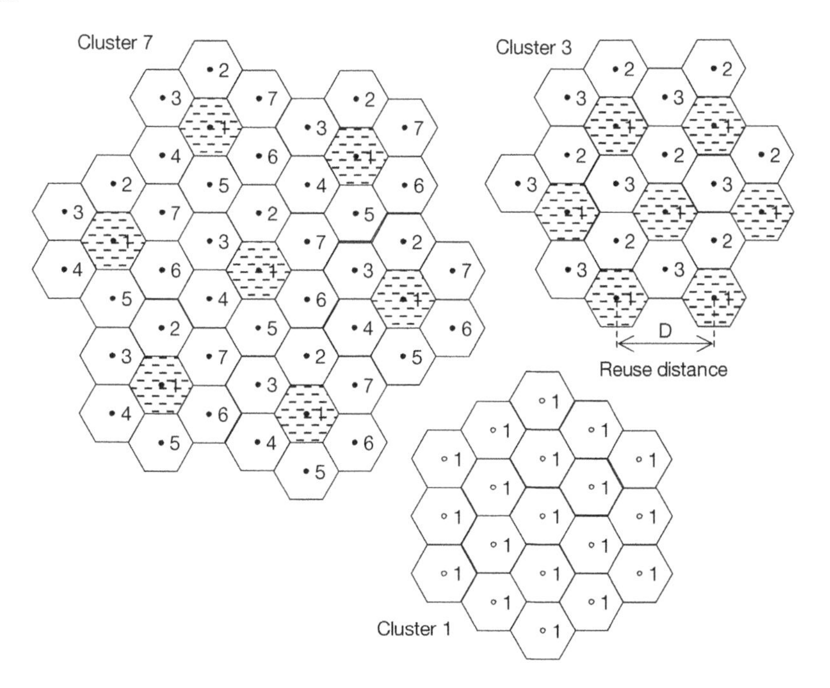

Figure 5.8 Frequency reuse in homogeneous hexagonal networks.

is only intercell interference in this subsection). Figure 5.9 shows examples of the *SIR* probability functions for $K = 1$, 3 and 9 and networks of omnicells as solid lines. Similar results for sectorized networks are also included in the figure and will be commented later.

The probability functions were obtained by Monte-Carlo simulations using the following assumptions:

- Mobile stations are spread randomly over the network.
- For each pair of MS and BS, the received level is calculated using $B = 40$ and $\sigma = 8$ dB as propagation parameter (the parameter A has no effect on the *SIR* since it cancels while taking the difference of the signal and interference level).
- Having calculated the received level with respect to each BS, an MS is assigned to the BS with the highest level that is taken as the carrier level for that connection. Subsequently, the interference level is calculated by summing up the interfering powers of all connections using the same frequency.

Figure 5.9 shows that for a transmission scheme, which needs an *SIR* of 10 dB, a sufficient coverage probability of $c_p = 0.97$ ($o_p = 0.03$) can be achieved using $K = 9$. A cluster 1 requires very robust transmission techniques – as for example, spreading – working at negative *SIR* values.

The network performance can be improved significantly by sectorization. Starting, for example – as shown in Figure 5.10 – with an omnicellular cluster of $K = 1$, one can separate the frequency set 1 into 3 disjoint subsets (denoted as 1a, 1b, 1c) and allocate them to

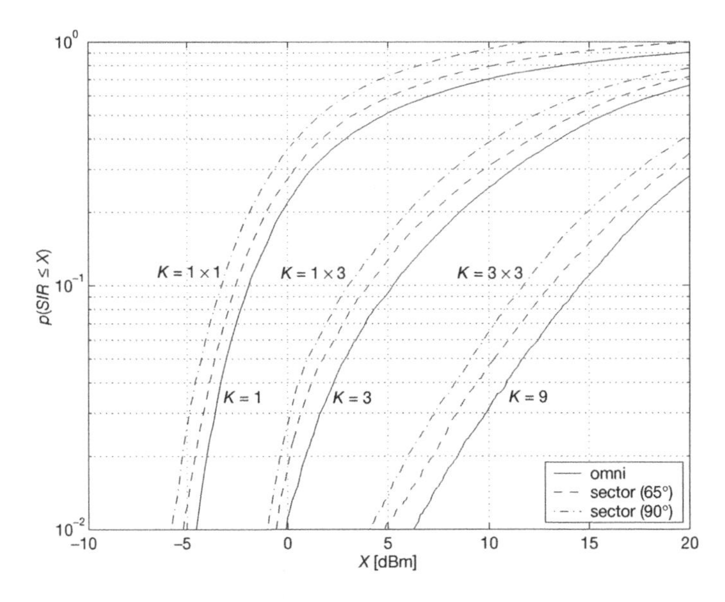

Figure 5.9 Probability functions of the *SIR* for different cluster sizes.

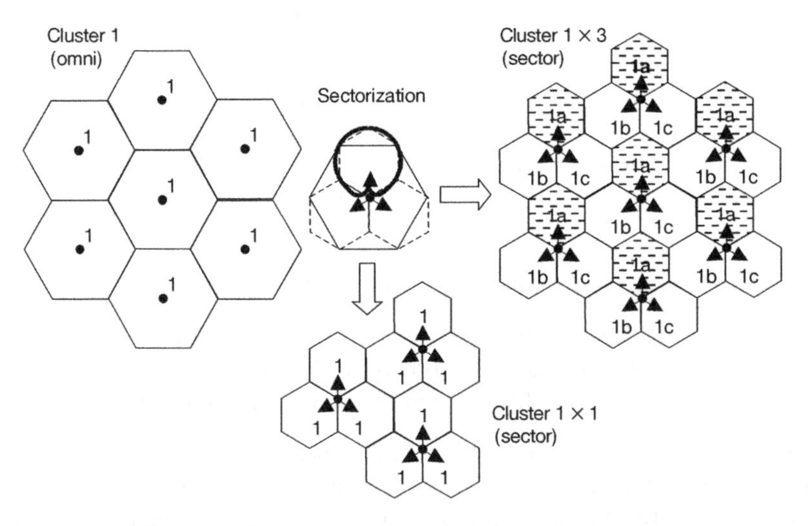

Figure 5.10 Cluster sizes for sectorized networks.

the three different sectors. This cluster is denoted as $K = 1 \times 3$. The number of channels per base station site is the same as for omnicells of $K = 1$, but from the point of view of interference it is very similar to the omnicellular cluster $K = 3$: the *SIR* is only 1 dB or 2 dB worse than that of the omnicellular structure depending on the horizontal half power beam width of the antennas (65° for the dashed lines, 90° for the dotted lines). The other possibility for performing sectorization that is often used in CDMA networks is to allocate the complete frequency set to each sector; this is denoted as cluster $K = 1 \times 1$. The number of channels is three times as high as for a cluster $K = 1$, whereas the *SIR* distributions are very similar.

Hence, one can conclude that by applying sectorization a gain of three times the number of channels per base station sites can be achieved at the expense of a moderate loss in *SIR* of about 1 or 2 dB.

5.1.3 Spreading codes and their properties

In this subsection, various types of codes are discussed presenting and summarizing the basic facts about spreading codes and their properties as far as they are needed for the following sections. More details as well as proofs for some propositions can be found, for example, in (Gibson 1999; Hanzo *et al.* 2003; Proakis 2001; Viterbi 1995).

Since spreading codes have different advantages and disadvantages, the appropriate type of code has to be selected for the relevant situation or application. The main criterion for selecting a set of codes is based upon the autocorrelation and cross-correlation functions of these codes. The autocorrelation function, defined as the scalar product of the code signals with code signal shifted by a delay τ, should be zero – at least approximately – for $\tau \neq 0$. Such a good autocorrelation property is especially required in radar or positioning systems where the autocorrelation process is used for an exact determination of the propagation delay; but it is also very useful in mobile communication systems for separating the different propagation paths and hence avoiding intersymbol interference. For CDMA systems, different code signals c and c' are used to distinguish different connections. The mutual interference between the connections is proportional to the scalar product of c and c'. Hence, for these applications of spreading one is tempted to require orthogonality of codes. However, orthogonality alone does not suffice for scenarios where the code signals are not transmitted synchronously or where a high delay spread due to multipath propagation occurs. In that case also the scalar product between c and c' shifted by the delay τ, that is, the cross-correlation function of c and c' has to be minimized. Each of the known types of codes fulfills one requirement to a higher and the other to a lower degree. Therefore, the codes giving the best compromise for the respective applications have to be selected.

Another method is to apply several spreading steps and use different types of codes for the separate steps.

Orthogonal variable spreading fact or codes

The orthogonal variable spreading factor codes (OVSF codes) look like the Walsh functions discussed in Section 1.1.4.

However, they are arranged and numbered in a different way, that is, according to a tree structure, which is illustrated in Figure 5.11.

For each spreading factor $SF = 1, 2, 4, \ldots$, which is a power of 2, there are $N = SF$ orthogonal codes obtained by the recursion relations:

$$c_{2SF,2m} = [c_{SF,m}, c_{SF,m}], \quad m = 0, 1, \ldots, SF - 1. \tag{5.3}$$

$$c_{2SF,2m+1} = [c_{SF,m}, -c_{SF,m}], \quad m = 0, 1, \ldots, SF - 1. \tag{5.4}$$

The symbol $[c, c]$ denotes the composition of codes. The orthogonality of these codes for a fixed spreading factor *SF* follows immediately from the defining equations. The notation

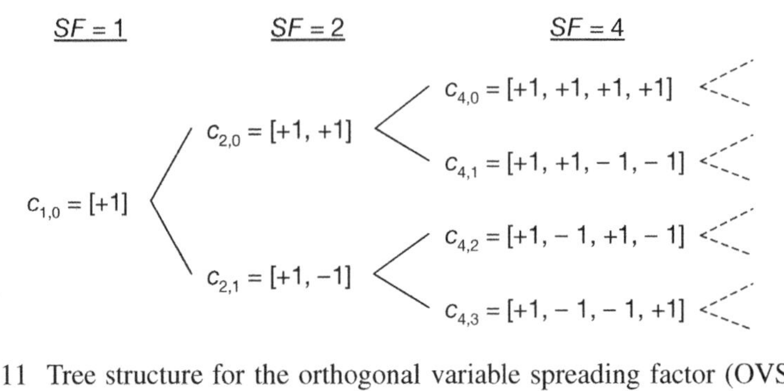

Figure 5.11 Tree structure for the orthogonal variable spreading factor (OVSF) codes.

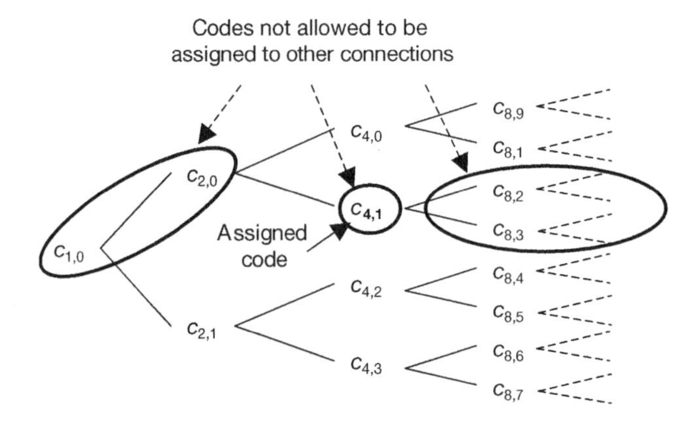

Figure 5.12 Illustration of code allocation rules for different data rates.

variable spreading factor results from the fact that they are applied in CDMA communication systems offering different data rates. For these systems, the chip rate and chip duration is kept fixed. Hence, different symbol rates can be realized by varying the spreading factor: $T_S = SF \cdot T_c$. Establishing connections with different data rates, some rules for selecting the corresponding codes have to be observed for maintaining the orthogonality: if a certain code $c_{SF,m}$ is already used for one connection, neither this code nor a code that is a descendant or an ancestor of this code is allowed to be used for another connection since these codes are not orthogonal to the already allocated one.

This rule is illustrated in Figure 5.12: if, for example, code $c_{4,1}$ is in use, another connection with a different data rate is not allowed to use the encircled codes, but all other codes. If, for example, the second connection has twice the data rate of the first one, it has to select the code $c_{2,1}$. Within the period of one data bit of connection 1, connection 2 transmits two data bits. The chip sequence corresponding to these data bits is either proportional to $c_{4,2}$ or $c_{4,3}$ and hence orthogonal to the one for connection 1. To summarize the discussion above: the OVSF codes can be applied to realize connections with different data rates by varying the spreading factor. Orthogonality is preserved exactly even between connections of different rates if some rules for code selection are observed.

This property makes OVSF codes very attractive for modern communication systems as, for example, UMTS. However, some disadvantages of these codes should also be mentioned which limit their application. First of all, OVSF codes have a poor autocorrelation property, which is obvious by looking, for example, at the codes $c_{SF,0}$. Furthermore, the orthogonality gets lost when there is no perfect synchronization, that is, there are high values for the cross correlation.

Pseudonoise sequences

Codes with a good autocorrelation property can be generated by linear feedback shift registers.

An example of a register of length $m = 5$ is shown in Figure 5.13. In each step, the content of the register is shifted one place to the right and it is also fed back to the left-most place by weighting the content by the coefficients a_n and summing up all results. The weights may take the value "0" and "1" and addition has to be performed as modulo-2 sum. Considering a register of length m, it possesses 2^m different states. The state where all places are set to "0" is not very exciting since it reproduces itself in each step. Starting with another rather than this trivial state, the register produces a sequence of "0" and "1" as an output whose period cannot exceed $M = 2^m - 1$. A generated sequence, which has exactly this maximum possible length M, is called *maximal length sequence* or – shortly *m-sequence*. The m-sequence of the example in Figure 5.13 has a length of $M = 31$. It has been generated using the values of the weights given in the brackets. Comments on the run length values are given below. The m-sequences are often referred as pseudonoise (PN) sequences since they have similar properties as random numbers generated by tossing a coin M times. Some of these properties are as follows:

- An m-sequence contains $\frac{1}{2}(M+1)$ ones and $\frac{1}{2}(M+1) - 1$ zeros, that is, for large M one has nearly the same number of ones and zeros.
- Defining a run as a maximal subsequence of consecutive identical numbers, one has 1 run of ones of length m, 1 run of zeros of length $m - 1$, and 2^{m-r-2} runs of zeros and 2^{m-r-2}, $r = 1, 2, \ldots, m - 2$ runs of ones, each of length r.

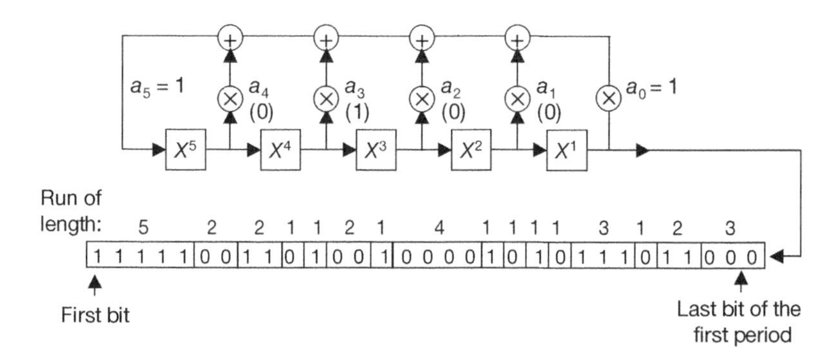

Figure 5.13 Example for an m-sequence generated by a linear feed back shift register of five stages.

The most interesting property from the point of view of spread spectrum applications concerns the autocorrelation function. Defining the code sequence c corresponding to an m-sequence by mapping each "0" to "+1" and each "1" to "−1", the periodic discrete normalized autocorrelation function is given by:

$$\phi(\nu) = \frac{1}{M} \sum_{\mu=0}^{M-1} c_{\mu+\nu} \cdot c_{\mu}$$

In an analogous way one can define the periodic discrete normalized cross-correlation function between two periodic codes c_i and c_j as

$$\phi_{i,j}(\nu) = \frac{1}{M} \sum_{\mu=0}^{M-1} c_{i,\mu+\nu} \cdot c_{j,\mu}$$

It can be shown that the autocorrelation function of m-sequences is only two-valued: $\phi(0) = 1$ and $\phi(\nu) = -\frac{1}{M}$, $\nu \neq 0$. Hence, a code given by an m-sequence has – for large M – nearly an ideal autocorrelation function like a sequence of independent Bernoulli random numbers. It should be noted that defining a continuous autocorrelation function instead of a discrete one by using integrals instead of sums, this continuous function is obtained from the discrete one by linear interpolation. Proofs of these and some more properties of m-sequences can be found in (Gibson 1999; Hanzo *et al.* 2003; Viterbi 1995). These proofs are based on the relation between a linear feedback shift register with coefficients $a_n (n = 0, 1, \ldots, m)$ and its corresponding generating polynomial, which is given by $P(X) = a_m X^m + a_{m-1} X^{m-1} + \cdots a_1 X + a_0$. This relationship also allows the classification of those registers that generate m-sequences: A linear feedback shift register of length m produces an m-sequence if and only if the corresponding generating polynomial of degree m is primitive. A polynomial of degree m is called *primitive* if it is irreducible, that is, if it cannot be factored, and if is a factor of $x^M + 1$, $M = 2^m - 1$.

The examples are as follows:

- The polynomial generating the m-sequence of Figure 5.13 given by $P(X) = X^5 + X^3 + 1$ is a primitive one.

- The simplest examples for primitive polynomials are the following two polynomials of degree 3:

  $$P_1(X) = X^3 + X^2 + 1, \ \ P_2(X) = X^3 + X^1 + 1.$$

  They are irreducible since they cannot be divided either by X or by $1 + X$. Furthermore, they fulfill the relation

  $$(1 + X) \cdot (X^3 + X^2 + 1) \cdot (X^3 + X^1 + 1) = X^7 + 1,$$

  hence they are primitive.

As mentioned above, m-sequences have a nearly optimal autocorrelation property. Unfortunately, there is some weakness concerning the cross correlation, whose peak values decrease quite slowly with increasing sequence length M. The peak value of the undesired elements in the cross- and autocorrelation functions is defined by

$$\phi_{\max} = \max(\phi_{i,j}(\nu)| \, (i = j \text{ and } \nu \neq 0) \, or \, (i \neq j \text{ and all } \nu)).$$

Though for a given M a subset of m-sequences with pairwise low cross correlation may be selected, this set is quite small. Therefore, other methods are used to derive codes with low cross correlation from the m-sequences.

Gold codes

A set of Gold codes of length M can be obtained by combining specific pairs of m-sequences c, c' which are called *preferred* m-sequences. This set of Gold codes Γ is given by c, c' and the modulo-2 sums of c and all M different cyclically shifted versions of c', hence it contains $M + 2$ elements. Another way to number the Gold codes generated by c, and c' is:

$$\Gamma = \{c_0, c_1, c_2, \ldots, c_M, c_{M+1}\}$$
$$c_0 = c, \;\; c_{M+1} = c', \;\; c_\mu = c + c'(\mu), \;\; \mu = 1, 2, \ldots, M,$$

where the sum has to be understood as a modulo-2 sum and $c'(\mu)$ is the m-sequence given by the binary representation of μ as the initial setting for the generating shift register. This process is illustrated in Figure 5.14 for the shift registers corresponding to the primitive polynomials of degree 3 mentioned above. The resulting nine Gold codes have a length of $M = 7$.

Concerning the decrease of auto- and cross-correlation functions Gold proofed the following proposition: the cross-correlation functions of Gold sequences take only the three values $-1/M, -t(M), t(M) - 2$, where $t(M)$ decreases as $2/\sqrt{M}$ for large even M and as $\sqrt{2/M}$ for large odd $M = 2^m - 1$. To summarize, for large M the peak values of the

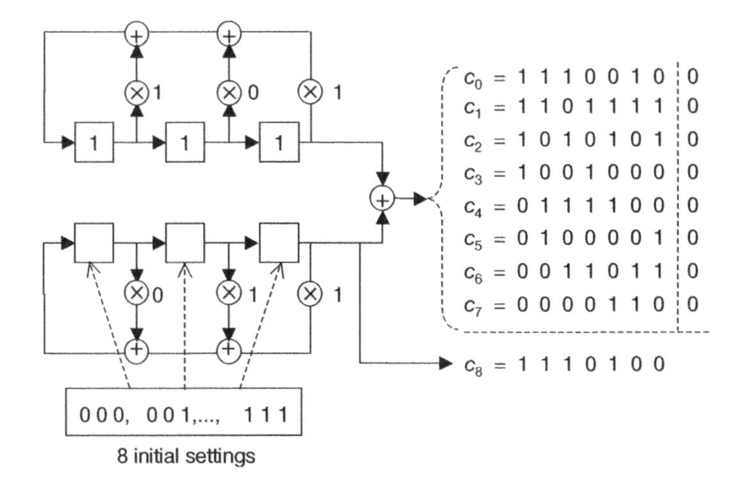

Figure 5.14 Example for a generation of Gold codes.

cross-correlation functions of Gold codes are much smaller than for the m-sequences, but at the expense of higher (but also decreasing) values of the autocorrelation functions. The combined codes in the set of Gold codes are no m-sequences.

Furthermore, it should be mentioned that the first eight Gold codes have even an optimal cross correlation, that is, the cross-correlation functions take only the value $-1/M = -1/7$. Hence, padding a "0" to each of these eight codes, one gets eight codes of length eight, which are exactly orthogonal. These orthogonal codes may be used as roots for code trees produced in the same way as the one for the OVSF codes discussed above. Compared to the Walsh codes, the orthogonal Gold codes have a better autocorrelation property.

Kasami codes

The most important property of Kasami codes is that the peak values of the cross-correlation functions are even smaller than for the Gold codes: $\phi_{\max}$ decreases as $1/\sqrt{M}$ for large M. The Kasami codes are also derived from m-sequences. The degree m of the corresponding polynomial has to be even: $m = 2k$. In that case, the length $M = 2^m - 1$ can be factorized: $M = (2^k - 1) \cdot (2^k + 1)$, $k = m/2$. Starting from an m-sequence c_0, the corresponding decimated sequence c_d is obtained by taking every dth chip from c_0 – where $d = 2^k + 1$ for a Kasami code generation – and repeating these $2^k - 1$ chips $2^k + 1$ times. The resulting code sequence c_d has the same length as c_0 but a period of $2^k - 1$. The set of Kasami codes is constructed in a similar way as the set of Gold codes by taking c_0 and the modulo-2 sum of c_0 and all $2^k - 1$ cyclically shifted versions of c_d.

Comparing Gold and Kasami codes of code length M, the peak value of cross correlation is lower for Kasami codes, whereas the cross-correlation functions of Gold codes take the lowest possible value $-1/M$ more often. Furthermore, there are more Gold than Kasami codes.

Figure 5.15 summarizes the essential facts concerning the number of codes and the peak values of their cross-correlation functions for m-sequences, Gold and Kasami codes.

Barker codes

Another type of codes used in spread spectrum systems without CDMA, that is, without the need for separating different users, is given by the so-called *Barker codes*. These are characterized by minimizing certain kinds of reduced autocorrelation functions defined by

$$\phi_r(\nu) = \sum_{\mu=0}^{M-1-\nu} c_{\mu+\nu} \cdot c_\mu, \quad 0 \le \nu < M$$

$$\phi_r(\nu) = \sum_{\mu=-\nu}^{M-1} c_{\mu+\nu} \cdot c_\mu, \quad -M < \nu < 0.$$

A code c consisting of $+1$ and -1 of length M is called a *Barker code* if and only if $|\phi_r(\nu)| \le 1$ for all ν with $0 < |\nu| < M$.

Correlating a Barker sequence b of odd length M with a concatenated Barker sequence of the form $[\sigma_1 \cdot b, \sigma_2 \cdot b, \ldots]$, $\sigma_i = -1, +1$, which occurs when spreading a sequence of data bits, the out-of-phase values of the discrete autocorrelation function take the values $1/M$ and $-1/M$, that is, the smallest possible values.

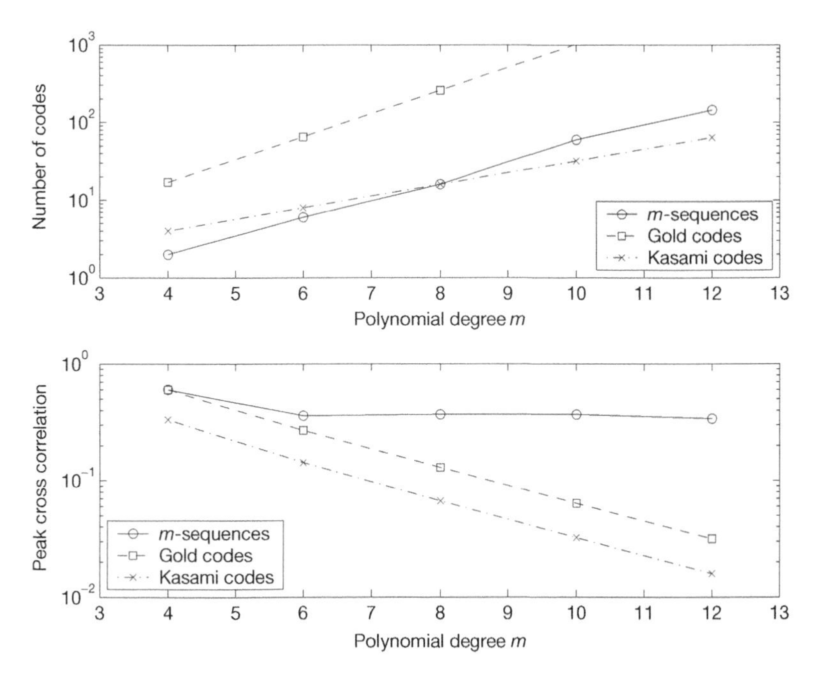

Figure 5.15 Comparison of m-sequences, Gold and Kasami codes.

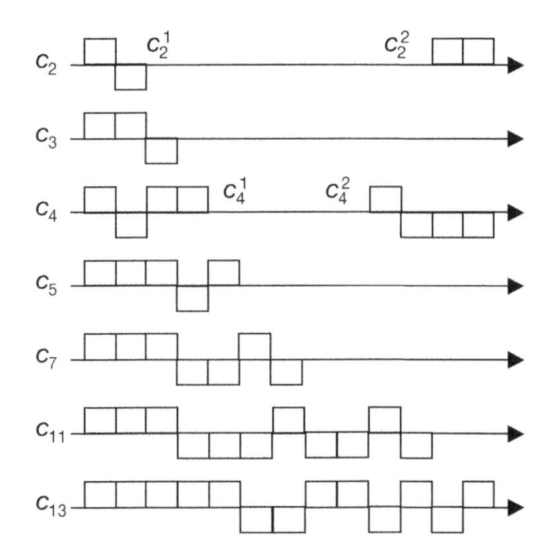

Figure 5.16 Barker codes.

Up to transformations of the type $c_\mu \to -c_\mu$, $c_\mu \to (-1)^\mu \cdot c_\mu$, $c_\mu \to -c_{M-\mu}$ there exist only the Barker codes shown in Figure 5.16. A Barker code of length $M = 11$ is used in Wireless LANs according to the standard IEEE 802.11 as will be explained in Subsection 5.5.1.

Usage of codes

The codes discussed above are used in different systems in different ways:

- For *modulation*, which may be called *code keying*, the different data symbols of one connection are mapped onto different codes of short length as the Walsh codes. This kind of modulation is used in the uplink of the cdmaOne (IS-95) mobile communications standard and has been discussed in Subsection 1.1.4 as Walsh modulation. It is also applied using complex-valued versions of the Walsh codes within Wireless LANs according to the IEEE 802.11b standard, which will be explained in Subsection 5.5.1.

- As a *direct sequence spread spectrum* method, all data symbols of one connection (or at least of one data packet) are spread using the same code; different codes are allocated to different connections. In this case the code length is either chosen equal to the period length of a data symbol or much longer. As an example for the first case, the spreading factors of the OVSF codes in UMTS are selected in exactly the way that the corresponding code length matches the period of the data symbols of the connection. In contrast, the precision code of the Global Positioning System (GPS) has a duration of nearly nine month partitioned into 37 one-week segments. This is of course longer than a data symbol. Also in mobile communication systems like UMTS or IS-95, quite long codes are applied. However, for UMTS the respective Gold codes corresponding to polynomials of degree $m = 25$ are truncated to the length of a physical data frame (10 ms corresponding to 38,400 chips). In this case, the correlation functions of the truncated code instead of the complete Gold codes have to be considered.

In CDMA mobile communication systems spreading is often performed in two steps: In a first step, each symbol of a connection is multiplied by a Walsh or OVSF code allocated to the respective connections. Different connections generated by one and the same source use orthogonal codes. The source may be a BS having connections to many MSs or an MS having several parallel connections (e.g. one connection for speech and another one for internet browsing). Since the signals are transmitted synchronously by one source, the corresponding codes are expected to keep their orthogonality – at least to a high degree – in environments with a low delay spread. Hence, the Walsh and OVSF codes are applied as channelization codes to separate several signals transmitted by one source. Signals transmitted by different sources cannot be assumed to be synchronized. Therefore, Walsh and OVSF codes are not the appropriate ones to separate these signals. Hence, in a second step the signals of different sources are multiplied by long m-sequences or Gold codes. These scrambling codes are allocated in a way to distinguish the signals of different sources – at least locally. Different MSs use different codes; a scrambling code allocated to one cell is not reused in a neighbor cell, but may be allocated to another cell within the network far away. This code allocation method is illustrated and explained in some more detail in Subsection 5.5.4.

5.1.4 Methods for handling interference in CDMA mobile radio networks

As discussed in the previous subsection, orthogonality of codes gets lost – at least to some degree – in mobile radio networks, since for the uplink direction one has no perfectly synchronized transmitters (the different mobile stations) and in both directions one has multipath propagation with a certain delay spread. Hence, without any additional measure a high level of intracell interference will occur, thereby reducing the performance of the respective CDMA mobile radio systems significantly. Therefore, methods for managing and reducing interference have to be applied, which are as follows:

- power control (PC)
- soft handover (SHO)
- interference cancellation (IC) and other multiuser detection receiver techniques
- smart antenna techniques (SAT).

Power control

A method that is absolutely necessary to achieve sufficient performance in UL direction is *transmission power control*. As demonstrated in Figure 5.6, the uplink received level from MSs at different positions varies in a range of about 50 dB or more. Without PC the result will be a probability function of the local mean *UL SIR* as shown by the solid line in Figure 5.17, where a cluster $K = 1$ and 10 active connections per carrier are assumed. A significant part of the connections experiences an *SIR* of -30 dB or less. This results from the so-called *near–far problem*: the transmitted signal by an MS *near* the BS will highly

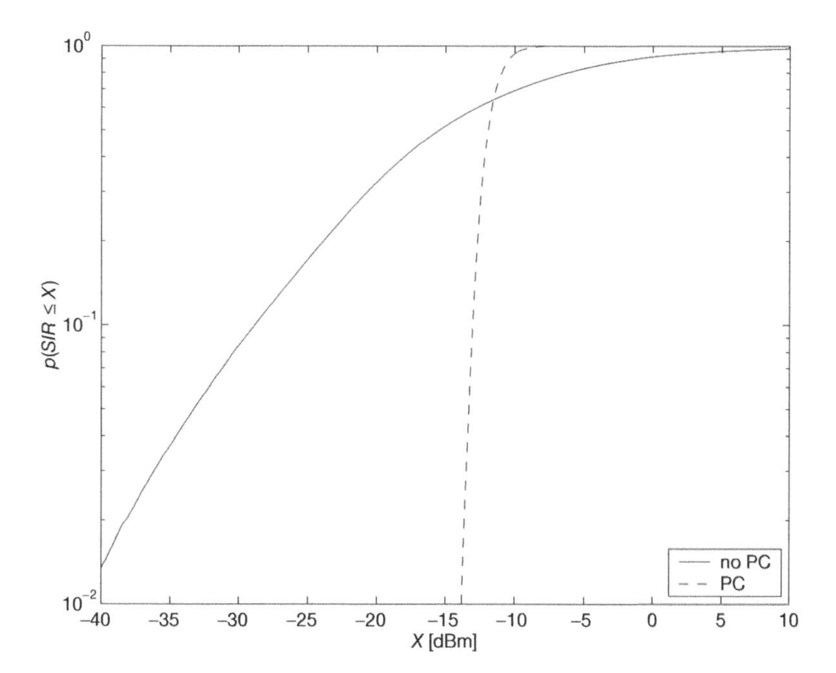

Figure 5.17 Probability function of the uplink *SIR* with and without power control.

dominate the signals of all other mobile stations *far* away from the BS. Hence, spreading factors higher than 1000 will be necessary to be able to separate the 10 connections. This will of course be a giant waste of bandwidth.

The situation gets much better applying PC in uplink direction as shown by the dashed curve in Figure 5.17: the probability distribution of the *SIR* with PC is very narrow, nearly all connections have an *SIR* better than −14 dB. It should be noted that for Figure 5.17 only the local mean value of the *SIR* has been considered. However, there is also a significant variation of the signal level due to short-term fading (depending on the ratio between signal and coherence bandwidth). Hence, PC has to be able to react also on short-term fading and therefore has to be very fast. For example, a vehicular mounted MS moving at a speed of $v = 90$ km/h = 25 m/s crosses a distance of 25 mm in 1 ms, which is a significant part of the wavelength of 15 cm at a frequency of 2 GHz. This rough estimation shows that PC actions have to be taken at a rate of about one per millisecond. For UMTS the power control rate is 1500 Hz, for IS-95 and cdma2000 it is 800 Hz.

Having estimated the power control rate, the next question is how to organize power control. Concerning the control loop, one distinguishes

- open-loop power control and
- closed-loop power control.

These concepts are illustrated in Figure 5.18 for power control in UL direction. Nevertheless, the concepts may be transferred directly to the DL by interchanging the roles of MS and BS. For the UL open loop, the MS adjusts its own transmit power on the basis of the received DL signal, whereas in a closed loop the BS measures the received signal strength and transmits a power control command to the MS. In consequence, the MS adjusts its transmit power on the basis of the received UL signal. The advantage of the open-loop

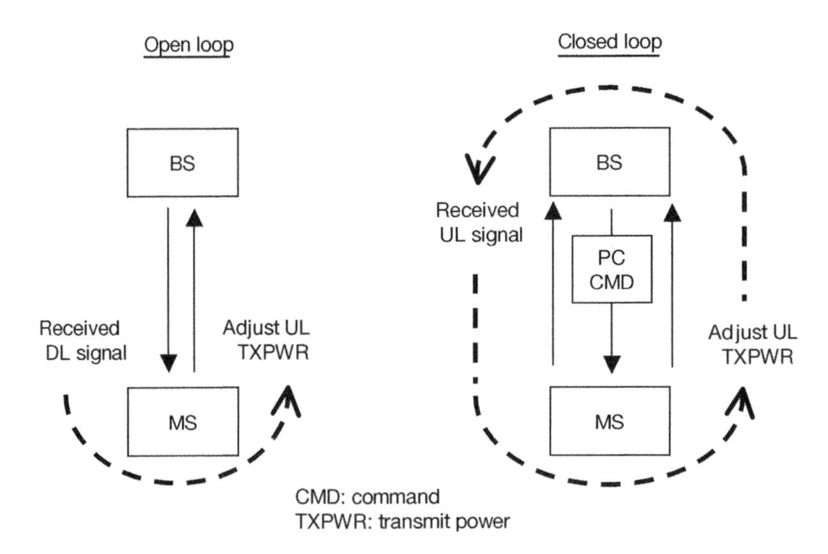

Figure 5.18 Open- and closed-loop power control.

concept is that no control channel for PC commands in the reverse direction is needed and that a PC action may be taken immediately without signaling delay. The essential disadvantage is that the reception conditions may be very different in UL and DL direction due to mainly two reasons:

- Short-term fading differs since it depends on frequency and changes rapidly with time. Within systems using frequency division duplex, UL and DL are usually separated by 20–200 MHz, while in time division duplex systems they are separated by a time period of some milliseconds.
- Interference differs in UL and DL direction.

In general, an open loop is applied to set the initial power at a call setup, thereafter a fast closed-loop power control is used.

Though in DL direction the near–far problem does not exist and therefore DL PC is not as necessary as UL PC, it is worthwhile to implement DL PC for the following reason: though intracell interference is constant in DL direction, intercell interference varies for different MSs, since they have different distances with respect to the disturbing base stations. DL PC can equalize the *SIR* where *I* is the sum of inter- and intracell interference. However, as long as intracell interference significantly contributes to the total interference, the *DL SIR* without PC varies in a relatively small range of less than 10 dB. Hence, the potential for DL PC is much smaller than for UL PC.

In systems like IS-95 and UMTS, power control is accomplished by setting a target SIR_{tar} for the required *SIR*. This is illustrated in Figure 5.19. If the *SIR* measured on

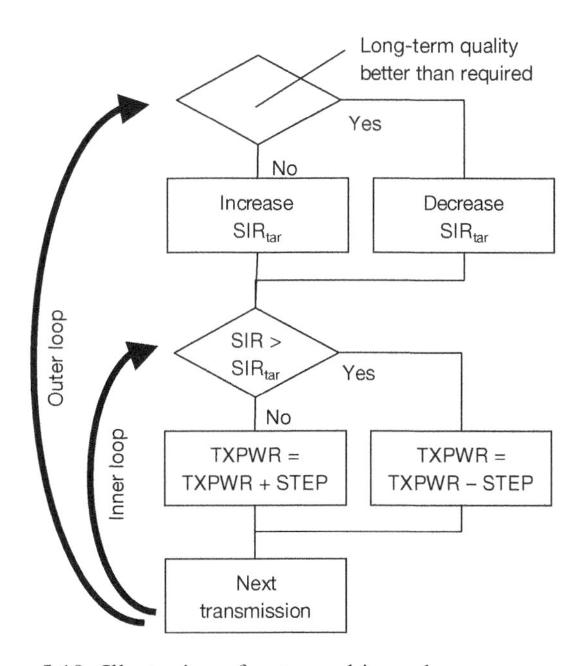

Figure 5.19 Illustration of outer and inner-loop power control.

a millisecond scale is above SIR_{tar}, the transmit power level *TXPWR* is decreased by a certain *STEP*, otherwise *TXPWR* is increased by *STEP*. The parameter *STEP* may be set by the operator to, for example, 0.5, 1, 2 or 3 dB. However, it should be noted that the appropriate target for the *SIR* depends on different parameters as the considered service, the velocity of the MS and the multipath profile. Hence, the *SIR* target also has to be set adaptively using a second control loop, the so-called *outer loop*. While the inner loop reacts on the measured *SIR* on a millisecond scale, the outer loop reacts on longer scales using averaged values of, for example, the following quantities as an input:

- the block error rate;
- the estimated bit error rate before channel decoding;
- soft output information of the channel decoder.

Having roughly explained the usefulness as well as some variants of PC, the performance and the gain achieved by fast UL PC is now quantified in some more detail using an example with the following assumptions:

- An UMTS-like closed-loop *PC* algorithm with $STEP = 1$ dB and $SNR_{\mathrm{tar}} = SIR_{\mathrm{tar}} =$ 0 dB is investigated.
- The signal S is only disturbed by a constant noise level N, interference I is neglected.
- The local mean signal level is assumed to be constant, the signal level only varies due to short-term fading which has been generated using the ITU channel models for the vehicular and indoor-to-outdoor-pedestrian environment described in Section 2.3. Resolvable propagation paths are combined within the BS RAKE receiver. Furthermore, a twofold antenna diversity is used to reduce the effect of fading.

The results of applying UL PC in such situations for a pedestrian ($v = 1$ m/s) and a vehicular ($v = 10$ m/s) subscriber are illustrated in Figure 5.20. Part (a) shows the short-term fading level, the (relative) transmit power level *TXPWR* as well as the resulting *SNR* as a function of time. For the pedestrian, PC is able to equalize the fading nearly completely: the deviation of received *SNR* from SNR_{tar} is less than 1 dB, which can be seen in part (c) of Figure 5.20. The deviation would be even smaller, if a smaller *STEP* had been selected. Also for the vehicular subscriber moving at moderate velocity, PC is able to follow the fast fading in general. However, there are situations where power control is to slow: About 10% of the SNR values deviate more than 1 dB from SNR_{tar}, even a deviation of 3 and 4 dB occurs. For a higher velocity the deviation becomes even larger. In UMTS or IS-95 networks power control is not able to follow the fast fading for a velocity above approximately 50 km/h. However, at high velocities interleaving becomes more efficient instead.

Figure 5.20(b) shows the distribution of the values of *TXPWR*. It should be noted that the mean transmit power for the pedestrian is about 2 dB higher than for the case of no PC, whereas the mean *SNR* is the same. This increase of the mean transmit power due to fast power control results in an increased interference to other connections. For the vehicular subscriber, this interference increase is lower (about 1 dB), partly because of the higher multipath diversity reducing the fading in the vehicular environment and partly because PC cannot follow fast fading to the same degree as for the pedestrian.

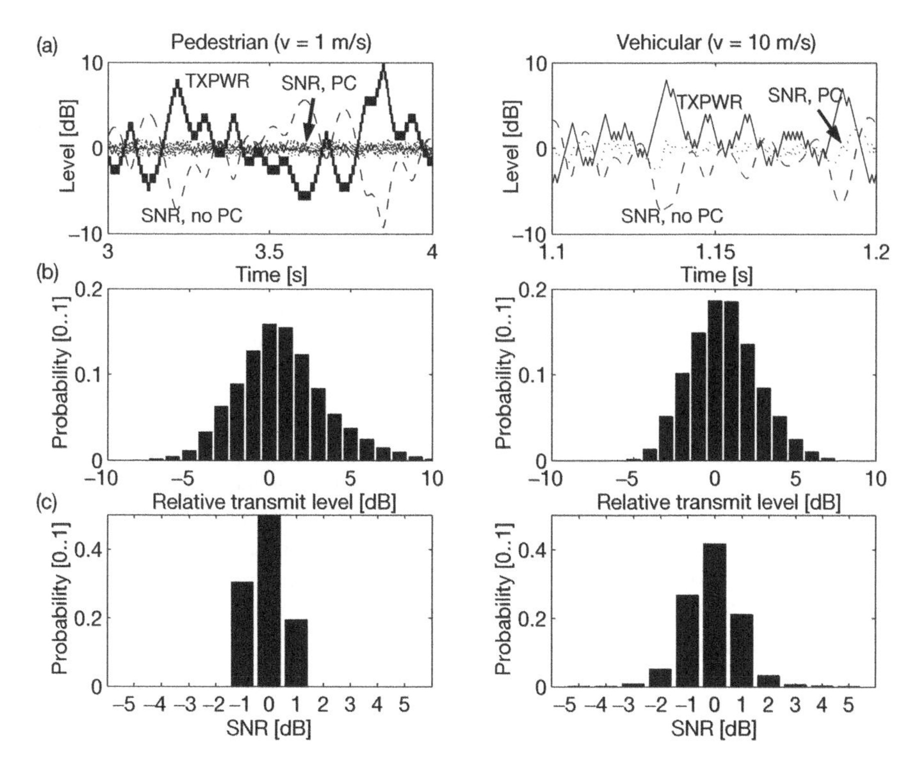

Figure 5.20 Illustration of the dynamics of fast UL PC and its effect to fading compensation.

Hence, while a slow power control algorithm removes the near–far problem and reduces the local mean intercell interference, a fast power control algorithm additionally may cancel the effect of the fast fading to a certain degree depending on the velocity of the MS. This results in a gain with respect to the required E_b/N_0 since the channel to be considered is approximately a pure Gaussian instead of a fading one. Simulation results reported in (Holma and Toskala 2001) show that for a pedestrian UMTS subscriber in a low delay spread environment the E_b/N_0 requirement for a speech-like service can be reduced by fast power control from 11.3 dB to 5.5 dB. On the other hand, an interference rise of about 2 dB is caused, reducing the effective gain from about 6 dB to about 4 dB. For a vehicular subscriber with a velocity of about 50 km/h and above, there is no additional gain by fast power control; however, interleaving becomes efficient reducing the effect of fast fading.

Summing up the discussion for PC, one obtains the following main results:

- UL PC removes the near–far problem caused by intracell interference and leads to a significantly improved probability distribution of the local mean *SIR* as illustrated in the example of Figure 5.17.
- Compensating the near–far effect by PC reduces not only the intracell, but also the intercell interference: an MS at a reduced power causes less interference in other cells. This effect is included in Figure 5.17.

- Fast UL PC can cancel fast fading nearly completely for pedestrian subscribers. Depending on the environment, this results in a gain of some decibels with respect to the required E_b/N_0 since the channel to be considered is approximately a pure Gaussian instead of a fading one.
- At higher velocities, fast fading may not be cancelled. However, interleaving becomes efficient instead so that nearly the same E_b/N_0 requirement is achieved as for a pedestrian subscriber.
- The signaling effort for fast power control depends on the fading rate. Within the present mobile communication systems it is about 1–2 kbit/s.
- DL PC reduces the intercell interference. Its potential is lower than that of UL PC.

In present TDMA-based systems like GSM, power control is implemented as slow PC with a control rate of about 1 Hz. By this method, an improvement for the local mean *SIR* can also be achieved. In principle, it would be possible to speed up power control. However, due to the TDMA nature, data and power control commands for one connection are not transmitted continuously, but only within the allocated time slot. Hence, (closed-loop) power control in TDMA systems is slower than in CDMA systems. The ratio mainly depends on the duration of a TDMA frame.

More details on power control in CDMA mobile radio systems can be found, for example, in (Holma and Toskala 2001; Viterbi 1995).

Frequency allocation and capacity estimation

In the following text, some simplified capacity estimations are presented to work out the optimum cluster size for CDMA networks and to discuss the benefits of these networks with respect to capacity. The estimations are based upon the following assumptions:

- The interference power $I = I_a + I_r$ composed of intracell interference I_a and intercell interference I_r is much larger than the noise power N so that N can be neglected.
- Interference affects the bit error rate in the same way as white noise, that is, the signal-to-interference ratio S/I is related to the E_b/N_0 by $E_b/N_0 = SF_{\text{eff}} \cdot S/I$, where SF_{eff} is the effective spreading factor.
- The number of active connections n is the same in each cell. All connections use the same effective spreading factor.
- An ideal UL PC is applied so that all connections within a cell are received at the BS with the same signal power S. Consequently, the relative intracell interference power is given by $I_a/S = \alpha \cdot (n - 1)$. The factor α, which ranges between 0 and 1, is called *orthogonality factor*. For ideal orthogonal codes or if intracell interference could be cancelled completely by special receiver techniques, α would be 0. However, without these techniques $\alpha = 1$ is a more realistic assumption for the UL.
- UL intercell interference is generated by a large number of independent sources, the MSs. Hence, it is reasonable to model the relative intercell interference power

I_r/S as a Gaussian random variable with a mean value $n \cdot q$ and standard deviation of $\sqrt{n} \cdot \sigma_I$, where q and σ_I are the mean value and the standard deviation of the interference distribution for one interferer per cell. These quantities mainly depend on the propagation parameters, the cluster size K and the used antenna configuration; examples are given below.

Composing all these assumptions, one obtains in a first step for the *median SIR*:

$$SIR_m = \frac{1}{\alpha \cdot (n-1) + q \cdot n} \geq \frac{E_b/N_0}{SF_{\text{eff}}}. \tag{5.5}$$

Hence, considering only median values, the maximum number of active connections per cell n_m is given by

$$n_m = \left(\frac{SF_{\text{eff}}}{(E_b/N_0)} + \alpha\right) \cdot \frac{1}{(\alpha + q)}. \tag{5.6}$$

However, these results for the median values are too optimistic. An operator usually has to guarantee a coverage probability of, for example, $c_p = 0.95$. Hence, in this case one has to consider the 95%-percentile of the relative interference:

$$(I/S)_{95} = \alpha \cdot (n-1) + n \cdot q + \sqrt{n} \cdot \sigma_I \cdot \gamma_{95}, \tag{5.7}$$

where $\gamma_{95} = 1.65$ is the 95%-percentile of the normalized Gaussian distribution. Requiring that the corresponding *SIR* shall be greater than or equal to the E_b/N_0 divided by the effective spreading factor, one obtains after some elementary algebraic manipulation for the maximum number of connections per cell (at 95% coverage probability):

$$n_{95} = n_m + 2b^2 - 2b \cdot \sqrt{n_m + b^2}, \quad b = \frac{\sigma_I \cdot \gamma_{95}}{2(\alpha + q)}. \tag{5.8}$$

Using Equation 5.6 and Equation 5.8, the effect of spreading and CDMA on radio network capacity can be discussed. For this reason, sectorized networks with 65° half power beam width antennas and cluster sizes $K = 1 \times 1$ and $K = 1 \times 3$ are considered as examples. Assuming a propagation parameter of $B = 40$ dB per decade and a standard deviation of the long-term fading of $\sigma = 8$ dB, the following values for the intercell interference parameters are derived by Monte-Carlo simulations:

- for $K = 1 \times 1$ one has $q = 0.79$ and $\sigma_I = 0.5$,
- for $K = 1 \times 3$ one has $q = 0.15$ and $\sigma_I = 0.19$.

Decreasing B, that is, the decay of the signal level as a function of the distance, increases the intercell interference and therefore q and σ_I. For a cluster 1 or 1×1, the intracell interference (given by α) and mean intercell interference (given by q) per connection are of the same order of magnitude.

Equation 5.6 and Equation 5.8 are illustrated in Figure 5.21 using the values above and $E_b/N_0 = 6$ dB as input values. Though the spreading factor in Figure 5.21 is varied, the E_b/N_0 is kept at a constant value, to highlight additional effects of spreading beside frequency diversity. The number of active connections per cell n has been normalized by the effective spreading factor SF_{eff} and the cluster size K to get a comparison of the network

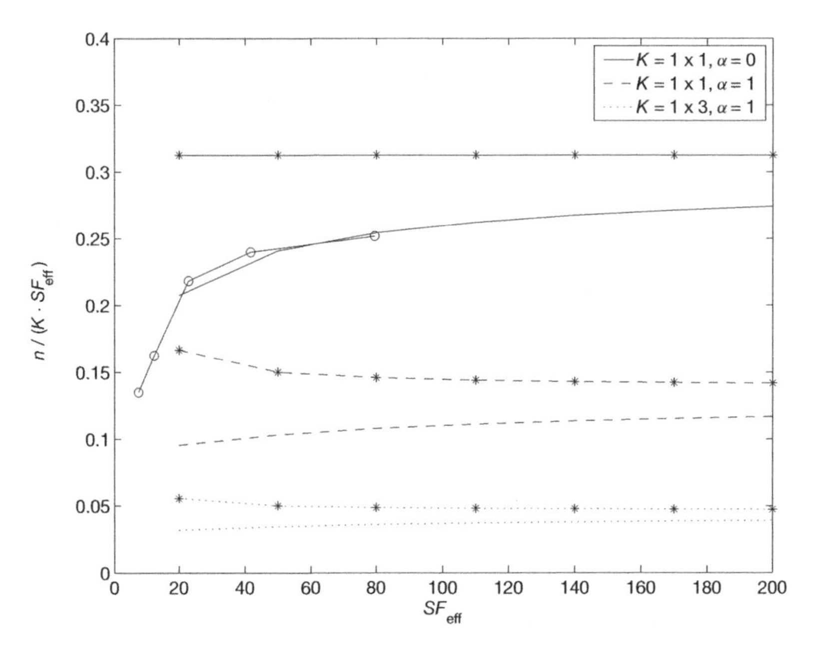

Figure 5.21 Network capacity per bandwidth as a function of the spreading factor for different interference scenarios.

capacity per totally allocated frequency spectrum; the total required spectrum is proportional to both quantities. Results corresponding to the median coverage are marked by the symbol "*", whereas the results corresponding to the 95%-percentiles are not marked. Considering first the cluster $K = 1 \times 1$ and the overidealized case of no intracell interference (solid line in Figure 5.21), the median value of the allowed number of connections n_m is proportional to SF_{eff}, that is, there is no gain of spreading with respect to the relative capacity as long as frequency diversity is neglected. However, when considering the 95% coverage probability, a gain is achieved due to the large number of users; this gain is called *interference averaging* or *interference diversity*. For high spreading factors, the value for the 95% coverage tends toward the median values, that is, a CDMA network operator may perform network planning based on nearly the mean values, whereas a TDMA operator has to consider the worst case scenarios. To highlight the effect of interference averaging, the results from Monte-Carlo simulations for small spreading factors and therefore small numbers of n are included in Figure 5.21 and they are marked by the symbol "o".

Taking into account intracell interference with $\alpha = 1$ (dashed lines), there is still some interference diversity gain. Since intracell interference has a strong impact on network performance, one gets a capacity loss for CDMA networks in this case (which however, may be compensated or even turned into a gain when considering frequency diversity as well). In the example of Figure 5.21, the capacity is reduced by about a factor of 2.1 since the ratio of total interference and intercell interference ($(\alpha + q)/q$) is approximately 2.1. This example shows also the potential for methods like interference cancellation, even if only intracell interference may be cancelled.

Reducing the intercell interference by increasing the cluster size to $K = 1 \times 3$, increases the SIR by about 2 dB for the considered example, but at the cost of requiring the three-fold bandwidth which is of course a bad deal. Hence, as long as there is a strong impact of intracell interference, the cluster $K = 1 \times 1$ is the optimal one from the point of view of network performance. For higher cluster sizes, the additional bandwidth costs are not compensated by the gain due to intercell interference reduction. However, if intracell interference can be reduced significantly (α below 0.1), higher cluster sizes may become more efficient depending on many parameters as, for example, the propagation parameters.

Until now only the UL capacity has been discussed. Though similar consequences can be drawn for the DL, there are some differences; since the signals for the different connections within one cell are transmitted synchronously, orthogonality of codes may be preserved at the receiving end, at least in environments with a low delay spread. Hence, the intracell interference for the DL is expected to be significantly lower than for the UL. In (Holma and Toskala 2001) orthogonality factors of $\alpha = 0.4$ and $\alpha = 0.1$ are reported for UMTS (chip duration: 260 ns) in a vehicular environment with a delay spread of 370 ns and in a pedestrian environment with a delay spread of about 50 ns, respectively. Hence, for the DL the potential of interference cancellation methods is expected to be lower than for the UL.

Intercell interference in DL direction is generated by the base stations, that is, by much less sources than for the UL. Hence, the interference averaging effect discussed for the UL reduces to zero. Nevertheless, also in DL direction one profits from interference averaging since the interference generated by one BS or even by one connection at a BS may vary – mainly for the following three reasons:

- Because of DL PC, different connections at one BS result in different values for the interference power.
- For speech services, discontinuous transmission (DTX) may be applied, that is, in phases of no speech activity the transmission for the respective connection is switched off reducing the corresponding interference power to zero.
- For inhomogeneous load conditions, the interference power generated by neighboring base stations may differ.

It should be mentioned that interference averaging with respect to the two last mentioned effects (which may also be present in UL direction) results in a higher gain than the one mentioned when discussing Figure 5.21 for the UL.

Because of the interference averaging effect and since the number of codes per cell in general puts no restriction to the number of connections per cell, CDMA networks are usually planned by the so-called *soft capacity* planning strategy, which is explained in the following text, in contrast to a *hard capacity* planning strategy.

- For a hard capacity planning, in general, a high value of the cluster size is used so that even if all installed channels are busy, the *SIR* stays above the required value (with a probability of e.g. 95%). The capacity of a cell is limited by the number of installed channels and that limit is a hard one, that is, if all channels are busy a new request is blocked.

- For a soft capacity planning, a much smaller cluster size (e.g. $K = 1$) is used. Having a smaller cluster size, more channels per cell exist. This means that in this case the capacity limit is not given by the maximum number of channels per cell, but by an upper limit on the total interference power. Therefore, powerful and reliable methods for controlling the interference within the network are required to reduce the risk that a small increase of interference causes a significant performance degradation for many connections; these methods are called *load* and *admission controls*.

  Furthermore, the interference experienced by different connections should be approximately the same and hence interference averaging is required. The soft capacity planning strategy is therefore usually applied in CDMA mobile radio network. But since recent years, more and more GSM networks using frequency hopping are also planned according to this strategy (see e.g. (Rehfuess and Ivanov 1999)).

The main advantage of the soft capacity planning strategy is that only the mean interference has to be controlled, that is, there may be a very high load in one cell as long as the load within other cells, and therefore the interference caused by these cells, is low. Hence, a network planned by this strategy is able to react automatically on inhomogeneous and time-dependent load conditions.

Another consequence that may be drawn from Equation 5.6 and Equation 5.8 is related to the effect of channel coding in CDMA networks. Because of these equations the number of active connections per cell is approximately proportional to the ratio between the effective spreading factor SF_{eff} and the E_b/N_0 (assuming a high bandwidth and spreading factor). Keeping the bandwidth at a constant value and introducing channel coding reduces the effective spreading factor. On the other hand, there is a coding gain (reduced E_b/N_0 requirements) turning the loss of the spreading factor into a gain for the number of active connections. This means that in a CDMA network the coding gain can be directly and continuously turned into a capacity gain. Though a TDMA network using the hard capacity planning strategy also profits from channel coding, it is more difficult to implement the corresponding capacity gain by reducing the cluster size and reworking the frequency plan. Furthermore, the reduction of cluster size can only be performed in discrete steps, which may be too high for a given coding gain.

As argued in the preceding text, a cluster 1 or 1×1 using the same frequency carriers in neighboring cells is in general the most efficient frequency allocation scheme in CDMA mobile radio networks, at least when cells within the same hierarchical level are considered. However, in regions of very high load as pedestrian area, railway stations or shopping malls, additional small sized cells called *microcells* may be implanted into an existing network of macrocells. While the high-power macrocells accomplish the overall coverage, the low power microcells are installed to serve most of the traffic, that is, cell selection between these two hierarchical levels is not primarily based upon the signal level, but on load conditions. Since these two types of cells are operating at unbalanced power levels, interference caused, for example, by the macrocells to the microcells cannot be controlled by the same methods (power control, soft handover) as the interference within the macrocell layer. Therefore, the cells in different levels of this hierarchical cell structure should use disjoint frequency carriers.

The discussion on frequency allocation and capacity is summarized as follows:

- The gain of CDMA networks with respect to capacity is not achieved by spreading itself, but by interference averaging. Especially, for a strongly varying and inhomogeneous network load, a high gain can be achieved. With respect to capacity there is no other benefit of spreading except for frequency diversity.
- Intracell interference strongly degrades the network performance, especially in UL direction. However, methods like interference cancellation are expected to reduce intracell interference and therefore enhance capacity significantly. The effect of intracell interference is expected to be lower in DL direction where orthogonality of codes is preserved to a certain degree.
- Owing to the strong impact of intracell interference, a cluster 1×1 leads to the highest capacity values. Using a cluster 1×1, the high effort for frequency planning can be avoided.
- The soft capacity planning strategy related to the cluster 1×1 frequency allocation and the interference averaging effect of CDMA allows an adaption of the network capacity to time-varying and inhomogeneous load conditions. Furthermore, a channel coding gain can be directly and continuously transferred into a capacity gain.
- Soft capacity planning is not only applicable in CDMA networks, but also in TDMA networks using frequency hopping.
- In a hierarchical cell structure, the different layers should use disjoint frequency carriers.

Soft handover

As argued earlier, CDMA mobile radio networks should be operated in general by using a cluster 1 or a cluster 1×1, that is, allocating the same frequencies in neighboring cells. However, this fact results in a high degree of intercell interference. Even a single MS near the cell border may disturb in UL direction all connections in a neighbor cell to a high degree if no special measure is taken. To illustrate this, consider an MS near the border of two cells called *cell 0 and cell 1*. The MS is assumed to be currently served and power controlled by cell 0, that is, the corresponding BS 0 is the one with the highest received level for that MS (higher than for e.g. BS 1) and the signal level is adjusted to the target level by PC. However, due to fading, the level with respect to BS 1 may temporarily become much larger than the target level within some milliseconds, that is, the interference by that MS may exceed the signal level of all other connections in cell 1 significantly. To avoid this undesired situation, a soft handover is required, that is, the MS has to be served and power controlled not only by BS 0 but also by BS 1 (and eventually further base stations receiving nearly the same signal level from the MS as BS 0 and BS 1).

In UL direction, soft handover is implemented usually in one of the two following ways:

- The signals processed by the RAKE receivers of all involved base stations are combined by maximum ratio combining.

- Each involved BS performs the channel decoding for the received signal and adds a frame reliable indicator to each decoded frame. The frames are transferred to the radio network controller which selects the most reliable one.

Obviously, the first method which is called *softer handover* in UMTS gives the highest performance; however, the highest data rate also is required to transfer all received signals to the combining element. Therefore, it is applied (e.g. in UMTS) only for base stations that are installed at the same site, that is, for a soft handover between sector cells served from the same site. A UL soft handover between base stations at different sites is usually managed by the second method, that is, by a selection combining of data frames that have a length of some tenth of milliseconds.

In DL direction, soft handover is performed by transmitting the same data to the MS from several base stations. Since a cluster 1 is used, each of the corresponding signals is sent on the same frequency and is spread by the codes of the respective cells. Furthermore, the transmitted signals are roughly synchronized (to an order of about some microseconds). Hence, from the point of view of the receiving MS, the different signals can be handled in nearly the same way as multipath components of one signal, that is, they can be combined by the RAKE receiver. The only modification is that correlation within the RAKE fingers has to be performed using the different codes corresponding to the involved base stations. Furthermore, it should be observed that the number of RAKE fingers in an MS is limited.

Having explained the general principles of soft handover, some comments on the gain that can be achieved by this method should be added:

The soft handover gain comprises

- a microdiversity gain against short-term fading
- and a macrodiversity gain against long-term fading.

Considering the macrodiversity gain, it is obviously profitable to switch the connection as fast as possible to the BS with the highest local mean received level. Also in the case of a hard handover the general strategy is usually to switch to the BS guaranteeing the best level. However, as mentioned in Subsection 5.1.3, in this case one aims to avoid many forward and backward handovers between different cells by basing the decision on an averaged level and by introducing a hysteresis margin of some decibels. This means that as long as the averaged receive level of the neighbor cell does not exceed the averaged level of the old cell by, for example, 4 dB, no hard handover is performed for the respective MS. Hence, for a hard handover there may be phases of some seconds where the MS is not served by the BS with the best local mean signal level, that is, where the performance is lower than for a soft handover. The performance difference between hard and soft handover with respect to macrodiversity depends on the averaging length and hysteresis margin, which on their part have to be selected on the basis of the MS velocity, the standard deviation and correlation length of the long-term fading and the tolerable rate of handovers. Though it is very difficult to quantify the macrodiversity gain exactly, some results from (Graf *et al.* 1997) are quoted to give an idea of the order of magnitude; for typical scenarios, the *SIR* at 95% coverage is improved by about 1–2 dB.

How much additional microdiversity gain can be achieved by soft handover depends on to what extent short-term fading has already been combatted by other means like antenna diversity and multipath combining within the RAKE receiver. A significant microdiversity gain by soft handover is only expected if the difference between the local mean signal levels of the involved signals is low. Furthermore, for the DL direction it should be observed that the number of RAKE fingers in an MS is limited to, for example, four. This means that if one or two of these fingers are needed for soft handover connections to additional base stations, multipath diversity combining is reduced. Hence, it depends on the multipath profile and the difference of the local mean values of the signal levels, whether a multipath or a soft handover combining is preferable. Since the gain of soft handover depends on many parameters, it requires thorough investigations to derive reliable and exact values. Nevertheless, a very simplified model is presented to give an idea of the order of magnitude of the gain. Concerning short-term fading the following assumptions are made leading to results of Figure 5.22:

- The RAKE receiver in the MS and BS is able to combine four propagation paths.
- The ITU channel model A for a vehicular environment (see Section 2.3) is taken as the multipath profile. The four fading paths have the relative mean power levels of 0, −1, −9 and −10 dB.
- Antenna diversity is implemented as a maximum ratio combining within the BS for the UL, but no antenna diversity is used in DL direction.

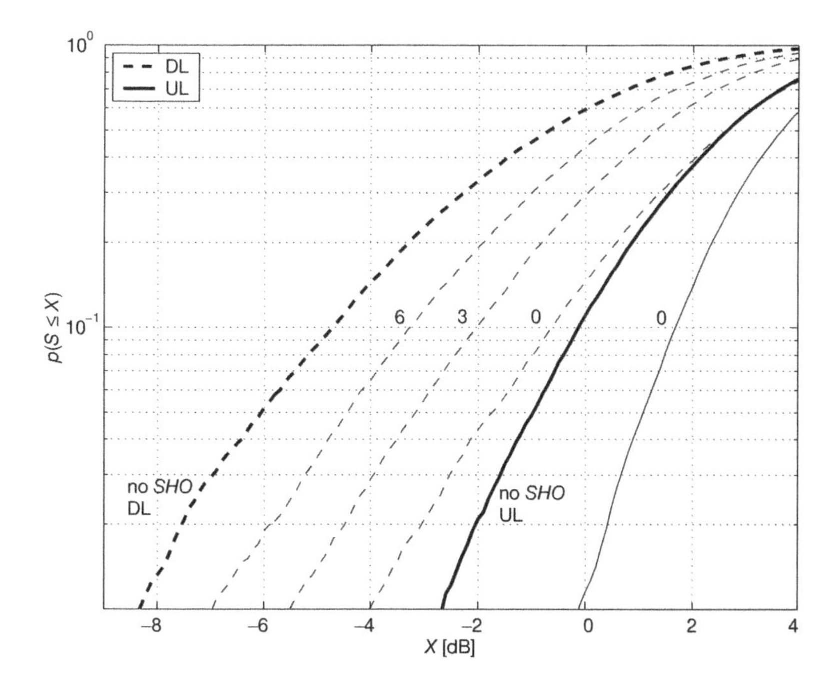

Figure 5.22 Illustration of the microdiversity gain by soft handover.

- A soft handover between two base stations is considered. It is modeled as a selection combining of the short-term fading values in the UL assuming the same local mean received signal level for both base stations (i.e. the optimum case).

- In DL direction, a maximum ratio combining of the signals transmitted by the two base stations is assumed, where the restriction that only four paths can be combined is observed. The difference of the local mean levels of both signals has been set to 0, 3 and 6 dB.

In Figure 5.22, the corresponding probability functions of the signal levels with and without soft handover are compared; the case of no soft handover is represented by the bold curves for the UL and DL, the difference (in decibels) of the local mean receive level between the strongest and the other connection is indicated by the numbers in the diagram. It should be noted that the received level is shown relative to the local mean level of the strongest BS. If the local mean levels with respect to both base stations are the same, the soft handover gain is about 2 dB for the UL and about 4–5 dB for the DL. If the level difference is 6 dB, the DL gain reduces to about 1.5 dB. For the DL, it should be observed that the gain is achieved by using twice the transmission power, that is, the original power is transmitted by both base stations. Hence, from the point of view of power efficiency, the DL soft handover curves have to be shifted by 3 dB to the left. Though BS transmission power itself is not the most critical parameter, twice the transmission power also means that the high DL gain for one connection can only be achieved at the expense of an increased interference level for other connections.

For this reason and other reasons to be discussed below, a BS should only be involved in a soft handover, if it contributes significantly to the totally received power. To check this condition, various algorithms are specified within the different CDMA systems. For example, in UMTS, a BS is included in the active set of base stations for a soft handover, only if its averaged received level exceeds $RXLEV_0 - H_{SHO}$, where $RXLEV_0$ is the strongest averaged received level and H_{SHO} is a hysteresis parameter. Looking at Figure 5.22, a hysteresis of about $H_{SHO} = 4$–5 dB seems to be reasonable.

Besides the increased transmission and interference power, there are two other aspects of soft handover causing additional effort:

- additional transmitter and receiver hardware within each BS;

- additional transmission lines between the base stations and the combining network elements.

Also for limiting these costs, the number of base stations involved in a soft handover should be kept small. To illustrate this soft handover effort, Figure 5.23 shows the fraction of connections involved in a soft (or softer) handover and the mean number of active base stations per connection as a function of the hysteresis H_{SHO}. The results have been obtained by Monte-Carlo simulations for a sectorized network using $B = 30$, $B = 40$ and $\sigma = 8$ dB as propagation parameters. Though one connection in soft handover mode may use even more than two base stations, the maximum number of used base stations has been restricted to four. For $H_{SHO} = 5$ dB, about 40–50% of the connections are involved in a soft handover and each connection uses on average about 1.6–1.9 base stations.

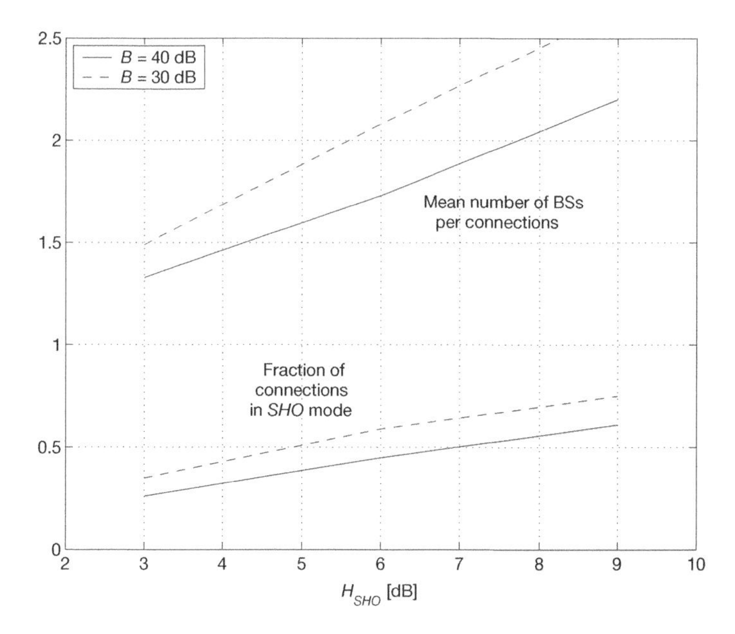

Figure 5.23 Soft handover probabilities as a function of the hysteresis margin.

Finally, it should be noted that a soft handover cannot be applied between cells using different frequency carriers, for example, between the different layers of a hierarchical cell structure. In these cases, a hard handover is required. While for a hard handover in TDMA systems the MS can perform neighbor cell measurements in time slots not used for data transmission or reception, a hard handover in a CDMA system requires some additional effort for the MS; to be able to perform neighbor cell measurement, the MS has to be equipped with an additional measurement receiver or a slotted transmission mode has to be used. Slotted mode means that the data to be transmitted are compressed, for example, by reducing the spreading factor or the channel coding rate for some period to obtain some time for neighbor cell measurements. Obviously, during the slotted mode phases the connection quality is reduced.

The summary of the discussion on soft handover is as follows:

- Soft handover is required in CDMA networks using a cluster 1 to control the intercell interference caused by MSs near the cell border.
- On the other hand, using a cluster 1 and CDMA, soft handover can be implemented in quite a simple way, for example, in DL direction it can be implemented using the RAKE receiver within the MS.
- Compared to a hard handover, a gain of several decibels (depending on many parameters) is achieved for the UL as well as for the DL.
- The gain is achieved at the expense of additional costs for transmission lines and BS transmitter and receiver hardware.

- Soft handover is not only restricted to CDMA systems, but may also be applied in other systems, where the MS is able to combine different propagation paths, for example, by using an equalizer.

- Switching between cells with different frequency carriers, for example, in a hierarchical cell structure, a hard handover has to be used requiring additional effort compared to a hard handover in TDMA systems.

More details on soft handover can be found, for example, in (Holma and Toskala 2001; Viterbi 1995).

The potential of multiuser detection and interference cancellation

The idea behind multiuser detection is to detect and to demodulate not only the useful signal, but also the interfering signals – at least some of the strongest ones. Having detected the dominant interferers, their undesired contribution may be removed from the total received signal using some sophisticated algorithms to obtain a less interfered signal. Since, for applying this method, the signals of multiple users have to be detected, the corresponding receiver structure is called a *multiuser detector receiver*. Some important multiuser detectors and their application areas are discussed in detail in Section 5.3 and Section 5.4. In this subsection, only some qualitative arguments concerning the potential of interference cancellation are presented.

First of all, it should be noted that for efficiently cancelling the interference caused by other connections the corresponding code signals have to be known and a connection individual channel estimation has to be performed.

As discussed above, the intracell and intercell interference power in UL direction is nearly the same for typical propagation parameters and a cluster 1 (or 1×1) network layout. Hence, cancelling the intracell interference reduces the overall interference by about a factor 2, which results in a doubling of network capacity. Since the BS knows all the codes allocated to the active MSs in its cell, one prerequisite for performing a multiuser detection is given. Furthermore, as explained in Section 5.5, connection-specific pilot symbols accomplishing channel estimation are included within the UL physical channels of modern CDMA systems like UMTS or cdma2000. Nevertheless, multiuser detection is a hard challenge since there may be a large number of intracell interferers and all of these contribute with nearly the same interference power due to power control, that is, there is no dominant one.

The gain that can be achieved by cancelling the intracell interference in DL direction depends on the environment where it is applied. As discussed above, in a low delay spread environment the orthogonality of codes is preserved to a high degree. Hence, in this case the contribution of intracell interference to the total interference power may be only about 10% or less, that is, the potential for intracell interference cancelling is low. However, environments with a higher delay spread and higher degree of DL intracell interference also exist. From the implementation point of view, it should be noted that the receiving MS needs some information on the allocated codes within the cell, which has to be signaled by the BS in DL direction. Since the data rates of the connections and therefore the code allocation may change very rapidly, a high overhead would be required to transfer this information. Using the tree structure of the OVSF codes, some proposals have been developed to reduce this overhead (see e.g. (Bing *et al.* 2000)).

A further gain can be achieved if not only the intracell interference, but also the intercell interference is cancelled. However, this would further increase the receiver complexity. Furthermore, information concerning the code allocation has to exchanged between cells.

Combining CDMA and TDMA

Combining CDMA with TDMA means that each radio carrier is divided into a certain number of time slots and each of these time slots is further subdivided into a number of code channels. Hence, the physical channel assigned to a connection is characterized by its time slot and code number. It should be noted that this method leads to another arrangement of physical channel, but not to an increase of the total number. For example, instead of having 256 orthogonal code channels per carrier in a pure CDMA systems, these channels may be rearranged into 16 time slots each separated into 16 code channels.

Though there is no difference with respect to the number of channels, there are mainly three benefits of combining CDMA and TDMA:

- Connections on different time slots do not interfere. Hence, intracell interference is only generated by connections using the same time slot, that is, the number of intracell interferers decreases. As a consequence, the effort for jointly detecting the interfering signals and cancelling the interference may be significantly reduced.
- Dynamic channel allocation can be applied in a cell with one frequency carrier, that is, a connection affected by strong interference may be handed over to a less interfered time slot.
- Because of the time slot structure, a TDD transmission mode can be implemented. Time division duplex (TDD) means that UL and DL use the same frequency carrier but different time slots. The TDD mode allows a flexible division of transmission capacity between UL and DL. Especially, if the network load is generated mainly by highly asymmetric services like internet browsing, it is recommendable to assign more time slots for the DL than for the UL.

Note that the TDD mode has also a drawback, at least when using nonsynchronized base stations. Since there is no frequency separation between UL and DL, there may be situations of severe interference between two base stations or between two MSs using adjacent carriers. For FDD systems, this interference can be neglected because of the large frequency duplex separation between UL and DL.

The method of combining CDMA and TDMA is applied within the TDD transmission mode of UMTS (see Subsection 5.5.5), where a frequency carrier is divided into 15 time slots. Within the TDD mode, a so-called *joint detection algorithm* is foreseen and may be implemented with moderated effort. This method allows the joint detection of all signals using one time slot and thereby reduces the intracell interference (see e.g. (Baier *et al.* 2000; Bing *et al.* 2000)).

Smart antenna techniques

Antenna systems that are able to automatically adapt their beam pattern or antenna characteristics to the reception conditions are usually denoted as adaptive, intelligent or smart antennas; throughout this section the name smart antennas is used.

Beam forming is accomplished by an array of antenna elements affected by individual complex weight factors or phase shifts. In general, these systems are applied at the BS, but not at the MS side. Forming narrow beams, one may profit in three different ways:

- The delay spread is reduced.
- The cell area may be increased due to an increase of the antenna gain.
- The SIR is improved since less interference is received in UL direction and less interference is spread in DL direction.

With respect to the topic of this subsection, the third item is the most important one: applying smart antenna techniques reduces the intracell as well as the intercell interference and thereby enhances the capacity. In a certain sense, smart antennas may be seen as a special approach for interference cancellation.

As to the implementation of smart antenna techniques one may distinguish the following methods illustrated in Figure 5.24:

- switched beams
- full adaptive beams

The switched beams approach may be seen as an enhancement of sectorization discussed in Subsection 5.1.2. Each cell – an omnicell or a sector cell – is divided into a certain number of subsectors with angular width of typically 5–30°. Hence, the antenna diagrams point to several fixed directions. An MS moving within such a cell is switched from beam to beam in a similar manner as for a soft or softer handover. However, it should be noted

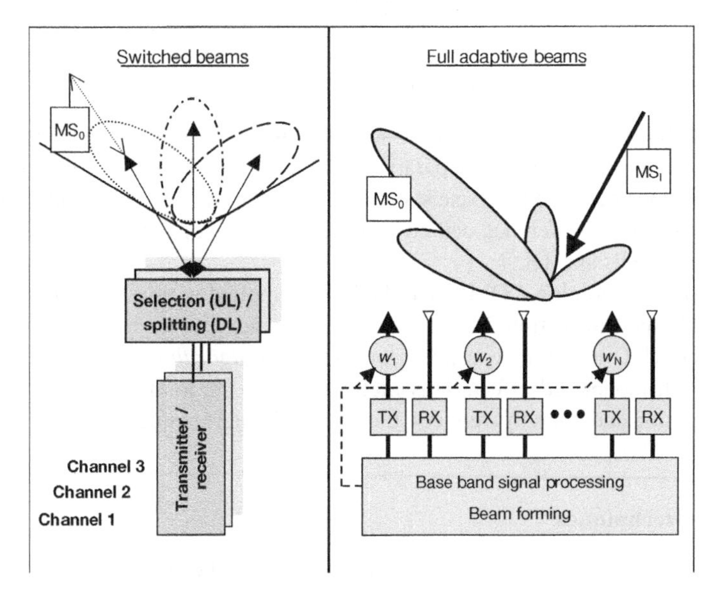

Figure 5.24 Smart antenna techniques.

that the subsectors are usually not handled as proper cells – they do not carry their own identification – but as parts of the corresponding cell. Hence, the soft handover is managed internally by the BS and the MS is not aware of this.

Forming an antenna diagram electronically in a full adaptive way, the beam may follow a moving MS to be served continuously. In this case, the BS equipment has to form the diagram for each MS individually. This technique promises to reduce interference not only by using narrow beams, but also by fading out the signal received from specific directions, namely, the directions of the strongest interferers. Since the signal and especially the interference level in UL and DL direction may differ significantly, it is questionable whether the additional gain of the full adaptive approach can be really achieved.

Considering the switched beams approach, measurements reported, for example, in (Mogensen *et al.* 1999) show the following: if the BS antenna is mounted above the rooftop level, by far the most signal energy is received within an angular interval of about 10–20° (depending on the antenna installation height) around the geometrical direction of the corresponding MS. Hence, dividing a cell into subsectors of about this angular width, either the best beam signal may be selected or the signals received via two beams may be combined in UL direction. In DL direction, the signal is transmitted using the same beams as selected for the UL.

To give a rough estimation of the potential of smart antenna techniques, one may say that reducing the beam width of the used antennas by a factor 2, reduces the interference power by a factor 2 and thereby increases the capacity by the same factor. More thorough investigations for applying smart antenna techniques for UMTS presented in (Monogioudis *et al.* 2004) confirm this argument. Using a base station site with six instead of three sectors increased the capacity by nearly 100%; applying even more sophisticated techniques, gain values of more than 200% were presented.

Comparing smart antennas with antenna diversity techniques, smart antennas require signals with a low angular spread and a coherent reception at all antenna elements, whereas antenna diversity techniques have their benefits in environments with a high angular spread resulting in uncorrelated fading values at the different antenna positions.

Because of the high potential of smart antennas, prerequisites for these techniques (as e.g. beam individual pilot channels) are foreseen in all modern CDMA systems.

More details on smart antenna techniques can be found, for example, in (Haardt and Alexiou 2004; Holma and Toskala 2001; Hottinen *et al.* 2004; Mogensen *et al.* 1999; Monogioudis *et al.* 2004).

Summary on interference handling in CDMA networks

In the following text, the discussion concerning the methods for handling interference in CDMA mobile radio networks is summarized.

A variety of profitable methods for achieving a performance gain and for simplifying network planning can be implemented within CDMA systems in a very natural way, namely,

- fast power control,
- soft handover,
- a cluster 1 network and
- soft capacity planning.

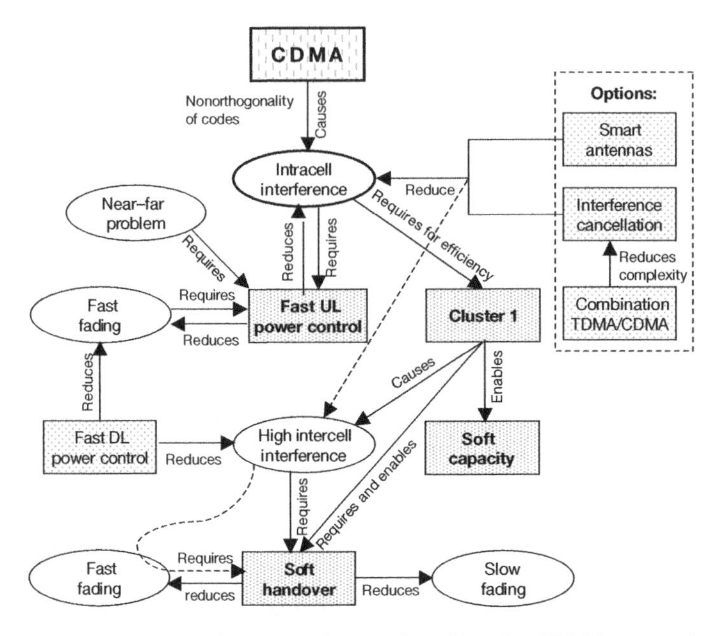

Figure 5.25 Overview: Interference handling in CDMA networks.

On the one hand, these methods may be viewed as a big advantage of CDMA networks, and on the other hand, it should be noted that these methods are required for CDMA mobile radio networks in any case to give an acceptable network performance. As shown in Figure 5.25, they are a direct or indirect consequence of intracell interference, that is, a consequence of the nonorthogonality of codes caused by nonsynchronized transmitters (mobile stations) and by multipath propagation.

While intercell interference and a widely varying received signal level due to the near–far effect and due to long and short-term fading have to be taken into account in any mobile radio network planning process, intracell interference represents a special challenge for CDMA networks.

The mentioned methods remove the undesired effects of intracell interference to an acceptable part and lead to some additional and significant benefits. Other methods like interference cancellation or smart antenna techniques are an option for CDMA systems for further increasing the network performance.

However, it should also be mentioned that all these methods require some additional effort in terms of signaling load, system complexity and hardware effort. Furthermore, their application is not restricted to only CDMA networks. To a certain degree, they may be – and in fact are – applied also in TDMA-based networks.

5.2 CDMA Transmission Channel Models

5.2.1 Representation of CDMA signals

For the theoretical analysis of the receiver structures and the performance of CDMA transmission, we need to introduce a suitable notation to describe the signals. As in the preceding chapters, we shall represent signals as vectors and look at them from a geometric point of view wherever this is possible.

In contrast to the signals investigated in the preceding chapters, we now have to deal with several users that share the same physical channel, that is, the same frequency band at the same time slot. We thus will have to introduce an additional index that numbers the user, and we have to deal with a signal in the air that is the superposition of the signals corresponding to the different users. The K users that share the same physical channel can be identified (and hopefully be separated at the receiver) because they use different complex baseband transmit pulses, $g_k(t)$ $(k = 1, \ldots, K)$, that are called *signature pulses* or *signature waveforms*. We normalize the pulses according to

$$\|g_k\|^2 = \int_{-\infty}^{\infty} |g_k(t)|^2 \, \mathrm{d}t = 1,$$

but, in general, we cannot assume that the signature pulses are orthogonal. We thus have a nonzero correlation coefficient

$$\rho_{ik} = \langle g_i, g_k \rangle = \int_{-\infty}^{\infty} g_i^*(t) g_k(t) \, \mathrm{d}t$$

between the signature pulses of two users indexed by i and k.

The special case where the signature pulses are orthogonal (i.e. where $\rho_{ik} = \delta_{ik}$ holds) seems to be desirable, but, as discussed above, there are often reasons given, which make nonorthogonal signature pulses a better choice. In Subsection 1.1.4, we discussed the orthogonal Walsh functions of length M as an example for a set of $K = M$ orthogonal signature pulses.

In that subsection, we have already introduced the notion of a *chip*. As done for the Walsh functions, we write any set of (in general, nonorthogonal) signature pulses as a superposition

$$g_k(t) = \sum_{i=1}^{N} \gamma_{ik} \psi_i(t) \tag{5.9}$$

of *chip pulses* $\psi_i(t)$. The chip pulses themselves are assumed to be an orthonormal base, that is, we assume that

$$\langle \psi_i, \psi_k \rangle = \int_{\infty}^{\infty} \psi_i^*(t) \psi_k(t) \, \mathrm{d}t = \delta_{ik}$$

holds. The coordinates

$$\gamma_{ik} = \langle \psi_i, g_k \rangle$$

of the user number k in Equation (5.9) may be grouped together to form the *signature vector*

$$\mathbf{g}_k = \begin{pmatrix} \gamma_{1k} \\ \vdots \\ \gamma_{Nk} \end{pmatrix}$$

of user number k. The signature vector characterizes the user. As mentioned above, PN sequences are often used as signature vectors. However, as we have seen for the example of the orthogonal Walsh functions, other choices are possible too. Typically, in DS-CDMA systems, the spreading sequence has a constant amplitude. For real valued sequences, due to the normalization of the signature pulses, we have

$$\gamma_{ik} = \pm\frac{1}{\sqrt{N}}.$$

However, complex γ_{ik} are possible as well. All the K signature vectors together are grouped to form a signature matrix

$$\mathbf{G} = \begin{pmatrix} \gamma_{11} & \cdots & \gamma_{1K} \\ \vdots & \ddots & \vdots \\ \gamma_{N1} & \cdots & \gamma_{NK} \end{pmatrix}.$$

For DS-CDMA, the chip pulses are a Nyquist base as defined in Subsection 1.1.3, that is,

$$\psi_i(t) = \psi(t - iT_c)$$

is the delayed version of a pulse $\psi(t)$ for which the convolved pulse $\psi^*(-t) * \psi(t)$ satisfies the well-known Nyquist condition for the *chip period* (or *chip duration*) T_c. The *symbol period* is given by $T_S = NT_c$. As we have discussed in that subsection, the Nyquist condition is just a special case of orthogonality. Figure 5.1 shows the situation for $N = 8$ and for the rectangular chip pulse

$$\psi(t) = \frac{1}{\sqrt{T_c}}\Pi\left(\frac{t}{T_c} - \frac{1}{2}\right).$$

For *multicarrier* (MC-) CDMA, the chip pulses are the base pulses of a multicarrier transmission scheme as discussed in Section 4.1. For a given time slot (i.e. one *OFDM symbol* in the terminology introduced there), the chip pulses are the frequency-shifted version of one base pulse $\psi(t)$ with Fourier transform $\Psi(f)$. In the frequency domain, the chip pulses are given by

$$\Psi_i(f) = \Psi(f - f_i).$$

The chip pulses in the time domain are then given by

$$\psi_i(t) = e^{j2\pi f_i t}\psi(t).$$

The *subcarrier frequencies* f_i are normally assumed to be equally spaced. OFDM with guard interval is the most important multicarrier scheme. As discussed in Subsection 4.1.4, the symbols of length $T_S = T + \Delta$ consist of a Fourier analysis window of length T and a guard interval of length Δ (see Figure 4.8). The subcarrier frequencies are given by

$$f_i = \frac{i}{T}.$$

We adopt the formalism from that subsection and define chip pulses by

$$\psi_i(t) = \frac{1}{\sqrt{T}}\exp(j2\pi f_i t)\,\Pi\left(\frac{t}{T} - \frac{1}{2}\right)$$

and the chip pulses with guard interval by

$$\psi_i'(t) = \frac{1}{\sqrt{T_S}} \exp(j2\pi f_i t)\, \Pi\left(\frac{t+\Delta}{T_S} - \frac{1}{2}\right).$$

The transmitter uses pulses with guard interval, whereas the receiver uses pulses without guard interval (see Figure 4.9). We have the orthogonality condition

$$\langle \psi_i, \psi_k' \rangle = \sqrt{\frac{T}{T_S}}\delta_{ik},$$

that is, there is an energy loss given by T/T_S, which is caused by the guard interval. For MC-CDMA (multicarrier CDMA) with OFDM and guard interval, we may write

$$g_k'(t) = \sum_{i=1}^{N} \gamma_{ik}\psi_i'(t) \tag{5.10}$$

instead of Equation (5.9). However, to keep the notation unified and simple, we will keep in mind the guard interval but we will not take it into account in the notation.

5.2.2 The discrete channel model for synchronous transmission in a frequency-flat channel

Until now, we only discuss K users that share one time slot. This is sufficient if we can assume ideal synchronous transmission for all users and we can neglect the time dispersion of the channel, that is, the channel can be assumed to be frequency flat over the signaling bandwidth. This is of course a very ideal situation, but it is the simplest to analyze and illustrates the basic properties of the most important receiver structures. We further assume that the channel is approximately constant during the transmission of the signature pulses during one time slot.

We note that we must exclude *wideband* DS-CDMA because, by definition, the corresponding channel is time dispersive. For the same reason, we must also exclude *wideband* MC-CDMA because the channel is frequency selective and different subcarriers are affected by different fading amplitudes.

With the assumptions made above, we may work with a channel model in which the receive signal is given by

$$r(t) = \sum_{k=1}^{K} c_k s_k g_k(t) + n(t). \tag{5.11}$$

Each user is affected by its own transmit channel represented by c_k as the complex fading amplitude corresponding to the channel of user number k. As done in the preceding chapters, the complex transmit symbols are denoted by s_k and $n(t)$ is the complex baseband AWGN with PSD N_0. For BPSK modulation, which is utilized in many CDMA systems, we have $s_k = \pm\sqrt{E_b}$, where E_b is the energy per bit.

To recover all available information at the receiver end, a base of detectors is necessary to guarantee sufficient statistics (see Subsection 1.4.1). Applying these results, we note that the *transmit space*, that is, vector space spanned by the transmit base $\{g_k(t)\}_{k=1}^{K}$, must be a subspace of the *receive space*, that is, the vector space spanned by the detector pulses. Obviously, the transmit base itself is a possible detector base. In that case, the detector outputs are sampled outputs of matched filters or correlators and we may speak of a *matched filter base* receiver. However, in general, the transmit base is not orthogonal. Sometimes it is convenient to choose an orthogonal base of detector pulses. In that case, we speak of the *orthogonal detector base* receiver.

A discrete model for the matched filter base

We first discuss the matched filter base, which uses the (nonorthogonal) transmit base $\{g_k(t)\}_{k=1}^{K}$ as detectors. The detector output for the kth user is given by

$$v_k = \mathcal{D}_{g_k}[r] = \int_{-\infty}^{\infty} g_k^*(t) r(t)\, \mathrm{d}t.$$

We may write

$$v_i = \sum_{k=1}^{K} c_k s_k \rho_{ik} + m_i,$$

where

$$m_i = \mathcal{D}_{g_i}[n] = \int_{-\infty}^{\infty} g_i^*(t) n(t)\, \mathrm{d}t$$

is correlated Gaussian noise with covariance given by

$$\mathrm{E}\left\{m_i m_k^*\right\} = N_0 \rho_{ik}.$$

We may rewrite the discrete transmission channel in vector notation as

$$\mathbf{v} = \mathbf{RCs} + \mathbf{m}, \tag{5.12}$$

where

$$\mathbf{R} = \begin{pmatrix} \rho_{11} & \cdots & \rho_{1K} \\ \vdots & \ddots & \vdots \\ \rho_{K1} & \cdots & \rho_{KK} \end{pmatrix}$$

is the correlation matrix of the signature vectors, and $\mathbf{C} = \mathrm{diag}\,(c_1, \ldots, c_K)$ is the channel matrix. $\mathbf{m} = (m_1, \ldots, m_N)^T$ represents the vector of noise samples.

The problem in using the discrete matched filter model for the theoretical performance analysis is that we have to deal with correlated noise $\mathbf{m}$ and thus the results obtained for AWGN cannot be applied. The same problem occurs in the analysis of equalizer structures where the channel matched filter introduces correlations into the noise. The solution in equalizer theory is the *whitening matched filter* that decorrelates the noise. The same methods may be applied here. In our model, noise whitening means that we have to multiply Equation (5.12) by an appropriate matrix so that the noise becomes white. However, this can be avoided by using an orthogonal detection base for the theoretical analysis, no matter what kind of base will be used in practice.

A discrete model for the orthogonal detector base

As a base of orthogonal detectors we may use the chip base $\{\psi_i(t)\}_{i=1}^N$. This is a very natural base, and it is close to implementation. Because of Equation (5.9), the matched filter base outputs can be obtained from the chip base outputs as described below. Other orthogonal bases than the chip base are possible. For instance, the sinc base with an appropriate sampling frequency is a possible choice for band-limited signals. Alternatively, one can always obtain an orthogonal base by applying the Gram–Schmidt algorithm to the base of signature pulses $g_k(t)$. If we have chosen an appropriate orthogonal detector base with sufficient statistics, we can express everything else in terms of this base. We thus emphasize that the concept of an orthogonal detector base is a very useful tool for the theoretical analysis even if another base may be implemented in practice.

We assume an orthogonal base of detector pulses $\{\psi_i(t)\}_{i=1}^N$ that provides a set of sufficient statistics. To avoid the introduction of new symbols, we use the same notation as for the chip base. The following treatment, however, applies to any orthonormal base. The set of sufficient statistics for the receive signal $r(t)$ is given by the detector outputs

$$r_i = \mathcal{D}_i[r] = \int_{-\infty}^{\infty} \psi_i^*(t) r(t)\, \mathrm{d}t$$

for all $i = 1, \ldots, N$. Using Equations (5.9) and (5.11), we may write

$$r_i = \sum_{k=1}^{K} c_k s_k \gamma_{ik} + n_i$$

with

$$\gamma_{ik} = \mathcal{D}_i[g_k] = \langle \psi_i, g_k \rangle = \int_{-\infty}^{\infty} \psi_i^*(t) g_k(t)\, \mathrm{d}t$$

and

$$n_i = \mathcal{D}_i[n] = \int_{-\infty}^{\infty} \psi_i^*(t) n(t)\, \mathrm{d}t.$$

We now rewrite the discrete transmission channel in vector notation as

$$\mathbf{r} = \mathbf{GCs} + \mathbf{n} = \sum_{k=1}^{K} c_k s_k \mathbf{g}_k + \mathbf{n}, \tag{5.13}$$

where $\mathbf{G} = (\mathbf{g}_1, \ldots, \mathbf{g}_K)$ is the matrix of signature vectors, $\mathbf{C} = \mathrm{diag}(c_1, \ldots, c_K)$ is the diagonal matrix of complex fading amplitudes and $\mathbf{s} = (s_1, \ldots, s_K)^T$ and $\mathbf{n} = (n_1, \ldots, n_N)^T$ are the vectors of transmit symbols and noise samples, respectively.

We note that the matched filter detector outputs $v_k = \mathcal{D}_{g_k}[r]$ can easily be expressed by the orthogonal detector outputs $r_i = \mathcal{D}_i[r]$. By using

$$g_k(t) = \sum_{i=1}^{N} \gamma_{ik} \psi_i(t) \tag{5.14}$$

with $\gamma_{ik} = \langle \psi_i, g_k \rangle$, we find that

$$\mathcal{D}_{g_k}[r] = \sum_{i=1}^{N} \gamma_{ik}^* \mathcal{D}_i[r]$$

and thus

$$v_k = \mathbf{g}_k^{\dagger}\mathbf{r}$$

or, in matrix notation,

$$\mathbf{v} = \mathbf{G}^{\dagger}\mathbf{r}. \tag{5.15}$$

Similarly, we obtain

$$\mathbf{m} = \mathbf{G}^{\dagger}\mathbf{n}. \tag{5.16}$$

Furthermore, we insert Equation (5.14) into

$$\rho_{ik} = \langle g_i, g_k \rangle$$

and obtain

$$\rho_{ik} = \sum_{i=1}^{N} \gamma_{ji}^{*}\gamma_{jk}$$

or, in matrix notation,

$$\mathbf{R} = \mathbf{G}^{\dagger}\mathbf{G}. \tag{5.17}$$

Equation (5.12) can be obtained from Equation (5.13) by multiplication with $\mathbf{G}^{\dagger}$ from the left. Thus, there is a straightforward connection between the matched filter base and the orthogonal detector representation.

In case we take the chip base as $\{\psi_i(t)\}_{i=1}^{N}$, the above vector-algebraic treatment is also of practical use. The above equations say that, for the matched filter receiver, we do not need to implement a (analog) matched filter for each signature pulse. We only need an (analog) filter that is matched to the chip pulse $\psi(t)$ and sample it with the chip clock period T_c to obtain the values for $r_i = \mathcal{D}_i[r]$. We may then obtain the K matched filter detector outputs $v_k = \mathcal{D}_{g_k}[r]$ by digital signal processing according to Equation (5.15).

For the theoretical analysis of the receivers, we may choose any orthonormal base $\{\psi_i(t)\}_{i=1}^{N}$ of sufficient statistics. Especially, it may be convenient to choose the base in such a way that

$$\psi_1(t) = g_1(t)$$

holds. For the geometrical visualization for two users, we then have a simple picture in the two-dimensional plane (see the following text).

5.2.3 The discrete channel model for synchronous wideband MC-CDMA transmission

In the preceding subsection, we assumed synchronous transmission and that the channel is the same for every chip. This is not the case for wideband CDMA (WCDMA). For wideband MC-CDMA, we may derive a synchronous channel model that is very similar to the one derived in the preceding subsection. We note that user synchronization even in the uplink is not such a severe problem for MC-CDMA. The time dispersion of the channel is typically relatively small compared to the symbol duration T_S. For OFDM, it will be absorbed by the guard interval. The synchronization inaccuracy is allowed to be in the order of the time dispersion of the channel, which typically is in the order of microseconds. Thus, the synchronization requirements are not more severe than for GSM. We assume that the synchronization inaccuracy and the time dispersion are small (or absorbed by the guard interval) so that we can write the received signal affected by multiplicative fading as

$$r(t) = \sum_{i=1}^{N} \sum_{k=1}^{K} c_{ik} \gamma_{ik} s_k \psi_i(t) + n(t). \tag{5.18}$$

The complex fading amplitude for user k at frequency index i is given by c_{ik}. The orthogonal base $\{\psi_i(t)\}_{i=1}^{N}$ is the chip base, that is, the base corresponding to the subcarrier pulses. As mentioned above, for simplicity, we do not take into account the guard interval in the notation. In that case, we will have to use $\left\{\psi_i'(t)\right\}_{i=1}^{N}$ at the transmitter and $\{\psi_i(t)\}_{i=1}^{N}$ at the receiver. The outputs corresponding to the orthogonal detector base $\{\psi_i(t)\}_{i=1}^{N}$ are given by the discrete transmission channel

$$r_i = \sum_{k=1}^{K} c_{ik} \gamma_{ik} s_k + n_i,$$

where n_i is the complex discrete AWGN. For OFDM, r_i is just the Fourier coefficient at frequency number i that has been obtained by the FFT analysis. We define the matrix

$$\mathbf{H} = \begin{pmatrix} \gamma_{11} c_{11} & \cdots & \gamma_{1K} c_{1K} \\ \vdots & \ddots & \vdots \\ \gamma_{N1} c_{N1} & \cdots & \gamma_{NK} c_{NK} \end{pmatrix},$$

which is formally the Hadamard product (i.e. elementwise product)

$$\mathbf{H} = \mathbf{G} \circ \mathbf{C}$$

between the signature matrix $\mathbf{G}$ and the matrix $\mathbf{C}$ of fading amplitudes.

We now rewrite the discrete transmission channel in vector notation as

$$\mathbf{r} = \mathbf{H}\mathbf{s} + \mathbf{n}, \tag{5.19}$$

where $\mathbf{s} = (s_1, \ldots, s_K)^T$ and $\mathbf{n} = (n_1, \ldots, n_N)^T$ are the vectors of transmit symbols and noise samples, respectively. This has formally the same structure as the vector channel given by Equation (5.13) with the signature matrix $\mathbf{G}$ replaced by $\mathbf{H}$ and the matrix $\mathbf{C}$ of fading amplitudes formally replaced by the identity matrix $\mathbf{I}$. We note that in this vector channel model, the fading amplitudes are now already included in the matrix $\mathbf{H}$. The receiver structures for this channel model can thus be derived quite like those that are derived for the synchronous CDMA vector channel in the next section. The main difference is that the column vectors $\mathbf{g}_k$ of the matrix

$$\mathbf{G} = \left[\mathbf{g}_1, \ldots, \mathbf{g}_K\right]$$

are normalized to length 1, while the column vectors $\mathbf{h}_k$ of the matrix

$$\mathbf{H} = [\mathbf{h}_1, \ldots, \mathbf{h}_K]$$

are not.

5.2.4 The discrete channel model for asynchronous wideband CDMA transmission

We now consider the general case of K asynchronous users with signals to be transmitted over K independent time-dispersive (i.e. frequency-selective) fading channels. This covers the case of wideband DS-CDMA. For simplicity, we still ignore the time variance of the channel. Let $c_k(\tau)$ be the (time-invariant) impulse response of the channel for user number k and $g_k(t)$ his signature waveform.

$$h_k(t) = c_k(t) * g_k(t)$$

is then the signature waveform for user number k as it is seen at the receiver in a noise-free and MAI-free channel. Note that the asynchronism is covered by the different channel impulse responses. Let s_k be the transmit symbol for user number k. For time slot zero, the transmit signal is then given by

$$\sum_{k=1}^{K} s_k h_k(t),$$

where s_k is the transmit symbol for user number k. Let l be the time index for the continuous data stream, that is, user number k transmits the symbol s_{kl} by

$$s_{kl} g_k(t - lT_S)$$

at time slot number l. The noisy signal at the receiver is given by

$$r(t) = \sum_{k=1}^{K} \sum_{l=0}^{L-1} s_{kl} h_k(t - lT_S) + n(t). \tag{5.20}$$

Here L is the number of time slots to be transmitted, for example, during one frame. It is not really a restriction to assume that L is finite because, even in the absence of a frame structure, any transmission will start and stop at some time instant. We note that Equation (5.20) stands for a very general setup with several asynchronous users transmitting over a time-dispersive channel. We also note that two special cases are covered by this model:

1. Synchronous nondispersive transmission as discussed above. In that case, $h_k(t) = c_k g_k(t)$ holds, and the only term in the second sum that needs to be considered is that for $l = 0$ and we have to sum only over the user index k, that is,

$$r(t) = \sum_{k=1}^{K} s_k h_k(t) + n(t), \tag{5.21}$$

   which is the same as Equation (5.11).

2. Single-user transmission over a time-dispersive channel. In this case, the only term in the first sum is that for $k = 1$ and we may drop the index k. The pulse shape of the whole channel (without noise) is $h(t) = c(t) * g(t)$, where $c(t)$ is the impulse response of the channel and $g(t)$ is the transmit pulse shape. We may assume that $g^*(-t) * g(t)$ fulfills the Nyquist condition but this is not necessary, especially

because the Nyquist condition is violated when the channel impulse response is included. We write $h_l(t) = h(t - lT_S)$ and get the channel model

$$r(t) = \sum_{l=0}^{L-1} s_l h_l(t) + n(t). \tag{5.22}$$

Receivers for such a transmission setup are known as *equalizers*. Equalizer structures are treated exhaustively in the literature (see e.g. (Benedetto and Biglieri 1999; Kammeyer 2004; Proakis 2001)). There are optimal equalizers like the MLSE receiver, suboptimal linear equalizers like the *zero-forcing equalizer* (ZFE) and the *minimum mean square error* (MMSE) equalizer, and suboptimal nonlinear equalizers like the *decision feedback equalizer* (DFE).

We note that Equation (5.21) and (5.22) are formally the same. The only difference is the interpretation of the index, which is a user index in the first case and a time index in the second case. Thus, it will not be surprising that the well-known equalizer structures find their correspondence in the theory of multiuser detection.

To avoid summation over two indices (which makes the treatment quite cumbersome), we introduce the multiindex $j = (k, l)$. We may as well regard j as a one-dimensional index that we obtain by writing the double-indexed quantities s_{kl} and $h_k(t - lT_S)$ in a matrix and read them out column-wise. Then $j \in \{1, 2, \ldots, KL\}$ will be the index of the vector obtained that way, that is, write

$$j = lK + k$$

and

$$s_j = s_{kl}, \quad h_j = h_k(t - lT_S).$$

We may now rewrite Equation (5.20) as

$$r(t) = \sum_{j=1}^{J} s_j h_j(t) + n(t) \tag{5.23}$$

with

$$J = KL.$$

We may introduce a discrete model by utilizing detector outputs of sufficient statistics for that signal. Again, we may alternatively use a matched filter base or an orthonormal base. As shown above, the detector outputs for the matched filter base can be obtained from those of an orthonormal base. We thus start with the orthonormal base.

Let

$$\{\phi_i(t)\}_{i=1}^{N}$$

be an orthonormal base for the vector space spanned by the nonorthogonal base of the pulses $h_j(t)$. We note that, because of the time dispersion of the channel, this vector space will in general be larger than the vector space spanned by the vectors $g_k(t - lT_S)$. For a complex baseband signal with a band-limited spectrum between $-B/2$ and $B/2$, we may choose the sinc base

$$\phi_i(t) = B \operatorname{sinc}(Bt - i)$$

as defined in Subsection 1.1.2. We define the detector outputs of the orthogonal base as

$$r_i = \mathcal{D}_{\phi_i}[r]$$

and obtain

$$r_i = \sum_{j=1}^{J} h_{ij} s_j + n_j,$$

where

$$h_{ij} = \langle \phi_i, h_j \rangle$$

defines the coordinate matrix $\mathbf{H}$ of the pulse vectors $h_j(t)$ corresponding to the base $\{\phi_i(t)\}_{i=1}^{N}$. We may write this in compact matrix notation as

$$\mathbf{r} = \mathbf{Hs} + \mathbf{n}. \tag{5.24}$$

Obviously, this is formally the same structure as obtained for the synchronous MC-CDMA model in the preceding subsection, but the interpretation is different. However, this formal similarity simplifies the derivation of receiver structures.

The representation in the matched filter base will be obtained as the output corresponding to the detector base $\left\{h_j(t)\right\}_{j=1}^{J}$ of the signal $r(t)$ in Equation (5.23). Note that $h_j(t)$ corresponds to the column vector $\mathbf{h}_j$ in the matrix

$$\mathbf{H} = [\mathbf{h}_1, \ldots, \mathbf{h}_J].$$

Thus, in the discrete model, this detection by $h_j(t)$ corresponds to multiplying Equation (5.24) by $\mathbf{h}^{\dagger}$ from the left, that is, the detector output corresponding to $h_j(t)$ is given by $v_j = \mathbf{h}_j^{\dagger}\mathbf{r}$. Thus, the vector of matched filter outputs $\mathbf{v} = (v_1, \ldots, v_J)^T$ is given by

$$\mathbf{v} = \mathbf{Rs} + \mathbf{m},$$

where

$$\mathbf{R} = \mathbf{H}^{\dagger}\mathbf{H}$$

is the autocorrelation matrix with components

$$(\mathbf{R})_{ij} = \langle h_i, h_j \rangle = \mathbf{h}_i^{\dagger}\mathbf{h}_j,$$

and $\mathbf{m}$ is the complex colored Gaussian noise with autocorrelation matrix given by

$$\mathrm{E}\left\{\mathbf{mm}^{\dagger}\right\} = N_0\mathbf{R}.$$

The autocorrelation matrix has the following block structure:

$$\mathbf{R} = \begin{pmatrix} \mathbf{R}[0] & \mathbf{R}^{\dagger}[1] & \mathbf{R}^{\dagger}[2] & \cdots & & \\ \mathbf{R}[1] & \mathbf{R}[0] & \mathbf{R}^{\dagger}[1] & \mathbf{R}^{\dagger}[2] & \cdots & \\ \mathbf{R}[2] & \mathbf{R}[1] & \mathbf{R}[0] & \mathbf{R}^{\dagger}[1] & \mathbf{R}^{\dagger}[2] & \cdots \\ \vdots & \mathbf{R}[2] & \mathbf{R}[1] & \mathbf{R}[0] & \mathbf{R}^{\dagger}[1] & \ddots \\ & \vdots & \mathbf{R}[2] & \mathbf{R}[1] & \mathbf{R}[0] & \ddots \\ & & \vdots & \ddots & \ddots & \ddots \end{pmatrix}, \tag{5.25}$$

with $K \times K$ submatrices $\mathbf{R}[l] = \mathbf{R}^{\dagger}[-l]$ that have elements $\rho_{ik}[l] = (\mathbf{R}[l])_{ik}$ given by

$$\rho_{ik}[l] = \int h_i^*(t - lT_S)h_k(t)\,\mathrm{d}t.$$

We have seen that the receiver structures for all the three models can be treated using the same formalism. The channel is always given by a matrix multiplication applied to the vector $\mathbf{s}$ of the transmit symbols. However, the interpretation of that matrix $\mathbf{GC}$ or $\mathbf{H}$ or $\mathbf{R}$ is quite different. Therefore, we prefer to split up the analysis of the receiver structures for the different models. First, because it is the simplest, we analyze the receiver structures for the synchronous model. For the other models, the derivation of receiver structures is formally the same, even though the interpretation is different.

5.3 Receiver Structures for Synchronous Transmission

The simplest receiver for CDMA systems is one that is matched to the user under consideration, regarding the other users as additive noise-like interference. This is called a single-user matched filter (SUMF) receiver. It has no other detectors except for that single user. This is the traditional CDMA receiver. The common argument why one can proceed this way is that the spreading sequences have good correlation properties. As a consequence, the signal of all other users can be regarded as approximately orthogonal to the signal under consideration and may thus be ignored. The residual interference term that is due to nonorthogonality is assumed to be small. If the number of the other users is high enough, due to the central limit theorem, this interference has Gaussian statistics and may be treated as Gaussian noise.

It is evident that the SUMF receiver cannot be optimal. Because the signals are not exactly orthogonal, any information about other users may improve the performance. Moreover, severe problems occur in the uplink if the receive power of the (distant) user under consideration is significantly below the power of another (nearby) user. This is known as the *near–far problem* that we discussed in Subsection 5.1.4.

In contrast to the SUMF receiver, a *multiuser detection* (MUD) receiver has detectors for (some or all) other users. It takes into account information about their signature waveforms and about the corresponding fading coefficients. In the preceding chapters, several general receiver structures have been derived for optimal bit decisions (the bitwise MAP receiver, see Subsection 3.1.5) and for optimal decisions about sequences (the MLSE receiver, see Subsection 1.4.2 and Subsection 2.4.1). These results can be applied to CDMA systems. However, the optimal receiver structures are often too complex to be applied in practice. Therefore, suboptimal (linear) MUD receivers as the *decorrelating receiver* and the *minimum mean square error* (MMSE) *receiver* are of high practical relevance.

In the following subsections, we will discuss and compare these various CDMA receiver structures. To make the ideas more clear, we start with the synchronous channel model. Moreover, we will explain the ideas in the example with a two-user scenario. This is because it is possible to visualize this case by using vectors in a two-dimensional plane, which makes the understanding much easier. The mathematical generalization to more users is then often straightforward.

5.3.1 The single-user matched filter receiver

We consider K users in a fading channel and a receiver that is matched to one user, that is, we treat the optimum receiver for a single user. The receiver is matched to the signature waveform (transmit pulse) $g_1(t)$ of this user number one and makes use of the corresponding complex fading amplitude c_1. The transmit pulses $g_k(t)$ and fading amplitudes c_k of the other users ($k > 1$) are unknown to the receiver. The receive signal in the synchronous channel model is given by Equation (5.11). We split up the receive signal into three terms

$$r(t) = c_1 s_1 g_1(t) + \sum_{k=2}^{K} c_k s_k g_k(t) + n(t),$$

where the first term represents the desired signal component, the second one comprises $K - 1$ interfering users, and the third one describes AWGN.

The SUMF detector output for user number one,

$$r_1 = \mathcal{D}_{g_1}[r] = \int_{-\infty}^{\infty} g_1^*(t) r(t)\,\mathrm{d}t,$$

is given by

$$r_1 = c_1 s_1 + \sum_{k=2}^{K} c_k s_k \rho_{1k} + n_1 \tag{5.26}$$

with $\rho_{1k} = \langle g_1, g_k \rangle$. The sum in this expression is the *multiple access interference* (MAI) caused by the other users. The noise term

$$n_1 = \mathcal{D}_{g_1}[n] = \int_{-\infty}^{\infty} g_1^*(t) n(t)\,\mathrm{d}t$$

is a complex Gaussian random variable with zero mean and variance

$$\mathrm{E}\left\{|n_1|^2\right\} = N_0.$$

For the geometric visualization, it is helpful to look at the situation in a finite-dimensional vector space and use the discrete channel model given by Equation (5.13). We choose the corresponding orthonormal base $\{\psi_i(t)\}_{i=1}^{N}$ in such a way that $\psi_1(t) = g_1(t)$ to get the notation consistent with the above formulas. We rewrite Equation (5.13) as

$$\mathbf{r} = c_1 s_1 \mathbf{g}_1 + \sum_{k=2}^{K} c_k s_k \mathbf{g}_k + \mathbf{n}.$$

We note that we get back to Equation (5.26) if we multiply this from the left by $\mathbf{g}_1^\dagger$. We further note that

$$\mathbf{g}_1 = \begin{pmatrix} 1 \\ 0 \\ \vdots \\ 0 \end{pmatrix}.$$

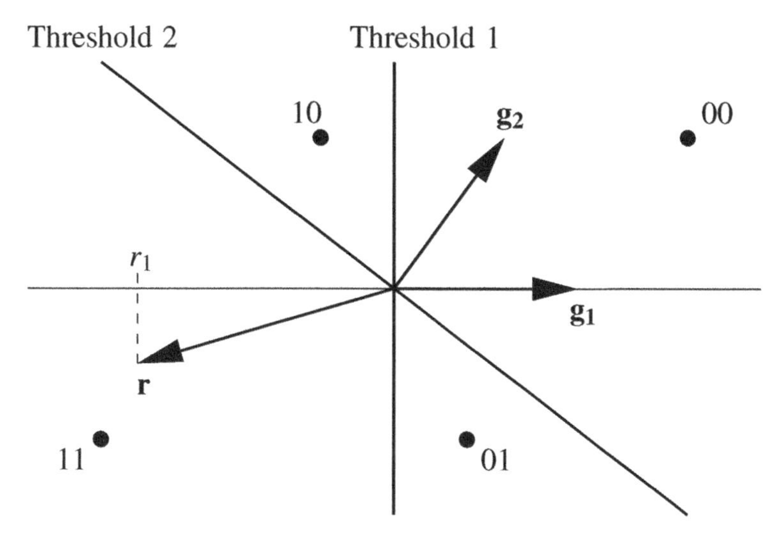

Figure 5.26 Two-user BPSK SUMF receiver for equal fading amplitudes and $\rho_{12} = 0.6$.

One can visualize the scenario in the two-dimensional plane for $K = 2$ and $N = 2$, real positive fading amplitudes and BPSK transmission. Figure 5.26 shows the situation for

$$\mathbf{g}_1 = \begin{pmatrix} 1 \\ 0 \end{pmatrix}, \quad \mathbf{g}_2 = \frac{1}{5}\begin{pmatrix} 3 \\ 4 \end{pmatrix}$$

corresponding to $\rho_{12} = 0.6$ and equal fading amplitudes $c_1 = c_2 = 1$. The four possible transmit signal vectors are given by

$$\mathbf{s}_{00} = +\mathbf{g}_1 + \mathbf{g}_2,$$

$$\mathbf{s}_{01} = +\mathbf{g}_1 - \mathbf{g}_2,$$

$$\mathbf{s}_{10} = -\mathbf{g}_1 + \mathbf{g}_2,$$

and

$$\mathbf{s}_{11} = -\mathbf{g}_1 - \mathbf{g}_2.$$

Here $\mathbf{s}_{b_1 b_2}$ is the transmit signal if user 1 transmits the bit b_1 and user 2 transmits the bit b_2. For simplicity, we have assumed $E_S = 1$ in this constellation example. The figure looks like a skew QPSK constellation with the square replaced by a parallelogram. The two-dimensional receive vector is given by

$$\mathbf{r} = s_1 \mathbf{g}_1 + s_2 \mathbf{g}_2 + \mathbf{n},$$

where $\mathbf{n}$ is the two-dimensional real AWGN. The SUMF output is given by the scalar product

$$r_1 = \mathbf{g}_1 \cdot \mathbf{r}$$

and can be interpreted as the orthogonal projection of the vector $\mathbf{r}$ on the vector $\mathbf{g}_1$. The receiver makes a decision based on the sign of that projection, that is, there is a decision threshold for user 1 which is orthogonal to $\mathbf{g}_1$ that separates the left and the right half plane. The decision threshold for user 2 is also drawn in the figure.

To illustrate the near–far problem, we consider the same signature vectors $\mathbf{g}_1$ and $\mathbf{g}_2$, but we assume that the fading amplitudes are given by $c_1 = 1$ and $c_2 = 2$. The receive vector is now given by

$$\mathbf{r} = s_1\mathbf{g}_1 + 2s_2\mathbf{g}_2 + \mathbf{n}.$$

This situation is depicted in Figure 5.27. It is obvious from this figure that the bit value of user 2 completely dominates the bit decision for user 1. Only if both bits have the same value, the SUMF for user 1 will make a correct decision in a noise-free channel. Thus, for equal probable bits in the noise-free channel, the bit error rate will equal the constant value 1/2.

From this figure, one easily concludes that correct SUMF decisions for user 1 are only possible if the projection of $c_2\mathbf{g}_2$ on $\mathbf{g}_1$ is smaller than the length of $c_1\mathbf{g}_1$. Here, we have assumed that everything is real. For the general case we note that the BPSK decision for user 1 is based on the sign of

$$\Re\left\{c_1^* r_1\right\} = \Re\left\{|c_1|^2 s_1 + c_1^* c_2 \rho_{12} s_2 + c_1^* n_1\right\}.$$

From this equation, we conclude that correct SUMF decisions for user 1 are only possible if

$$\left|\Re\left\{c_1^* c_2 \rho_{12}\right\}\right| < |c_1|^2 \tag{5.27}$$

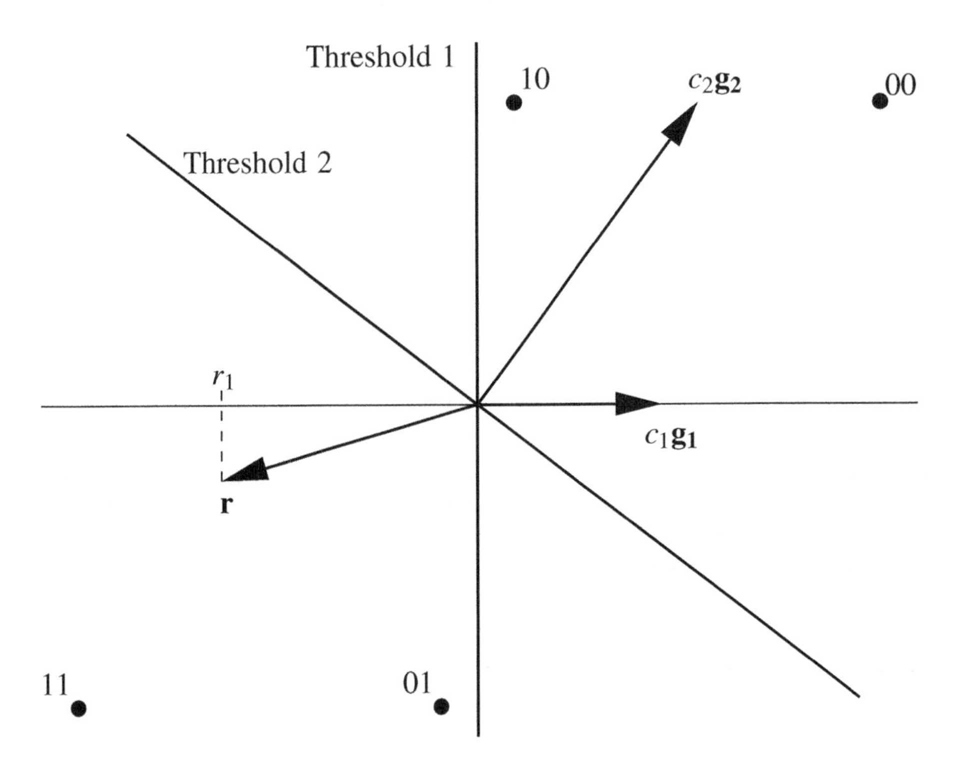

Figure 5.27 Two-user BPSK SUMF receiver for unequal fading amplitudes and $\rho_{12} = 0.6$.

holds. For the general case of K users, we have the condition

$$\sum_{k=2}^{K} \left| \Re\left\{c_1^* c_k \rho_{1k}\right\}\right| < |c_1|^2 . \tag{5.28}$$

Gaussian channel performance

We consider again two users and BPSK transmission. We assume that the condition of Equation (5.27) for possible correct SUMF receiver decisions is fulfilled. We assume fixed fading amplitudes so that we have an AWGN channel. From the geometrical constellation of Figure 5.26, we easily conclude how the bit error probability for user 1 can be obtained. We have to calculate the probability that user 1 has transmitted the bit $b_1 = 0$ but the receiver decides for $b_1 = 1$. In Figure 5.26, this is the case if one of the constellation points in the right half plane is transmitted but the receive signal is in the left half plane. We denote the distance of the point corresponding to the bit pair 00 to the threshold by Δ_0 and the distance corresponding to 01 by Δ_1. Assuming that both points are equal probables, the bit error probability for user 1 is then given by

$$P_b = \frac{1}{2}\left(\mathrm{Q}\left(\frac{\Delta_0}{\sigma}\right) + \mathrm{Q}\left(\frac{\Delta_1}{\sigma}\right)\right)$$

with $\sigma^2 = N_0/2$. A simple upper bound is given by

$$P_b \le \mathrm{Q}\left(\frac{\Delta}{\sigma}\right) \tag{5.29}$$

with $\Delta = \min(\Delta_0, \Delta_1)$. Using geometrical arguments one can easily check that distances are given by

$$\Delta_{0,1} = \sqrt{E_b}\left(|c_1| \pm \frac{1}{|c_1|}\Re\left\{c_1^* c_2 \rho_{12}\right\}\right)$$

(see Problem 1). Thus we have

$$\Delta = \sqrt{E_b}\left(|c_1| - \frac{1}{|c_1|}\left|\Re\left\{c_1^* c_2 \rho_{12}\right\}\right|\right)$$

and

$$P_b \le \frac{1}{2}\mathrm{erfc}\left(\sqrt{\left(|c_1| - \frac{1}{|c_1|}\left|\Re\left\{c_1^* c_2 \rho_{12}\right\}\right|\right)^2 \frac{E_b}{N_0}}\right). \tag{5.30}$$

For the general case with K users, Equation (5.29) holds with

$$\Delta = \sqrt{E_b}\left(|c_1| - \frac{1}{|c_1|}\sum_{k=2}^{K}\left|\Re\left\{c_1^* c_k \rho_{1k}\right\}\right|\right),$$

that is,

$$P_b \le \frac{1}{2}\mathrm{erfc}\left(\sqrt{\left(|c_1| - \frac{1}{|c_1|}\sum_{k=2}^{K}\left|\Re\left\{c_1^* c_k \rho_{1k}\right\}\right|\right)^2 \frac{E_b}{N_0}}\right). \tag{5.31}$$

Rayleigh channel performance

We consider a SUMF receiver and K users in a Rayleigh fading channel with BPSK transmission. The complex baseband receive signal is given by

$$r(t) = \sum_{k=1}^{K} c_k s_k g_k(t) + n(t).$$

In that equation, $s_k = \pm\sqrt{E_b}$ denotes the BPSK symbols, and $n(t)$ is the complex baseband AWGN with power density N_0. In a Rayleigh fading channel, the fading amplitudes c_k are complex Gaussian random variables with variances

$$\mathrm{E}\left\{|c_k|^2\right\} = \alpha_k^2$$

and statistically independent real and imaginary parts (see Subsection 2.2.5). The power transfer of the user under consideration is normalized to $\alpha_1^2 = 1$, and the fading of different users is assumed to be independent. The SUMF detector output according to Equation (5.26) is given by

$$r_1 = c_1 s_1 + \sum_{k=2}^{K} c_k s_k \rho_{1k} + n_1.$$

The first term corresponds to the useful signal, the second term is the multiple access interference (MAI). Because each c_k is a Gaussian random variable, the MAI term has Gaussian statistics, and it can be treated as an additional noise term. We therefore expect that the bit error rates are the same as in a Rayleigh fading channel, but with a noise enhancement that is caused by MAI. In the following discussion, we will formally substantiate this heuristic argument.

An independent multiplicative random sign does not change the statistics of a Gaussian random variable. Thus, each of the $K-1$ MUI terms $c_k s_k \rho_{1k}$ $(k > 1)$ is a complex mean zero Gaussian random variable with variance $\sigma_k^2 = \alpha_k^2 |\rho_{1k}|^2 E_b$. We may thus rewrite Equation (5.26) as

$$r_1 = c_1 s_1 + n_c, \tag{5.32}$$

where n_c is the complex one-dimensional AWGN with variance

$$\sigma^2 = N_0 + E_b \sum_{k=2}^{K} \alpha_k^2 |\rho_{1k}|^2 .$$

Equation (5.32) describes BPSK in a Rayleigh fading channel with enhanced noise. The bit error rate has been derived in Subsection 2.4.3. In that subsection, we have obtained the formula

$$P_b = \frac{1}{2}\left[1 - \sqrt{\frac{\frac{E_b}{N_0}}{1 + \frac{E_b}{N_0}}}\right]$$

for the bit error rate. We replace N_0 by σ^2 and get the expression

$$P_b = \frac{1}{2}\left[1 - \sqrt{\frac{E_b}{E_b + N_0 + E_b \sum_{k=2}^{K} \alpha_k^2 |\rho_{1k}|^2}}\right].$$

Obviously, the MAI disturbs the signal like an additional Gaussian noise source. Thus, the system performance will be severely degraded unless the MAI variance σ^2 is significantly below E_b. An interferer close to the receiver will thus severely disturb the reception of a distant user, which is known as the *near–far problem*.

We finally note that the above treatment can also cover the case of asynchronous transmission because the synchronous model may be regarded as a worst case. For rectangular transmission pulses, this can easily be proven (see Problem 2). Heuristically, one should expect the same behavior also for other pulse shapes.

5.3.2 Optimal receiver structures

In this subsection, we study receiver structures that are optimal in a certain sense. The *maximum likelihood sequence estimation* (MLSE) receiver is *jointly* optimum for estimating the most probable sequence of transmit symbol estimation for *all* users together. The *bitwise maximum a posteriori* (MAP) receiver is optimum for estimating the most probable individual bit transmitted by one user, thereby taking into account the available information about all other users. Both types of receivers have already been derived and discussed in some detail in preceding chapters (Chapters 1, 2, and 3). Here, we have only another interpretation for the same formalism. As discussed in these chapters, both receivers may lead to different bit decisions. However, this is a very rare event except when E_b/N_0 is extremely small.

Let us consider the discrete-time model for the synchronous channel corresponding to a base $\{\psi_i(t)\}_{i=1}^N$ of orthonormal detectors. The vector $\mathbf{r} = (r_1, \ldots, r_N)^T$ of detector outputs is given by

$$\mathbf{r} = \mathbf{GCs} + \mathbf{n}. \tag{5.33}$$

Here, $\mathbf{G} = (\mathbf{g}_1, \ldots, \mathbf{g}_K)$ is the matrix of signature vectors, $\mathbf{C} = \mathrm{diag}(c_1, \ldots, c_K)$ is the diagonal matrix of complex fading amplitudes and $\mathbf{s} = (s_1, \ldots, s_K)^T$ and $\mathbf{n} = (n_1, \ldots, n_N)^T$ are the vectors of transmit symbols and noise samples, respectively.

The maximum likelihood receiver

The MLSE receiver for a discrete-time fading channel has already been discussed in detail in Subsection 2.4.1. Comparing Equations (2.27) and (5.33), we note that the only formal difference is that we formally have to replace $\mathbf{Cs}$ in

$$\mathbf{r} = \mathbf{Cs} + \mathbf{n}$$

by $\mathbf{GCs}$. The interpretation, however, is different. In Subsection 2.4.1, we treated a single user, and we wanted to estimate the sequence of transmit symbols for all K time slots. Here we have K users, and we want to estimate the sequence of transmit symbols for all users corresponding to one time slot. Keeping this in mind, we may now adopt the results of that subsection and find that the most likely vector of transmit symbols is given by

$$\hat{\mathbf{s}} = \arg\min_{\mathbf{s}} \|\mathbf{r} - \mathbf{GCs}\|^2 \tag{5.34}$$

(see Equation (2.28)). Instead of Equation (2.30), we now have

$$\hat{\mathbf{s}} = \arg\max_{\mathbf{s}} \left(\Re\left\{\mathbf{s}^\dagger \mathbf{C}^\dagger \mathbf{G}^\dagger \mathbf{r}\right\} - \frac{1}{2}\|\mathbf{GCs}\|^2 \right). \tag{5.35}$$

Using Equations (5.15) and (5.17), we may write this as

$$\hat{\mathbf{s}} = \arg\max_{\mathbf{s}} \left(\Re\left\{\mathbf{s}^\dagger \mathbf{C}^\dagger \mathbf{v}\right\} - \frac{1}{2}\mathbf{s}^\dagger \mathbf{K}\mathbf{s} \right), \tag{5.36}$$

where we have defined

$$\mathbf{K} = \mathbf{C}^\dagger \mathbf{R} \mathbf{C}.$$

We note that the first term in Equation (5.36) is a cross correlation. $\mathbf{C}^\dagger \mathbf{v}$ is the matched filter output vector $\mathbf{v}$ with the components v_i back rotated in phase and weighted with the fading channel amplitude. This has to be correlated with all possible transmit sequences $\mathbf{s}$.

Again, we can visualize the scenario in the two-dimensional plane for $K = 2$, real positive fading amplitudes and BPSK transmission. Figure 5.28 shows the situation for

$$\mathbf{g}_1 = \begin{pmatrix} 1 \\ 0 \end{pmatrix}, \quad \mathbf{g}_2 = \frac{1}{5}\begin{pmatrix} 3 \\ 4 \end{pmatrix}$$

and equal fading amplitudes $c_1 = c_2 = 1$. The receiver decides for the transmit vector $\mathbf{s}_{b_1 b_2} \in \left\{\pm c_1 \sqrt{E_b}\, \mathbf{g}_1 \pm c_2 \sqrt{E_b}\, \mathbf{g}_2\right\}$, which has the smallest distance to the receive vector $\mathbf{r}$. This condition divides the two-dimensional plane into four regions corresponding to four possible decisions for the two-user bit pair $(b_1 b_2)$. The regions are separated by pieces of straight lines that have equal distances to two closest possible transmit vectors. Comparing Figure 5.28 with Figure 5.26, we observe that there are large regions where the decisions

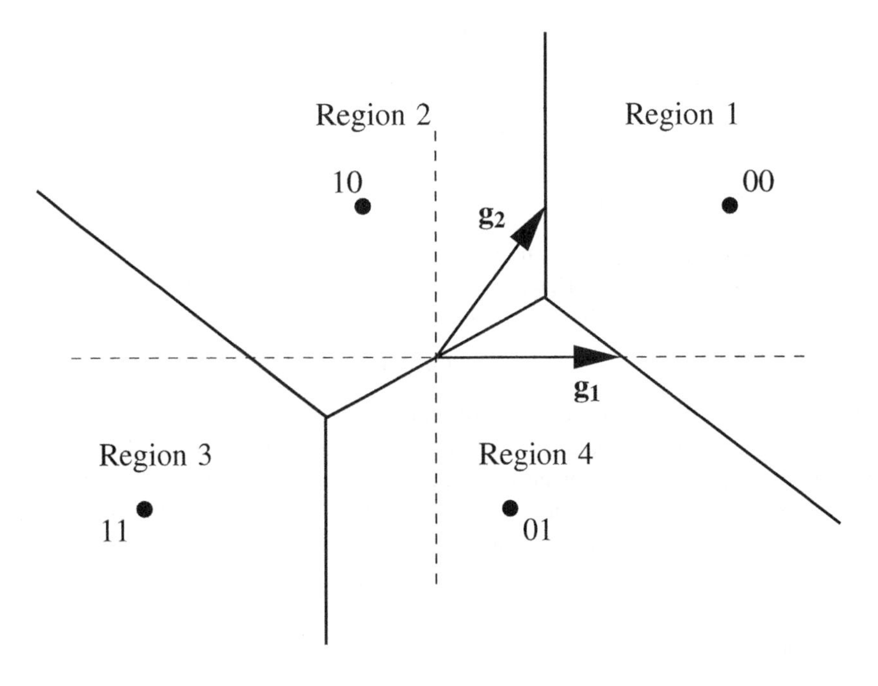

Figure 5.28 Two-user BPSK MLSE receiver for equal fading amplitudes.

for user 1 differ. For example, the receive vector

$$\mathbf{r} = \frac{1}{5}\begin{pmatrix} 1 \\ 4 \end{pmatrix},$$

the SUMF receiver will decide for $b_1 = 0$, but the MLSE receiver will decide for the transmit vector $\mathbf{s}_{10}$ and thus for $b_1 = 1$.

To illustrate the resistance of the MLSE receiver against the near–far problem, we consider the same signature vectors $\mathbf{g}_1$ and $\mathbf{g}_2$, but we assume that the fading amplitudes are given by $c_1 = 1$ and $c_2 = 2$. The receive vector is now given by

$$\mathbf{r} = s_1\mathbf{g}_1 + 2s_2\mathbf{g}_2 + \mathbf{n}.$$

This situation is depicted in Figure 5.29. It is obvious from the figure that correct decisions will be made for user 1 as long as the noise does not exceed the distance to the decision threshold, which is the piecewise linear curve that separates the union of Region 1 and Region 4 from the union of Region 2 and Region 3.

Now, we evaluate the error probability for user 1. Assume that user 1 transmits $b_1 = 0$, that is, one of the transmit vectors $\mathbf{s}_{00}$ or $\mathbf{s}_{01}$. An error occurs if the receive vector $\mathbf{r}$ lies in

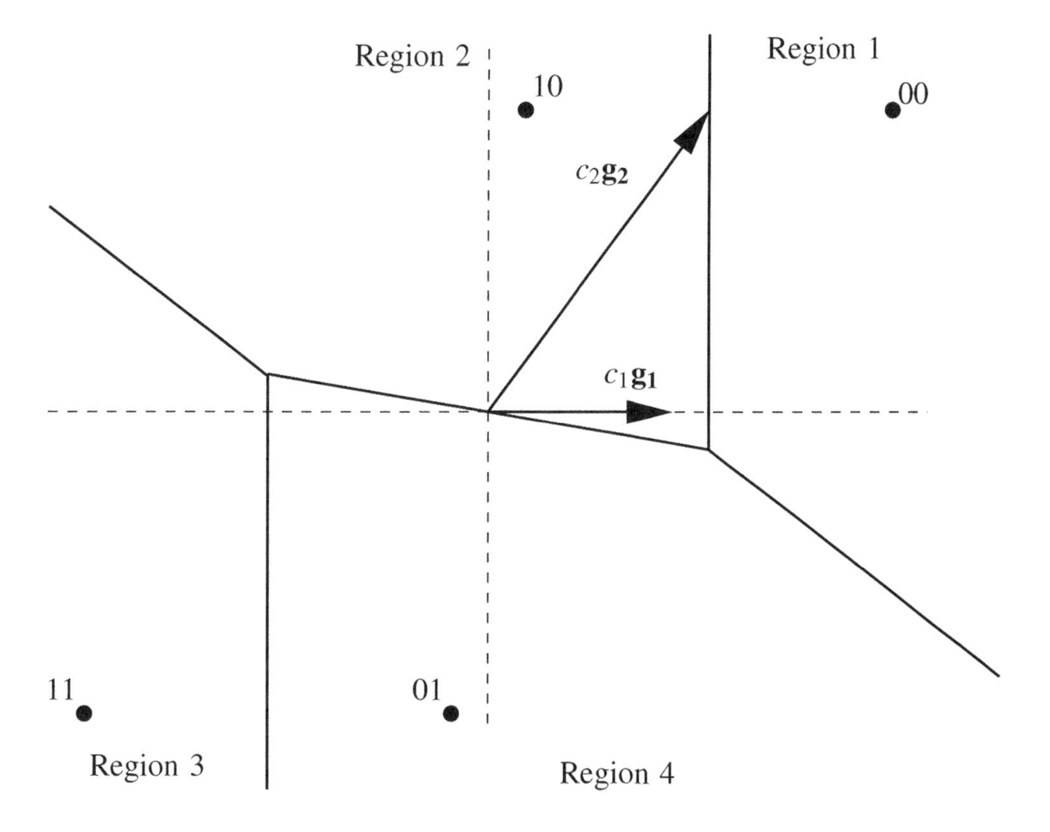

Figure 5.29 Two-user BPSK MLSE receiver for unequal fading amplitudes.

Region 2 or Region 3. First of all, we calculate the four distances between $\mathbf{s}_{0b_2}$ and $\mathbf{s}_{1b_2}$ as

$$\begin{aligned}
\|\mathbf{s}_{00} - \mathbf{s}_{10}\|^2 &= 4E_b\, |c_1|^2\,,\\
\|\mathbf{s}_{00} - \mathbf{s}_{11}\|^2 &= 4E_b \left(|c_1|^2 + |c_2|^2 + 2\Re\left\{c_1^* c_2 \rho_{12}\right\}\right),\\
\|\mathbf{s}_{01} - \mathbf{s}_{10}\|^2 &= 4E_b \left(|c_1|^2 + |c_2|^2 - 2\Re\left\{c_1^* c_2 \rho_{12}\right\}\right),\\
\|\mathbf{s}_{01} - \mathbf{s}_{11}\|^2 &= 4E_b\, |c_1|^2\,.
\end{aligned}$$

For our analysis, we assume that

$$\|\mathbf{s}_{00} - \mathbf{s}_{11}\|^2 > \|\mathbf{s}_{01} - \mathbf{s}_{10}\|^2\,,$$

that is,

$$\Re\left\{c_1^* c_2 \rho_{12}\right\} > 0.$$

This is consistent with the situation in Figure 5.28 and Figure 5.29. Otherwise, the treatment is similar. For $\rho_{12} = 0$, the SUMF is already optimal.

We recall that the pairwise error probabilities are given by

$$P\left(\mathbf{s}_{b_1 b_2} \mapsto \mathbf{s}_{\hat{b}_1 \hat{b}_2}\right) = \mathrm{Q}\left(\frac{1}{2\sigma}\left\|\mathbf{s}_{b_1 b_2} - \mathbf{s}_{\hat{b}_1 \hat{b}_2}\right\|\right).$$

We now assume that $\mathbf{s}_{00}$ is the transmit signal. Because Region 2 and Region 3 are completely contained in the half plane left from the decision threshold between $\mathbf{s}_{00}$ and $\mathbf{s}_{10}$, the error probability is upper bounded by the pairwise error probability $P\left(\mathbf{s}_{00} \mapsto \mathbf{s}_{10}\right)$. We then assume that $\mathbf{s}_{01}$ is the transmit signal. The probability that $\mathbf{r}$ lies in Region 2 or Region 3 can be upper bounded by $P\left(\mathbf{s}_{01} \mapsto \mathbf{s}_{11}\right) + P\left(\mathbf{s}_{01} \mapsto \mathbf{s}_{10}\right)$. The bit error probability is then bounded by

$$P_b \le \frac{1}{2} P\left(\mathbf{s}_{00} \mapsto \mathbf{s}_{10}\right) + \frac{1}{2}\left(P\left(\mathbf{s}_{01} \mapsto \mathbf{s}_{11}\right) + P\left(\mathbf{s}_{01} \mapsto \mathbf{s}_{10}\right)\right).$$

Using $P\left(\mathbf{s}_{00} \mapsto \mathbf{s}_{10}\right) = P\left(\mathbf{s}_{01} \mapsto \mathbf{s}_{11}\right)$, we get

$$P_b \le \mathrm{Q}\left(\frac{1}{2\sigma}\|\mathbf{s}_{00} - \mathbf{s}_{10}\|\right) + \frac{1}{2}\mathrm{Q}\left(\frac{1}{2\sigma}\|\mathbf{s}_{01} - \mathbf{s}_{10}\|\right).$$

We may express this as

$$P_b \le \frac{1}{2}\operatorname{erfc}\left(\sqrt{|c_1|^2 \frac{E_b}{N_0}}\right) + \frac{1}{4}\operatorname{erfc}\left(\sqrt{\left(|c_1|^2 + |c_2|^2 - 2\Re\left\{c_1^* c_2 \rho_{12}\right\}\right)\frac{E_b}{N_0}}\right). \qquad (5.37)$$

For

$$\Re\left\{c_1^* c_2 \rho_{12}\right\} < 0,$$

the derivation is similar and we get another sign under the square root for the second term. Thus, both cases are covered by

$$P_b \le \frac{1}{2}\operatorname{erfc}\left(\sqrt{|c_1|^2 \frac{E_b}{N_0}}\right) + \frac{1}{4}\operatorname{erfc}\left(\sqrt{\left(|c_1|^2 + |c_2|^2 - \left|2\Re\left\{c_1^* c_2 \rho_{12}\right\}\right|\right)\frac{E_b}{N_0}}\right). \qquad (5.38)$$

To illustrate the performance gain of the MLSE receiver compared to the SUMF receiver, we have evaluated this expression for

$$\mathbf{g}_1 = \begin{pmatrix} 1 \\ 0 \end{pmatrix}, \ \mathbf{g}_2 = \frac{1}{5}\begin{pmatrix} 3 \\ 4 \end{pmatrix}$$

and equal fading amplitudes $c_1 = c_2 = 1$, that is, the situation of Figure 5.28. Figure 5.30 shows the bit error rate for the MLSE receiver according to the above formulas compared with the bit error rate for the MLSE receiver according to Equation (5.30) and ideal (single-user) BPSK. The SUMF receiver shows a severe degradation compared to the ideal BPSK (approximately 8 dB at 10^{-5}), while the MLSE receiver shows only a loss of approximately 1 dB. For the same signature vectors and $c_2 = 2c_1$, the performance curves are depicted in Figure 5.31. Surprisingly, the higher interference power of user 2 leads to a better performance for user 1 which even approaches the BPSK curve. This can be explained by the fact that for a joint detection receiver, user 2 must not be regarded as an interferer. In fact, user 2 helps user 1 by increasing the relevant distances. To explain this, we note that the first term in Equation (5.37) corresponding to the distances $\|\mathbf{s}_{00} - \mathbf{s}_{10}\|$ and $\|\mathbf{s}_{01} - \mathbf{s}_{11}\|$ equals the BPSK performance curve. The second term corresponds to the distance $\|\mathbf{s}_{01} - \mathbf{s}_{10}\|$. Looking at Figure 5.29, we observe that the distance $\|\mathbf{s}_{01} - \mathbf{s}_{10}\|$ now is smaller than the other distances and thus the second term can be neglected for high values of E_b/N_0. This is in contrast to the situation in Figure 5.28 where that term dominates the performance curve.

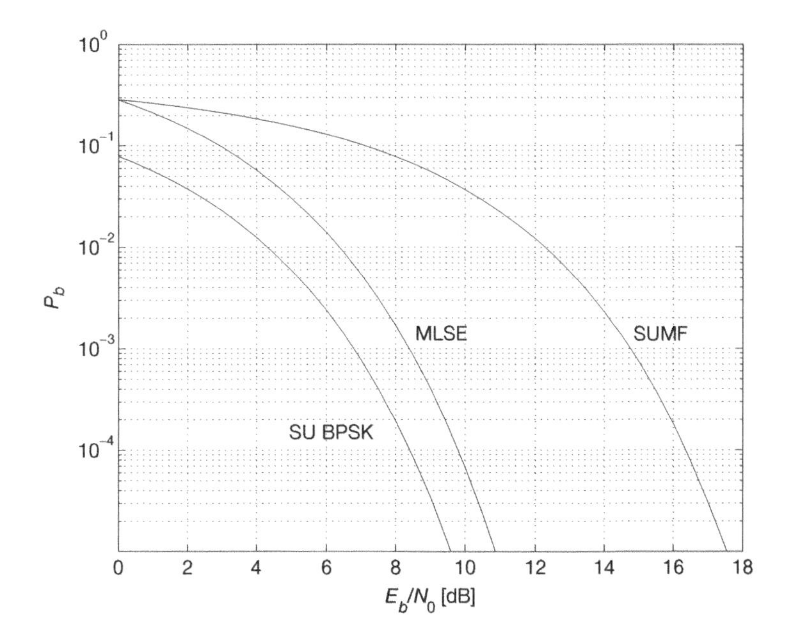

Figure 5.30 The performance of the MLSE receiver compared with the SUMF receiver and ideal *single-user* (SU) BPSK for equal real fading amplitudes and $\rho_{12} = 0.6$.

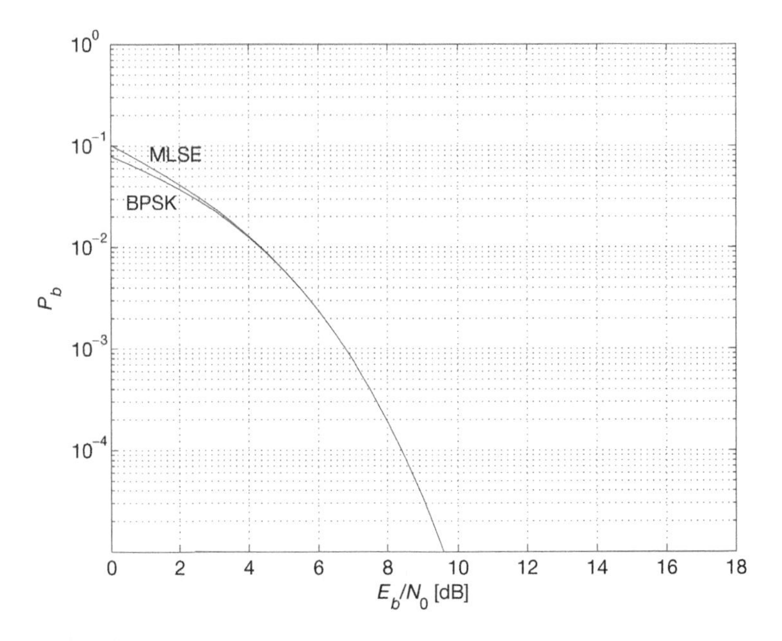

Figure 5.31 The performance of the MLSE receiver compared with ideal (single-user) BPSK for unequal real fading amplitudes and $\rho_{12} = 0.6$.

The bitwise MAP receiver

The bitwise MAP receiver is derived and discussed in Subsection 3.1.5. Given a receive vector $\mathbf{r}$, it provides us with the *log-likelihood ratio* (LLR) $L(b_k = 0|\mathbf{r})$ of one individual bit in the transmit sequence. As discussed in that subsection, the sign of $L(b_k = 0|\mathbf{r})$ gives the hard bit decision ($b_k = 0$ for positive LLR, $b_k = 1$ otherwise). The absolute value is a measure for the reliability. Consider the bit b_1 corresponding to user 1. We do not necessarily need to assume BPSK, so other bits of user 1 can be transmitted during the same time slot. Let us denote the set of all transmit symbol vectors $\mathbf{s}$ with $b_1 = 0$ by $\mathcal{S}_0$ and the set of all transmit symbol vectors $\mathbf{s}$ with $b_1 = 1$ by $\mathcal{S}_1$. If all vectors $\mathbf{s}$ have equal *a priori* probability, the LLR value for b_1 can be obtained by the method of that subsection as

$$L(b_1 = 0|\mathbf{r}) = \log\left(\frac{\sum_{\mathbf{s}\in\mathcal{S}_0} \exp\left(-\frac{1}{2\sigma^2}\|\mathbf{r} - \mathbf{GCs}\|^2\right)}{\sum_{\mathbf{s}\in\mathcal{S}_1} \exp\left(-\frac{1}{2\sigma^2}\|\mathbf{r} - \mathbf{GCs}\|^2\right)}\right). \tag{5.39}$$

In this expression, $\sigma^2 = N_0/2$ is the noise variance per real dimension. The expression for the LLR can equivalently be written as

$$L(b_1 = 0|\mathbf{r}) = \log\left(\frac{\sum_{\mathbf{s}\in\mathcal{S}_0} \exp\left(\frac{1}{\sigma^2}\left(\Re\left\{\mathbf{s}^\dagger\mathbf{C}^\dagger\mathbf{v}\right\} - \frac{1}{2}\mathbf{s}^\dagger\mathbf{Ks}\right)\right)}{\sum_{\mathbf{s}\in\mathcal{S}_1} \exp\left(\frac{1}{\sigma^2}\left(\Re\left\{\mathbf{s}^\dagger\mathbf{C}^\dagger\mathbf{v}\right\} - \frac{1}{2}\mathbf{s}^\dagger\mathbf{Ks}\right)\right)}\right). \tag{5.40}$$

For $K = 2$ and BPSK transmission, this expression can be evaluated further. We have

$$\mathbf{K} = \begin{pmatrix} |c_1|^2 & c_1^* c_2 \rho_{12} \\ c_1 c_2^* \rho_{21} & |c_2|^2 \end{pmatrix},$$

and thus

$$\mathbf{s}^\dagger \mathbf{K} \mathbf{s} = \sqrt{E_b}\, \left(|c_1|^2 + |c_2|^2\right) + 2 s_1 s_2 \Re\left\{c_1^* c_2 \rho_{12}\right\}.$$

Furthermore,

$$\mathbf{s}^\dagger \mathbf{C}^\dagger \mathbf{v} = s_1 c_1^* v_1 + s_2 c_2^* v_2.$$

We insert this into the exponential expressions in Equation (5.40) and get

$$L(b_1 = 0|\mathbf{r}) = \log\left(\frac{\mathrm{e}^{\frac{1}{\sigma^2}\Re\{c_1^* v_1\}} \cosh\left(\frac{1}{\sigma^2}\Re\left\{c_2^* v_2 - c_1^* c_2 \rho_{12}\right\}\right)}{\mathrm{e}^{-\frac{1}{\sigma^2}\Re\{c_1^* v_1\}} \cosh\left(\frac{1}{\sigma^2}\Re\left\{c_2^* v_2 + c_1^* c_2 \rho_{12}\right\}\right)}\right),$$

and, finally,

$$L(b_1 = 0|\mathbf{r}) = \frac{2}{\sigma^2}\Re\left\{c_1^* v_1\right\} + \log\left(\frac{\cosh\left(\frac{1}{\sigma^2}\Re\left\{c_2^* v_2 - c_1^* c_2 \rho_{12}\right\}\right)}{\cosh\left(\frac{1}{\sigma^2}\Re\left\{c_2^* v_2 + c_1^* c_2 \rho_{12}\right\}\right)}\right).$$

Figure 5.32 depicts this quantity as a contour plot for the curves of equal LLR with the parameters of Figure 5.28 and $E_b/N_0 = 6$ dB. The bold curve corresponds to LLR $= 0$ and has to be interpreted as the *smooth threshold* that separates the union of Regions 1 and 4

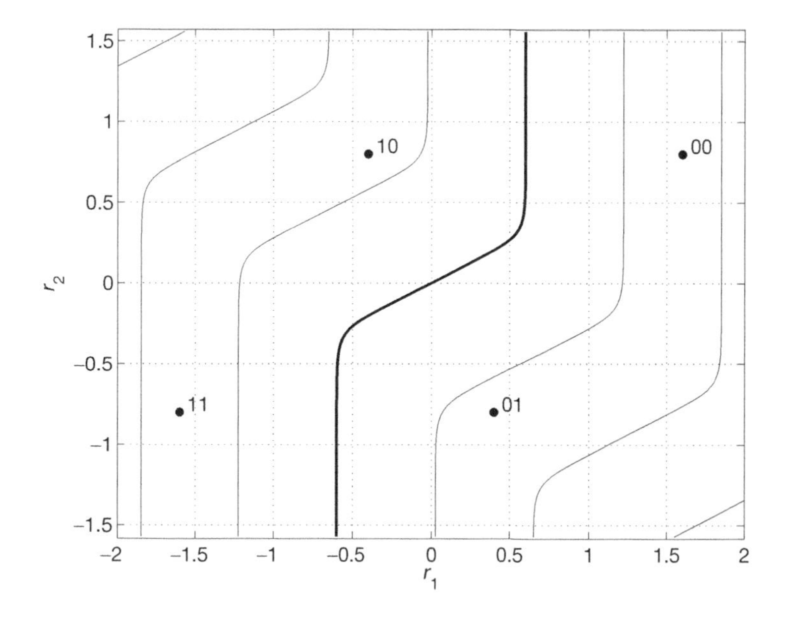

Figure 5.32 Smooth threshold for the MAP receiver ($E_b/N_0 = 6$ dB).

from the union of Regions 2 and 3. For high values of E_b/N_0, it comes closer and closer to the piecewise linear MLSE threshold depicted in Figure 5.28.

We note that for sufficiently high values of E_b/N_0, the *maxlog* MAP becomes a very good approximation (see Subsection 3.5.1). We then have the approximate expression for the LLR

$$L(b_k = 0|\mathbf{r}) \approx \frac{1}{2\sigma^2}\left(\min_{\mathbf{s}\in\mathcal{S}_1} \|\mathbf{r} - \mathbf{GCs}\|^2 - \min_{\mathbf{s}\in\mathcal{S}_0} \|\mathbf{r} - \mathbf{GCs}\|^2\right).$$

This approximate expression leads to the same hard decisions as the MLSE receiver. However, it provides us with reliability information that can be used by an outer decoder.

5.3.3 Suboptimal linear receiver structures

The complexity of the optimal receivers discussed above grows exponentially with the number of users. In this subsection, we will discuss suboptimal linear receiver structures with a complexity that grows only linearly with the number of users.

The decorrelating receiver

As discussed above, the SUMF receiver suffers from the near–far problem. In the presence of strong interferers, the receiver may produce erroneous bit decisions even in a noise-free channel.

We now derive a linear receiver that always perfectly recovers the information in the absence of noise. In that case, the synchronous discrete model receive signal is given by

$$\mathbf{r} = \sum_{k=1}^{K} c_k s_k \mathbf{g}_k.$$

Here $\mathbf{g}_k$ is the signature vector for user number k, and s_k and c_k are the corresponding transmit symbol and complex fading amplitude, respectively. We define

$$u_k = c_k s_k$$

and write

$$\mathbf{r} = \sum_{k=1}^{K} u_k \mathbf{g}_k.$$

In that equation, the values u_k can be interpreted as the *coordinates* of the vector $\mathbf{r}$ corresponding to the base of the signature vectors $\mathbf{g}_k$. We recall that the coordinates of any vector that lies in the hyperplane spanned by that base are uniquely defined. Therefore, if $\mathbf{n}$ is a noise vector that lies in the hyperplane spanned by the base of signature vectors $\mathbf{g}_k$, the coordinates x_k of the vector

$$\mathbf{r} = \sum_{k=1}^{K} u_k \mathbf{g}_k + \mathbf{n}$$

are uniquely defined by

$$\mathbf{r} = \sum_{k=1}^{K} x_k \mathbf{g}_k.$$

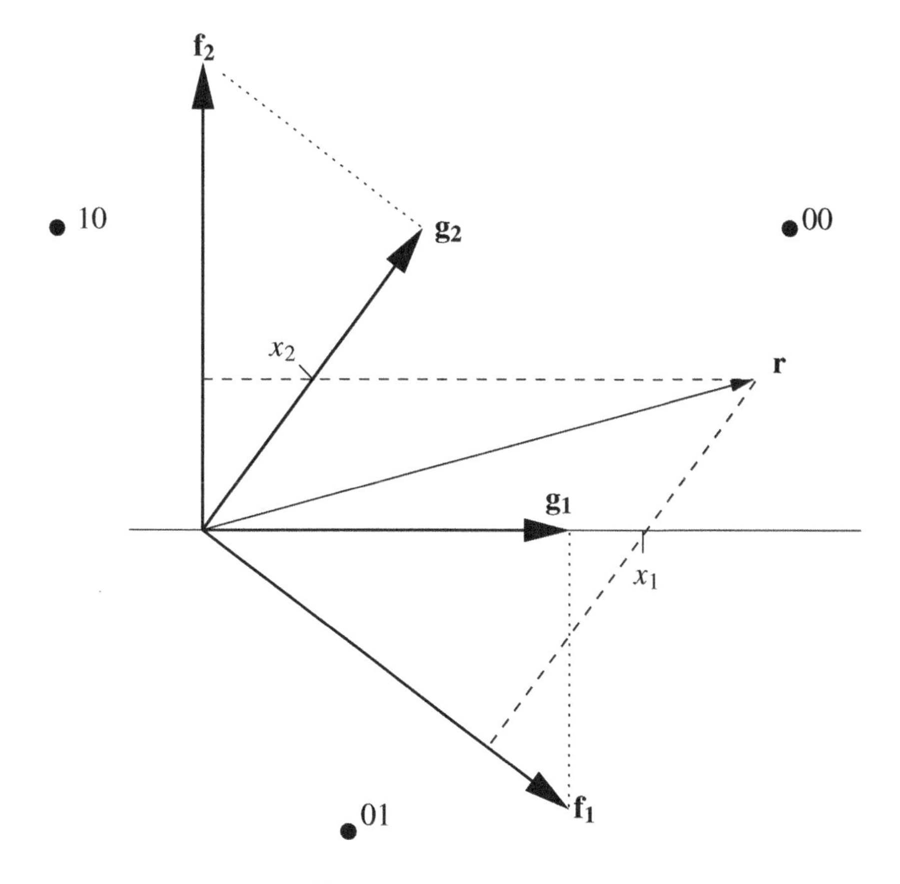

Figure 5.33 Two-user decorrelating receiver.

These coordinates x_k will serve as estimates for the coordinates u_k of the transmit signal. For the moment, we have assumed that the vector $\mathbf{r}$ lies in the hyperplane spanned by the base of signature vectors $\mathbf{g}_k$. We will later see that we can remove this restriction.

The calculation of the coordinates is standard linear algebra. Before doing this, however, we want to illustrate the geometrical structure for the case of two users, real fading amplitudes and BPSK transmission. This situation is depicted in Figure 5.33. The receive vector is a vector sum

$$\mathbf{r} = x_1 \mathbf{g}_1 + x_2 \mathbf{g}_2,$$

which can geometrically be visualized by the diagonal of the parallelogram spanned by the vectors $x_1 \mathbf{g}_1$ and $x_2 \mathbf{g}_2$. The coordinates x_1 and x_2 can be interpreted as the lengths of the edges of the parallelogram (note that $\|\mathbf{g}_1\| = \|\mathbf{g}_2\| = 1$). They can geometrically be obtained as the skew projections (Eldar 2002) on the base: x_1 is the projection of $\mathbf{r}$ on $\mathbf{g}_1$ along the direction of $\mathbf{g}_2$, and x_2 is the projection of $\mathbf{r}$ on $\mathbf{g}_2$ along the direction of $\mathbf{g}_1$. To express the skew projections by orthogonal projections by means of scalar products, we introduce vectors $\mathbf{f}_1$ and $\mathbf{f}_2$ in the plane with the property

$$\mathbf{f}_i \cdot \mathbf{g}_k = \delta_{ik},\ i, k \in \{1, 2\},$$

that is, $\mathbf{f}_1 \perp \mathbf{g}_2$ and $\mathbf{f}_2 \perp \mathbf{g}_1$. Now $\mathbf{r}$ and $x_1\mathbf{g}_1$ have the same projection on the direction of $\mathbf{f}_1$, and $\mathbf{r}$ and $x_2\mathbf{g}_2$ have the same projection on the direction of $\mathbf{f}_2$. The first fact means that

$$|\mathbf{f}_1|^{-1}\mathbf{f}_1 \cdot \mathbf{r} = |\mathbf{f}_1|^{-1}\mathbf{f}_1 \cdot (x_1\mathbf{g}_1),$$

which leads to

$$x_1 = \mathbf{f}_1 \cdot \mathbf{r}.$$

Similarly, we obtain

$$x_2 = \mathbf{f}_2 \cdot \mathbf{r}.$$

We now generalize to the case that $\mathbf{r}$ does not lie in the plane spanned by $\mathbf{g}_1$ and $\mathbf{g}_2$ or, equivalently, by $\mathbf{f}_1$ and $\mathbf{f}_2$. The vector $\mathbf{r}$ can then be uniquely decomposed according to

$$\mathbf{r} = \mathbf{r}_\diamond + \mathbf{r}_\perp,$$

where $\mathbf{r}_\diamond$ lies in the plane and $\mathbf{r}_\perp$ is orthogonal to that plane. Thus, $\mathbf{f}_k \cdot \mathbf{r}_\perp = 0$ and

$$x_k = \mathbf{f}_k \cdot \mathbf{r} = \mathbf{f}_k \cdot \mathbf{r}_\diamond$$

are the correct expressions for the coordinates of the orthogonal projection $\mathbf{r}_\diamond$ of $\mathbf{r}$ onto the plane.

We formalize the above treatment for K users. Let $\mathcal{V} = \mathrm{span}(\mathbf{g}_1, \ldots, \mathbf{g}_K)$ be the vector space (hyperplane) spanned by the base of signature vectors $\{\mathbf{g}_k\}_{k=1}^K$. There exists a uniquely defined base $\{\mathbf{f}_k\}_{k=1}^K$ called the *reciprocal base* with the property

$$\mathbf{f}_i^\dagger \mathbf{g}_k = \delta_{ik},\ i, k \in \{1, \ldots, K\}.$$

As depicted in Figure 5.34, we decompose again

$$\mathbf{r} = \mathbf{r}_\diamond + \mathbf{r}_\perp,$$

where $\mathbf{r}_\diamond$ lies in $\mathcal{V}$ and $\mathbf{r}_\perp$ is orthogonal to $\mathcal{V}$, that is,

$$\mathbf{f}_i^\dagger \mathbf{r}_\perp = 0,$$

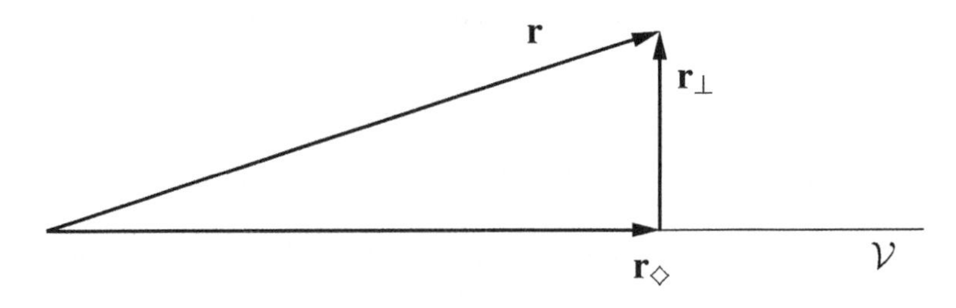

Figure 5.34 The orthogonality condition.

or

$$\mathbf{F}^{\dagger}\mathbf{r}_{\perp} = 0$$

with

$$\mathbf{F} = [\mathbf{f}_1, \ldots, \mathbf{f}_K] .$$

The orthogonal component $\mathbf{r}_{\perp}$ of the receive vector is not relevant to the decision. With

$$\mathbf{r}_{\diamond} = \sum_{k=1}^{K} x_k \mathbf{g}_k = \mathbf{G}\mathbf{x}$$

and $\mathbf{x} = (x_1, \ldots, x_K)^T$, we have

$$\mathbf{F}^{\dagger}(\mathbf{r} - \mathbf{G}\mathbf{x}) = 0.$$

From the orthogonality

$$\mathbf{F}^{\dagger}\mathbf{G} = \mathbf{I}$$

between the two bases, we obtain

$$\mathbf{x} = \mathbf{F}^{\dagger}\mathbf{r}. \tag{5.41}$$

There are standard methods to calculate the reciprocal base. Here, however, we will show how to derive the coordinate vector by the orthogonality condition which again gives some insight into the geometrical structure.

We look for a vector $\mathbf{x}$ that minimizes the quadratic distance (i.e. the mean square error) between the receive vector $\mathbf{r}$ and $\mathcal{V}$, that is,

$$\mathbf{x} = \arg\min_{\mathbf{u}} \|\mathbf{r} - \mathbf{G}\mathbf{u}\|^2 . \tag{5.42}$$

This is equivalent to the orthogonality condition

$$\mathbf{g}_k^{\dagger}(\mathbf{G}\mathbf{x} - \mathbf{r}) = 0$$

that must hold for all vectors $\mathbf{g}_k$ (see Figure 5.34) with $\mathbf{G}\mathbf{x} = \mathbf{r}_{\diamond}$. In matrix notation, we can write this as

$$\mathbf{G}^{\dagger}(\mathbf{G}\mathbf{x} - \mathbf{r}) = 0$$

or

$$\mathbf{G}^{\dagger}\mathbf{G}\mathbf{x} = \mathbf{G}^{\dagger}\mathbf{r}.$$

If the signature matrix $\mathbf{G}$ has maximal rank, we may invert $\mathbf{G}^{\dagger}\mathbf{G}$ and eventually find the desired solution

$$\mathbf{x} = \left(\mathbf{G}^{\dagger}\mathbf{G}\right)^{-1}\mathbf{G}^{\dagger}\mathbf{r}. \tag{5.43}$$

The matrix

$$\mathbf{G}^{-} = \left(\mathbf{G}^{\dagger}\mathbf{G}\right)^{-1}\mathbf{G}^{\dagger}$$

is called the *Moore–Penrose pseudoinverse* of the matrix $\mathbf{G}$. Since $\mathbf{x} = \mathbf{G}^{-}\mathbf{r} = \mathbf{F}^{\dagger}\mathbf{r}$, we have shown that

$$\mathbf{G}^{-} = \mathbf{F}^{\dagger}.$$

We note that the same formula can easily be obtained in the discrete matched filter model

$$\mathbf{v} = \mathbf{RCs} + \mathbf{m}$$

with $\mathbf{v} = \mathbf{G}^{\dagger}\mathbf{r}$. In the noise-free channel, $\mathbf{x} = \mathbf{Cs}$ holds and we simply obtain $\mathbf{x}$ by applying $\mathbf{R}^{-1}$, that is,

$$\mathbf{x} = \mathbf{R}^{-1}\mathbf{v},$$

which is equivalent to Equation (5.43). Thus, the decorrelating receiver just extends the method of simple matrix inversion to the noisy channel.

An estimate for the transmit signal vector $\mathbf{s}$ can be obtained by channel inversion as

$$\hat{\mathbf{s}} = \mathbf{C}^{-1}\mathbf{x} = \mathbf{C}^{-1}\mathbf{R}^{-1}\mathbf{v}.$$

For user number k, we get the estimate

$$\hat{s}_k = c_k^{-1}\left(\mathbf{R}^{-1}\mathbf{v}\right)_k, \tag{5.44}$$

where the index k indicates the kth element of the vector. Thus, we need only to know the signature waveforms of all the other users. It is not necessary to know their fading amplitudes.

The discrete transmission model including the decorrelation receiver is given by

$$\mathbf{x} = \mathbf{Cs} + \tilde{\mathbf{m}}$$

with $\tilde{\mathbf{m}} = \mathbf{R}^{-1}\mathbf{m}$. The autocorrelation matrix of the noise vector $\tilde{\mathbf{m}} = (m_1, \ldots, m_K)^T$ is given by

$$\mathrm{E}\left\{\tilde{\mathbf{m}}\tilde{\mathbf{m}}^{\dagger}\right\} = N_0\mathbf{R}^{-1}.$$

For user number k, we thus have the transmission model

$$x_k = c_k s_k + \tilde{m}_k$$

with noise variance given by

$$\mathrm{E}\left\{|\tilde{m}_k|^2\right\} = N_0\left(\mathbf{R}^{-1}\right)_{kk},$$

where $(\mathbf{R})_{kk}$ denotes the kth diagonal element of the matrix $\mathbf{R}^{-1}$. For BPSK transmission, the BER performance of user k for fixed channel amplitudes is then given by

$$P_b^{(k)} = \frac{1}{2}\mathrm{erfc}\left(\sqrt{\frac{|c_k|^2}{\left(\mathbf{R}^{-1}\right)_{kk}}\frac{E_b}{N_0}}\right).$$

We now evaluate this BER performance in an alternative way that gives some insight into the geometrical structure. We recall that the receiver calculates the coordinates of the relevant part $\mathbf{r}_{\diamond}$ of the receive vector $\mathbf{r}$ with respect to the signature base vectors $\mathbf{g}_k$. Without losing generality, we consider user number 1. An error occurs if the first coordinate of the transmit vector corresponding to the base vector $\mathbf{g}_1$ has a positive sign, but the first coordinate of the receive vector has a negative sign (or vice versa). We can visualize the

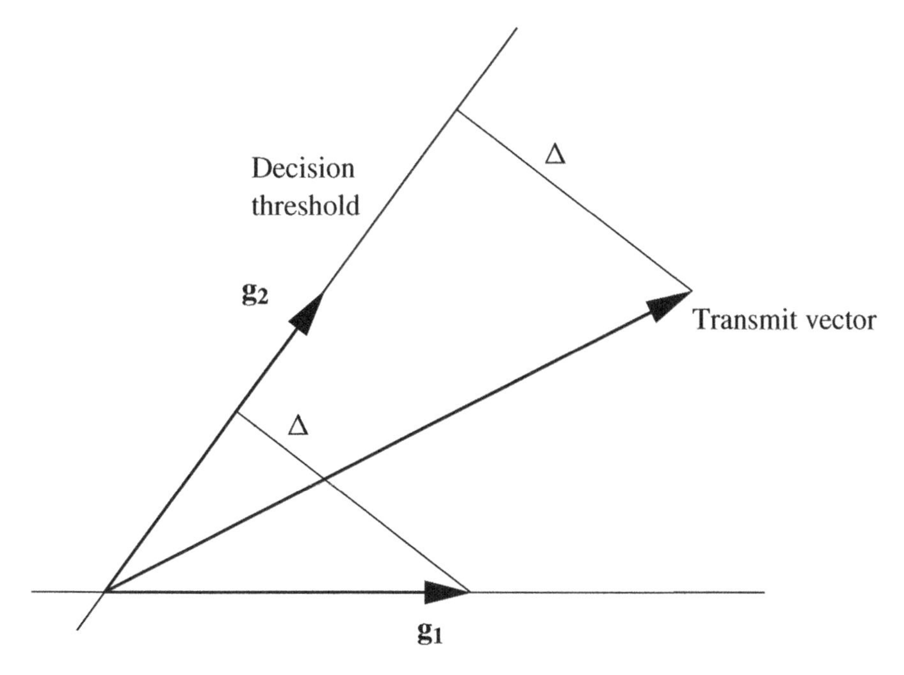

Figure 5.35 Distance to the decision threshold for the decorrelator for $c_1 = c_2 = 1$.

geometrical situation as follows. The $K-1$-dimensional hyperplane spanned by $\mathbf{g}_2, \ldots, \mathbf{g}_K$ divides the K-dimensional hyperplane given by the vector space $\mathcal{V} = \mathrm{span}(\mathbf{g}_1, \ldots, \mathbf{g}_K)$ into two half planes. An error occurs if the (relevant) receive vector $\mathbf{r}_\diamond$ lies in the other half plane of $\mathcal{V}$ than the transmit vector. For $K = 2$ and $c_1 = c_2 = 1$, the situation is depicted in Figure 5.35 We may thus regard the $K-1$-dimensional hyperplane $\mathrm{span}(\mathbf{g}_2, \ldots, \mathbf{g}_K)$ as a decision threshold. Let Δ be the distance between the transmit vector and this threshold. The bit error probability for user 1 is then given by

$$P_b = \mathrm{Q}\left(\frac{\Delta}{\sigma}\right).$$

We note that the coordinates of all users are independent and uniquely defined, and the decision about the transmit symbol s_1 of user 1 does not depend on the values of the other transmit symbols $s_2, \ldots, s_K$. For the performance analysis, we may thus set $s_2 = \cdots = s_K = 0$. This means that the distance of the transmit vector to the threshold is the same as the distance of $c_1\mathbf{g}_1$ to the threshold (see Figure 5.35). It is evident from the figure and the Pythagorean theorem that

$$\Delta^2 = |c_1|^2 \sqrt{E_b}\left(1 - \|\mathbf{P}\mathbf{g}_1\|^2\right),$$

where the vector $\mathbf{P}\mathbf{g}_1$ is the orthogonal projection of the base vector $\mathbf{g}_1$ onto the threshold. For two dimensions, we know from elementary vector geometry that

$$\mathbf{P}\mathbf{g}_1 = (\mathbf{g}_2^\dagger \mathbf{g}_1)\mathbf{g}_2 = \rho_{12}\mathbf{g}_2.$$

In that case, the *projector* $\mathbf{P}$ is given by the matrix

$$\mathbf{P} = \mathbf{g}_2\mathbf{g}_2^\dagger.$$

For the general case of more dimensions, we calculate Δ by using the orthogonality principle. The formalism is similar to the derivation of the decorrelating receiver. We define

$$\mathbf{G}_1 = [\mathbf{g}_2, \ldots, \mathbf{g}_K].$$

Let $\mathbf{k}$ be the vector of coordinates that minimizes the distance between $\mathbf{g}_1$ and the threshold, that is,

$$\mathbf{k} = \arg\min_{\mathbf{u}} \|\mathbf{g}_1 - \mathbf{G}_1\mathbf{u}\|^2.$$

From the orthogonality condition

$$\mathbf{G}_1^\dagger(\mathbf{g}_1 - \mathbf{G}_1\mathbf{k}) = 0,$$

we obtain

$$\mathbf{k} = (\mathbf{G}_1^\dagger\mathbf{G}_1)^{-1}\mathbf{G}_1^\dagger\mathbf{g}_1.$$

We thus have

$$\Delta^2 = |c_1|^2\sqrt{E_b}\,\|\mathbf{g}_1 - \mathbf{G}_1\mathbf{k}\|^2 = |c_1|^2\sqrt{E_b}\left\|\mathbf{g}_1 - \mathbf{G}_1(\mathbf{G}_1^\dagger\mathbf{G}_1)^{-1}\mathbf{G}_1^\dagger\mathbf{g}_1\right\|^2.$$

We note that

$$(\mathbf{G}_1^\dagger\mathbf{G}_1)^{-1}\mathbf{G}_1^\dagger = \mathbf{G}_1^-$$

is the pseudoinverse of $\mathbf{G}_1$ and

$$\mathbf{P} = \mathbf{G}_1\mathbf{G}_1^-$$

is the *orthogonal projector* onto the hyperplane span$(\mathbf{g}_2, \ldots, \mathbf{g}_K)$. An orthogonal projector $\mathbf{P}$ is a matrix with the following properties: it is Hermitian, that is,

$$\mathbf{P} = \mathbf{P}^\dagger$$

and has the property

$$\mathbf{P}^2 = \mathbf{P}.$$

Since $\|\mathbf{g}_1\|^2 = 1$, we readily get the expression

$$\Delta^2 = |c_1|^2\sqrt{E_b}\left(1 - \mathbf{g}_1^\dagger\mathbf{P}\mathbf{g}_1\right)$$

or, equivalently,

$$\Delta^2 = |c_1|^2\sqrt{E_b}\left(1 - \|\mathbf{P}\mathbf{g}_1\|^2\right),$$

which is just the Pythagorean Theorem. The bit error rate is then given by

$$P_b = \frac{1}{2}\operatorname{erfc}\left(\sqrt{|c_1|^2\left(1 - \|\mathbf{P}\mathbf{g}_1\|^2\right)\frac{E_b}{N_0}}\right). \tag{5.45}$$

As a secondary result, we have thus seen that

$$\frac{1}{\left(\mathbf{R}^{-1}\right)_{11}} = \left(1 - \left\|\mathbf{P}\mathbf{g}_1\right\|^2\right).$$

The quantity

$$\left\|\mathbf{P}\mathbf{g}_1\right\|^2 = (\mathbf{G}_1^\dagger \mathbf{g}_1)^\dagger (\mathbf{G}_1^\dagger \mathbf{G}_1)^{-1} (\mathbf{G}_1^\dagger \mathbf{g}_1)$$

has the following structure. The matrix

$$\mathbf{R}_1 = \mathbf{G}_1^\dagger \mathbf{G}_1$$

is just the autocorrelation matrix $\mathbf{R}$ with the first row and the first column deleted. The vector

$$\mathbf{v}_1 = \mathbf{G}_1^\dagger \mathbf{g}_1,$$

is just the column of the $K-1$ detector outputs of $\mathbf{g}_1$ for all other detectors $\mathbf{g}_2, \ldots, \mathbf{g}_K$.

From Equation (5.45), we see that the performance compared to ideal single-user BPSK is degraded by the channel attenuation $|c_1|^2$ and the geometrical factor $\left(1 - \left\|\mathbf{P}\mathbf{g}_1\right\|^2\right)$. For Rayleigh fading, we simply average over the fading amplitude c_1 as shown in Subsection 2.4.3 to get

$$P_b = \frac{1}{2}\left[1 - \sqrt{\frac{\left(1 - \left\|\mathbf{P}\mathbf{g}_1\right\|^2\right)\frac{E_b}{N_0}}{1 + \left(1 - \left\|\mathbf{P}\mathbf{g}_1\right\|^2\right)\frac{E_b}{N_0}}}\right]. \tag{5.46}$$

We now evaluate the expression (5.45) with $\left\|\mathbf{P}\mathbf{g}_1\right\|^2 = |\rho_{12}|^2$ for the two-user performance in the AWGN channel and compare it with the SUMF receiver. We set the fading amplitude $c_1 = 1$. Figure 5.36 shows the bit error curves for both receivers for equal fading amplitudes $c_1 = c_2 = 1$ and different values of the correlation coefficient ρ_{12}. We see that for these relatively high correlation coefficients, the decorrelator performs several decibels better than the SUMF receiver. The difference becomes smaller for small correlations, but the decorrelator is always better. In fact, one can show that for real-valued coefficients c_1, c_2 and ρ_{12}, the decorrelator is better than the SUMF receiver (see Problem 3). For complex values of the fading amplitudes, it may occasionally happen that $\left|\Re\left\{c_1^* c_2 \rho_{12}\right\}\right|$ in Equation (5.30) becomes smaller than $|c_1 \rho_{12}|^2$ in Equation (5.45) because of favorable channel phases. Comparing the performance of the decorrelator for $\rho_{12} = 0.6$ with MLSE curve for the same ρ_{12} in Figure 5.30, we note that the loss is always less than 1 dB. Figure 5.37 shows the bit error curves for both receivers for a fixed correlation coefficient $\rho_{12} = 0.6$ and the different values of the fading amplitudes. For small interferer amplitudes $|c_2| \ll |c_1|$, the performance curves of the SUMF receiver approach the limit of the ideal single-user (SU) BPSK curve. However, for this relatively high correlation coefficient, the degradation becomes severe if the interferer power is of the same order (or even higher) as that of the user under consideration. The curve for the decorrelator does not depend on the amplitude of the interferer. It has a fixed degradation (caused by the geometry) of approximately 2 dB compared to the ideal BPSK. Therefore, for very low values of the interferer amplitude, the SUMF receiver performs better.

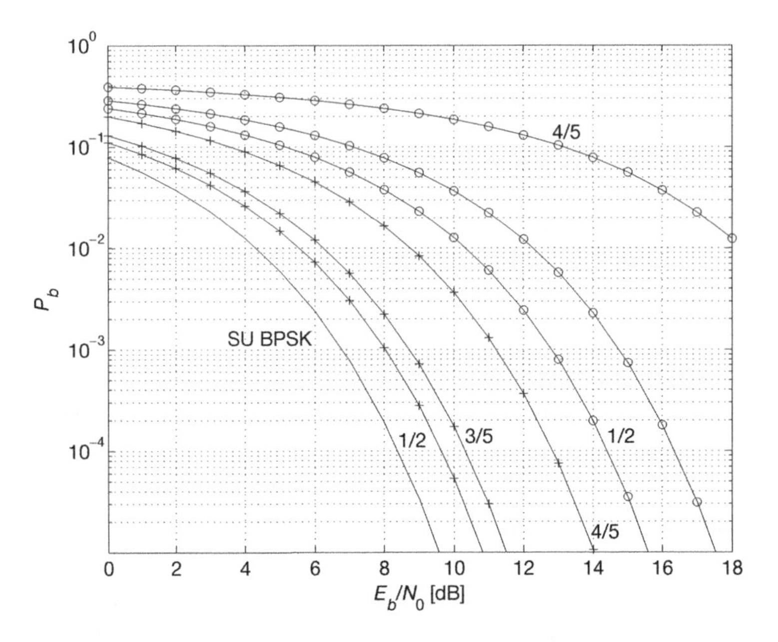

Figure 5.36 Bit error curves for the SUMF receiver (—o—) and the decorrelator (— + —) for equal fading amplitudes $c_1 = c_2 = 1$ and $\rho_{12} = 1/2,\ 3/5,\ 4/5$.

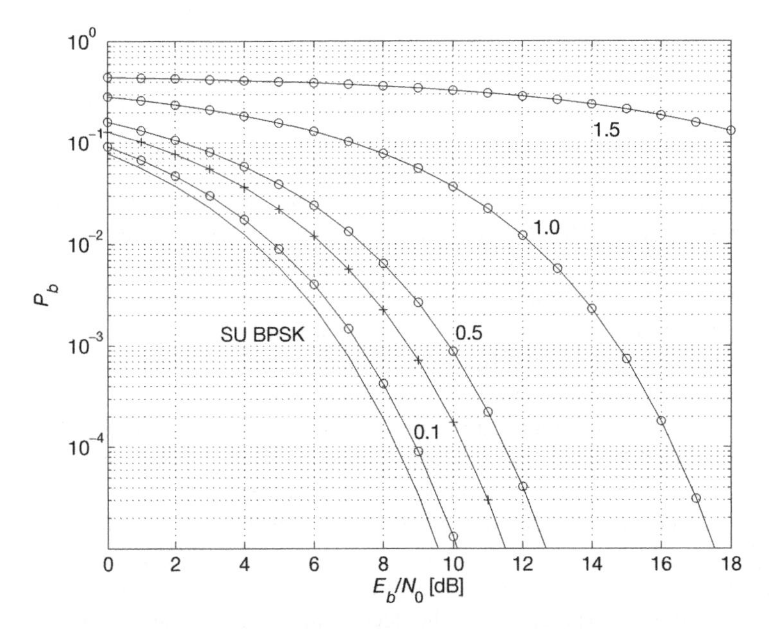

Figure 5.37 Bit error curves for the SUMF receiver (—o—) and the decorrelator (— + —) for $\rho_{12} = 3/5$ and fading amplitudes $c_1 = 1$ and $c_2 = 0.1,\ 0.5,\ 1,\ 1.5$.

The MMSE receiver

If an estimate for the SNR is available, a linear receiver that is better than the decorrelator can be obtained from the *minimum mean square error* (MMSE) condition. We consider the orthonormal base discrete channel model

$$\mathbf{r} = \mathbf{GCs} + \mathbf{n}$$

We look for a linear estimate $\hat{\mathbf{s}} = (\hat{s}_1, \ldots, \hat{s}_K)^T$ for $\mathbf{s} = (s_1, \ldots, s_K)^T$ that minimizes the mean square error

$$\mathrm{E}\left\{|\hat{s}_k - s_k|^2\right\}$$

for all k. A linear estimator is given by

$$\hat{s}_i = \sum_{k=1}^{K} b_{ik} r_k$$

or, in matrix notation, by

$$\hat{\mathbf{s}} = \mathbf{Br}$$

with $(\mathbf{B})_{ik} = b_{ik}$. The method to find the matrix $\mathbf{B}$ is quite similar to the derivation of the Wiener estimator in Subsection 4.3.3. We apply the orthogonality theorem of probability theory that says that the MMSE condition is equivalent to the orthogonality condition

$$\mathrm{E}\left\{(\hat{s}_i - s_i)\, r_k^*\right\} = 0.$$

In matrix notation, this can be written as

$$\mathrm{E}\left\{(\hat{\mathbf{s}} - \mathbf{s})\, \mathbf{r}^\dagger\right\} = 0.$$

We insert the linear estimator for $\hat{\mathbf{s}}$ and get

$$\mathbf{B}\mathrm{E}\left\{\mathbf{rr}^\dagger\right\} = \mathrm{E}\left\{\mathbf{sr}^\dagger\right\}.$$

Using the fact that the signal and the noise are uncorrelated and that the symbols are independently modulated, the left hand side can be evaluated as

$$\mathrm{E}\left\{\mathbf{rr}^\dagger\right\} = \mathbf{GC}\mathrm{E}\left\{\mathbf{ss}^\dagger\right\}(\mathbf{GC})^\dagger + \mathrm{E}\left\{\mathbf{nn}^\dagger\right\} = E_S\mathbf{GC}\,(\mathbf{GC})^\dagger + N_0\mathbf{I}.$$

For the right-hand side, we get

$$\mathrm{E}\left\{\mathbf{sr}^\dagger\right\} = E_S\,(\mathbf{GC})^\dagger.$$

We thus have

$$\mathbf{B}\left(E_S\mathbf{GC}\,(\mathbf{GC})^\dagger + N_0\mathbf{I}\right) = E_S\,(\mathbf{GC})^\dagger$$

or

$$\mathbf{B} = (\mathbf{GC})^\dagger\left(\mathbf{GC}\,(\mathbf{GC})^\dagger + \frac{N_0}{E_S}\mathbf{I}\right)^{-1}.$$

We note that for a matrix $\mathbf{A}$ the relation

$$\mathbf{A}^\dagger\left(\mathbf{AA}^\dagger + t\mathbf{I}\right)^{-1} = \left(\mathbf{A}^\dagger\mathbf{A} + t\mathbf{I}\right)^{-1}\mathbf{A}^\dagger \tag{5.47}$$

holds for any real number t (if the inverses mentioned above exist) (see Problem 4). We can thus write

$$\mathbf{B} = \left((\mathbf{GC})^\dagger \mathbf{GC} + \frac{N_0}{E_S}\mathbf{I} \right)^{-1} (\mathbf{GC})^\dagger .$$

Using elementary matrix operations, we can write this as

$$\mathbf{B} = \mathbf{C}^{-1} \left(\mathbf{G}^\dagger \mathbf{G} + \frac{N_0}{E_S} \left(\mathbf{CC}^\dagger\right)^{-1} \right)^{-1} \mathbf{G}^\dagger .$$

We thus have

$$\hat{\mathbf{s}} = \mathbf{C}^{-1} \left(\mathbf{G}^\dagger \mathbf{G} + \frac{N_0}{E_S} \left(\mathbf{CC}^\dagger\right)^{-1} \right)^{-1} \mathbf{G}^\dagger \mathbf{r}. \tag{5.48}$$

We note that in the limit $N_0 \to 0$ this approaches the decorrelating receiver. For very noisy channels it approaches the SUMF receiver.

The treatment is similar for the discrete matched filter model

$$\mathbf{v} = \mathbf{RCs} + \mathbf{m}.$$

We look for a linear estimator

$$\hat{\mathbf{s}} = \mathbf{Av}.$$

The orthogonality principle in matrix notation reads as

$$\mathrm{E}\left\{ (\hat{\mathbf{s}} - \mathbf{s})\, \mathbf{v}^\dagger \right\} = 0.$$

We insert the linear estimator for $\hat{\mathbf{s}}$ and get

$$\mathbf{A}\mathrm{E}\left\{ \mathbf{vv}^\dagger \right\} = \mathrm{E}\left\{ \mathbf{sv}^\dagger \right\}.$$

The left-hand side can be evaluated as

$$\mathrm{E}\left\{ \mathbf{vv}^\dagger \right\} = \mathbf{RC}\mathrm{E}\left\{ \mathbf{ss}^\dagger \right\} (\mathbf{RC})^\dagger + \mathrm{E}\left\{ \mathbf{mm}^\dagger \right\} = E_S \mathbf{RCC}^\dagger \mathbf{R}^\dagger + N_0 \mathbf{R}.$$

For the right-hand side, we get

$$\mathrm{E}\left\{ \mathbf{sv}^\dagger \right\}$$

We thus have

$$\mathbf{A}\left(E_S \mathbf{RCC}^\dagger \mathbf{R} + N_0 \mathbf{R} \right) = E_S \mathbf{C}^\dagger \mathbf{R},$$

where we have used the fact that $\mathbf{R}$ is Hermitian. We assume that the matrices $\mathbf{R}$ and $\mathbf{C}$ are regular, and multiply from the right-hand side by $\mathbf{R}^{-1} \left(\mathbf{C}^\dagger\right)^{-1} \mathbf{C}^{-1}$ and obtain

$$\mathbf{A}\left(E_S \mathbf{R} + N_0 \left(\mathbf{CC}^\dagger\right)^{-1} \right) = E_S \mathbf{C}^{-1}$$

and, finally,

$$\mathbf{A} = E_S \mathbf{C}^{-1} \left(E_S \mathbf{R} + N_0 \left(\mathbf{C}^\dagger \mathbf{C}\right)^{-1} \right)^{-1}.$$

The estimate for the transmit vector is then given by

$$\hat{\mathbf{s}} = \mathbf{C}^{-1} \left(\mathbf{R} + \frac{N_0}{E_S} \left(\mathbf{CC}^\dagger\right)^{-1} \right)^{-1} \mathbf{v}. \tag{5.49}$$

5.3.4 Suboptimal nonlinear receiver structures

In this subsection, we present the method of *successive interference cancellation* (SIC) (see e.g. (Frenger *et al.* 1999)). There are two variations of that theme. The *serial* SIC should be preferred if the different users have very different power levels as is the case in the absence of power control. We see that, in this case, the SUMF receiver may totally fail for the users with low signal level. The *parallel* SIC is the better choice if the different users have approximately the same power levels as it is the case in the presence of power control. For this method, it is necessary that the SUMF receiver gives correct results at least for a significant part of the users.

Serial successive interference cancellation

The idea is simple: if the reception is corrupted by strong interferers with known signature pulses and known fading amplitudes, one can combat the near–far problem by adjusting the decision threshold to the situation.

We illustrate the idea for the case of two users with the same parameters as in Figure 5.27 and Figure 5.29. In those figures, the fading amplitudes are $c_1 = 1$ and $c_2 = 2$, and the transmit symbol energy has been normalized to $E_S = 1$. The signature vectors are given by

$$\mathbf{g}_1 = \begin{pmatrix} 1 \\ 0 \end{pmatrix}, \; \mathbf{g}_2 = \frac{1}{5}\begin{pmatrix} 3 \\ 4 \end{pmatrix}.$$

The situation is depicted in Figure 5.38. In the beginning, we do not have any knowledge about the BPSK transmit symbols s_1 and s_2. The receive vector is given by

$$\mathbf{r} = s_1\mathbf{g}_1 + 2s_2\mathbf{g}_2 + \mathbf{n},$$

where $\mathbf{n}$ is the two-dimensional AWGN. Because the signal of user 2 is stronger than that of user 1, the SUMF receiver will work quite well for that user by utilizing Threshold 2 in Figure 5.38 to provide us with a reliable estimate

$$\hat{s}_2 = \operatorname{sign}(\mathbf{g}_2 \cdot \mathbf{r})$$

of symbol s_2. If s_2 has been correctly decided as $\hat{s}_2 = +1$, we may now decide whether

$$\mathbf{s}_{00} = +\mathbf{g}_1 + 2\mathbf{g}_2$$

or

$$\mathbf{s}_{10} = -\mathbf{g}_1 + 2\mathbf{g}_2$$

has been transmitted. These two cases corresponding to Region 1 and Region 2 in Figure 5.38 are separated by Threshold 1a which is a straight half line with the same distance between $\mathbf{s}_{00}$ and $\mathbf{s}_{10}$. The half line ends at Threshold 2. For $\hat{s}_2 = -1$, we have to decide whether

$$\mathbf{s}_{01} = +\mathbf{g}_1 - 2\mathbf{g}_2$$

or

$$\mathbf{s}_{11} = -\mathbf{g}_1 - 2\mathbf{g}_2$$

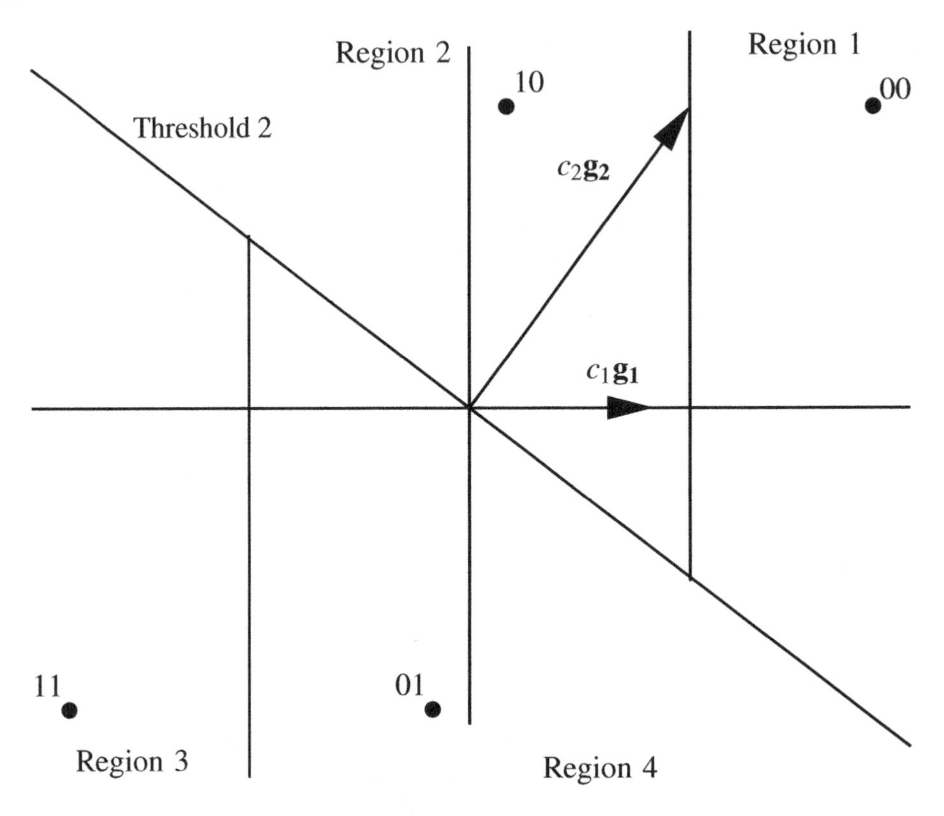

Figure 5.38 Two-user BPSK SIC receiver for unequal fading amplitudes.

corresponding to Region 3 and Region 4, respectively. These regions are separated by Threshold 1b. Formally, the decision on s_1 is given by

$$\hat{s}_1 = \operatorname{sign}(\mathbf{g}_1 \cdot (\mathbf{r} - 2\hat{s}_2\mathbf{g}_2)).$$

If s_2 has been correctly decided, this corresponds to the maximum likelihood decision for s_1. However, the estimate $\hat{s}_2$ may be wrong. Comparing Figure 5.29 and Figure 5.38, we note that there are two stripe-like shaped regions where the decisions on s_2 differ. However, there are only two small triangular regions where this affects the decision $\hat{s}_1$. Therefore, s_1 may be correctly decided even if the decision on s_2 is erroneous. This correct decision $\hat{s}_1$ now provides us with a better estimate for s_2, which is given by

$$\hat{s}_2 = \operatorname{sign}(\mathbf{g}_2 \cdot (\mathbf{r} - \hat{s}_1\mathbf{g}_1)).$$

Thus, this successive interference cancellation also improves the decisions for the stronger signal.

We now formalize the procedure and generalize to K users. For simplicity, we restrict ourselves to BPSK. The generalization on QPSK is straightforward. The receive signal is then given by

$$\mathbf{r} = \sum_{k=1}^{K} c_k s_k \mathbf{g}_k + \mathbf{n}.$$

Without losing generality, we assume that the signals are ordered according to their receive powers, that is,

$$|c_1|^2 \leq |c_2|^2 \ldots \leq |c_K|^2 .$$

We start with the estimate for the strongest signal and get

$$\hat{s}_K = \text{sign}\left(c_K^* \mathbf{g}_K^\dagger \mathbf{r}\right).$$

We now make decisions for the other $K - 1$ symbols $s_{K-1}, s_{K-2}, \ldots, s_1$ according to the descending signal strength order as

$$\hat{s}_k = \text{sign}\left(c_k^* \mathbf{g}_k^\dagger \left(\mathbf{r} - \sum_{i=k+1}^{K} c_i \hat{s}_i \mathbf{g}_i\right)\right).$$

After this first iteration, estimates $\hat{s}_1, \ldots, \hat{s}_K$ for all K symbols are available. We may now start a second iteration. We replace the old estimate $\hat{s}_K$ with the new one

$$\hat{s}_K^{(\text{new})} = \text{sign}\left(c_k^* \mathbf{g}_k^\dagger \left(\mathbf{r} - \sum_{i=1}^{K-1} c_i \hat{s}_i \mathbf{g}_i\right)\right),$$

thereby using the information available about all other symbol decisions obtained in the first iteration. We now make decisions for the other $K - 1$ signals according to the descending signal strength order as

$$\hat{s}_k^{(\text{new})} = \text{sign}\left(c_k^* \mathbf{g}_k^\dagger \left(\mathbf{r} - \sum_{i=1}^{k-1} c_i \hat{s}_i \mathbf{g}_i - \sum_{i=k+1}^{K} c_i \hat{s}_i^{(\text{new})} \mathbf{g}_i\right)\right).$$

Note that the estimates $\hat{s}_i$, $i < k$ of the last iteration are used together with the already available new estimates $\hat{s}_i^{(\text{new})}$, $i > k$. When all estimates are calculated in descending order, we may start a third iteration, and so forth.

The method described above is called *serial* SIC because the estimates are obtained serially in an order given by the signal strengths. It is especially useful if there are significant differences in the signal strengths. This is the case in the absence of power control. In case of an efficient power control, all signal powers are of the same order, and there is no reason for one of these signals to be privileged. In that case, we should apply the parallel SIC receiver.

Parallel successive interference cancellation

In a parallel SIC receiver, all iterations are performed in parallel. Again, we consider BPSK transmission and the discrete channel model

$$\mathbf{r} = \sum_{k=1}^{K} c_k s_k \mathbf{g}_k + \mathbf{n}.$$

In the first iteration, the SUMF receiver outputs for *all users* are calculated *in parallel* to obtain the estimates

$$\hat{s}_k = \text{sign}\left(c_k^* \mathbf{g}_k^\dagger \mathbf{r}\right)$$

for $k = 1, \ldots, K$. After this first iteration, estimates $\hat{s}_1, \ldots, \hat{s}_K$ for all K symbols are available. We may now start a second iteration. We replace the old estimate $\hat{s}_k$ by the new one

$$\hat{s}_k^{(\text{new})} = \text{sign}\left(c_k^* \mathbf{g}_k^\dagger \left(\mathbf{r} - \sum_{i \neq k}^{K} c_i \hat{s}_i \mathbf{g}_i\right)\right).$$

Again, this is done in parallel. The iteration can be repeated several times. At all times, a new estimate $\hat{s}_k^{(\text{new})}$ will be obtained from the estimates $\hat{s}_i$, $i \neq k$ for all other users from the last iteration step.

It is obvious that this method needs a reliable first SUMF estimate for most users. It will fail if most users have a power level far below the level of one or more high-power level users.

5.4 Receiver Structures for MC-CDMA and Asynchronous Wideband CDMA Transmission

5.4.1 The RAKE receiver

As already pointed out at the beginning of this chapter, spreading by itself does not improve the power efficiency (i.e. the BER performance as a function of E_b/N_0) in a Gaussian channel. It only uses a spectrally inefficient pulse shape that spreads the same power over a wider transmission band. However, in a fading environment, the receiver may take advantage from this higher transmission bandwidth. As we have seen in Subsection 4.4.3, a wideband channel has an inherent diversity degree that is significantly greater than 1 if the transmission bandwidth B significantly exceeds the correlation bandwidth f_{corr}. Since $B \approx 1/T_c$, this means that the delay spread $\Delta\tau$ must significantly exceed the chip duration T_c. The interpretation is that the signal arrives at the receiver via several uncorrelated or only weekly correlated transmission paths corresponding to different delays. We may thus speak of *path diversity* of the channel. In the heuristic discussion in Subsection 5.1.1, we introduced the RAKE receiver as a device that can exploit this diversity. In the following discussion, we shall derive this receiver structure as a special case of the maximum ratio combiner if the following model assumptions are made:

- The signal arrives at the receiver via L discrete transmission paths with delays τ_l, $l = 1, \ldots, L$.
- We may consider the signal of one time slot separately. Intersymbol interference may be neglected due to orthogonality or may be regarded as an additional noise-like disturbance. MAI is treated similarly.
- We neglect the time variance of the channel during the time slot under consideration.

For reasons of receiver implementation, we will assume at a later stage that each τ_l is an integer multiple of the chip duration T_c. We should keep in mind that the assumption of L discrete transmission paths is a very rough reflection of a real physical channel impulse response and that there is some arbitrariness in the model as to how these delays are chosen.

We now assume that user number k with the signature pulse $g_k(t)$ transmits a BPSK symbol $s_k \in \{\pm\sqrt{E_b}\}$. The transmit pulse is normalized to

$$\|g\|^2 = \int_{-\infty}^{\infty} |g(t)|^2 \, \mathrm{d}t = 1.$$

Because we only look at this one user number k, we may drop the index k and write $sg(t)$ instead of $s_k g_k(t)$. The signal will be transmitted over a multipath channel with complex fading amplitudes c_l corresponding to the delays τ_l. The receive signal is then given by

$$r(t) = \sum_{l=1}^{L} c_l s g(t - \tau_l) + n(t),$$

where $n(t)$ is the complex AWGN. We may write this as

$$r(t) = sh(t) + n(t),$$

where

$$h(t) = \sum_{l=1}^{L} c_l g(t - \tau_l)$$

is the convolution of the transmit pulse with the impulse response of the multipath channel and can be interpreted as a composed transmit pulse that includes the multipath channel. We may interpret $h(t)$ as a one-dimensional transmit base. Sufficient statistics is trivially given by the detector output

$$v = \mathcal{D}_h[r] = \int_{-\infty}^{\infty} h^*(t) r(t) \, \mathrm{d}t \tag{5.50}$$

corresponding to $h(t)$. This can be written as

$$v = \sum_{l=1}^{L} c_l^* \int_{-\infty}^{\infty} g^*(t - \tau_l) r(t) \, \mathrm{d}t = \sum_{l=1}^{L} c_l^* \mathcal{D}_l[r], \tag{5.51}$$

where

$$\mathcal{D}_l[r] = \int_{-\infty}^{\infty} g^*(t - \tau_l) r(t) \, \mathrm{d}t$$

is the detector for the delayed signature pulse $g(t - \tau_l)$. Thus, the receiver multiplies the detector outputs for all the delayed pulses by the complex conjugate channel amplitudes c_l^* and sums them up (*combines them*) (see Figure 5.39).

The combiner output is given by

$$v = \|h\|^2 s + m, \tag{5.52}$$

where

$$m = \mathcal{D}_h[n]$$

is a complex zero mean Gaussian random variable with variance $\sigma^2 = \|h\|^2 N_0$. The most probable transmit symbol $\hat{s}$ is obtained as

$$\hat{s} = \arg\min_s \left|v - \|h\|^2 s\right|^2 .$$

For BPSK, this is simply the sign[3] decision

$$\hat{s} = \text{sign}\left(\Re\{v\}\right).$$

For QPSK, we have

$$\hat{s} = \text{sign}\left(\Re\{v\}\right) + j\text{sign}\left(\Im\{v\}\right).$$

This so-called RAKE receiver is depicted in Figure 5.39 for BPSK. We note that the RAKE receiver is nothing but a special case of the MRC receiver discussed in Subsection 2.4.5. The L diversity branches corresponding to the detector outputs for the L paths are combined by Equation (5.51) in the same way as, for instance, the detector outputs obtained from different antennas.

We note that

$$\mathcal{D}_l[r] = g^*(-t) * r(t)\big|_{t=\tau_l}$$

is just the output of one filter (i.e. a correlator) matched to the signature pulse and sampled at the times $t = \tau_1, \ldots, \tau_L$. Thus, we do not have to implement a bank of L correlators as depicted in Figure 5.3. We only need one correlator (matched filter) to store the matched filter outputs and combine them with the appropriate multiplicative factors. If the delays τ_l are integer multiples of the chip duration T_c, we may implement the RAKE receiver by a filter matched to the chip pulse followed by a digital FIR (finite impulse response) filter structure.

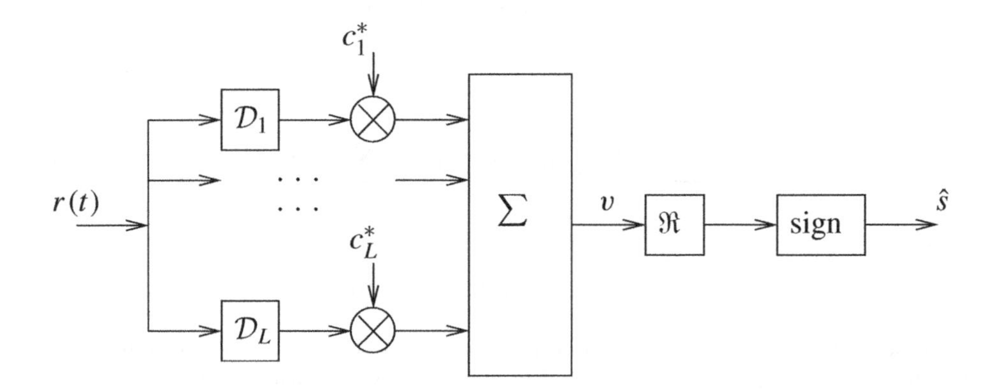

Figure 5.39 Block diagram for the BPSK RAKE receiver.

[3] Up to a factor that can be ignored for that decision.

Performance of the RAKE receiver

To evaluate the performance of the RAKE receiver, we divide Equation (5.52) by $\|h\|$ to obtain the normalized combiner output $u = v\,\|h\|^{-1}$ as

$$u = \|h\|\, s + n,$$

where the noise term n is a complex mean zero Gaussian random variable with variance $\sigma^2 = N_0$. We assume independent Rayleigh fading for the different transmission paths. Then the coefficients c_l are independent complex Gaussian random variables with mean zero. The respective transfer powers of the paths are given by the variances

$$\mathrm{E}\left\{|c_l|^2\right\} = \alpha_l^2.$$

To ensure that the energy per bit at the receiver end is given by E_b, we must normalize the signal components with respect to the fading coefficients such that

$$\alpha_1^2 + \cdots + \alpha_L^2 = 1.$$

To find an expression for the BER performance we assume pairwise orthogonality between the delayed versions of the signature pulse, that is, we assume

$$\int_{-\infty}^{\infty} g^*(t - \tau_i) g(t - \tau_l)\, \mathrm{d}t = \delta_{il}.$$

This assumption is not more severe than the other orthogonality assumptions above. However, we must keep in mind that our treatment is based on approximations. We may now write

$$\|h\|^2 = \sum_{l=1}^{L} |c_l|^2$$

and define the composed fading amplitude

$$a = \sqrt{\sum_{l=1}^{L} |c_l|^2 = \|h\|^2},$$

so the combiner output is then given by

$$u = as + n. \tag{5.53}$$

We thus have familiar BPSK (or QPSK) with a composed fading amplitude with power equal to the sum of the transmit powers of the different transmission paths. This is the same situation as for the MRC receiver for conventional diversity as discussed in Subsection 2.4.5. We may thus refer to the detailed analysis given in Subsection 2.4.6. For BPSK (or QPSK) transmission, Equation (2.38) simplifies to the expression for the bit error probability

$$P_b = \mathrm{E}_{c_1,\ldots,c_L}\left\{\frac{1}{2}\mathrm{erfc}\left(\sqrt{\sum_{i=1}^{L} |c_i|^2 \frac{E_b}{N_0}}\right)\right\},$$

where $\mathrm{E}_{c_1,\ldots,c_L}\{\cdot\}$ means averaging over the fading amplitudes. The same derivation as presented in that subsection now results in the integral expression

$$P_b = \frac{1}{\pi}\int_0^{\pi/2}\prod_{i=1}^{L}\frac{1}{1+\frac{\alpha_i^2}{\sin^2\theta}\frac{E_b}{N_0}}\,\mathrm{d}\theta,$$

which can easily be evaluated by numerical methods.

We finally note that the well-known formulas for L-fold diversity cited at the end of subsection 2.4.6 cannot be applied here because the transmit powers of the different paths typically are (in general) different. A formula for diversity with different powers can be found in (Proakis 2001). However, that formula can only be applied if no pair of diversity branches has the same power. The formula presented above is more flexible because it can be applied in any case.

The RAKE receiver and MAI

In the above derivation, we treated the RAKE receiver as a SUMF receiver with a matched filter $h^*(-t)$ that includes the impulse response of the channel. The RAKE receiver is the optimal receiver under the assumptions made at the beginning of the subsection. Even for one user only, there is the approximation that the signal for only one time slot and its delayed versions have to be taken into account. *Intersymbol interference* (ISI) from neighboring signals has been ignored. This may be justified by good correlation properties of the sequence.

For the RAKE receiver as a SUMF receiver, we may regard the MAI due to other users as a noise-like interference with Gaussian statistics. As already discussed for the synchronous case, this can be justified for the case of Rayleigh fading. The near–far problem is still present. If there is a *fast* power control mechanism that is able to follow the time variance of the channel, we do not have Rayleigh fading any more and the situation comes closer to a Gaussian channel.

Even though the RAKE receiver is essentially a detector matched to one user (including its signature waveform and the corresponding channel impulse response), we may think of multiuser detection for K users by using K RAKE receivers matched to all these users (including their signature waveforms and their respective channels impulse responses). If we restrict this treatment again to the signal for only one time slot and its delayed versions, we may write the receive signal as

$$r(t) = \sum_{k=1}^{K} s_k h_k(t) + n(t),$$

where s_k is the transmit symbol of user number k and $h(t)$ is the signature waveform convolved with the channel impulse response for user number k. The RAKE combiner output for user number i is then given by

$$v_i = \sum_{k=1}^{K} \rho_{ik} s_k + m_i$$

with $\rho_{ik} = \langle h_i, h_k \rangle$ and detector outputs m_i corresponding to the noise or, in vector notation,

$$\mathbf{v} = \mathbf{R}\mathbf{s} + \mathbf{m}$$

with $(\mathbf{R})_{ik} = \rho_{ik}$. This is just like the synchronous matched filter model, and we may thus apply any of the MUD receivers discussed in the preceding, e.g. successive interference cancellation. Optimality will never be achieved as other time slots are ignored, but some improvements are possible. This may help to reduce the effect of the near–far problem, but the problem is not resolved because the other time slots of the unwanted strong signal are not taken into account. This may severely influence the detector output for the wanted week signal. An optimal receiver must take all the time slots into account. We discuss such a receiver in the following subsection.

5.4.2 Optimal receiver structures

In this subsection, we study optimal receiver structures. The *maximum likelihood sequence estimation* (MLSE) for *jointly* optimum decisions and the *bitwise maximum a posteriori* (MAP) receiver for estimating the most probable individual bit transmitted by one user have already been discussed in some detail. The formalism can directly be applied to a channel that is formally described by

$$\mathbf{r} = \mathbf{H}\mathbf{s} + \mathbf{n}.$$

This channel model applies to synchronous wideband MC-CDMA and asynchronous wideband DS-CDMA as well. However, the interpretation of matrix $\mathbf{H}$ is different and both cases have to be discussed separately.

The optimal receiver for asynchronous DS-CDMA

We recall that the above compact notation for the discrete transmission model includes two degrees of freedom (or dimensions), and all quantities are labeled by two indices. One is the user index k that already occurs in the synchronous model. The other one is the time index l. Thus, the complexity of an exhaustive search for a maximum likelihood receiver would grow exponentially in the number of users K and with the frame duration L. Fortunately, one can usually assume that the time dispersion of the channel is finite. For single user transmission over a channel with finite time dispersion, an MLSE receiver can be implemented by using the Viterbi algorithm (see e.g. (Benedetto and Biglieri 1999; Kammeyer 2004; Proakis 2001)). This receiver type is sometimes called *Viterbi equalizer*. We may generalize that receiver structure for multiuser detection. Here, we note that the general case of different time-dispersive channels for different users already includes any case of asynchronous transmission.

The maximum likelihood receiver has to find the most probable transmit sequence $\hat{\mathbf{s}}$ that is given by

$$\hat{\mathbf{s}} = \arg\min_{\mathbf{s}} \|\mathbf{r} - \mathbf{H}\mathbf{s}\|^2$$

or, equivalently,

$$\hat{\mathbf{s}} = \arg\max_{\mathbf{s}} \left(\Re\left\{\mathbf{s}^\dagger \mathbf{H}^\dagger \mathbf{r}\right\} - \frac{1}{2} \|\mathbf{H}\mathbf{s}\|^2 \right).$$

We define

$$\mathbf{v} = \mathbf{H}^\dagger \mathbf{r}$$

and write

$$\hat{\mathbf{s}} = \arg\max_{\mathbf{s}} \left(\Re\left\{\mathbf{s}^\dagger \mathbf{v}\right\} - \frac{1}{2}\mathbf{s}^\dagger \mathbf{R}\mathbf{s} \right)$$

with the autocorrelation matrix defined by

$$\mathbf{R} = \mathbf{H}^\dagger \mathbf{H}.$$

Let us first consider the single-user case in a time-dispersive channel, that is, we rederive the well-known MLSE equalizer structure for that channel. The autocorrelation matrix $\mathbf{R}$ has elements given by

$$(\mathbf{R})_{lm} = \int h^*(t - lT_S)h(t - mT_S)\,\mathrm{d}t,$$

that only depend on the difference $l - m$ so $(\mathbf{R})_{lm} = \rho[l - m]$ with[4]

$$\rho[l] = \int h^*(t - lT_S)h(t)\,\mathrm{d}t.$$

The task now is to maximize the metric given by

$$\mu(\mathbf{s}) = \sum_l \Re\left\{ s_l^* \left(v_l - \frac{1}{2}\sum_m \rho[l - m]s_m \right) \right\}.$$

We assume that the time-dispersive channel has finite memory M, hence $\rho[m] = 0$ for $m > M$. The metric can now be written as

$$\mu(\mathbf{s}) = \sum_l \mu_l$$

with

$$\mu_l = \Re\left\{ s_l^* \left(v_l - \frac{1}{2}\sum_{m=-M}^{M} \rho[m]s_{l-m} \right) \right\},$$

where

$$\sum_{m=-M}^{M} \rho[m]s_{l-m}$$

is a convolution and can be interpreted as a digital *finite impulse response* (FIR) filter described by a shift register[5]. Assuming BPSK transmission, the shift register has 2^{2M} states characterized by

$$(s_{l-M+1}, s_{l-M+2}, \ldots, s_{l+M}),$$

and the MLSE receiver can easily be implemented by the Viterbi algorithm in the same way as for convolutional codes as described in Subsection 3.2.2, but with the metric defined above.

[4]We write ρ instead of ρ_{11} for a single user.

[5]We ignore that the filter is noncausal. This can easily be removed by a time shift.

The complexity of the trellis can be further reduced by noting that the matrix $\mathbf{R}$ is Hermitian. Consider the simplest case $M = 1$. The autocorrelation matrix is then given by

$$\mathbf{R} = \begin{pmatrix} \rho[0] & \rho^*[1] & 0 & 0 & 0 & \\ \rho[1] & \rho[0] & \rho^*[1] & 0 & 0 & \\ 0 & \rho[1] & \rho[0] & \rho^*[1] & 0 & \\ 0 & 0 & \rho[1] & \rho[0] & \rho^*[1] & \ddots \\ 0 & 0 & 0 & \rho[1] & \rho[0] & \ddots \\ & & & \ddots & \ddots & \ddots \end{pmatrix}.$$

For BPSK with $s_i = \pm\sqrt{E_b}$, we obtain[6]

$$\mathbf{s}^\dagger\mathbf{R}\mathbf{s} = E_b L \rho[0] + 2\Re\{\rho[1]\}(s_1 s_2 + s_2 s_3 + s_3 s_4 + \cdots).$$

Omitting the first term that is not relevant for the decision, we obtain

$$\mu(\mathbf{s}) = \Re\left\{\sum_l (s_l v_l - s_l s_{l+1}\rho[1])\right\},$$

and, with an appropriate renumbering,

$$\mu_l = \Re\{s_{l-1}(v_{l-1} - s_l\rho[1])\}.$$

Thus, the trellis diagram has two states labeled by the signs of s_{l-1}. The actual input bit is s_l.

For $M = 2$, the autocorrelation matrix has the following structure:

$$\mathbf{R} = \begin{pmatrix} \rho[0] & \rho^*[1] & \rho^*[2] & 0 & 0 & 0 & 0 & \\ \rho[1] & \rho[0] & \rho^*[1] & \rho^*[2] & 0 & 0 & 0 & \\ \rho[2] & \rho[1] & \rho[0] & \rho^*[1] & \rho^*[2] & 0 & 0 & \\ 0 & \rho[2] & \rho[1] & \rho[0] & \rho^*[1] & \rho^*[2] & 0 & \\ 0 & 0 & \rho[2] & \rho[1] & \rho[0] & \rho^*[1] & \rho^*[2] & \\ & & & \ddots & \ddots & \ddots & \ddots & \ddots \end{pmatrix},$$

and the metric increment is given by

$$\mu_l = \Re\{s_{l-2}(v_{l-2} - s_{l-1}\rho[1] - s_l\rho[2])\}.$$

Thus, we have a trellis diagram with four states labeled by the signs of (s_{l-1}, s_{l-2}). The actual input bit is s_l.

For general values of M, we readily find the expression

$$\mu_l = \Re\left\{s_{l-M}\left(v_{l-M} - \sum_{m=1}^{M} s_{l-M+m}\rho[m]\right)\right\},$$

which is described by a trellis diagram with 2^M states.

[6]Here and in the following text, we start with the index $l = 1$.

Since

$$v_l = \mathbf{h}_l^\dagger \mathbf{r} = \int h^*(t - lT_S) r(t)\, \mathrm{d}t$$

is already the combiner output of the RAKE receiver for time slot l, this single-user MLSE receiver is a generalization of the former because it takes intersymbol interference into account. The RAKE receiver by itself detects only one symbol. However, for a single user this intersymbol interference will not be severe if sequences with good correlation properties are chosen. Problems will occur because of intersymbol interference by a stronger user.

For K asynchronous users, the autocorrelation matrix is given by the general expression (5.25). As for the single-user case, for a channel with finite memory, the matrix elements will be zero outside the inner subdiagonals. We may then proceed in a similar fashion as for the single-user case and calculate $\mathbf{s}^\dagger \mathbf{R} \mathbf{s}$ to obtain metric expressions corresponding to that case. In contrast to the single-user case the subdiagonals do not consist of identical elements, and thus the metric increment calculation will be different from time-step to time-step, but it will be repeated after K time-steps. For $K = 2$, there will be two different expressions for the metric calculation, one for even and one for odd time indices.

We shall now evaluate the metric expression for $K = 2$ and finite memory such that $\mathbf{R}[l] = 0$ for $l > 1$. Thus we have to consider only the submatrices $\mathbf{R}[0]$ and $\mathbf{R}[1]$ with

$$\mathbf{R}[0] = \begin{pmatrix} \rho_{11}[0] & \rho_{12}[0] \\ \rho_{12}^*[0] & \rho_{22}[0] \end{pmatrix},$$

and

$$\mathbf{R}[1] = \begin{pmatrix} \rho_{11}[1] & \rho_{12}[1] \\ 0 & \rho_{22}[1] \end{pmatrix},$$

that is, we have assumed for simplicity that $\rho_{21}[1] = 0$. This is the case if the time dispersion is small. We note that for the asynchronous case without time dispersion, the diagonal elements of $\mathbf{R}[1]$ vanish.

We now write the matrix $\mathbf{R}$ as

$$\mathbf{R} = \begin{pmatrix} \rho_{11}[0] & \rho_{12}[0] & \rho_{11}^*[1] & 0 & 0 & 0 & 0 & \\ \rho_{12}^*[0] & \rho_{22}[0] & \rho_{12}^*[1] & \rho_{22}^*[1] & 0 & 0 & 0 & \\ \rho_{11}[1] & \rho_{12}[1] & \rho_{11}[0] & \rho_{12}[0] & \rho_{11}^*[1] & 0 & 0 & \\ 0 & \rho_{22}[1] & \rho_{12}^*[0] & \rho_{22}[0] & \rho_{12}^*[1] & \rho_{22}^*[1] & 0 & \\ 0 & 0 & \rho_{11}[1] & \rho_{12}[1] & \rho_{11}[0] & \rho_{12}[0] & \rho_{11}^*[1] & \\ & & \ddots & \ddots & \ddots & \ddots & \ddots & \ddots \end{pmatrix}.$$

This matrix shows some similarities to the matrix for the single-user case and channel memory $M = 2$. There is the main diagonal and two nonvanishing subdiagonals both in the upper and the lower triangle. Thus, we will also have a trellis with four states. However, the elements of the diagonals are different for even and for odd row indices. For BPSK with $s_i = \pm\sqrt{E_b}$, we evaluate the energy term as

$$\begin{aligned} \mathbf{s}^\dagger \mathbf{R} \mathbf{s} = {} & E_b L \left(\rho_{11}[0] + \rho_{22}[0]\right) \\ & + 2\Re\left\{\rho_{12}[0]\right\}(s_1 s_2 + s_3 s_4 + \cdots) + 2\Re\left\{\rho_{11}[1]\right\}(s_1 s_3 + s_3 s_5 + \cdots) \\ & + 2\Re\left\{\rho_{12}[1]\right\}(s_2 s_3 + s_4 s_5 + \cdots) + 2\Re\left\{\rho_{22}[1]\right\}(s_2 s_4 + s_4 s_6 + \cdots) \end{aligned}$$

Ignoring the first term that is not relevant for the decision, the metric increment expressions are

$$\mu_l = \Re\{s_{l-2}(v_{l-2} - s_{l-1}\rho_{12}[0] - s_l\rho_{11}[1])\}$$

for odd values of l and

$$\mu_l = \Re\{s_{l-2}(v_{l-2} - s_{l-1}\rho_{12}[1] - s_l\rho_{22}[1])\}$$

for even values of l.

The expressions for the metric increments for more users and more channel memory can be derived in a similar manner. Given these metric increment expressions, the MLSE receiver can be implemented by means of the Viterbi algorithm (see Subsection 3.2.2). Alternatively, the bitwise MAP receiver can be implemented by the BCJR algorithm (see Subsection 3.2.4). The first receiver makes joint optimal decisions, and the second one makes individually optimal bit decisions. The decisions may be different, but typically this is a very rare event. The MAP algorithm is computationally more complex, but it provides reliability information that may be used by an outer decoder. Alternatively, the suboptimal SOVA (see Subsection 3.2.3) may be used to obtain such reliability information.

We note that, since the complexity of both optimal receivers grows exponentially with the number of users, it can only be applied if K is not too large. However, one may think of a receiver that utilizes only the most significant entries in the matrix $\mathbf{R}$, which will then be an intermediate solution between the SUMF receiver (implemented by a RAKE receiver) and the optimal receiver. The most significant matrix entries (outside the main diagonal) correspond to the strongest interferers. Since the main task of multiuser detection is to mitigate the near–far problem, such a strategy will certainly improve the performance significantly.

The optimal receiver for synchronous wideband MC-CDMA

For synchronous MC-CDMA, the discrete transmission model is formally the same as above and thus the maximum likelihood estimate is formally the same, that is, it is given by

$$\hat{\mathbf{s}} = \arg\min_{\mathbf{s}} \|\mathbf{r} - \mathbf{H}\mathbf{s}\|^2,$$

or, equivalently, by

$$\hat{\mathbf{s}} = \arg\max_{\mathbf{s}} \left(\Re\left\{\mathbf{s}^\dagger\mathbf{v}\right\} - \frac{1}{2}\mathbf{s}^\dagger\mathbf{R}\mathbf{s}\right),$$

with

$$\mathbf{v} = \mathbf{H}^\dagger\mathbf{r}$$

and

$$\mathbf{R} = \mathbf{H}^\dagger\mathbf{H}.$$

However, the interpretation is different. Since $\mathbf{H} = \mathbf{C} \circ \mathbf{G}$, the column vector $\mathbf{h}_k$ of $\mathbf{H}$ is the signature vector of user k with random signs in frequency direction multiplied with the fading amplitudes for that user at the respective subcarrier frequencies. The component

$$v_k = \mathbf{h}_k^\dagger\mathbf{r}$$

of $\mathbf{v}$ is the detector output corresponding to that composed pulse shape. It acts as a back rotation of the channel phase and the PN sequence and weights with the fading amplitude.

In contrast to the DS-CDMA time-dispersive channel, there is no trellis structure involved here. Hence, the MLSE must be implemented by exhaustive search. Thus, the complexity grows exponentially with the number of users and with the length of the spreading sequence. Therefore, the MLSE receiver seems to be too complex for implementation in most cases. The same holds for the bitwise MAP receiver.

5.5 Examples for CDMA Systems

As mentioned in Section 5.1, the origin of spread spectrum is in the field of military applications and radar systems. Today, there are also some specialized markets for CDMA as wireless microphones. However, in this section we focus on the main commercial applications as wireless local area networks, satellite navigation and mobile radio networks.

5.5.1 Wireless LANs according to IEEE 802.11

In 1997, the Institute of Electrical and Electronics Engineers (IEEE) released the standard IEEE 802.11 for wireless local area network (WLAN) applications. Systems according to this standard offer a wireless access to existing local area computer networks via the so-called *access points* as well as direct wireless interconnections within a small group of computers.

Within the standard, three transmission modes are specified:

- an infrared (IR) mode,
- a frequency hopping (FH) mode,
- a direct sequence spread spectrum (DSSS) mode.

However, only products using the DSSS mode have been established on the market. The DSSS mode is operating in the ISM frequency band 2400–2485 MHz. ISM means that this band is free for unlicensed wireless systems for industrial, medical and scientific applications. Hence, it is not exclusively reserved for IEEE 802.11. Another prominent and important system operating in the same band is Bluetooth. Therefore, the spread spectrum technique is used within IEEE 802.11 to lower the power spectral density to reduce the interference to other systems and as a kind of *antijamming* method with respect to interference experienced from other systems with a smaller bandwidth. Bluetooth, for example, is using frequency hopping instead. It should be noted that IEEE 802.11 is not a CDMA system; separation of different connections and radio resource allocation is managed by a carrier sense multiple access (CSMA) scheme, which is known for wired Ethernet (IEEE 802.3) networks and which is adapted to the special needs of wireless networks.

Within the DSSS mode, two transmission schemes are defined working at data rates of 1 Mbit/s and 2 Mbit/s. Spreading within IEEE 802.11 is performed using an 11-chip Barker code word at a chip rate of 11 Mchip/s. Spreading and modulation is illustrated in Figure 5.40: for the data rate of 1 Mbit/s the bits are modulated by a DBPSK changing the

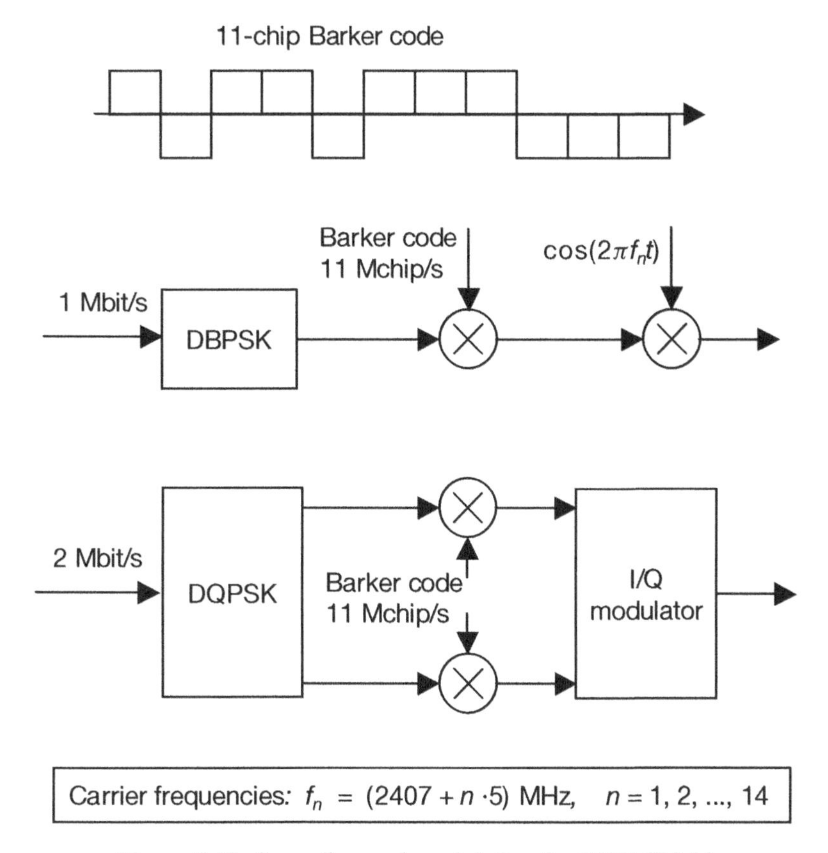

Figure 5.40 Spreading and modulation for IEEE 802.11.

phase by either 0 or π from bit to bit. For the data rate of 2 Mbit/s a DQPSK with symbols consisting of dibits is used; the phase is changed by a multiple of $\pi/2$ from dibit to dibit.

No details of the pulse shaping of the signal, but a spectrum mask is specified within IEEE 802.11, as illustrated in Figure 5.41. Within the ISM band, 14 carrier frequencies with a separation of 5 MHz are defined. However, in some countries or regions of the world, only a subset of them is allowed to be used. Furthermore, it should be noted that a carrier separation of 5 MHz is too low when operating, for example, two access points in one area. To reduce the adjacent channel interference to an acceptable level, the carriers used in the nearby access points should have a distance of at least 25 MHz. Hence, effectively there are only three to four carriers for IEEE 802.11, which are nearly disjoint with respect to the adjacent channel interference.

In 1999, the IEEE 802.11b standard (see (IEEE 802.11b 1999)) was released as an addition to the original standard specifying two new transmission modes with higher data rate, namely, 5.5 Mbit/s and 11 Mbit/s. These transmission modes are based on a code keying method named complementary code keying. The used 64 codes may be interpreted as complex-valued variants of the Walsh–Hadamard codes or OVSF codes discussed in Subsections 1.1.4 and 5.1.3, respectively. Each code consists of eight complex chips. The

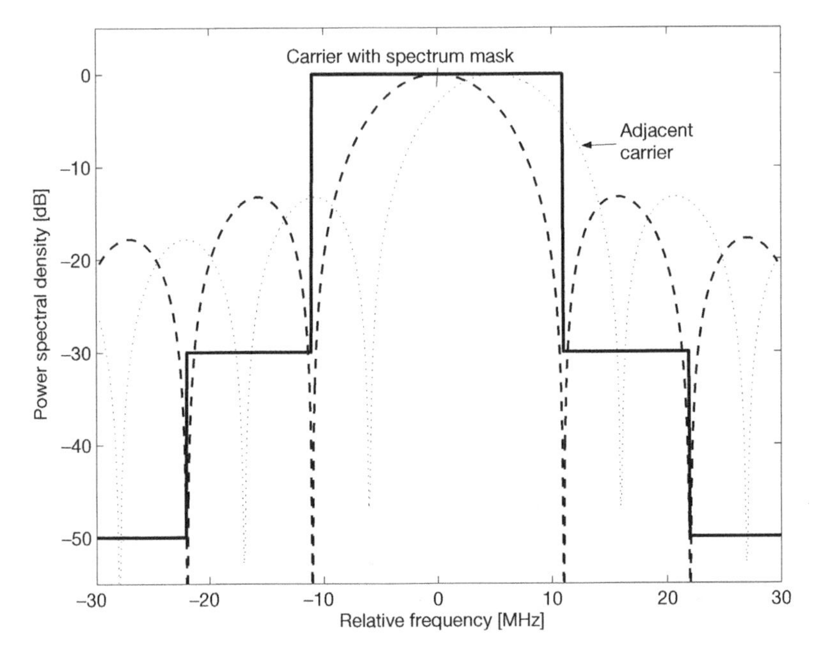

Figure 5.41 Spectrum of an IEEE 802.11b carrier.

chip rate has been chosen to be 11 Mchip/s as for the original transmission modes in order to preserve the spectral properties of a carrier. For the data rate of 11 Mbit/s, the mapping of data symbols to codes is performed in the following manner: a data symbol consists of eight bits that are grouped into four dibits (d_0, d_1, d_2, d_3). Using the last three dibits, one of the 64 8-chip codes is generated according to the tree structure shown in Figure 5.42; dibit d_1 affects each odd chip, d_2 each odd pair of chips and d_3 the first four chips (the odd quads of chips). The overall phase of the code is fixed dibit d_0 according to a DQPSK.

For the data rate of 5.5 Mbit/s, a symbol consists of two dibits (d_0, d_1), where the role of d_0 is the same as before, that is, fixing the overall phase of the code. Dibit d_1 selects one out of four 8-chip codes which form a subset of the 64 codes shown in Figure 5.42. The four codes in the subset are pairwise orthogonal.

Modulation for the additional IEEE 802.11b transmission modes is illustrated in Figure 5.43. A so-called *cover code* flipping the fourth and seventh chip is applied to optimize the autocorrelation properties and to minimize DC offsets in the codes.

Systems according to IEEE 802.11b use link adaption, that is, one out of the four transmission schemes discussed above is selected automatically based on the measured link quality. It should be mentioned that to handle transmission errors of data packets, an Automatic Repeat reQuest (ARQ) scheme, not a forward error correction, is applied.

Further types of IEEE 802.11 – the types IEEE 802.11a and IEEE 802.11g mentioned in Subsection 4.6.3 – use OFDM to handle data rates of up to 54 Mbit/s.

More details on the IEEE 802.11 standard can be found, for example, in (O'Hara and Petrick 1999).

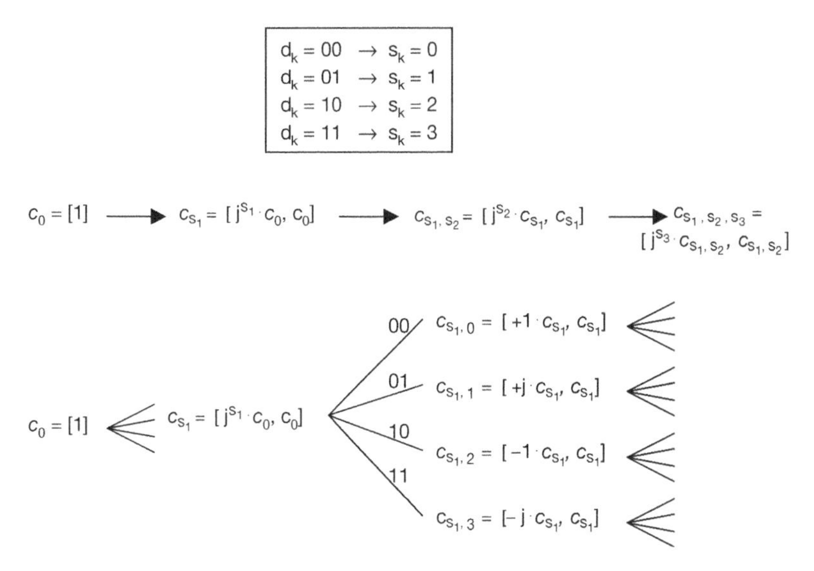

Figure 5.42 Generation of complementary codes.

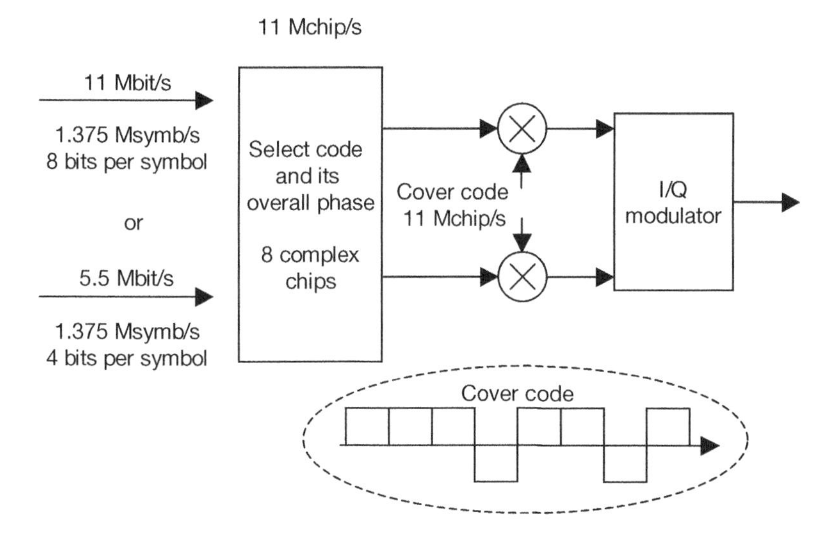

Figure 5.43 Modulation and spreading for IEEE 802.11b.

5.5.2 Global Positioning System

The core of the Global Positioning System (GPS) is the space segment consisting of 24 operational satellites surrounding the earth on six different orbits (i.e. four satellites per orbit) at a distance of about 20,000 km. Each satellite is equipped with a very precise atomic clock generating the fundamental frequency $f_0 = 10.23$ MHz from which all other frequencies as the carrier frequencies, the chip rates and the data bit rates are derived. The development of GPS was initiated by the US Department of Defense for military applications, but later was opened to civil applications. Correspondingly, GPS offers two kinds of services: a precise positioning service (PPS) – primarily for military and other authorized applications – and a less precise standard positioning service (SPS) for everyone.

Positioning is based on the principle of trilateration: measuring the propagation delay of the signals transmitted by three satellites, the corresponding distances can be calculated using values of the velocity of electromagnetic waves in free space and in the ionosphere and atmosphere. Knowing the position of the satellites, the position of the GPS receiver is obtained as the intersection of the spheres with the derived radii around the three satellites. Since the time when the satellite transmits the signal cannot be determined exactly by the receiver, only relative propagation delays can be measured. Therefore, an exact positioning requires the reception of the signals of at least four satellites.

Satellite signals are transmitted on two frequency bands, namely, on the so-called *L1 band* with a carrier frequency of $f_1 = 154 \cdot f_0 = 1575.42$ MHz and on the L2 band with a carrier frequency of $f_2 = 120 \cdot f_0 = 1227.60$ MHz. For the SPS only the first carrier is needed, whereas the PPS requires both carriers, since reception on both carriers allows estimation and elimination of the effect of the ionospheric refraction which is an essential source for errors.

Spreading and CDMA is performed in GPS by two types of codes:

1. The coarse/acquisition or clear/access codes (C/A codes) with a chip rate of $r_{C/A} = f_0/10 = 1.023$ Mchip/s.

2. The precision or protected code (P code) with a chip of $r_P = f_0 = 10.23$ Mchip/s.

The C/A codes are Gold codes generated by two 10-stage shift registers as illustrated in Figure 5.44. Hence, the codes have a period of 1023 chips corresponding to a duration of 1 ms. The chip length is about 1 μs, so that the time of arrival may be measured by code correlation with an accuracy of about some tenth of microsecond corresponding to a position accuracy of about 100 m. A higher resolution can be achieved by tracking not only the code, but also the phase of the carrier. Using the two 10-stage shift registers, 1023 different Gold codes can be generated by adding shifted versions of the m-sequence of the second register to the (unshifted) m-sequence of the first register as explained in Subsection 5.1.3. In GPS, only a part of these Gold codes are used which are identified by a so-called *PRN number* (pseudorandom noise sequence number). Shifted versions of the second m-sequence are obtained by extracting the register content at two variable stages S_1 and S_2 and by combining the corresponding two outputs. Examples for the relation between the PRN number, the output stages and the corresponding shift are given in the table at the bottom of Figure 5.44. A unique C/A code is assigned to each satellite. The used code spreads the proper navigation message transmitted by the respective satellite.

The navigation message contains, for example, the system time, the orbit data of the satellite, the satellite status and its signal reliability as well as data for correcting propagation delay measurements. A satellite transmits a navigation message every 30 seconds at a bit rate of 50 kbit/s. Hence, within one bit period a C/A code is repeated 20 times. Since each

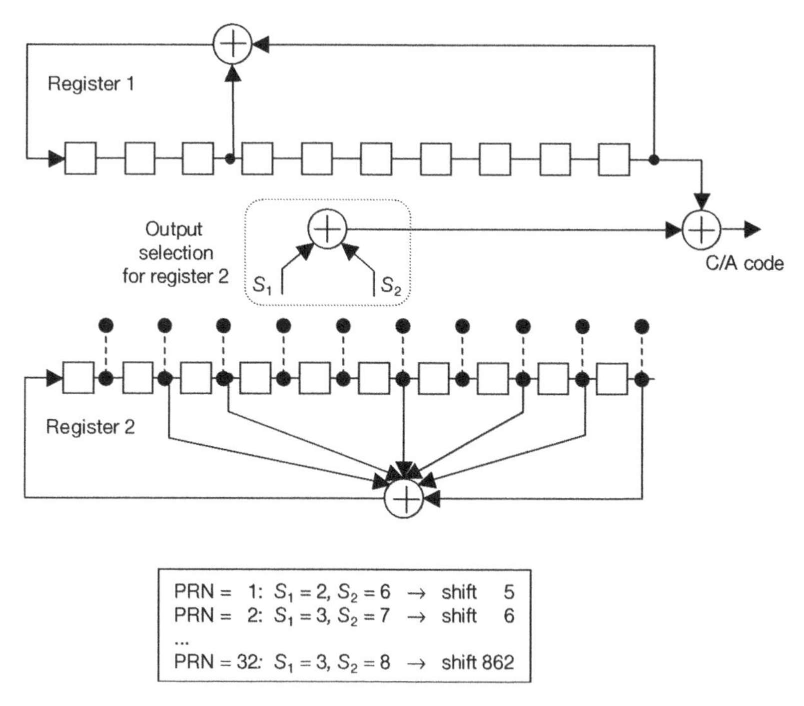

Figure 5.44 Generation of the C/A codes using two 10-stage shift registers.

satellite uses its individual code, the satellite signals can be separated and a parallel decoding of the corresponding navigation messages can be accomplished. A GPS receiver has to be able to manage at least four parallel code channels. However, 12-channel receivers are also available. Though the process of receiving messages from several satellites via different code channels has some similarities to a soft handover, there is an essential difference: for a soft handover the same data are transmitted by the involved base stations, while the satellites transmit different navigation messages.

The second type of code, the P code, is used on the L1 and the L2 carriers mainly for military applications. The very long P code of a length of about $2.4 \cdot 10^{14}$ chips (about 266 days) is generated at a rate of $r_P = f_0 = 10.23$ Mchip/s, that is, at a rate that is 10 times the rate of the C/A code and therefore results in better time resolution. The P code may be encrypted by a so-called secret *W code* to form the Y code which can only be decoded by authorized users. The P or Y code is separated into 37 unique one-week segments. Each satellite uses a satellite-specific segment that is restarted at the beginning of every GPS week (Saturday midnight).

The overall spreading modulation process is illustrated in Figure 5.45. More details on GPS can be found, for example, in (Hofmann-Wellenhof *et al.* 2001).

5.5.3 Overview of mobile communication systems

Before discussing some CDMA mobile communication systems in more detail, a general overview is presented to explain the evolution of the most important standards.

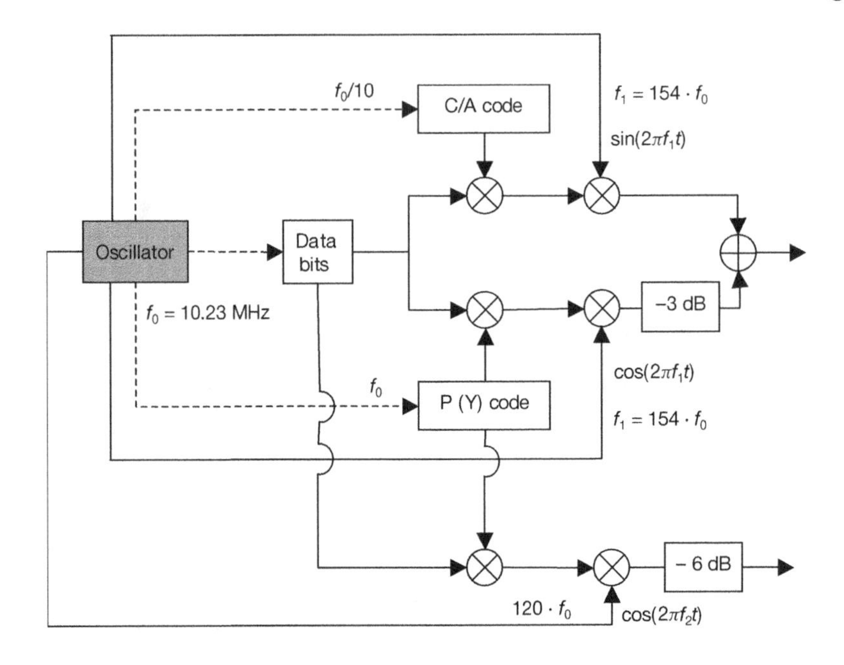

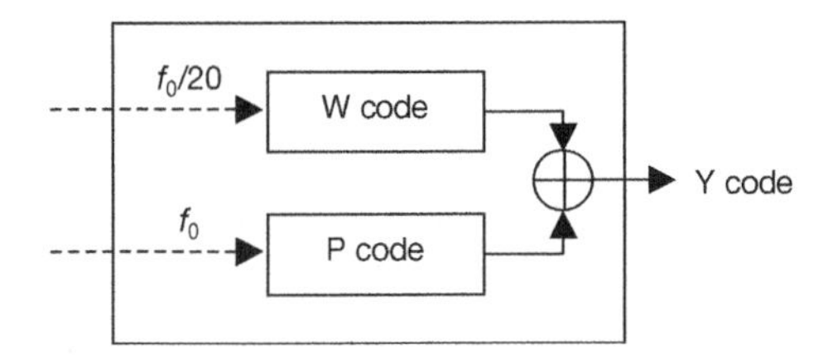

Figure 5.45 Spreading and modulation for GPS.

Mobile communication systems are usually classified into a certain number of generations as illustrated in Figure 5.46. The first generation (1G) comprised of several incompatible systems using analog transmission techniques for mainly speech services. In general, a roaming between different countries was not possible. To overcome these as well as security and capacity problems, standardization bodies in Europe, America and Japan developed the following standards for public mobile communication networks:

- Global System for Mobile Communications (GSM, Europe)
- Digital American Mobile Phone System (DAMPS) or Interim Standard 136 (IS-136, USA)
- Interim Standard 95 (IS-95), nowadays called *cdmaOne* (USA)
- Personal Digital Cellular (PDC, Japan).

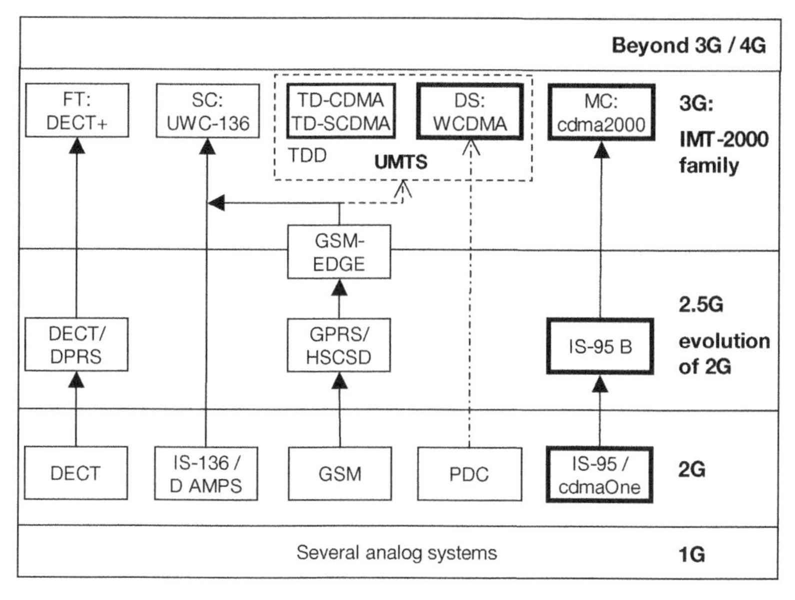

Figure 5.46 Overview of mobile communication systems.

The first releases of these standards were published between 1989 and 1994. Since then, several hundred networks according to these standards were established all over the world serving more than 1.3 billion subscribers (June 2004) as illustrated by Figure 5.47[7]. During the same time, standards for cordless telecommunications as the Digital Enhanced Cordless Telecommunications (DECT) or the Personal Handy-phone System (PHS) evolved. The main common features of these systems of the second generation (2G) are

- a digital transmission technique
- an enhanced network capacity
- speech and data services with data rates of about 10 kbit/s
- security mechanisms
- international roaming capabilities.

Between 1995 and 1998, the main focus for new releases of these standards was on creating new data services allowing higher rates and a packet switched mode. The increase of the data rate was accomplished by allocating several channels for one connection and by using

[7]The data of the Figure have been extracted from the internet pages of the GSM Association (www.gsmworld.com) and of the CDMA Development Group (www.cdg.org) where the interested reader may find the recent statistics.

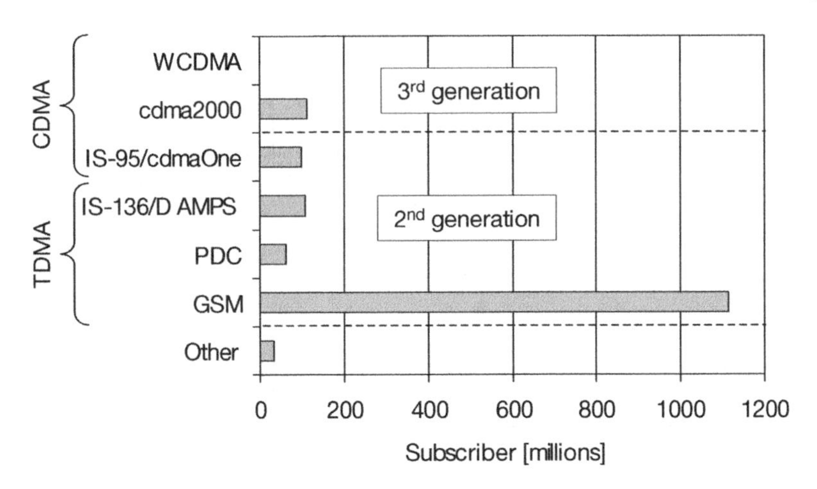

Figure 5.47 Subscriber statistics (June 2004).

flexible coding schemes as well as sophisticated ARQ schemes. For example, the GSM standard was enhanced by two types of bearer services called *High-speed Circuit Switched Data* (HSCSD) and *General Packet Radio Service* (GPRS) that offer data rates of up to 160 kbit/s (see e.g. (Steele *et al.* 2001)). Using higher-order modulation schemes as defined in the EDGE mode (Enhanced Data rates for GSM Evolution) of GSM, even data rates of up to 500 kbit/s can be achieved. 2G systems with these increased data rate capabilities are called *systems* of generation 2+ or generation 2.5, indicating that they represent an evolutionary step from 2G to 3G systems.

For systems of the third generation (3G), the International Telecommunications Union (ITU) set the following main requirements:

- capabilities for offering mixed services with various quality of service requirements;
- an enhanced network capacity with an efficient usage of the frequency spectrum;
- data rates of up to 2 Mbit/s.

It should be noted that the data rate of 2 Mbit/s has to be achieved only in indoor environments; for pedestrian environments (city centers) and for a wide area coverage at a high velocity of the mobile station data rates of 384 kbit/s and 144 kbit/s are required, respectively (see e.g. (ETSI TR 101 112 1998)).

These requirements initiated a system selection process which proceeded in several steps: in a first step, different proposals (partly based on funded research projects) were submitted to the standardization bodies of the respective regions where a preselection was performed. These preselected proposals were further submitted to the ITU. After a harmonization phase, a multimode concept was defined selecting basically five transmission modes that formed the IMT-2000 family of systems. IMT-2000 means International Mobile Telecommunications in frequency bands at about 2000 MHz. Figure 5.48 shows the frequency bands allocated to IMT-2000 by the ITU and their present usage in different regions of the world. It should be noted that in the meantime further frequency bands have been

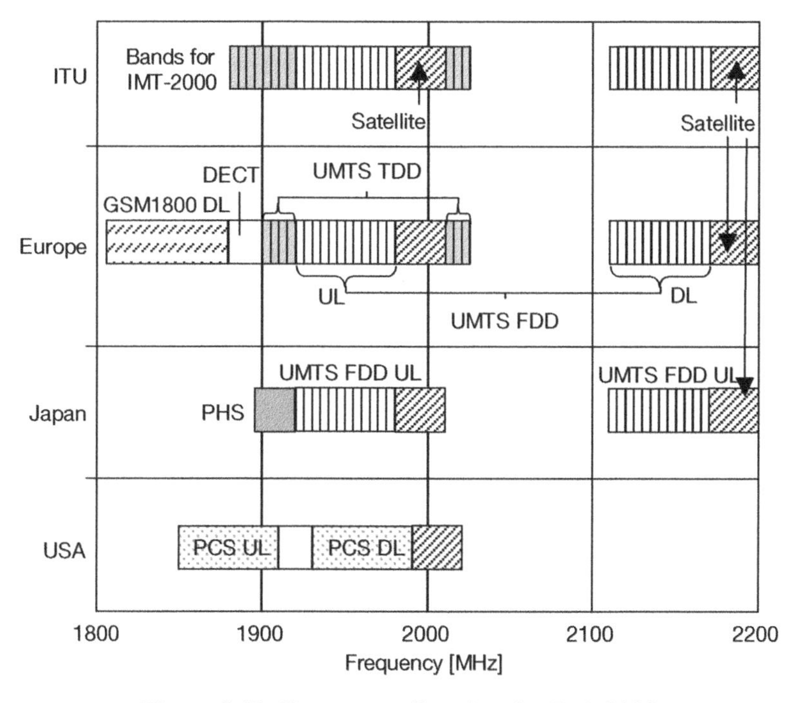

Figure 5.48 Frequency allocation for IMT-2000.

identified by the ITU for systems of the IMT-2000 family. The systems of the IMT-2000 family are listed on the top of Figure 5.46.

The Universal Mobile Telecommunication System (UMTS) originally has been proposed by the European Telecommunications Standards Institute (ETSI). However, since countries outside Europe also decided to introduce UMTS-like networks, the 3rd Generation Partnership Project (3GPP) was created in 1999 to harmonize the standardization activities. Beside the ETSI, standardization organizations from the United States, Japan, Korea and China are involved in the 3GPP. Within this specification group, the radio access technology is referred to as UTRA (Universal Terrestrial Radio Access). For UTRA two different transmission modes have been specified: a frequency division duplex (FDD) mode using pure CDMA and a time division duplex (TDD) mode using a combination of CDMA and TDMA as the multiple access scheme. The FDD mode is based on a proposal originally called the *Wideband CDMA* (WCDMA), whereas the TDD mode is based on the so-called *TD-CDMA* proposal. Within the TDD mode a second submode has been specified, which was submitted to the ITU by the Chinese standardization organization CWTS as Time Division Synchronous CDMA (TD-SCDMA). While the WCDMA and TD-CDMA-based modes use a chip rate of 3.84 Mchip/s and a carrier separation of about 5 MHz, the chip rate for the TD-SCDMA mode is 1.28 Mchip/s at a carrier separation of about 1.6 MHz. In the first phase, UMTS networks are operated mainly in the FDD mode. UMTS networks are and will be established in Europe and other countries with GSM as the 2G technology as well as in Japan. Though the multiple access scheme and many details of the transmission techniques of UMTS differ significantly from those of GSM, some frame periods were adapted to facilitate a handover between UMTS and GSM networks. Furthermore, some packet switching schemes and concepts for the core network were harmonized. In this sense, UMTS may be viewed as a kind of evolution of GSM. More details concerning UMTS can be found in Subsections 5.5.4 and 5.5.5.

In the United States, the preconditions for 3G mobile radio networks are more complicated: 2G networks according to the three standards, GSM, IS-95 and IS-136 have been established. Furthermore, these networks are operated for a main part in the frequency bands that were originally allocated to IMT-2000 by the ITU. For this reason, concepts for a continuous *reforming* of the already used frequency bands have been considered within the 3G standardization process. One evolutionary path starts from the IS-136 and the GSM/EDGE technology to create the UWC-136 standard (UWC: Universal Wireless Communications) using TDMA as the multiple access scheme. The starting point for the CDMA evolution path to 3G is cdmaOne (IS-95) using a chip rate of 1.2288 Mchip/s and a carrier spacing of 1.25 MHz. For the corresponding IMT-2000 member cdma2000, a variety of transmission modes is specified, which is based on these parameters. Some transmission modes use the same chip rate (and carrier spacing) as cdmaOne, but different spreading factors and channel coding schemes. For especially high rate services, the threefold chip rate (and carrier separation) may be used in the UL. For the DL, a multicarrier concept has been defined allowing the transmission on three (or in future versions even more) parallel cdmaOne carriers. Further details on cdmaOne and cdma2000 can be found in Subsections 5.5.6 and 5.5.7, respectively. Standardization of cdma2000 is performed in a second 3GPP group called *3GPP2*, which is formed by standardization bodies from the United States, Korea, China and Japan.

To complete the discussion on the IMT-2000 family, it should be mentioned that an evolution of the DECT standard also has been included as a member. However, DECT is not intended to be used for a wide area coverage, but for indoor coverage where it offers the required data rate of 2 Mbit/s.

Looking once more at Figure 5.46 and Figure 5.47, one observes that TDMA is by far the dominating multiple access technology in 2G while for 3G networks CDMA plays the dominant role. CDMA is used in 3G in three different modes:

- the classical direct sequence mode for UTRA FDD;
- a combination of TDMA and CDMA for UTRA TDD;
- as CDMA with an integrated multicarrier concept for cdma2000.

Within the IMT-2000 family, these modes are therefore denoted as DS-CDMA, TDD-CDMA and MC-CDMA. Since the evolutionary path from cdmaOne to cdma2000 was significantly smoother than the one from GSM to UMTS, in 2004 more subscribers could be found in cdma2000 than in UMTS networks.

More details on the standardization process for 3G mobile communication systems can be found, for example, in (Holma and Toskala 2001; Konhäuser *et al.* 2000; Steele *et al.* 2001).

5.5.4 Wideband CDMA

Within this subsection, an overview of the physical layer of the Wideband CDMA system, which is specified as the UTRA FDD mode, is given. More details can be found, for example, in (Holma and Toskala 2001) or within the standardization documents quoted in the following text.

Frequency allocation

As illustrated in Figure 5.48, the frequency bands 1920–1980 MHz (UL) and 2110–2170 MHz (DL) are allocated to the UTRA FDD mode based on WCDMA. Hence, totally there are 12 frequency blocks of 5 MHz each for the UL and the DL. Since, in general, there are several operators per network, typically only 1–3 duplex blocks are available per network. For example, in Germany each operator has two duplex blocks. Within each block one duplex carrier can be realized; hence, the standard carrier spacing is 5 MHz. However, the standard allows some fine-tuning by adjusting the carrier separation in steps of 0.2 MHz (see e.g. (3GPP TS 25.104 2004; 3GPP TS 25.101 2004)). For example, an operator may choose a more narrow carrier spacing of 4.4 MHz between his own carriers and therefore a wider spacing with respect to carriers of other operators, as it is easier to control the interference from adjacent carriers within the same network than that from foreign networks.

For a network rollout, the operators start in general with one duplex carrier per cell, that is, the same frequency carrier is reused in each cell so that a soft handover is possible within the complete network. To enhance the capacity, the operator can allocate a second carrier to cells in regions with high load or he can build up a hierarchical network allocating the second carrier to the additional microcells in hot spots. A typical frequency allocation pattern is illustrated in Figure 5.49.

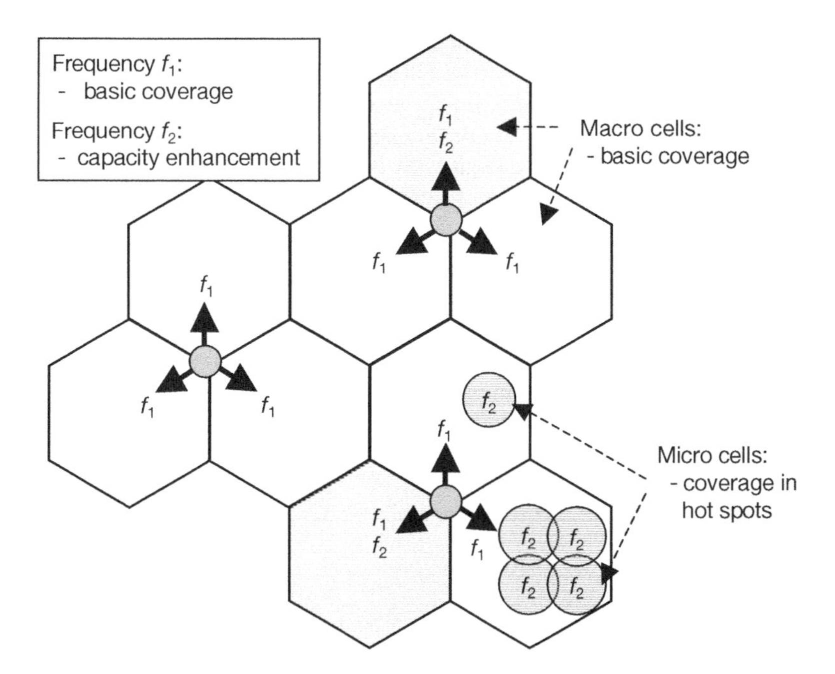

Figure 5.49 Typical UTRA FDD frequency plan using two carriers.

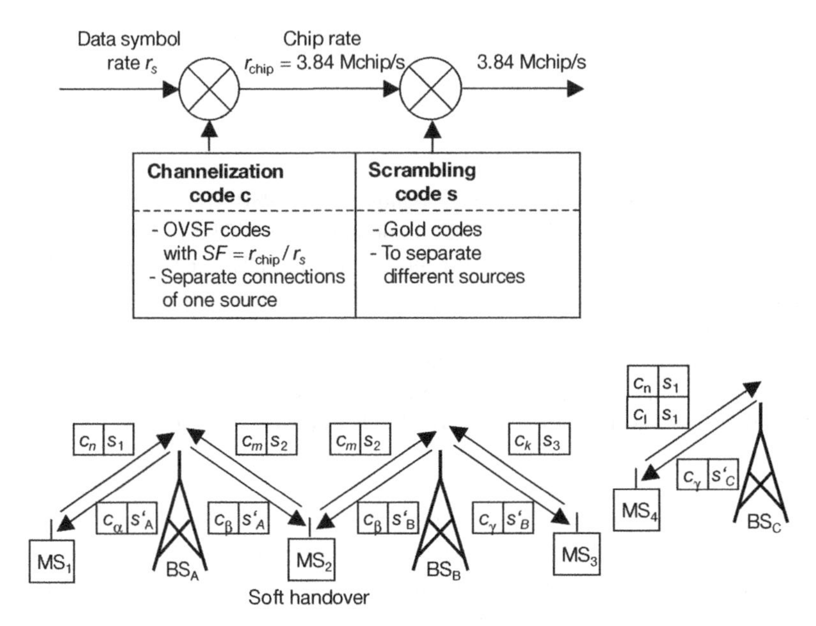

Figure 5.50 The principle of spreading and code allocation within UTRA FDD.

Code allocation

Before explaining some more details on spreading and modulation within UTRA FDD, the general principles on code allocation are sketched (see e.g. (3GPP TS 25.213 2004)).

UTRA FDD uses a fixed chip rate of $r_{\text{chip}} = 3.84$ Mchip/s; spreading is performed in two steps as shown in Figure 5.50. The OVSF codes described in Subsection 5.1.3 are used as channelization codes, that is, to separate different connections transmitted by one and the same source. The source may be one BS having connections to many MSs or one MS having several parallel connections; for example, the MS transmits user data and control data on different code channels or it may use two or more code channels to enhance the data rate. OVSF codes are allocated according to the rules mentioned in Subsection 5.1.3, where the code length or spreading factor is chosen to match the length of one data symbol to be spread. Hence, a rate matching of data symbols has to be performed (e.g. by puncturing or symbol repetition) so that the relation $r_s = (3.84\,\text{Mchip/s})/SF$ for the symbol rate r_s can be fulfilled. In UTRA FDD, the following spreading factors and symbol rates can be applied:

- UL: $SF = 2^m,\ r_s = 2^{-m} \cdot 3840$ ksym/s $m = 2, 3, \ldots, 8$;
- DL: $SF = 2^m,\ r_s = 2^{-m} \cdot 3840$ ksym/s $m = 2, 3, \ldots, 9$.

Since OVSF codes can only separate signals transmitted synchronously by one source, a second spreading step is introduced to separate different sources. In the UL direction, essentially segments of Gold codes generated by shift registers corresponding to the following polynomials of degree 25 are used:

- $P_{UL,1}(X) = X^{25} + X^3 + 1$;
- $P_{UL,2}(X) = X^{25} + X^3 + X^2 + X + 1$.

Segments of the corresponding m-sequences are combined to give a complex-valued sequence, which is subsequently truncated to the basic physical frame length of 10 ms, that is, to 38,400 chips. The resulting sequences are called *long scrambling* sequences.

Alternatively, short scrambling sequences consisting of 256 chips may be allocated in UL direction. These complex-valued sequences, which are derived from quaternary codes, are specified to be used only in connection with interference cancellation to reduce its algorithmic complexity (see e.g. (Bing *et al.* 2000)).

In both cases, there are many millions of UL scrambling codes.

Scrambling in DL direction is also performed using Gold codes, where the degree of the generator polynomials is 18 as in this case:

- $P_{DL,1}(X) = X^{18} + X^7 + 1$;
- $P_{DL,2}(X) = X^{18} + X^{10} + X^7 + X^5 + 1$.

Also for the DL two segments of each Gold code are combined to give a complex-valued sequence, which is subsequently truncated to the basic physical frame length of 10 ms, that is, to 38,400 chips. Not all, but only a part of these sequences consisting of $2^{13} = 8192$ codes is used for DL scrambling. These $8192 = 64 \cdot 8 \cdot 16$ codes are divided into 64 scrambling code groups each consisting of eight scrambling code sets. Hence, totally there are 512 scrambling code sets which comprise 16 codes as illustrated in Figure 5.51. The first code in each set is called the primary scrambling code (PSC), the other ones the secondary scrambling codes (associated with the respective PSCs). To each cell one and only one scrambling code set is allocated in a way that avoids the usage of the same set in neighboring cells, that is, code sets are planned with a high reuse distance. The grouping of sets is

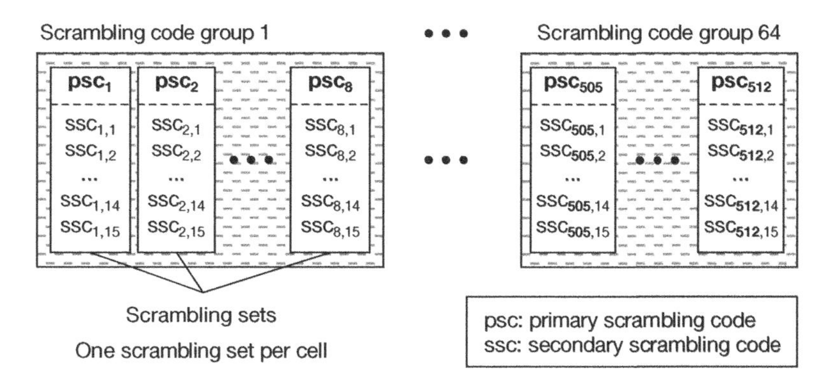

Figure 5.51 Division of scrambling sequences into different groups and sets.

important for cell selection, which will be explained in the following text. In a standard configuration, only the primary scrambling code is used within a cell; in any case it has to be allocated to control channels that are broadcasted within the complete cell. One or more secondary scrambling codes may be used in situations where the interference is low, for example, when smart antennas are applied. In this case, the number of connections allowed by the interference conditions may exceed the number of available OVSF codes within one tree. Additional OVSF trees in one cell can be generated using the secondary scrambling codes as a kind of root. Obviously, there is no orthogonality between different trees.

A peculiarity has to be observed for a soft handover: in UL direction, of course, only one channelization and one scrambling code is used, which is known and adjusted in the RAKE receivers of all involved base stations. However, in DL direction the involved base stations use the same channelization code indeed, but different scrambling codes, namely, their specific primary or secondary scrambling codes. Hence, the RAKE fingers of the MS in soft handover mode have to be fed with these different scrambling codes. Furthermore, UTRA requires no synchronization of the base station transmission. Though the RAKE receiver is able to handle moderate delays with respect to the different propagation paths, at least a synchronization up to a level of some data symbols is necessary. On the other hand, to keep the orthogonality of the channelization codes, the transmission is only allowed to be shifted in steps of the maximum code length. Since OVSF codes of $SF = 512$ are expected to be used very rarely, the step size for shifting the transmission in soft handover mode has been set to 256 chips and some exemptions have been specified for connections operating at an $SF = 512$.

Data transmission, channel coding and multiplexing

UTRA has been designed to allow a large variety of services with very different quality requirements. Furthermore, besides the proper user data, a lot of system control data have to be transmitted in an efficient way. Therefore, various channel types and coding and multiplexing schemes adapted to the individual needs have been defined (see e.g. (3GPP TS 25.211 2004)). From the point of view of communication layers, one distinguishes transport channels and physical channels. Transport channels are characterized by the way the data are transmitted, whereas the physical channels are characterized mainly by the spreading scheme and the physical information they carry. Within the transport channels, one distinguishes, for example,

- channels that carry broadcast data as system information (e.g. network and cell identity);
- channels that are used as common control channels for establishing a connection (paging, random access, channel allocation);
- dedicated channels that are allocated for an individual connection;
- shared common channels that are used for packet transmission modes using special medium access and collision resolution schemes.

On the physical layer many different tasks have to be fulfilled, for example,

- transmission of pilot symbols for channel estimation;

- dedicated transmission of user data or higher-layer control data;
- dedicated transmission of physical layer control data (e.g. power control commands, individual pilot symbols);
- synchronization;
- shared transmission of packet data;
- providing capabilities for broadcast transmission and paging;
- providing capabilities for random access.

Totally, more than 10 types of physical channels have been defined partly using very special spreading schemes. In this section, we focus on the main channels.

First, the dedicated transmission of user and control data is considered, which is performed via the dedicated channel (DCH) as the transport channel and via the dedicated physical data or control channel (DPDCH or DPCCH) as the underlying physical channel. It should be emphasized that data from different applications may be multiplexed on one DPDCH. On the other hand, one higher-layer connection may use more than one DPDCH to increase the data rate.

As a first step in the multiplexing and channel coding chain, CRC bits may be attached to the higher-layer data frames where block codes with 24, 16, 12 and 8 CRC bits are defined. Subsequently, convolutional coding or turbo coding is applied. Convolutional coders of rate $1/2$ and $1/3$ and constraint length 9 are used. Turbo coding of rate $1/3$ is accomplished by a parallel concatenated convolutional code (compare Subsection 3.2.5) with two 8-state constituent encoders and internal interleaver (see e.g. (3GPP TS 25.212 2004)).

Having adapted the data volume to the size of radio frames, an interleaving over 20, 40 or 80 ms frames depending on the delay constraints of the respective services is performed. Before multiplexing various connections to one physical channel, a rate matching mechanism taking into account the different quality of service requirements is necessary, since all symbols are transmitted at the same power. Rate matching is performed by three methods:

- code puncturing (DL and UL)
- repetition of symbols (DL and UL)
- discontinuous transmission (only DL).

Subsequently, the matched and multiplexed data are mapped to one or more DPDCHs using a segmentation into and a second interleaving over the basic radio frames of 10 ms. The data structure for the dedicated channel on a radio frame is illustrated in Figure 5.52. There is a remarkable difference between UL and DL: while in DL direction user data and physical control data are time multiplexed using the same channelization code, in UL direction these data are separated by using different channelization codes. The reason for this is explained later. Each radio frame is subdivided into 15 slots, which are not used for time multiplexing of several connections but to achieve a high transmission rate of 1500 Hz for the physical control data (see e.g. (3GPP TS 25.211 2004)). These control data consist of

- transmission power control (TPC) bits indicating whether the transmission power shall be increased or decreased for the next slot;

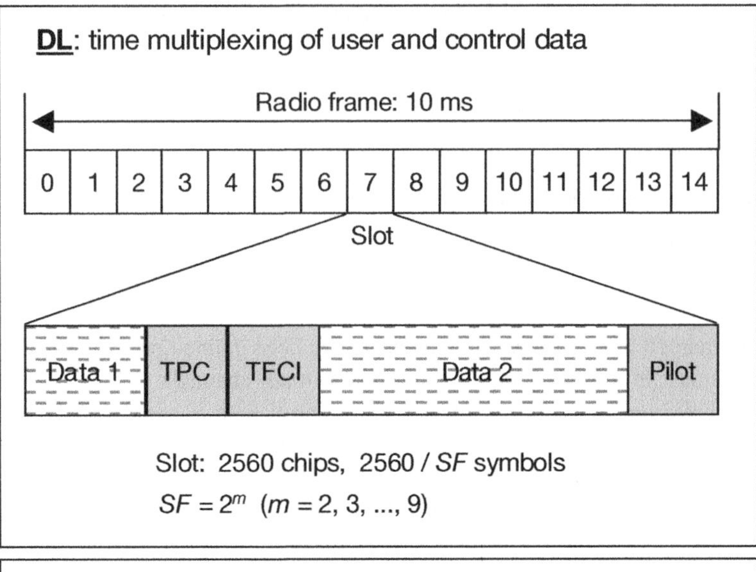

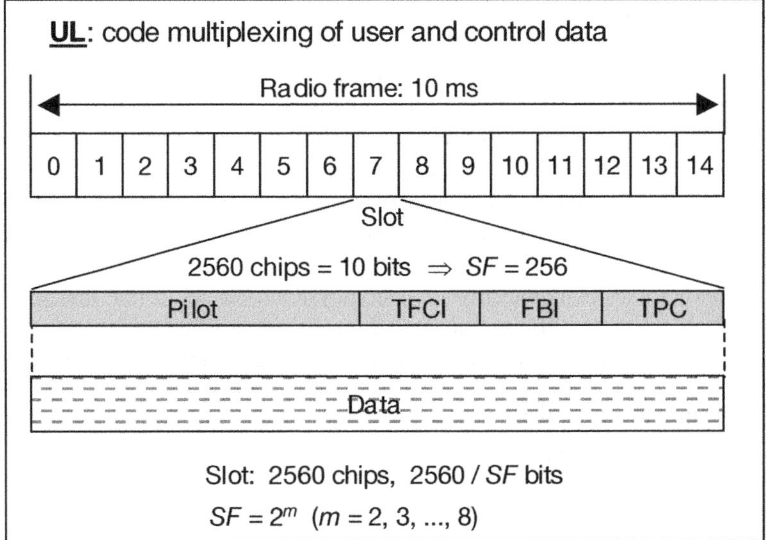

FBI: feed back information

TFCI: transport format combination indicator

TPC: transmission power control

Figure 5.52 Data structure of a UTRA FDD radio frame.

- pilot symbols for an individual channel estimation;
- feedback information (FBI) that allows a closed-loop transmit diversity scheme (additionally defined besides the open-loop scheme explained in Subsection 2.4.7);

- a transport format combination indicator (TFCI), which provides information for the receiver on the used rate matching and multiplexing scheme.

It should be noted that in DL direction there are also common pilot channels (CPICH). The primary common pilot channel (P-CPICH) is transmitted within the whole cell

- to support phase tracking and channel estimation for dedicated channels;
- to serve as a phase reference and to allow channel estimation for common channels not carrying individual pilot symbols;
- to serve as a beacon for neighbor cell measurements (handover).

In addition, secondary common pilot channels may be allocated if the cell is divided into several subsectors using switched beams techniques, that is, different secondary common pilot channels are assigned to different subsectors. Scrambling of the common pilot channels is performed by using the primary and secondary scrambling sequences, respectively. The channel-specific DL pilot symbols are needed when full adaptive smart antenna techniques are applied resulting in individual propagation channels for the different connections.

Spreading and modulation of dedicated channels

Spreading and modulation for UTRA FDD is specified in (3GPP TS 25.213 2004). As mentioned above, there are some differences between data transmission in UL and DL directions, not only with respect to multiplexing of user and control data, but also with respect to spreading and modulation.

The main reason for this is that in UL direction a nearly equal transmission power is desirable to avoid disturbing pulsed transmissions with audio frequencies and to relax the requirements on the linearity of the MS power amplifier. This is accomplished within two steps as shown in Figure 5.53: Multiplexing user and physical control data on different codes in a first step guarantees that there is at least a continuous transmission of control data even in phases where no user data have to be transmitted, for example, in phases of no speech activity. In a second step, the control data are phase shifted by 90° (multiplication by j) and added to the user data stream; the sum is then multiplied by a complex-valued scrambling code. By this method, control and user data symbols are distributed to both the I- and the Q-branch of the subsequent I/Q modulator (chip modulation). Since the complex-valued scrambling code is constructed in such a way that there is at most a phase change of $+90^\circ$ or -90° between consecutive chips, zero transitions are largely avoided except for a symbol change. Finally, some notes on UL spreading and modulation for the dedicated channel are listed as follows:

- The data channels and the control channels may use different power weights.
- For the control data, the spreading factor $SF = 256$ and the channelization code $c_{256,0}$ are used.
- The spreading factor of the DPDCH may change from frame to frame between 4 and 256. However, the channelization codes shall be taken from a fixed branch of the code tree disjoint from the 0-branch occupied by the control channel.

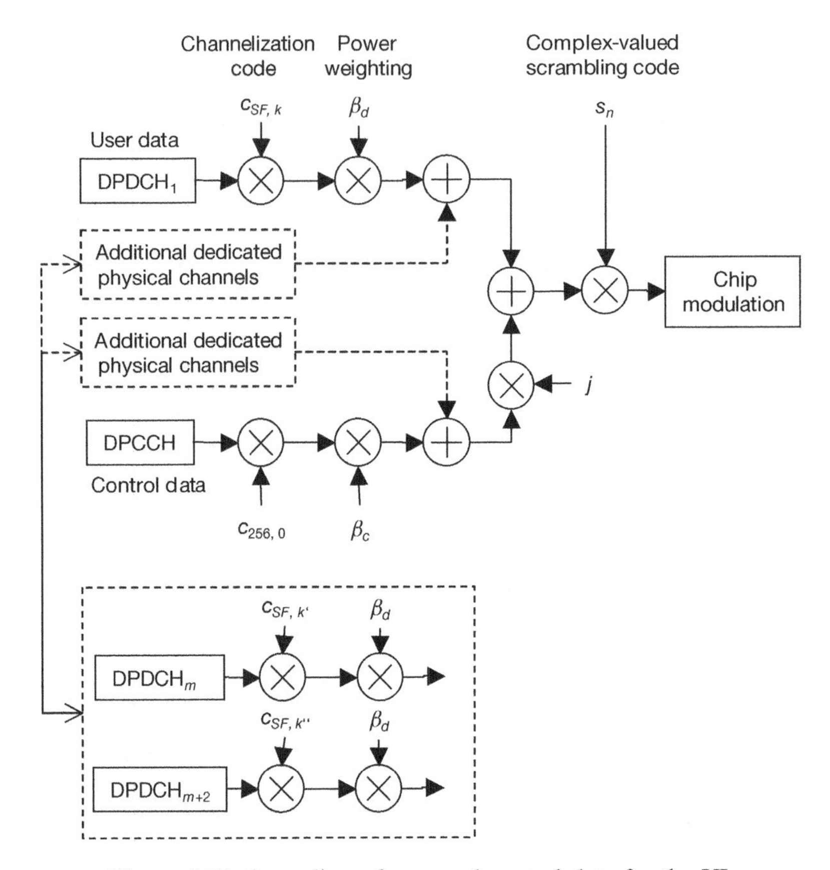

Figure 5.53 Spreading of user and control data for the UL.

- Up to five additional DPDCHs may be allocated to one connection to increase the data rate if all other methods for an enhancement are exhausted; these channels are mapped to I- and Q-branch in an alternating way. In this case, only the spreading factor $SF = 4$ is allowed with the channelization codes $c_{4,1}, c_{4,2}, c_{4,3}$.

- In a recent version of UTRA FDD, also an additional control channel corresponding to high-speed packet data channels can be transmitted.

- For pulse shaping a root-raised cosine filter with a roll-off factor of 0.22 is used (see e.g. (3GPP TS 25.104 2004; 3GPP TS 25.101 2004)).

In DL direction, in general, many channels are transmitted by the BS. Hence, even if discontinuous transmission is applied, transmission power is expected to vary only smoothly due to averaging effects. Furthermore, using an extra channelization code for the control channel of each connection would set too many constraints to code allocation. Hence, in DL direction a different spreading and modulation strategy illustrated in Figure 5.54 is chosen. The first step is a modulation mapping: for a standard dedicated channel (DCH), pairs of subsequent data bits are mapped to the I- and Q-branch according to a QPSK;

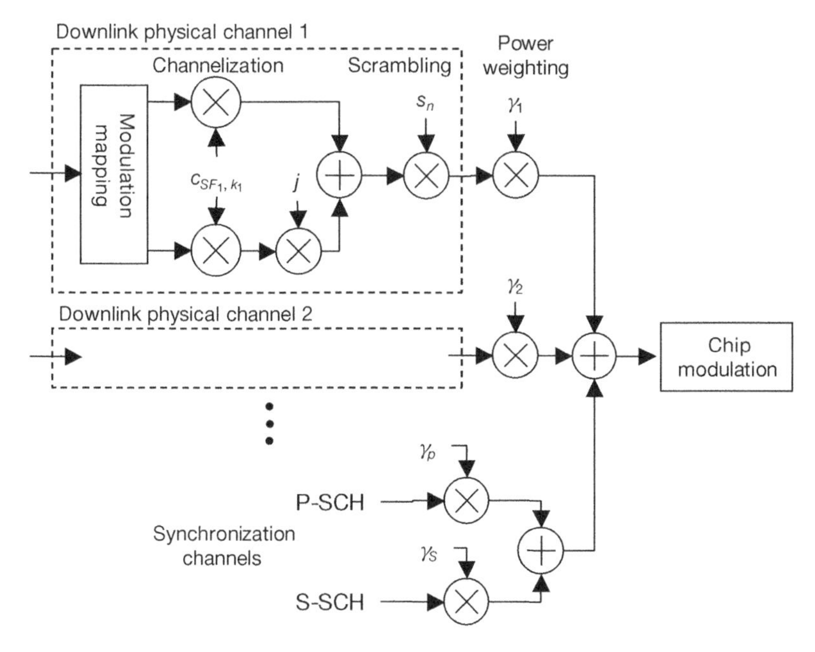

Figure 5.54 Downlink spreading for UTRA FDD.

for a high-speed packet data channel, quadruples of bits and a mapping according to 16-QAM may be used. The I- and Q-branch is then spread by the same channelization code. Afterwards, the signal is scrambled with a cell-specific complex-valued scrambling code. Chip modulation is performed in the same way as for the UL.

- Each channel transmitted by the BS can be weighted with an individual power value.
- Some channelization codes are reserved for special channels as $c_{256,0}$ for the common pilot channel and $c_{256,1}$ for the common control physical channel (system information, paging, channel assignment). All other channelization codes can be allocated in a flexible way.
- Since a coordination of many connections is required for the DL, the spreading factor cannot not changed on a frame-by-frame basis as in UL direction, but at a lower rate.
- The channels needed for synchronization of the MSs and for cell search are handled in a special way.

Cell search and synchronization

The first task an MS has to perform after it has been switched on is to search for an appropriate cell to camp on. As a precondition the MS has to synchronize to the frame and slot period of the respective BS and it has to find out the allocated scrambling code to be able to decode the system information sent via the broadcast channel. Since it is very time consuming to test and correlate to hundreds of scrambling codes, the cell search is divided

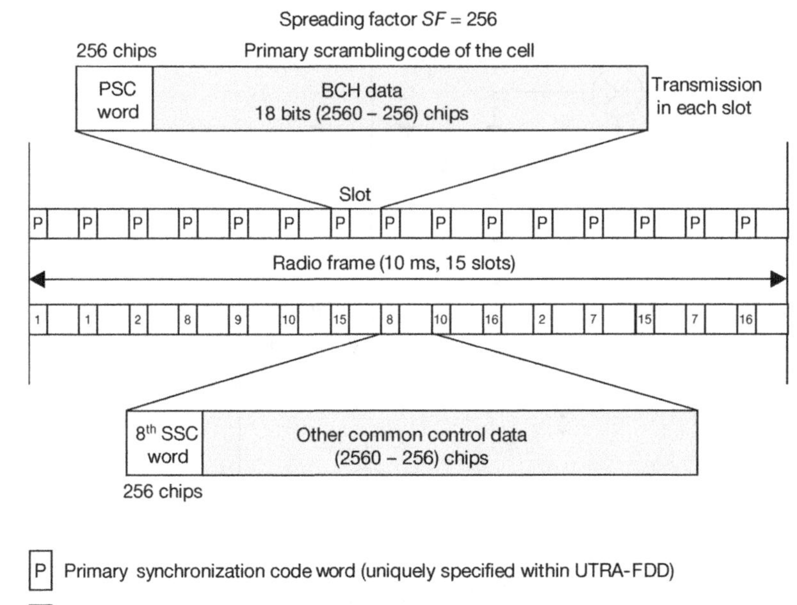

Figure 5.55 Code words for synchronization and cell search.

into several steps to speed it up. Furthermore, it should be noted that the transmissions of base stations at different sites are not synchronized in general. For this reason, two types of synchronization code words consisting of 256 chips have been defined as illustrated in Figure 5.55:

- The primary synchronization code word (PSCW) is a code word with good aperiodic correlation properties, which is uniquely defined for UTRA FDD (see e.g. (3GPP TS 25.213 2004)). It is classified as a generalized hierarchical Golay code. The PSCW is transmitted in each slot without scrambling in front of the broadcast channel data.

- The 16 secondary synchronization code words (SSCW) are derived from Walsh–Hadamard sequences (see e.g. (3GPP TS 25.213 2004)). These SSCWs are transmitted without scrambling in each slot in front of the forward access channel data in a characteristic order (the forward access channel is used to assign a specific code at a connection setup). 64 disjoint sequences, each consisting of 15 SSCWs, are defined and are in one-to-one correspondence with the 64 primary scrambling code groups defined above. The respective sequence of SSCWs is repeated in each frame (consisting of 15 slots) of the cell.

As a first step within the cell search procedure (see e.g. (3GPP TS 25.214 2004)), the MS detects the PSCWs and synchronizes to the slot period of the cell with the strongest received signal. Subsequently, the sequence of SSCWs is detected. Since the sequences are constructed in a way that also a cyclic shift of one sequence is disjointed from a cyclic shift of any other sequence, the mobile can derive both the start of the frame within the cell and the PSC group the cell belongs to. Hence, the MS has to correlate only to the eight primary scrambling codes of that group to find out the scrambling code allocated to that cell. Subsequently, the system information can be descrambled and decoded.

Random access

To setup a connection, the MS has to request for transmission resources, especially for a scrambling code to be allocated to the UL of that connection. The procedure for establishing the first contact with the network has to handle the following challenges:

- Because of the propagation delay the MS is not completely synchronized to the BS in UL direction, that is, a data packet received from the MS may not exactly match the frame or slot period of the BS.
- Since the MSs try to access the network in an uncoordinated way (random access), channel requests may collide and a collision resolution strategy has to be defined.
- Before a successful random access no specific code is allocated to the MS. Furthermore, a random access signal interferes with other already established connections in the cell. Hence, its signal power has to be on the one hand high enough to be detected, but on the other hand low enough to restrict the interference to other connections.
- Since in a CDMA network frequencies are reused generally in neighboring cells, care has to be taken that the random access signal is only interpreted by the appropriate BS.

While the first two challenges also exist in TDMA networks, the last two are CDMA-specific ones. The UTRA FDD solution uses the following ingredients (see e.g. (3GPP TS 25.214 2004)):

The random access message is split into a preamble and a proper message part containing information on the requested resources. The preamble – carrying no higher-layer information – consists of 4096 chips called the *preamble signature sequence*. Totally there are 16 signature sequences obtained by repeating Walsh–Hadamard codes of length 16; the repetition factor is 256. For signature scrambling, a subset of the long scrambling sequences specified for the UL is used. Since a one-to-one correspondence between signature scrambling sequences and DL primary scrambling sequences has been fixed, the destination (cell) of a preamble can be derived. Preambles are only allowed to be transmitted in certain periods given by the BS; such a random access period consists of two slots ($2 \cdot 2560$ chips). For the first time, the MS transmits the preamble with a randomly selected signature sequence at a low power value, which is derived by an open-loop power control procedure based on the received common pilot channel signal and other parameters. If no acquisition indication from the BS is received within a certain time interval, the MS repeats the preamble transmission with a power increased by a certain step. Otherwise, the acquisition indication (echoing the preamble signature) is interpreted by the MS as a kind of acknowledgment that the BS has decoded the preamble and is ready to receive the proper random access message part. Transmitting the message, only specific channelization codes and scrambling sequences are allowed to enable the BS to despread the message.

The channelization code may only be selected from a branch related to the signature of the preamble; a correspondence between signatures and branches of the code tree is defined in the standard. The scrambling sequence is derived from the same long sequence as applied for the preamble. However, another segment of this sequence is used.

Obviously, there is a limitation on the maximum number of retransmissions of the preamble as well as on the maximum allowed transmission power. If one of these limits is reached, the access procedure is aborted.

Power control

Within UTRA FDD a fast closed-loop power control with a rate 1500 Hz is specified for the UL as well as for the DL (see e.g. (3GPP TS 25.214 2004)). On the basis of the measurements of the SIR, a certain SIR target (set by a slow outer loop mechanism) is tried to be achieved. Depending on whether the received SIR within a slot is lower or higher than the target, the transmitter is commanded (using the PCI bits in the next slot) to increase or decrease its transmit power by a certain PC step. The effective PC step per slot may be 0.5, 1.0 or 2.0 dB and is set by higher-layer information fields.

For the call setup phase (random access) and the transmission of short packets, an open-loop PC has been defined. Furthermore, special rules are specified for the slotted mode and the soft handover. For a soft handover, different PC commands may be transmitted by the involved base stations and the MS has to resolve this situation.

Obviously, fast PC is not allowed on common channels like, for example, the common pilot channel or the broadcast channel.

Handover

Within UTRA FDD, two main types of handover are specified:

- a soft handover between cells using the same frequency carriers;
- a hard handover between cells using different frequency carriers.

For a hard handover, one further has to distinguish between a hard handover within UTRA FDD, for example, in a hierarchical cell structure, and a hard handover between cells of different systems, for example, between UTRA FDD and GSM or UTRA TDD cells.

As mentioned in Subsection 5.1.4, for a soft handover one has to find a compromise between its gain and its effort by restricting the so-called *active set*, that is, the set of cells involved in a handover for a certain connection. In UTRA FDD, the number of cells in the active set can be limited. Furthermore, a new cell is only included in the set of active cells if the corresponding averaged received signal level is not far below the best averaged level. The meaning of *not far below* is fixed by a hysteresis margin. Level measurements are performed using the primary pilot channels of the respective cells.

A soft handover in DL direction results in an increased interference level for other connections if power is transmitted from all base stations in the active set. To reduce the interference, a method called *site selection transmit diversity power control* is foreseen in UTRA FDD as an option: the MS measures the signal levels with respect to all primary pilot channels of the active set. The cell with the best level is indicated to all base stations of the active set by “misusing” the FBI field of the dedicated physical control channel. Having received this information, the transmission of the dedicated channel is switched off by all cells of the active set except for the strongest one. Obviously, this feature can only be active if no closed-loop transmission diversity requiring the FBI is used.

While a soft handover can be managed using the multiple fingers of the MS RAKE receiver, additional means are required for a hard handover between different frequency carriers. To perform measurements, either an additional measurement receiver is needed or the so-called *slotted or compressed* mode has to be applied. In this mode, data transmission is compressed to shorter intervals to gain some time for measurements on carrier frequencies different from the present one. Compression can be achieved by

- a reduction of the data rate by higher layers,
- transmitting the data at twice the original rate using half the spreading factor and
- increasing the puncturing.

By these methods, several idle slots per frame can be produced.

5.5.5 Time Division CDMA

A part of the spectrum allocated to IMT-2000 by the ITU consists of unpaired frequency bands and is therefore not suited for operation within the FDD mode. For this reason, also a TDD mode for UTRA has been specified. The TDD mode has been derived from a system proposal called *TD-CDMA* that combines time division and code division multiple access. The original idea behind this proposal submitted to the ETSI was to reduce the number of intracell interferers by dividing a frequency carrier into several disjoint time slots and thereby having only a few code channels per time slot. Since there are only a few interfering connections within a cell, these can be jointly detected and separated with a moderate calculation complexity. Since the specific *joint detection* (JD) algorithm was an essential ingredient of the system proposal, it also was called *JD-CDMA*. Having selected TD-CDMA as a TDD mode for IMT-2000, it has been harmonized to WCDMA, for example, the chip rate has been chosen to be the same (3.84 Mchip/s). There are many further similarities which are mentioned below.

Besides the ETSI contribution TD-CDMA, another proposal called *TD-SCDMA* (Time Division Synchronous CDMA) using a lower chip rate has been submitted to the ITU by the Chinese standardization organization. After a harmonization phase, this proposal has been included in the UTRA TDD standard as a second option with a chip rate of 1.28 Mchip/s. Besides this lower chip rate, a slightly different TDMA frame structure has been defined including periods for special physical channels to accomplish a fast random access and a synchronization for the UL transmissions.

However, except for the different chip rate, channel coding, spreading and modulation are very similar in both submodes. Therefore, the following discussion focuses on the transmission mode with 3.84 Mchip/s.

Within this subsection, an overview of the physical layer of the UTRA TDD mode is given. More details can be found, for example, in (Baier *et al.* 2000; Holma and Toskala 2001; Steele *et al.* 2001) or within the standardization documents quoted in the following text.

Frequency allocation

The spectrum allocated to IMT-2000 contains 11 unpaired blocks of 5 MHz, namely, in the bands 1880–1920 MHz and 2010–2025 MHz (see e.g. (3GPP TS 25.102 2004; 3GPP TS 25.105 2004)). A part of them is foreseen for licensed operation in public networks, another part for unlicensed operation in private networks.

In Germany, for example, (nearly) each UTRA operator has acquired one TDD carrier besides his two FDD duplex carriers. Hence, he may use this additional TDD carrier for capacity enhancement, especially in regions with a highly asymmetric data traffic demand. The division of a carrier into UL and DL time slots can be adapted to the corresponding capacity requirements.

However, as mentioned in Subsection 5.1.4, a synchronization of nearby TDD base stations is needed to avoid very critical interference situations. Within the network of one operator, synchronization may be accomplished via the radio network controller (RNC) using some special protocols defined for the TDD mode. However, synchronization of base stations of different operators or of base stations for unlicensed operation is not achievable. Therefore, at least for these situations a dynamic channel selection feature (see e.g. (Bing *et al.* 2000; Holma and Toskala 2001)) similar to the one specified for DECT is needed.

Code allocation

As for the FDD mode, code allocation is performed in two steps using OVSF channelization codes and cell-specific scrambling sequences (see e.g. (3GPP TS 25.223 2004)). OVSF codes are selected according to the code tree of Subsection 5.1.3. 128 scrambling sequences consisting of 16 complex-valued chips are defined by code tables. The sequences are divided into 32 groups and are allocated to cells in a way that nearby cells use a scrambling sequence from different groups. Hence, scrambling codes are much shorter than for the FDD mode. Furthermore, there are much less scrambling sequences that allow only a separation of different cells, but not a separation of different connections. Hence, in UL direction the OVSF codes together with the joint detection mechanism also have to be used to distinguish different connections. In UL direction, OVSF codes with spreading factors $SF = 1, 2, 4, 8, 16$ are applied observing the allocation rules mentioned in Subsection 5.1.3. In DL, only the spreading factors $SF = 16$ (standard case) and $SF = 1$ (high data rates) are used. However, multiple (orthogonal) code channels with $SF = 16$ can be allocated to one connection in order to achieve higher data rates.

Data transmission, channel coding and multiplexing

Similar types of transport and physical channels as for the FDD mode are defined for the TDD mode. The main difference is that there is no common pilot channel since channel estimation is performed using training sequences that are included within the data bursts. As to multiplexing and channel coding, the same principles and methods as for the FDD mode are applied (see e.g. (3GPP TS 25.221 2004; 3GPP TS 25.222 2004)).

Also the length of the basic physical frame has been fixed to the same value, namely, 10 ms. A frame is divided into 15 time slots, which are used to separate connections, that is, for the time division multiple access scheme. Connections using the same time slots can be further separated by different codes as explained below. As illustrated in Figure 5.56

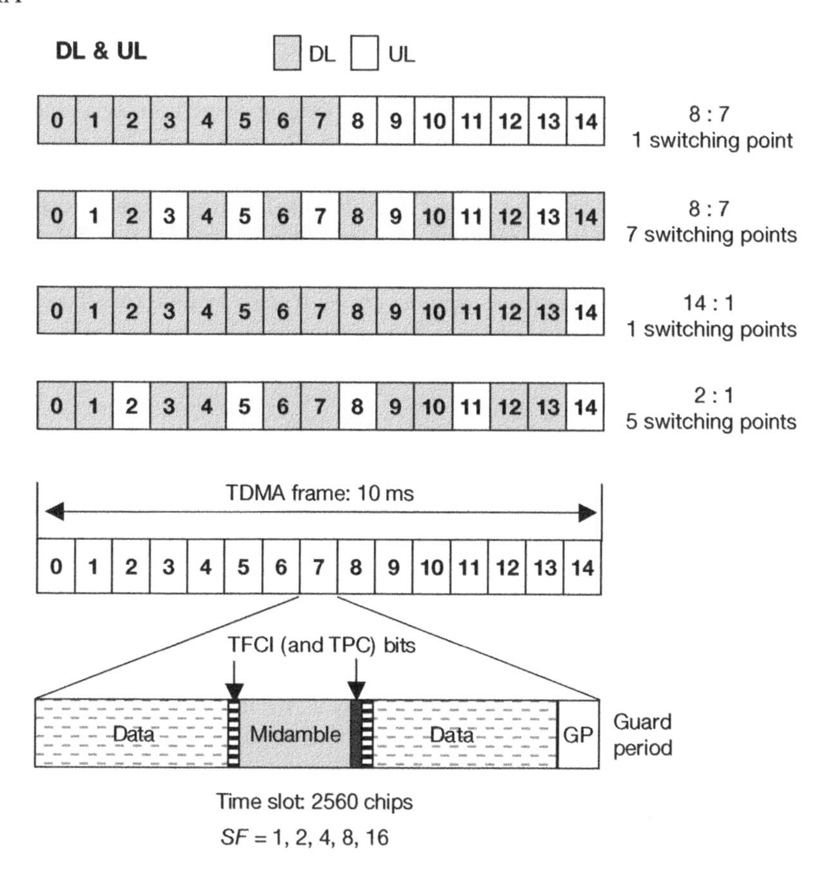

Figure 5.56 Variable time duplex division scheme and data structure for UTRA TDD.

time slots can be divided into UL and DL slots in a very flexible way using even multiple switching points between UL and DL.

Within one time slot of length 2560 chips, data are transmitted as so-called *bursts*. A burst consists of three parts:

- a guard period of a length of 96 or 192 chips where transmission is switched off to separate connections using different time slots;
- a midamble or training sequence of 256 or 512 chips for channel estimation;
- a data part including user data and physical layer control data as the transport format combination indicator (TFCI) or the transmit power control bits.

The longer guard period of 192 chips is used in UL direction for a random access or a handover access to a new base station, that is, before the MS transmission is exactly synchronized to the respective BS.

Within UTRA TDD, 128 long (512 chips) and 128 short (256 chips) training sequences have been defined, which are in one-to-one correspondence with the scrambling sequences, that is, the short and long training sequences corresponding to the cell-specific scrambling code is allocated to the respective cells. MSs within one cell transmitting on the same time slot use cyclically shifted versions of the respective long or short training sequences. A cyclic correlator allows a joint channel estimation of all UL connections on a time slot and this is the first step for the joint detection algorithm (see e.g. (Baier *et al.* 2000; Bing *et al.* 2000)). Obviously, the shift has to be greater than the maximum propagation delay value expected within the operational environment. In an environment with a low delay spread, up to 16 MSs may transmit on the same time slot using the long training sequence. In DL direction, the identical training sequence (without any shift) may be used for all connections on a time slot in a cell. However, if smart antenna techniques are applied, beam individual training sequences have to be assigned.

Spreading and modulation of dedicated channels

While in the FDD mode slightly different spreading and modulation schemes are applied for the UL and the DL, the TDD mode uses essentially the FDD DL scheme in both directions (see e.g. (3GPP TS 25.223 2004)), that is, data bits are grouped into pairs (or quadruples) and are mapped according to a QPSK (or a 16-QAM for high-speed packet data transmission) to the I- and Q-branch of a modulator, where they are first multiplied by a (real) OVSF code as a channelization code and afterwards by a complex multiplier depending on the selected channelization code. Subsequently, scrambling with a cell-specific code is performed. The resulting chip sequence is modulated in the same way as for the FDD mode.

Cell search and synchronization

Cell search is based upon the same code words that are also used in the FDD mode, namely, the unique system-specific primary and 16 secondary synchronization code words (see e.g. (3GPP TS 25.223 2004; 3GPP TS 25.224 2004)). However, owing to the TDMA nature of the TDD mode they are applied in a slightly different way.

- They are not transmitted within each slot, but only within one or two slots of a frame which we call the synchronization channel (SCH) time slots. Note that the code words have a length of 256 chips, whereas the length of a time slot corresponds to 2560 chips.

- A certain time offset between the start of the time slot and the start of the code words is used, which is related to the code group of the respective cells. Since cells in a public TDD network should be synchronized to avoid critical BS–BS or MS–MS interference situations, an SCH in one cell may mask the SCH in a neighboring cell if both channels are transmitted completely synchronously. To reduce this effect, the time offset is introduced.

- Parallel to the primary synchronization code word, not only one but three out of the 16 secondary synchronization code words are transmitted. Each of these three code words is multiplied by a complex-valued data symbol. Detecting the code words and the corresponding data symbols, the scrambling code group of the cell, the time offset, the slot number of the SCH and the slot number of the broadcast control channel carrying further system information can be derived.

In a first step, the MS detects the strongest cell and its timing using, for example, a matched filter that is matched to the primary synchronization code word. On the basis of this timing information, the received signal is correlated to the SSCWs. By this process, the scrambling code group of the cell and some further information mentioned above can be derived. In a next step, the MS is able to descramble the broadcast channel correlating the signal to the four scrambling codes of the known code group.

Random access

At least one time slot has to be allocated in UL direction for the random access procedure (see e.g. (3GPP TS 25.224 2004)). In contrast to the UTRA FDD mode, no preamble but only the proper random access message is sent. The message is spread by OVSF codes of spreading factor $SF = 8$ or $SF = 16$; which OVSF codes are allowed is broadcasted by the BS. Furthermore, there is a one-to-one correspondence between the used channelization code and the cyclic shift of the training sequence. For transmitting a random access message, the MS selects an allowed OVSF code and a related midamble shift value randomly. The message is transmitted at a power level derived by an open-loop process without timing advance. The message does not arrive at the BS at the beginning of the random access time slot, but with a certain propagation delay that can be measured by the BS. Having received and decoded the message, the BS allocates a dedicated channel. Furthermore, the MS may be commanded to adjust its transmission timing in steps of four chips (about 1 μs) to synchronize the UL transmission for allowing shorter guard periods.

Power control

As mentioned in Subsection 5.1.4, a CDMA system affected by a high degree of intracell interference requires a fast and exact UL power control. Since in a TDD system there is no continuous transmission in UL and DL direction, the high-power control rate of 1500 Hz of the FDD mode cannot be preserved in the TDD mode. In an adverse configuration, the rate may reduce to 100 Hz (one PC action per TDMA frame) with a high delay between the measurement and the corresponding power control command. To circumvent this problem, the joint detection mechanism is foreseen, which allows a separation of different code channels even if there is no perfect fast UL PC. On the other hand, since UL and DL use the same carrier frequency and there is a control delay anyway, an open-loop PC can be applied in UL direction (see e.g. (3GPP TS 25.224 2004)). Therefore, the MS measures the received signal level from the broadcast channel, which is transmitted at a constant known power one or two times a physical frame. From these values, the path loss can be derived. Knowing also the UL interference level (whose value is broadcasted by the BS), the MS is able to adjust its transmit power to approximately achieve a target *SIR*. In DL direction, open-loop PC cannot be applied since there is no unique UL reference transmitter. Therefore, a closed-loop PC analogous to the one for the FDD mode (but with a lower rate) is used.

Handover

Within UTRA TDD, handover is managed as a hard handover, that is, the active set consists of only one cell. However, due to the TDMA structure, the MS is able to perform neighbor cell measurements without requiring an additional measurement receiver or a compressed mode – at least for low and medium rate services where only a part of the time slot is needed for data transmission and reception. Hence, though no soft handover is applied and no signals are combined, the hard handover can be performed very fast on the basis of continuous measurements by the MS.

5.5.6 cdmaOne

The first cellular mobile communication system based on CDMA technology was designed by a single company, Qualcomm Incorporated. This system has been further developed to become the Interim Standard number 95 (IS-95) in 1994 in the United States. Networks according to this standard are operated mainly within two frequency bands: in the so-called *US cellular band* at about 850 MHz and in the so-called *PCS* (personal communication system) band at about 1900 MHz. In 1997, IS-95 was rebranded as cdmaOne. It should be noted that there are some (small) differences of the cdmaOne versions for the cellular and the PCS band.

Within this subsection, an overview of the physical layer of the cdmaOne system is given. More details can be found, for example, in (Steele *et al.* 2001) or within the standardization documents (J-STD-007 1999; TIA/EIA-95 1993).

Frequency allocation

As mentioned above two bands are used by cdmaOne in North America:

- the cellular band at about 824–849 MHz in UL and 869–894 MHz in DL direction[8];
- the PCS band at about 1850–1910 MHz in UL and 1930–1990 MHz in DL direction.

These bands are divided into subbands and have to be shared by several operators so that each operator may assign about 10 duplex carriers to his network. The typical carrier spacing is 1.25 MHz.

Hence, the number of carriers per operator is in general significantly higher than in UMTS networks discussed above. Therefore, a cdmaOne operator may install several carriers per cell or may even use a cluster size higher than $K = 1$. Furthermore, there are enough carriers for implementing hierarchical cell structures. It should be mentioned, that – though a cluster with $K > 1$ reduces the overall interference – a soft handover cannot be performed between cells using different frequencies.

A main architectural difference compared to UMTS is that in cdmaOne networks the base station controllers (BSC) are not interconnected. Hence, a soft handover is only possible between base stations belonging to the same controller, but not between base stations belonging to different controllers. Hence, at the border between two BSC areas different frequency carriers have to be allocated to avoid a strong mutual interference that cannot be managed by a soft handover.

[8]To be consistent with the other sections and subsections, the notation downlink (DL) and uplink (UL) is used throughout this and the next subsection though the standardization documents for cdmaOne and cdma2000 use the notation forward link and reverse link instead.

Code allocation

Within cdmaOne, three different types of codes are used:

- Walsh–Hadamard codes of fixed length 64 as channelization codes for the DL and as a basis for a Walsh modulation in UL direction;
- long PN sequences (m-sequences) corresponding to a 42-stage shift register to separate different UL connections and to cipher the data stream in UL and DL direction;
- a pair of PN sequences corresponding to two 15-stage shift registers to provide a cell-specific scrambling.

The chip rate for all these codes is $r_{\text{chip}} = 1.2288$ Mchip/s.

In contrast to UMTS, no variable spreading factors are used. Therefore, instead of using the OVSF codes, Walsh–Hadamard codes of fixed length 64 are applied in cdmaOne as channelization codes for the DL. In the UL, they have a completely different role: they serve as a basis for 64-ary orthogonal Walsh modulation of the data symbols.

In UL direction, the signals transmitted by different MS are separated by long PN sequences generated by a 42-stage shift register. The register is initialized by the so-called *channel mask* which includes a kind of channel identification number and a permuted version of the equipment registration number of the respective MS (in case of a traffic channel). Additionally, this PN sequence may be viewed as a kind of ciphering stream providing protection against eavesdropping. For this reason, the long PN sequences are also applied in DL direction.

A cell-specific scrambling is achieved on the basis of a pair of sequences generated by two 15-stage shift registers. One sequence is applied to the I-branch, the other to Q-branch of a QPSK modulator; hence we call them s_I and s_Q. The two corresponding m-sequences have a length of $2^{15} - 1$ chips. An additional zero is added to the unique maximum run of zeros to obtain a sequence length of $2^{15} = 64 \cdot 512$. Observing that the chip rate may be written as $r_{\text{chip}} = 2^{15} \cdot 75/2$ Mchip/s, 75 periods of these sequences match exactly to 2 s.

This is not an accident but related to the fact that cdmaOne base stations are synchronized to GPS timing. Using synchronized base stations facilitates the cell search and soft handover procedure compared to UMTS which needs no synchronization. On the other hand, the cdmaOne base stations have to be able to receive the signals from the GPS satellites, which might be difficult in street canyons or within buildings. On the basis of synchronization a cell-specific scrambling is accomplished by using shifted versions of the sequences s_I and s_Q. The shift has been defined in multiples of 64 chips, which corresponds to approximately 80 μs. Hence, by this method 512 different scrambling sequences are derived, which are allocated to cells in a way that nearby cells use different sequences. We recall that in UTRA FDD also 512 primary scrambling sequences have been defined. However, in addition, each primary sequence has 15 associated secondary sequences to support smart antenna techniques.

Data transmission, channel coding and multiplexing

Within cdmaOne the following types of channels have been defined:

- A pilot channel for channel estimation and coherent detection in DL direction and serving as a beacon for neighbor cell measurements.
- A synchronization channel transmitting, for example, the system time and the cell-specific offset (shift) of the scrambling codes s_I and s_Q in DL direction.
- A paging channel transmitting paging messages as well as system information and channel assignment messages in DL direction.
- A (random) access channel used by the MSs in UL direction for transmitting channel request messages at a call initiation.
- Traffic channels for user data transmission with different data rates.

Explaining the channel coding and multiplexing, we focus on the traffic channel; channel coding for the paging channels is quite similar. The other channels, their format and their role will be discussed in the following text.

A traffic channel carries user data as well as signaling data. Signaling and user data – even from different services as speech and FAX – may be multiplexed on one physical traffic channel by using different parts of a data frame.

The cdmaOne standard offers two rate sets for the traffic channels. As shown in Figure 5.57 the corresponding data rates are some multiples of 1.2 kbit/s and 1.8 kbit/s, respectively. The maximum data rate is 9.6 kbit/s in set 1 and 14.4 kbit/s in set 2.

In DL direction, a convolutional code of constraint length 9 and of rate $R_c = 1/2$ is applied, which is punctured for the channels of the rate set 2. Subsequently, a bit repetition is performed in a way that a rate of 19.2 kbit/s is achieved for all traffic channels. To equalize the transmitted energy per bit E_b for all channels of a rate set, the transmission power will be reduced proportional to the repetition rate. Interleaving is performed over a block of 384 bits corresponding to the basic physical frame length of 20 ms. As illustrated at the bottom of Figure 5.57 each frame is divided into 16 power control groups (similar to the 15 slots of an UMTS frame). In each power control group, a power control (PC) symbol is transmitted replacing two coded data bits. The PC symbols indicate whether the MS shall increase or decrease its transmit power. They are transmitted without power reduction at quasirandom positions. The position in one group is derived from the long PN sequence segment used in the preceding group.

In UL direction, convolutional codes of constraint length 9 and of rate $1/2$ and $1/3$ are applied to the channels of the rate set 1 and 2, respectively. Subsequently, a bit repetition is performed in a way that a rate of 28.8 ksym/s is achieved for all traffic channels. The symbols are interleaved over the frame length of 20 ms. Subsequently, a Walsh modulation is performed mapping a group of six symbols to a Walsh–Hadamard code of length 64, which results in a chip rate of 307 kchip/s. To compensate the bit repetition, not a continuous power reduction is used as in the UL, but transmission is switched off completely for certain parts of the frame. This means that only eight, four or two power control groups per frame may be filled with data bits to be transmitted. The corresponding groups are selected in a quasirandom way. The interleaving is adapted to this mechanism.

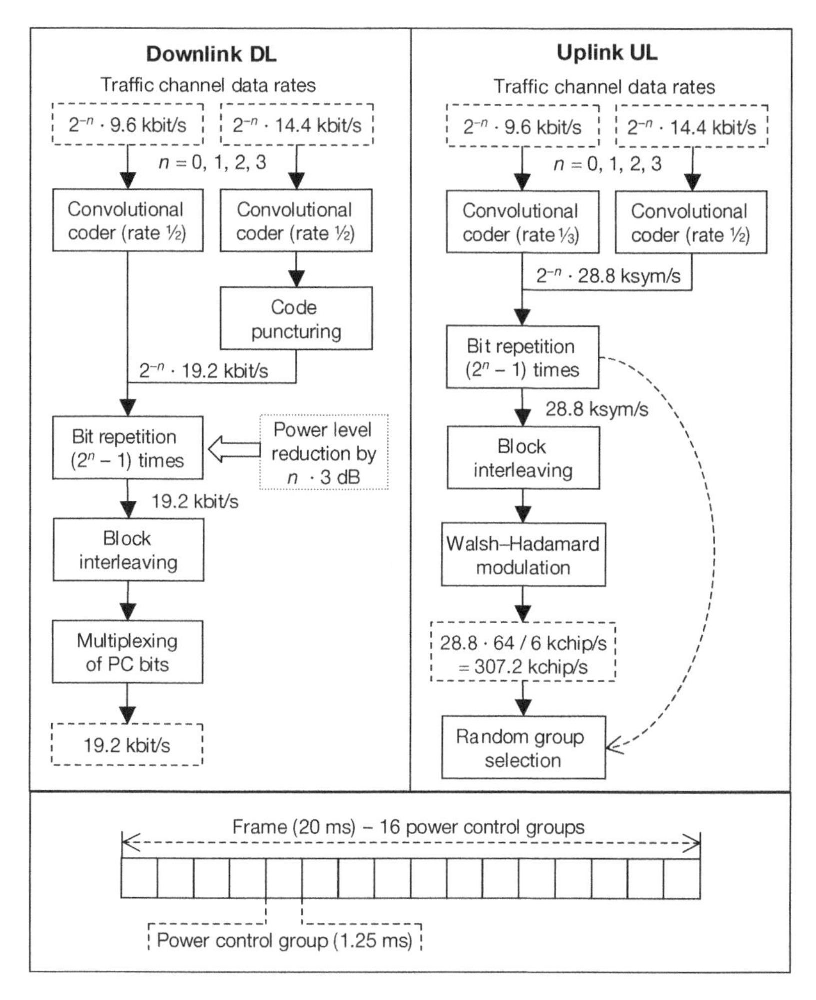

Figure 5.57 Channel coding, rate matching and frame structure for cdmaOne.

Spreading and modulation

In UL direction, the traffic channel data are first spread by the MS-specific long PN sequence, which is generated at a chip rate of 1.2288 Mchip/s as shown in Figure 5.58. The resulting chip sequence is transferred to both the I- and the Q-branch of a quadrature modulator where it is scrambled in a unique system-specific way using the scrambling sequences s_I and s_Q without any offset. Furthermore, the Q-branch is delayed by half the chip duration $T_c/2$ to avoid the amplitude of the carrier signal passing through zero during phase changes. Pulse shaping is specified by a spectrum mask. It should be noted that no pilot symbols are transmitted, so that incoherent detection is used in UL direction.

Figure 5.59 illustrates that the channels in DL direction are separated by Walsh codes of length 64 as channelization codes. The constant bit sequence of the pilot channel is multiplied by the (constant) Walsh code W_0 and is subsequently spread by the cell-specific

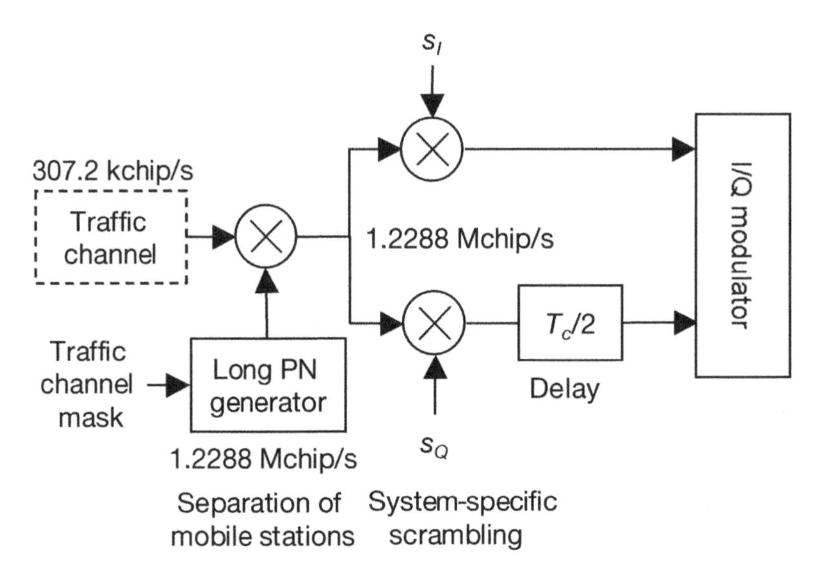

Figure 5.58 UL spreading and modulation for cdmaOne.

shifted version of the scrambling sequences s_I and s_Q in a quadrature modulator, that is, effectively, only the cell-specific scrambling sequences are modulated. The synchronization channel uses the Walsh code W_{32}. Furthermore, there may be up to seven paging channels. The remaining Walsh codes can be used by the traffic channels. Hence, the number of traffic channels per cell is restricted to about 60 depending on the respective configurations. However, as discussed in Subsection 5.1.4, the number of active connections is usually not limited by the number of codes, but by the interference within the network. Before being multiplied by a Walsh code, the traffic channel data stream is scrambled using the long PN sequence not to spread but to encrypt the data. The PN sequence is generated at a rate of 1.2288 Mchip/s. However, only each 64th chip is taken to match the data rate of 19.2 kbit/s, that is, the sequence is decimated. Cell-specific scrambling is performed by using shifted versions of s_I and s_Q in the two branches of a quadrature modulator. All channels are individually power weighted, combined and modulated in the same way as for the UL. On the basis of the pilot channel, a coherent detection is possible in DL direction. However, in cdmaOne there is only one common pilot channel in contrast to UTRA FDD where secondary and individual pilot channels allow an antenna beam–specific phase tracking and channel estimation.

Cell search and synchronization

In contrast to UTRA FDD where a cell may be characterized by one of 512 different primary scrambling sequences, only two scrambling sequences are used in cdmaOne. Different cells can be distinguished by different offsets of these sequences s_I and s_Q. For an offset 0, the start of each 75th period of the sequence matches exactly with the start of an even (GPS) second. An MS is able to lock to a cell by using a correlator fed by these codes and by varying the offset. The strongest peak of the correlator output is related to the cell best

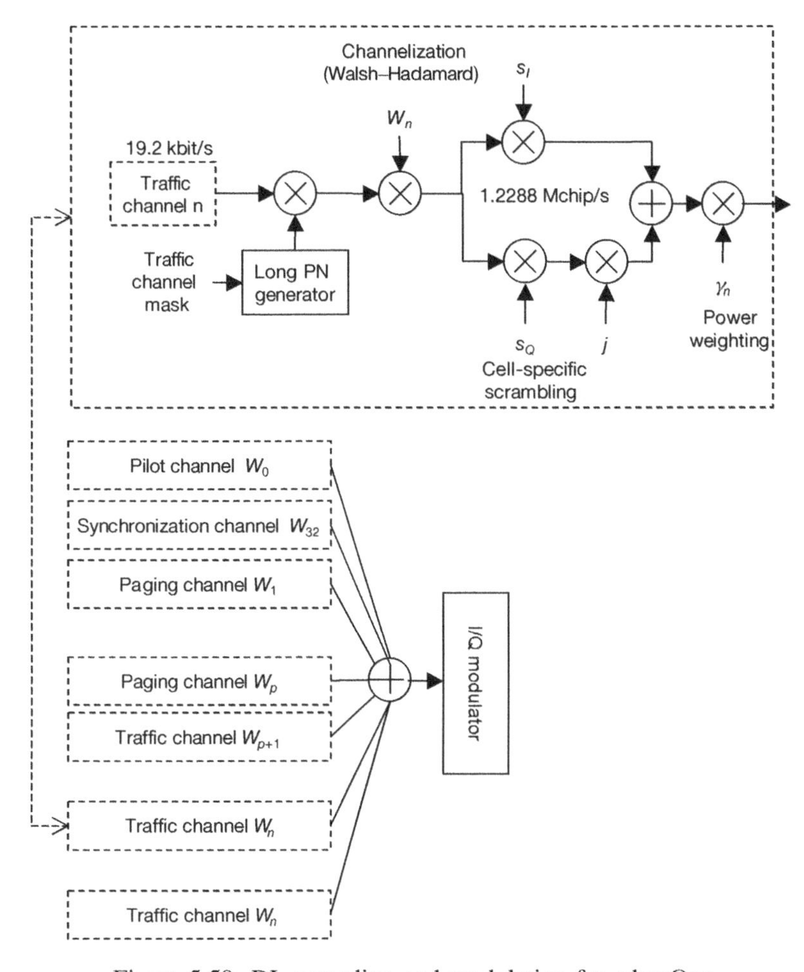

Figure 5.59 DL spreading and modulation for cdmaOne.

suited to camp on. The corresponding offset (assigned to all DL channels of that cell) can be used in the next step to descramble and decode the synchronization channel that carries information on the system and network identification, the system time, the absolute offset value and the state of the long PN code generator. Especially, the last information is needed to decode the paging channel to obtain further system information.

Random access

An access message, which may have a length of several physical frames, is coded, spread and modulated in a way analogous to the traffic channel data. The proper message is preceded by several preamble frames consisting of logical zeros, which serve as a kind of training sequence. The channel mask for the long code generator contains the cell-specific pilot channel offset, a BS identification number and an identification number of the used access subchannel. Hence, the BS is able to separate access attempts generated in its own cell area from attempts generated in neighboring cells.

To keep interference caused by access attempts (which may have a duration of more than 100 ms) at a low level, a mechanism similar to the one described for UTRA FDD is used: the first access probe is sent at a relatively low transmission power derived from a signal strength measurement on the DL pilot channel. If no acknowledgment is received, the message is sent again with a power level increased by a certain step. This procedure is repeated until the maximum allowed power level is reached.

Power control

As mentioned in the preceding section, a power control symbol is inserted in DL direction in each power control group indicating whether the corresponding MS has to increase or decrease its transmit power level by 1 dB. Hence, in UL direction a fast closed-loop power control at a rate of 800 Hz is implemented on the basis of *SIR* measurements by the BS. Furthermore, there is an open outer loop controlling the fast inner loop.

In DL direction, there are no power control symbols per group. However, the MS may command the BS to increase or decrease the power level by inserting the corresponding information into a data frame. Hence, the maximum rate for DL closed-loop power control is 50 Hz. DL power control is based on measurements of the frame erasure rate.

At the initiation of a call, an open-loop power control is used.

Handover

A handover decision is mainly based on signal strength measurements performed by the MS on the pilot channels of the serving as well as of the neighboring cells.

The preferred handover type in cdmaOne is the soft handover which can be applied between cells using the same frequencies. In DL direction, it is managed by combining the signals transmitted by the involved base stations within the RAKE receiver of the MS. In UL direction, a selection on a frame basis is performed if the involved base stations are located at different sites. A softer handover is implemented between collocated sector cells by a maximum ratio combining of the received signals. With respect to their role in a soft handover procedure, neighboring cells are separated into candidate, active and remaining sets. To be a member of the candidate set, the received strength of the pilot channel of that cell has to be above a certain level. A cell of the candidate set becomes a member of the active set if the corresponding pilot signal strength exceeds that of another cell in the active set by a certain margin.

A hard handover is required between cells at the border of a BSC area, in a hierarchical cell structure or between cells of different systems (e.g. between cdmaOne and DAMPS). In this case, the BS commands the MS to switch to the new frequency in the new cell. The BS may require some assistance in the form of measurement reports for the candidate cells. To perform these measurements, the MS needs an additional measurement receiver or it loses a data frame during the measurements. No compressed mode is specified for cdmaOne.

5.5.7 cdma2000

As explained in Subsection 5.5.3, cdma2000 is a direct evolution of cdmaOne fulfilling the ITU requirements for 3G mobile communication systems. To offer data rates of some Mbit/s and mixed heterogeneous services with very different quality requirements, a variety of coding, modulation and spreading schemes has been defined while preserving some fundamental methods and parameters of cdmaOne to guarantee a backward compatibility.

Within this subsection, an overview of the physical layer of the cdma2000 system is given. More details can be found, for example, in (Etemad 2004; Steele *et al.* 2001) or within the standardization documents (3GPP2 C.S0002 2004).

Frequency allocation

Networks according to the cdma2000 standard are and will be operated partly in bands that are currently allocated to cdmaOne networks. To allow a simple refarming[9] of these frequency bands, the chip rate and nominal carrier separation for cdma2000 have been defined as multiples of 1.2288 Mchip/s and 1.25 MHz, respectively. As mentioned in Subsection 5.5.3, cdma2000 is denoted as the multicarrier CDMA mode of the IMT-2000 family. This notation is related to the fact that a DL data stream may be multiplexed not only on one, but also on 3, 6, 9 or even 12 cdmaOne carriers. The multicarrier concept of cdma2000 should be clearly distinguished from what is usually understood as multicarrier CDMA within scientific publications (see e.g. (Hanzo *et al.* 2003)). In a proper multicarrier CDMA system, spreading sequences are applied in frequency domain and the chips are mapped to individual OFDM subcarriers. Within cdma2000 no OFDM is applied and the multiple DL carriers use the same spreading sequences.

For the current first version of cdma2000, only the single and threefold carrier variant have been specified, which are called the 1X and 3X mode, respectively. It should be noted that the multicarrier mode is only applied in DL direction, whereas the wideband mode in UL direction is based on a direct sequence spreading using a multiple of 1.2288 Mchip/s as the chip rate. This is illustrated in Figure 5.60.

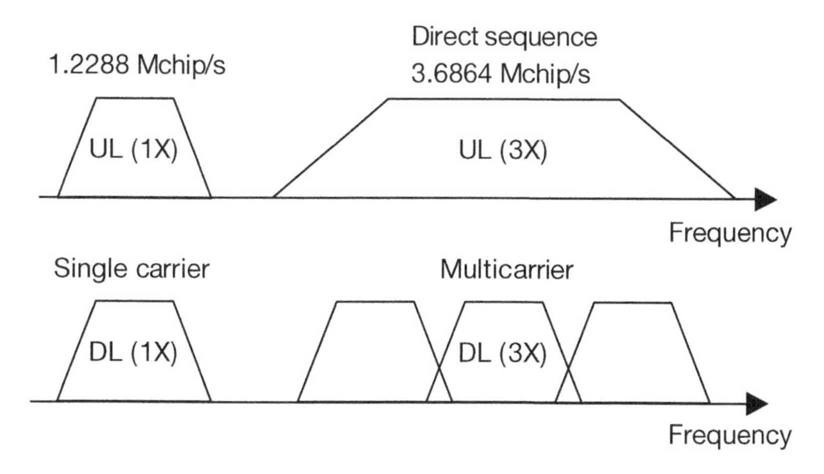

Figure 5.60 Transmission modes for cdma2000.

[9]Refarming of frequency bands means that frequency carriers of that band originally used by one system are continuously transferred to a newer mobile communication system.

Code allocation

Spreading within cdma2000 essentially is based on the same type of codes as used in cdmaOne, namely,

- Walsh–Hadamard codes;
- long PN sequences (m-sequences) corresponding to a 42-stage shift register to separate different MSs and to cipher the data stream in UL and DL direction;
- a pair of PN sequences corresponding to two 15-stage shift registers to provide a cell-specific scrambling.

However, some modifications are introduced to offer more flexibility with respect to the data rate and the chip rate.

In contrast to cdmaOne, the Walsh codes are not of fixed but of variable length similar to the OVSF codes of UMTS to offer a variety of data rates. Furthermore, in UL direction they are not only used for Walsh modulation but also to separate different physical channels transmitted by one MS.

To accomplish a spreading in UL direction at the threefold chip rate, the long PN code generator has been modified as shown in Figure 5.61. Furthermore, at this chip rate the pair (s_Q, s_I) of PN sequences is generated in another way by using two disjoint segments each of length $3 \cdot 2^{15}$ of one m-sequence which is derived from one 20-stage shift register. Using the threefold length at the threefold chip rate means that 75 periods of these sequences have exactly a length of 2 s as in the case of cdmaOne.

Hence, base stations in cdma2000 networks are also synchronized to GPS timing in order to distinguish cells by a PN sequence offset and to accomplish an easy cell search and soft handover mechanism.

Data transmission, channel coding and multiplexing

All of the channels defined for cdmaOne are also found in cdma2000. For the basic channel configuration they are used with the same transport format, spreading and modulation scheme as in cdmaOne to offer backward compatibility. However, some additional channels

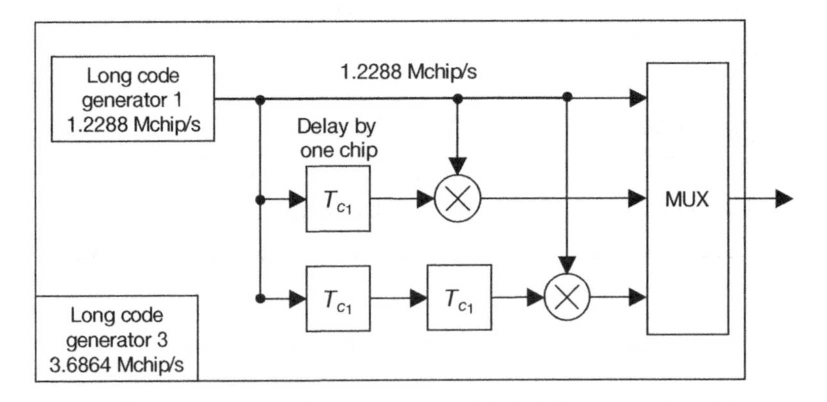

Figure 5.61 Long code generator for the threefold chip rate in cdma2000.

and transport formats have been specified to improve the link performance, to offer efficient packet switched modes and to increase the data rate.

For example, UL data transmission is accompanied by a pilot channel (similar to WCDMA) to allow a coherent detection. Furthermore, power control indicator bits may be multiplexed at a rate of 800 Hz to the UL pilot channel to accomplish a fast closed-loop DL power control. In DL direction, several pilot channels per cell may be allocated, which can be individually assigned to different beams in a smart antenna solution.

To increase the data rate, several traffic channels may be assigned to one connection. For a connection, one and only one so-called *fundamental channel* (FCH) is needed. This fundamental channel can be identified with a traffic channel defined for cdmaOne. Hence, it offers low data rates of up to 14.4 kbit/s. A data rate enhancement is achieved by additionally assigning one or more so-called supplemental channels (SCH) to the connection. For these supplemental channels, a variety of transport formats (channel coding and puncturing schemes, interleaving length, spreading and modulation schemes) has been specified to allow a multiplexing of services with very different quality and bandwidth requirements. The maximum data rate on one supplemental channel is 307.2 kbit/s for the 1X mode and 1036.8 Mbit/s for the 3X mode. Special packet channel may achieve even higher data rates of about 2–3 Mbit/s.

Within cdma2000, the following channel coding methods are defined:

- block codes (Reed–Solomon codes in DL direction, mapping to Walsh codes in UL direction);
- convolutional codes of constraint length 9 and of coding rates 1/2, 1/3, 1/4, 1/6;
- turbo codes of coding rate 1/2, 1/3, 1/4 and 1/5.

The coded bits may be punctured using a certain number of puncturing schemes. Interleaving is performed over frames of length 5, 10, 20, 40 or 80 ms in contrast to cdmaOne where only 20 ms frames are used.

Spreading and modulation

Different spreading and modulation schemes are defined for cdma2000. Using the basic channel configuration, spreading and modulation are performed in the same way as for cdmaOne. However, some enhancements are introduced to achieve higher data rates.

In DL direction, different channels are separated using Walsh codes of variable spreading factors SF (length) as shown in Figure 5.62. It should be noted that instead of duplicating the data symbols to the I- and Q-branch as for the basic configuration, symbols are mapped to the I- and Q-branch according to a QPSK; for packet data channels even a mapping according to an 8-PSK or 16-QAM may be used. Cell-specific scrambling is performed in the same way as for cdmaOne using the pair (s_I, s_Q) of PN sequences with a certain offset.

For the multicarrier mode (3X) the data symbols are cyclically demultiplexed to three symbol streams corresponding to the three carriers as shown in Figure 5.63. Spreading of the data symbols of one transport channel is performed by using the same Walsh code and spreading sequence on each carrier. The three carrier signals may be transmitted via separated antennas to achieve an additional diversity gain. A pilot channel is assigned to each carrier.

In UL direction, spreading for the basic channel configuration is performed in the same way as for cdmaOne. For an enhanced channel configuration spreading is similar to UTRA

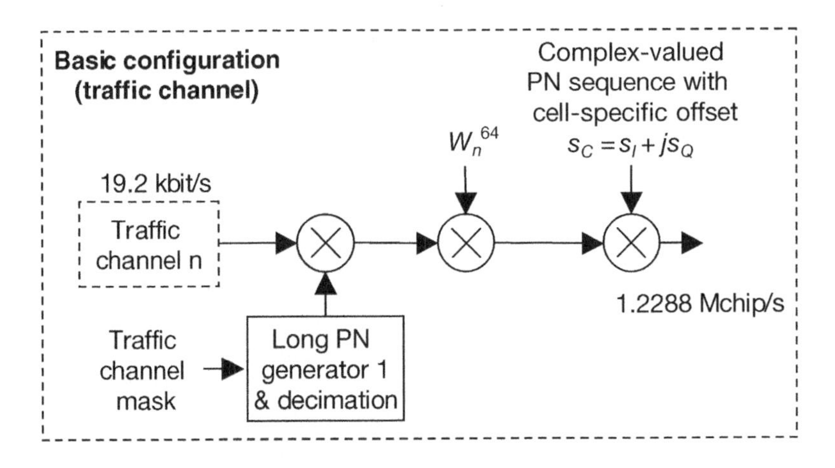

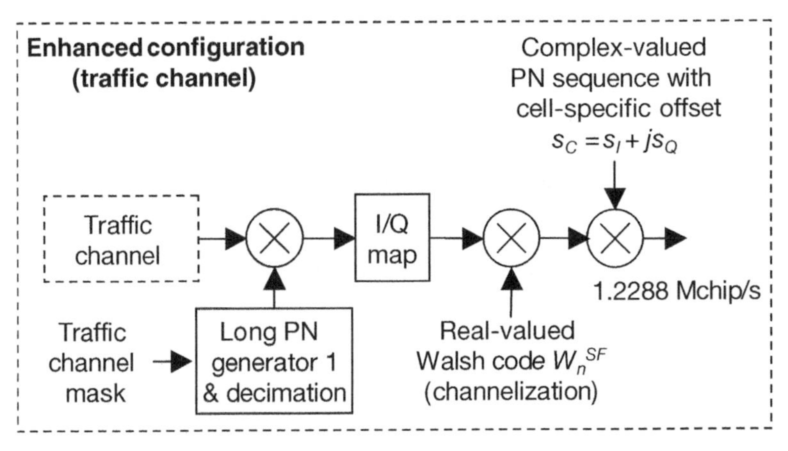

Figure 5.62 Downlink spreading for the single carrier mode.

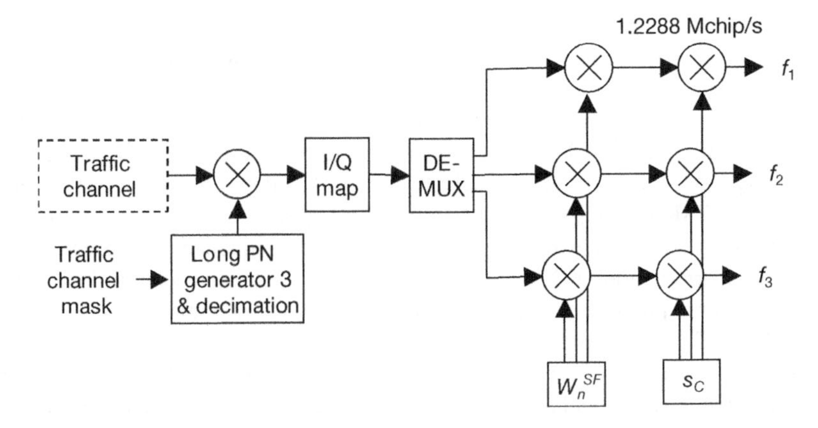

Figure 5.63 Downlink spreading and multiplexing for the multicarrier mode.

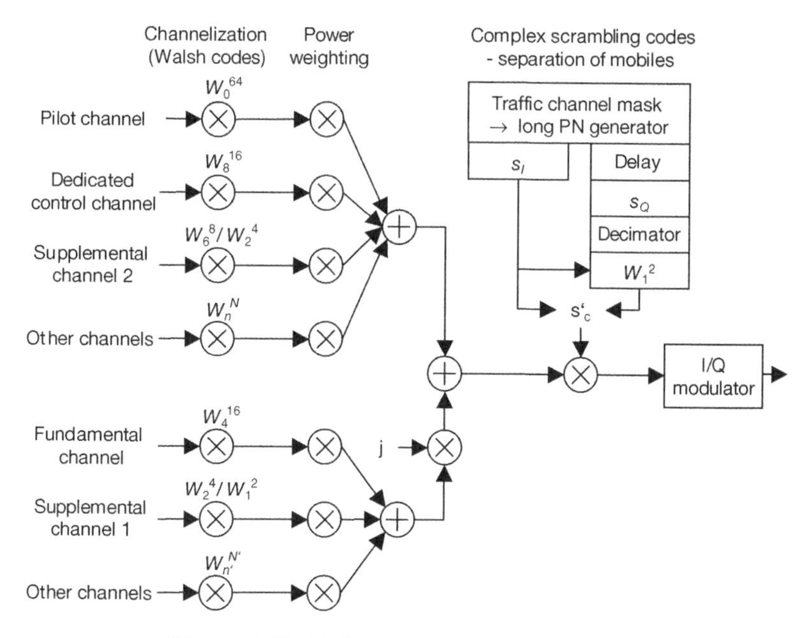

Figure 5.64 Uplink spreading for cdma2000.

FDD as explained in Subsection 5.5.4: One MS may transmit on several channels separated by different Walsh codes or by the two branches of the I/Q modulator as shown in Figure 5.64. There is at least one fundamental (traffic) channel of low data rate and a pilot channel offering coherent detection in UL direction. The pilot channel also contains power control bits for a fast closed-loop DL power control, that is, the pilot channel is very similar to the dedicated physical control channel of the UTRA FDD mode. Additionally, one or two supplemental channels for a data rate increase or a dedicated control channel or other channels (packet data) may be assigned. These channels are separated by orthogonal Walsh codes of fixed length. To adapt the input data symbol rate to the chip rate, the corresponding Walsh code may have to be repeated during one data symbol. Source (MS)-specific scrambling is performed using a complex-valued scrambling sequence, which is derived from the long code generator and the pair (s_I, s_Q) of PN sequences. The complex-valued sequence is constructed in such a way that there is a continuous transmission with, on average, the same power on the I- and Q-branch even if no traffic data have to be transmitted. Furthermore, it has been selected to avoid transitions through zero during phase changes as far as possible.

Cell search and synchronization

Cell search and synchronization is performed in the same way as in a cdmaOne network based on the pair (s_I, s_Q) of PN sequences with their cell-specific offset and on the fact that base stations in a cdma2000 network are also synchronized to GPS timing.

Random access

In addition, the basic random access procedure is very similar to the one for cdmaOne.

Power control

The main enhancement of cdma2000 compared to cdmaOne concerning power control is that a fast closed-loop DL power control mechanism at a rate of 800 Hz has been specified.

Handover

Soft and hard handover are performed in a way very similar to cdmaOne. For supporting a hard handover, a kind of compressed mode has been defined: the MS increases the transmission power within some time slots before suspending the data transmission to switch to another frequency for neighbor cell measurements.

5.6 Bibliographical Notes

The idea of spreading the signal spectrum is quite old. It was originally intended to hide the signal and thus it was very interesting for military applications. Some basic ideas even date back to the Second World War. The first important system that was available for civil applications (besides the military ones, and only with intentionally reduced performance) was the global positioning system GPS (see (Fazel and Kaiser 2003)). The idea to use spread spectrum techniques for multiple access is also not so new, but it became a serious candidate for mobile radio systems only when A. Viterbi and his Qualcomm Company promoted these ideas that resulted in the IS-95 standard ((J-STD-007 1999; Steele *et al.* 2001; TIA/EIA-95 1993)) with a high effort both in the field of scientific discussion and commercial marketing. During the 1990s many researchers started to work on this field, and more sophisticated receiver structures like multiuser detection were developed. For a detailed discussion of these techniques, we refer to (Verdu 1998). Besides efficient receiver structures, further methods for handling the interference in CDMA mobile radio networks are required to obtain an acceptable network performance. Methods like fast power control, soft handover, a cluster 1 soft capacity planning, and so on, are discussed in detail in (Holma and Toskala 2001; Steele *et al.* 2001; Viterbi 1995).

During the selection and standardization process for third generation mobile communication systems (which is reported in detail in (Holma and Toskala 2001)), CDMA was established as the dominating multiple access technology: systems like cdma2000 and UMTS with the two transmission modes Wideband CDMA and Time Division CDMA became global standards. While the physical layer specifications for cdma2000 can be found in one large document ((3GPP2 C.S0002 2004)), the corresponding specifications for UMTS released by the 3rd Generation Partnership Project Group are separated into many documents. Some essential specification documents are quoted as 3GPP TS 25.xxx within Subsection 5.5.4 and Subsection 5.5.5 while discussing the corresponding topics. A detailed description of these systems can also be found in (Etemad 2004; Holma and Toskala 2001; Steele *et al.* 2001).

Multicarrier CDMA, which combines CDMA techniques with OFDM, seems to be a promising candidate for future mobile radio systems. For a detailed discussion, we refer to (Fazel and Kaiser 2003; Hanzo *et al.* 2003).

The main parameters of the 2G and 3G CDMA mobile communication systems are listed in Table 5.1. Hence, this table serves as summary of the Subsections 5.5.4 to 5.5.7 where these systems are discussed in detail.

Table 5.1 Comparison of CDMA mobile communication systems

	UTRA FDD	UTRA TDD	IS-95/cdmaOne	cdma2000
Chip rate [Mchip/s]	3.84	3.84 (and 1.28)	1.2288	1.2288 and 3.6864
Duplex	FDD	TDD	FDD	FDD
Synchronization of BSs	Not required	Recommended, not required	via GPS	via GPS
Convolutional codes of rate	1/2, 1/3	1/2, 1/3	1/2, 1/3	1/2, 1/3, 1/4, 1/6
Constraint length	9	9	9	9
Turbo codes of rate	1/3	1/3	–	1/2, 1/3, 1/4, 1/5
Channelization codes DL (separation of MSs)	OVSF codes SF = 4, ..., 512	OVSF, SF = 1, 16 code pooling	Walsh codes SF = 64	Walsh codes of length 2, 4, 8, ..., 256
Channelization codes UL (separation of connections)	OVSF codes SF = 4, ..., 256	OVSF codes, SF = 1, 2, 4, 8, 16	–	Walsh codes of length 2, 4, 8, ..., 64
Scrambling codes DL (separation of cells)	Gold code segments: 38,400 chips, 16 codes per cell	128 sequences defined by code tables, 4 sequences per cell	512 shifted versions of one PN sequence, 1 code per cell	512 shifted versions of one PN sequence, 1 code per cell
Scrambling codes UL (separation of MSs)	Gold codes segments opt.: short codes (256 chips)	128 sequences defined by tables, 4 sequences per cell	Long PN sequences	Long PN sequences
Realization of variable data rates	Variable spreading & transport formats	As UTRA FDD, time slot aggregation	Bit repetition	Variable spreading & transport formats
DL pilot	Several common pilots, connection individual pilots	Midambles per burst	One pilot per cell	Several common pilots
UL pilot	Connection individual pilots	Midambles per burst	–	Connection individual pilots
Max. UL PC rate	1500 Hz	100, 200 Hz, open loop	800 Hz	800 Hz
Max. DL PC rate	1500 Hz	100–750 Hz	50 Hz	800 Hz
Soft handover	Yes	No	Yes	Yes
Interfrequency HO support	By compressed mode	TDMA	Additional receiver	Compressed mode

5.7 Problems

1. Show that
$$\Delta_{0,1} = \sqrt{E_b}\left(|c_1| \pm \frac{1}{|c_1|}\Re\left\{c_1^* c_2 \rho_{12}\right\}\right)$$
holds (see Figure 5.26).

2. For rectangular transmit pulses, show that the performance for synchronous transmission treated in Subsection 5.2.1 is the worst case compared to asynchronous transmission.

3. Show that for two users with real-valued ρ_{12} and real-valued coefficients $c_1 = c_2$, the decorrelator is better than the SUMF receiver.

4. Show that
$$\mathbf{A}^\dagger\left(\mathbf{A}\mathbf{A}^\dagger + t\mathbf{I}\right)^{-1} = \left(\mathbf{A}^\dagger\mathbf{A} + t\mathbf{I}\right)^{-1}\mathbf{A}^\dagger \tag{5.54}$$
holds for any real number t (if the inverses above exist).

5. Convert the free-space loss formula of Problem 1.9 into a decibel notation and show that one obtains a formula of the form of Equation (5.2). Show that for $f = 2$ GHz the parameters A and B take the values $A = 98.4$ dB and $B = 20$.

6. Consider Equation (5.2) with propagation parameters $A = 130$ dB and $B = 35$. Let us assume that the receiver sensitivity of the MS and the BS is improved by 3 dB or 6 dB (e.g. by channel coding or antenna diversity). Consequently, the cell radius may be increased. Calculate the factor for the increase of the radius. What does it mean for the number of required cells?

7. Consider the UTRA FDD mode and two services with user data rates of 80 kbit/s and 320 kbit/s which are protected by a convolutional code of rate 1/3. Let us assume that both services are transmitted at the same power and require the same E_b/N_0, namely, $E_b/N_0 = 6$ dB.

   (a) Which spreading factors SF are required for the UL, and which ones for the DL?

   (b) Which are the effective spreading factors?

   (c) Furthermore, consider the propagation conditions of Problem 6. By which factor does the maximum range (radius) of both services differ?

   (d) Assume that an OVSF code $c_{SF,3}$ is allocated to the 80 kbit/s service in DL direction. Which OVSF codes are allowed for the other service?

   (e) Calculate (separately for each service) the maximum number of allowed active connections per cell (median values) in UL direction using a cluster 1×1 with a mean relative intercell interference of $q = 0.8$ and orthogonality factors of $\alpha = 1$ and $\alpha = 0.1$. What would be the gain of an intracell interference cancellation reducing α from 1 to 0.1?

(f) Repeat the calculation for the maximum number of active connection for a cluster 1×3 assuming a mean relative intercell interference of $q = 0.15$.

(g) Compare the network capacity per required bandwidth for both clusters.

8. Consider a fictitious CDMA system in a frequency band at 10 GHz and a moving MS at a speed of $v = 20$ m/s. Is a power control rate of 1 kHz sufficient to follow fast fading? What order of magnitude of the interleaving depth (in milliseconds) would be required for an efficient interleaving?

9. Consider the long PN sequences of cdmaOne. How many different PN sequences may be generated by the corresponding shift register? What is the period length (in seconds) of these sequences?

10. Consider the UTRA TDD mode with the 2:1 separation between DL and UL as shown in Figure 5.56.

    (a) What is the maximum rate for a closed-loop DL power control?

    (b) What is the maximum allowed distance between MS and BS at the random access using the guard period of 192 chips?

    (c) What is the number of available code channels of $SF = 16$ in UL direction?

11. Consider a service of data rate 230.4 kbit/s within cdma2000, which is transmitted via a supplemental channel in UL and DL direction. The data are protected by a convolutional code of rate 1/4. Code puncturing is applied for the 1X mode, but not for the 3X mode.

    (a) For the UL 1X mode Walsh codes of length 2 are used. Which code rate is required after puncturing?

    (b) Which Walsh code length is required for the 1X mode in DL direction (using the same puncturing as in the UL)?

    (c) Which Walsh code length is required for the 3X mode in UL and DL direction?

Bibliography

3GPP TS 25.104 *Base Station (BS) Radio Transmission and Reception (FDD)*, 3rd Generation Partnership Project (3GPP).

3GPP TS 25.101 *User Equipment (UE) Radio Transmission and Reception (FDD)*, 3rd Generation Partnership Project (3GPP).

3GPP TS 25.102 *User Equipment (UE) Radio Transmission and Reception (TDD)*, 3rd Generation Partnership Project (3GPP).

3GPP TS 25.105 *Base Station (BS) Radio Transmission and Reception (TDD)*, 3rd Generation Partnership Project (3GPP).

3GPP TS 25.211 *Physical Channels and Mapping of Transport Channels onto Physical Channels (FDD)*, 3rd Generation Partnership Project (3GPP).

3GPP TS 25.212 *Multiplexing and Channel Coding (FDD)*, 3rd Generation Partnership Project (3GPP).

3GPP TS 25.213 *Spreading and Modulation (FDD)*, 3rd Generation Partnership Project (3GPP).

3GPP TS 25.214 *Physical Layer Procedures (FDD)*, 3rd Generation Partnership Project (3GPP).

3GPP TS 25.221 *Physical Channels and Mapping of Transport Channels onto Physical Channels (TDD)*, 3rd Generation Partnership Project (3GPP).

3GPP TS 25.222 *Multiplexing and Channel Coding (TDD)*, 3rd Generation Partnership Project (3GPP).

3GPP TS 25.223 *Spreading and Modulation (TDD)*, 3rd Generation Partnership Project (3GPP).

3GPP TS 25.224 *Physical Layer Procedures (TDD)*, 3rd Generation Partnership Project (3GPP).

3GPP2 C.S0002 *Physical Layer Standard for cdma2000 Spread Spectrum Systems*, 3rd Generation Partnership Project 2 (3GPP2).

Alamouti SM 1998 A simple transmit diversity technique for wireless communications. *IEEE J. Select. Areas Commun.* **16**, 1451–1458.

Alard M and Lassalle R 1987 Principles of modulation and channel coding for digital broadcasting for mobile receivers. *EBU Techn. Review* **224**, 47–68.

Bahl LR, Cocke J, Jelinek F and Raviv J 1974 Optimal decoding of linear codes for minimizing symbol error rate. *IEEE Trans. Inf. Theory* **20**, 284–287.

Baier PW, Bing T and Klein A 2000 TD-CDMA. In *Third Generation Mobile Communication Systems* (eds. Prasad R, Mohr W and Konhäuser W), pp. 25–72. Artech House Publishers.

Banelli P and Baruffa G 2001 Mixed BB-IF predistortion of OFDM signals in nonlinear channels. *IEEE Trans. Broadcast.* **47**, 137–146.

Bello PA 1963 Characterization of randomly time-variant linear channels. *IEEE Trans. Commun.* **11**, 360–393.

Benedetto S and Biglieri E 1999 *Principles of Digital Transmission. With Wireless Applications*, Kluwer Academic / Plenum Publishers, New York.

Theory and Applications of OFDM and CDMA Henrik Schulze and Christian Lüders

Berrou C, Glavieux A and Thitimajshima P 1993 Near Shannon limit error correcting coding and decoding. In *Proc. IEEE International Conference on Communications 1993*, 1064–1070. Geneva, Switzerland.

Biglieri E, Divsalar D, McLane PJ and Simon MK 1991 *Introduction to Trellis-Coded Modulation with Applications*, Macmillan, New York.

Bing T, Dahlhaus D, Latva-aho M and Nasshan M 2000 Advanced receiver algorithms. In *Third Generation Mobile Communication Systems* (eds. Prasad R, Mohr W, and Konhäuser W), pp. 91–132. Artech House Publishers.

Bingham JAC 1990 Multicarrier modulation for data transmission: An idea whose time has come. *IEEE Communications Magazine* **28**, 5–14.

Blahut RE 1983 *Theory and Practice of Error Control Codes*, Addison-Wesley, Reading, Massachusetts.

Blahut RE 1990 *Digital Transmission of Information*, Addison-Wessley.

Bossert M 1999 *Channel Coding for Telecommunications*, Wiley.

Bracewell RN 2000 *The Fourier Transform and its Applications*, McGraw-Hill.

Caire G, Taricco G and Biglieri E. 1998 Bit-interleaved coded modulation. *IEEE Trans. Inf. Theory* **44**, 927–946.

CCSDS 1987 *Recommendations for Space Data System Standard: Telemetry Channel Coding*, Blue Book Issue 2 CCSDS 101.0-B2. Consultative Committee for Space Data Systems.

Chang RW 1966 Synthesis of band-limited orthogonal signals for multi-channel data transmission. *Bell Labs Tech. J.* **45**, 1775–1796.

Chang RW and Gibby RA 1968 A theoretical study of performance of an orthogonal multiplexing data transmission scheme. *IEEE Trans. Commun.* **16**, 529–340.

Cimini LJ 1985 Analysis and simulation of a digital mobile radio channel using orthogonal frequency division multiplexing. *IEEE Trans. Commun.* **33**, 665–675.

Clark GC and Cain JB 1988 *Error-Correction Coding for Digital Communications*, Plenum, New York.

D'Andrea AN, Lottici V and Reggianini R 1996 RF power amplifier linearization through amplitude and phase predistortion. *IEEE Trans. Commun.* **44**, 1477–1484.

Eldar Y 2002 On geometric properties of the decorrelator. *IEEE Comm. Lett.* **6**, 16–18.

EN101475 2001 *Broadband Radio Access Networks (BRAN); HIPERLAN type 2; Physical (PHY) Layer*, ETSI, Sophia-Antipolis.

EN300401 2001a *Radio Broadcasting Systems: Digital Audio Broadcasting (DAB) to Mobile, Portable and Fixed Receivers*, ETSI, Sophia-Antipolis.

EN300744 2001b *Digital Broadcasting Systems for Television, Sound and Data Services; Framing Structure, Channel Coding and Modulation for Digital Terrestrial Television*, ETSI, Sophia-Antipolis.

Etemad K 2004 *cdma2000 Evolution*, Wiley.

ETSI TR 101 112 1998 *Universal Mobile Telecommunications Systems (UMTS); Selection Procedures for the Choice of Radio Transmission Technologies of the UMTS*, European Telecommunications Standards Institute (ETSI).

Fazel K and Kaiser K 2003 *Multi-Carrier and Spread Spectrum Systems*, Wiley.

Feller W 1970 *An Introduction to Probability Theory and its Applications*, 3rd edn. revised printing Wiley.

Forney GD 1966 *Concatenated Codes*, MIT Press, Cambridge, Massachusetts.

Forney GD 1970 Convolutional codes I: algebraic structure. *IEEE Trans. Inf. Theory* **16**, 720–738.

Forney GD 1973 The Viterbi algorithm. *Proc. IEEE* **61**, 268–278.

Frenger PK, Orten P and Ottoson T 1999 Code-spread CDMA with interference cancelation. *IEEE J. Select. Areas Commun.* **17**, 2090–2095.

Frenger PK, Orten P and Ottoson T 2000 Code-spread CDMA using maximum free distance low-rate convolutional codes. *IEEE Trans. Commun.* **48**, 135–144.

Gibson JD (ed.) 1999 *The Mobile Communications Handbook*, 2nd edn. CRC and IEEE Press.

Gitlin RD, Hayes JF and Weinstein SB 1993 *Data Communication Principles*, Plenum.

Graf F, Fischer P, Humburg E and Lüders C 1997 Technical potential and economic efficiency of macro diversity in GSM mobile radio networks. In *Proceedings of the 2nd European Mobile Communications Conference (EPMCC '97)*, pp. 127–134. VDE Verlag.

Haardt M and Alexiou A 2004 Smart antenna technologies for future wireless systems: trends and challenges. *IEEE Commun. Mag.* **42**, 90–97.

Hagenauer J 1982 Fehlerkorrektur und diversity in Fading-Kanälen. *Archiv für elektronische Übertragungstechnik (AEÜ)* **36**, 337–344.

Hagenauer J 1988 Rate compatible punctured convolutional codes (RCPC codes) and their applications. *IEEE Trans. Commun.* **36**, 389–400.

Hagenauer J 1995 Source-controlled channel coding. *IEEE Trans. Commun.* **43**, 2449–2457.

Hagenauer J and Hoeher 1989 A Viterbi algorithm with soft-decision outputs and its applications. In *Proc. GLOBECOM 1989* 47.1.1–47.1.7, Dallas, Texas.

Hagenauer J, Offer E and Papke L 1996 Iterative decoding of block and convolutional codes. *IEEE Trans. Inf. Theory* **42**, 429–445.

Hagenauer J and Offer E 2001 Matching Viterbi decoders and Reed–Solomon decoders in a concatenated system. In: (Wicker and Bhargava 2001) *Reed-Solomon Codes and Their Applications.* Wiley.

Hagenauer J, Seshadri N and Sundberg CEW 1990 The performance of rate-compatible punctured convolutional codes for mobile radio. *IEEE Trans. Commun.* **38**, 966–980.

Hanzo L, Münster M, Choi BJ and Keller T 2003 *OFDM and MC-CDMA for Broadband Multi-User Communications, WLANs and Broadcasting*, Wiley and IEEE Press.

Haykin S 1996 *Adaptive Filter Theory*, 3rd ed. Prentice-Hall.

Hill TL 1956 *Statistical Mechanics*, McGraw-Hill.

Hirosaki B 1981 An orthogonally multiplexed QAM system using the discrete Fourier transform. *IEEE Trans. Commun.* **29**, 982–989.

Hoeg W and Lauterbach T 2003 *Digital Audio Broadcasting. Principles and Applications of Digital Radio*, 2nd edn. Wiley.

Hoeher P 1991 TCM on frequency-selective land-mobile fading channels. *Proc. 5th Tirrenia International Workshop on Digital Communication*, Tirrenia

Hoeher P 1992 A statistical discrete-time model for the WSSUS multipath channel. *IEEE Trans. Veh. Technol.* **41**, 461–468

Hoeher P, Hagenauer J, Offer E, Rapp C and Schulze H 1991 Performance of an RCPC-Coded OFDM-based Digital Audio Broadcasting (DAB) system. *Proc. IEEE Globecom 1991* **1**, pp. 40–46.

Hoeher P, Kaiser S and Robertson P 1997 Two-dimensional pilot-symbol-aided channel estimation by Wiener filtering. *IEEE International Conference on Acoustics, Speech, and Signal Processing ICASSP-97*, Munich (Germany) **3**, pp. 1845–1848

Hofmann-Wellenhof B, Lichtenegger H and Collins J 2001 *GPS: Theory and Practice*, Springer.

Holma H and Toskala A (eds.) 2001 *WCDMA for UMTS*, Revised edn. Wiley.

Horn RA and Johnson CR 1985 *Matrix Analysis*, Cambridge University Press.

Hottinen A, Turkkonen O and Wichman R 2004 *Multi-antenna Transceiver Techniques for 3G and Beyond*, Wiley.

IEEE Standard 802.11a *Wireless LAN Medium Access Control (MAC) and Physical Layer (PHY) specifications: High Speed Physical Layer in the 5 GHz Band* Institute of Electrical and Electronics Engineers (IEEE).

IEEE Standard 802.11b *Wireless LAN Medium Access Control (MAC) and Physical Layer (PHY) specifications: Higher-Speed Physical Layer Extension in the 2.4 GHz Band* Institute of Electrical and Electronics Engineers (IEEE).

Jakes WC 1975 *Microwave Mobile Communications*, Wiley.

Jamali SH and Le-Ngoc T 1994 *Coded-Modulation Techniques for Fading Channels*, Kluwer.

J-STD-007 *PCS 1900 - Air Interface Specification (ANSI/J-STD-007)* Telecommunications Industry Association Standards and Engineering Publications.

Kammeyer KD 2004 *Nachrichtenübertragung*, 3rd edn. Teubner Verlag, Stuttgart, Germany.

Kammeyer KD, Tuisel U, Schulze H and Bochmann H 1992 Digital multicarrier transmission of audio signals over mobile radio channels. *Eur. Trans. Telecomm.* **3**, 243–254.

Konhäuser W, Mohr W and Prasad R 2000 Introduction. In *Third Generation Mobile Communication Systems* (eds. Prasad R, Mohr W and Konhäuser W), pp. 11–24. Artech House Publishers.

Landau LD, Lifshitz EM 1958 *Statistical Physics, Vol. 5 of Course of Theoretical Physics*, Addison–Wesley.

Lee AL and Messerschmidt DG 1994 *Digital Communication*, 2nd edn. Kluwer.

Lin S and Costello DJ 1983 *Error Control Coding. Fundamentals and Applications*, Prentice-Hall.

Massey JL 1984 The how and why of channel coding. In *Proceedings of the 1984 Zurich Seminar on Digital Communications.* Zurich (Switzerland), 67–73.

Massey JL 1992 Deep space communications and coding: a marriage made in heaven. In *Lecture Notes in Control and Information Sciences*, **182**, Springer–Verlag.

McEliece RJ and Swanson L 2001 Reed–Solomon codes and the exploration of the solar system. In: (Wicker and Bhargava 2001) *Reed-Solomon Codes and Their Applications.* Wiley.

Milstein LB 2000 Wideband code division multiple access. *IEEE J. Select. Areas Commun.* **18**, 1344–1364.

Mogensen PE, Espensen P, Pedersen KI and Zetterberg P *et al* 1999 Antenna arrays and space division multiple access. In *GSM - Evolution Towards 3rd Generation Systems* (eds. Zvonar Z, Jung P and Kammerlander K), pp. 117–152. Kluwer Academic Publishers.

Monogioudis P, Conner K, Das D, Gollamudi S, Lee JAC, Moustakas AL, Nagaraj S, Rao AM, Soni RA and Yuan Y *et al* 2004 Intelligent antenna solutions for UMTS: Algorithms and simulation results. *IEEE Commun. Mag.* **42**, pp. 28–38.

Neeser FD and Massey JL 1993 Proper complex random processes with applications to information theory. *IEEE Trans. Inf. Theory* **39**, 1293–1302.

O'Hara B and Petrick A 1999 *IEEE 802.11 Handbook: A Designer's Companion*,IEEE Press.

Papoulis A 1991 *Probability, Random Variable, and Stochastic Processes*, McGraw-Hill.

Parsons JD 2000 *The Mobile Radio Propagation Channel*, Wiley.

Proakis JR 2001 *Digital Communications*, 4th edn. McGraw-Hill.

Reed IS, Solomon G 1960 Polynomial codes over certain finite fields. *J. SIAM* **8**, 300–304.

Reed M and Simon B 1980 *Methods of Modern Mathematical Physics I: Functional Analysis*,Revised and enlarged edition Academic Press.

Rehfuess U and Ivanov K 1999 Comparing frequency planning against 1x3 and 1x1 reuse in real frequency hopping networks. In *Proceedings of the 49th IEEE Vehicular Technical Conference*, IEEE Press.

Saltzberg BR 1967 Performance of an efficient parallel data transmission system. *IEEE Trans. Commun. Tech.* **15** 805–811.

Schmidt H 2001 *OFDM für die drahtlose Datenübertragung Innerhalb von Gebäuden*, Ph. D. Thesis, University Bremen (Germany).

Schulze H 1988 Stochastische Methoden und digitale Simulation von Mobilfunkkanälen. *Kleinheubacher Berichte* **32**, 473–483.

Schulze H 2001 The performance of multicarrier CDMA for the correlated Rayleigh fading channel. *Int. J. Electron. Commun. (AEÜ)* **55**, 88–94.

Schulze H 2003a Geometrical properties of orthogonal space-time codes. *IEEE Comm. Lett.* **7**, 64–66.

Schulze H 2003b System design for bit interleaved coded QAM with iterative decoding in a Rician fading channel. *European Trans. Telecomm.* **14**, 119–129.

Schulze H 2003c Performance analysis of concatenated space-time coding with two transmit antennas. *IEEE Trans. Wireless Comm.* 2, 669–679.

Simon MK and Alouini MS 2000 *Digital Communications over Fading Channels*, Wiley.

Simon MK and Divsalar D 1998 Some new twists to problems involving the Gaussian probability integral. *IEEE Trans. Commun.* **46**, 200–210.

Steele R, Lee CC and Gould P 2001 *GSM, cdmaOne and 3G*, Wiley.

Therrien CW 1992 *Discrete Random Signals and Statistical Signal Processing*, Prentice-Hall.

TIA/EIA-95 1993 *Mobile Station - Base Station Compatibility Standard for Spread Spectrum Cellular Systems. (ANSI/TIA/EIA-95)*, Telecommunications Industry Association Standards and Engineering Publications.

TS101475 2001 *Broadband Radio Access Networks (BRAN); HIPERLAN Type 2; Physical (PHY) layer*, ETSI, Sophia-Antipolis.

Ungerboeck 1982 Channel codes with multilevel/multiphase signals. *IEEE Trans. Inf. Theory* **28**, 55–67.

van Kampen NG 1981 *Stochastic Processes in Physic and Chemistry*, North-Holland.

Van Trees HL 1967 *Detection, Estimation, and Modulation Theory I*, Wiley.

Verdu S 1998 *Multiuser Detection*, Cambridge University Press.

Viterbi AJ 1967 Error bounds for convolutional codes and an asymptotically optimum decoding algorithm. *IEEE Trans. Inf. Theory* **13**, 260–269.

Viterbi AJ 1995 *CDMA. Principles of Spread Spectrum Communications*, Addison-Wesley.

Weinstein SB and Ebert PM 1971 Data transmission by frequency-division multiplexing using the discrete Fourier transform. *IEEE Trans. Commun.* **19**, 628–634.

Wicker SB 1995 *Error Control Systems for Digital Communication and Storage*, Prentice-Hall.

Wicker SB and Bhargava VK 2001 *Reed–Solomon Codes and Their Applications*, Wiley.

Wozencraft JM and Jacobs IM 1965 *Principles of Communication Engineering*, Wiley.

Index

Theory and Applications of OFDM and CDMA Henrik Schulze and Christian Lüders

Printed and bound in the UK by
CPI Antony Rowe, Eastbourne

Printed and bound by CPI Group (UK) Ltd, Croydon, CR0 4YY

16/04/2025

14658558-0001